Benchmark Papers
in Ecology

Series Editor: Frank B. Golley
University of Georgia

Related Titles in BENCHMARK PAPERS IN BEHAVIOR Series

**Benchmark Papers
in Ecology / 13**

A BENCHMARK® Books Series

DIVERSITY

Edited by

RUTH PATRICK

The Academy of Natural Sciences of Philadelphia

Hutchinson Ross Publishing Company

Stroudsburg, Pennsylvania

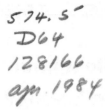
LIBRARY OF CONGRESS CATALOGING IN PUBLICATION DATA
Main entry under title:
Diversity
   (Benchmark papers in ecology; 13)
   Includes indexes.
   1. Variation (Biology)—Addresses, essays, lectures.
2. Biotic communities—Addresses, essays, lectures.
3. Species—Addresses, essays, lectures.   I. Patrick,
Ruth.   II. Series: Benchmark papers in ecology; v. 13.
QH401.D58   1983   574.5   83-4365
ISBN 0-87933-420-7

Distributed worldwide by Van Nostrand Reinhold Company Inc.,
135 W. 50th Street, New York, NY 10020.

CONTENTS

Contents

SERIES EDITOR'S FOREWORD

Ecology—the study of interactions and relationships between living systems and environment—is an extremely active and dynamic field of science. The great variety of possible interactions in even the most simple ecological system makes the study of ecology compelling but difficult to discuss in simple terms. Further, living systems include individual organisms, populations, communities, and ultimately the entire biosphere; there are thus numerous subspecialties in ecology. Some ecologists are interested in wildlife and natural history, others are intrigued by the complexity and apparently intractable problems of ecological systems, and still others apply ecological principles to the problems of man and the environment. This means that a Benchmark Series in Ecology could be subdivided into innumerable volumes that represent these diverse interests. However, rather than take this approach, I have tried to focus on general patterns or concepts that are applicable to two particularly important levels of ecological understanding: the population and the community. I have taken the dichotomy between these two as the major organizing concept in the series.

In a field that is rapidly changing and evolving, it is often difficult to chart the transition of single ideas into cohesive theories and principles. In addition, it is not easy to make judgments as to the benchmarks of the subject when the theoretical features of a field are relatively young. These twin problems—the relationship between interweaving ideas and the elucidation of theory, and the youth of the subject itself—make development of a Benchmark series in the field of ecology difficult. Each of the volume editors has recognized this inherent problem, and each has acted to solve it in his or her unique way. Their collective efforts will, we anticipate, provide a survey of the most important concepts in the field.

The Benchmark series is especially designed for libraries of colleges, universities, and research organizations that cannot purchase the older literature of ecology because of cost, lack of staff to select from the hundreds of thousands of journals and volumes, or the unavailability of the reference materials. For example, in developing countries where a science library must be developed *de novo,* I have seen where the Benchmark series can provide the only background literature available to the students and staff. Thus, the intent of the series is to provide an authoritative selection of literature, which can be read in the original form, but that is cast in a matrix of thought provided by the editor. The volumes are designed to explore the historical development of a concept in ecology and point the way toward new developments, without being a historical study. We hope that even though the Benchmark Series in Ecology is a library-oriented series and bears an appropriate cost it will also be of sufficient utility so that many professionals will place it in their personal libraries. In a few cases the volumes have even been used as textbooks for advanced courses. Thus we expect that the Benchmark Series in Ecology will be useful not only to the student who seeks an authoritative selection of original literature but also to the professional who wants to quickly and efficiently expand his or her background in an area of ecology outside his or her special competence.

In the present volume, Ruth Patrick has brought her immense experience in field biology to bear on the topic of diversity. Students of diversity must combine a skill in taxonomy and a knowledge of ecology in order to explore the relationships between species in a community. Dr. Patrick, who is Emeritus Director of the Department of Limnology, Philadelphia Academy of Sciences, has extensive knowledge of taxonomy, especially of the diatoms, and is a distinguished leader in ecological science. She has contributed directly to the study of natural diversity in freshwater and was among the first to employ diversity measures in assessment of the condition of streams and lakes. From this background, she has chosen a very special set of papers to represent several of the subsets that fall within the diversity problem. It is especially useful that she has examined diversity both from the taxonomic (kind of organism) and ecological (habitat) perspective.

FRANK B. GOLLEY

PREFACE

Over the past several decades a great deal has been written about ecological diversity, or as others term it, species diversity. At first the efforts were primarily focused on describing diversity of organisms in specific habitats or among certain taxonomic groups. Later the efforts shifted to explanation of differences among the observed patterns, focusing on how factors such as competition, predation pressure, invasion rate, and species pool develop and maintain diversity in communities. These studies continue today. Nevertheless, it seems a useful time to sum up and recapture, in a small way, some of the steps leading to our current concept of diversity.

In this collection of benchmark papers I have taken a biological, in contrast to a theoretical or mathematical, approach to the literature. As characterized by my own research work, I have stressed how diversity is expressed in various groups of organisms in the natural world and the factors important in developing and maintaining a diversified community.

Given the size of the literature and its complexity it is impossible to collect in one place all of the key or benchmark papers. Thus, a selection must be restrictive. For example, I have had to omit papers dealing with effects of perturbation on diversity of aquatic communities. Rather than select papers over the entire spectrum and be very shallow, I have attempted to stress general principles and key habitats and groups of organisms, under the assumption that if a student grasps the general pattern and concept he or she can understand specific topics and apply the concept in the field.

RUTH PATRICK

CONTENTS BY AUTHOR

DIVERSITY

INTRODUCTION

The word *diversity* denotes variety or multiformity, a condition of being different in character and quality. An appreciation that organisms in the natural world are diverse predates the writing of Aristotle, as the recognition of different kinds of animals is set forth in the Book of Genesis. This appreciation is also evident in the illumination of manuscripts of the fourteenth century. The Cocharelli manuscript, for example, is decorated with many kinds of invertebrates (Hutchinson, 1974). It is hard to comprehend the evaluation of diversity at that time. Clearly the numbers of new forms of life discovered in the Age of Exploration and through technical developments such as the microscope produced an explosion in biological knowledge. The attempt to classify such morphologically diverse organisms led to the system of binomial nomenclature produced by Linnaeus in 1758. Now biologists estimate that the total diversity of the biosphere may be several million species of which about half are known to science. This diversity of life is truly one of the most important features of the Planet Earth.

For many years routine ecological work focused on collecting, identifying, and then listing the species of organisms encountered in a habitat. These lists described the structure of the community. This activity was an essential element in the training of ecologists up to about 1960. The comparison of species lists from various habitats, at different times, and under various stresses caused ecologists to ask questions about why there were so many species present, why they occurred in such different numbers and sizes, and how these patterns related to other features of the ecological community.

Such questions were not new. For example, in 1858 Wallace and Darwin pointed out the importance of large numbers of species in developing the complex food webs observed in nature. The coupling of morphological diversity with functional diversity formed a basis for Darwin's theory of survival of the fittest (1859).

A bit later Forbes (1887) studied the interrelations among species and recognized the role of natural selection in adjusting the rates of multiplication and destruction of species to "best represent the common interests of the community." He realized the importance of diversity of prey and large populations of species for the maintenance of predators, and that certain predators have a greater selectivity of food than others. Commenting on the role of competition in the development of the foodweb, Forbes pointed out that this competition may exist among species of the same general systematic group, such as fish, or it may exist among members of very different taxonomic groups, such as insects, crustacea, and fish that feed upon *Entomostraca*.

Thus, well over a hundred years ago biologists recognized that the rich floras and faunas they encountered in habitats both were maintained by interactions among members of the flora and fauna and created certain recognizable features of ecological communities. In other words, diversity was both the resultant of actions of the biota and the environment, and a cause of community properties.

Later ecologists explored other elements of diversity. First, the relationships between the size or scale of the species in a community were studied, and then the functional attributes of the biota. For example, Elton (1930) discussed the importance of diversity of size of species in foodwebs and why there are many more small species than large species in natural systems. He pointed out that prey organisms were smaller than predators, and that more small organisms characteristically existed near the base of the foodweb. Hutchinson (1941), in his research on the cycling of phosphorus in lakes, noted that many species are involved in the cycling of this nutrient, and that the inorganic characteristics of a lake are important in regulating the numbers of species and sizes of their populations living within the lake. He drew attention to the fact that the overall functioning of a lake is similar to the functioning of an organism: The cycling of nutrients in lakes is accomplished by various organisms and in an organism, by various organs.

In 1949 Patrick showed that there was a similar pattern in the diversity of species in natural streams not perturbed by pollution. This pattern shifted in various ways as a result of the effects of different pollutants.

In 1957 Hutchinson considered the question of how so many organisms can live together and developed the theory of a niche as a hypervolume consisting of many resource parameters. He differentiated between the theoretical niche, which was a hypervolume in which a species might live, and the realized niche, which was the space the species occupied in competition with other species. Hutchinson's ideas provoked a series of papers dealing with the

2

question as to how so many species can live together. One of the researchers in this field was Robert MacArthur. He was most interested in developing mathematical models as to how species can divide resources and live compatibly with each other. In 1957 and 1958 he developed his theories concerning the fine-graining of the environment and the relationship of stability to complexity of the foodweb. These theories led MacArthur to consider the factors limiting diversity as well as those (MacArthur and Wilson, 1967) that increase diversity (1960). He continued his considerations with E. O. Wilson as set forth in their studies of island biogeography. Concurrently Odum (1969) and his students were investigating the diversity of the uses of energy in a community, and how the complexity of the foodweb allowed many more pathways for energy utilization because each species utilized the resources in a somewhat different way. He, as others, concluded that diversity was important in stable or steady-state systems.

The concern as to how species subdivide the environment and increase diversity led to a better understanding of predator/prey relations and, through the work of Jansen et al. (1976), the role of allelochemicals in regulating predation. Similarly, this concern has led to a more in-depth understanding of the various ways in which species compete: for example, in discovering the role of excreted chemicals by one species to control the growth of another (Proctor, 1957; Keating, 1976).

A desire to model these species relationships and to develop indices of diversity has resulted in a series of mathematical formulations. The two models that have been used most extensively are the "broken stick" by MacArthur (MacArthur, 1957, 1960; King, 1964; Deevey, 1969; Goulden, 1969) and the "truncated normal curve" (Preston, 1948; Williams, 1953, 1964; Patrick et al., 1954; Whittaker, 1965; May, Paper 19; Sugihara, Paper 20).

Several different indices have been developed. The most straightforward and oldest is simply counting the number of species, which indicates the number of niches available at any point in time. Simpson's (1949) equation is another relatively simple index, which emphasizes the dominance of the few species. The Shannon/Weiner index was first introduced in the field of ecology by MacArthur (Paper 1). It emphasizes the similarity or equibility of species population; that is, communities with many species with equal-size populations have the highest index. At about the same time, Whittaker (1965, Paper 9) developed "importance value curves," ranking species according to their commonness in a community in terms of the sample size. This index has the advantage over the Shannon/Weiner index in giving more information concerning the structure of the community.

One cannot separate a discussion of diversity from a discussion of the niche. Therefore, it is important that the students reading this volume should also refer to *Niche Theory and Application* by Whittaker and Levin, eds. (1975).

In conclusion, the word diversity, as used now, has many different connotations. One should be careful to define the type of characters of the species or of the community being considered when discussing "diversity" as the factors governing the diversity of species in various groups of organisms differ. In this book the various types of diversified communities and factors contributing to the diversity in various phylogenetic groups of plants and animals are discussed.

REFERENCES

Darwin, C., 1859, *On the Origin of Species by Means of Natural Selection, or the Preservation of Favoured Races in the Struggle for Life,* 1 vol., J. Murray, London, 502p.

Deevey, E. S., Jr., 1969, Specific Diversity in Fossil Assemblages, in *Diversity and Stability in Ecological Systems,* G. M. Woodwell and H. H. Smith, eds., Brookhaven Symp. Biol. No. 22, U.S. Dept. Commerce, Springfield, Va., pp. 224–241.

Elton, C., 1930, *Animal Ecology and Evolution,* Clarendon Press, London, 96p.

Forbes, S. E., 1887, The Lake as a Microcosm, *Sci. Assoc. Peoria, Ill. Bull.,* pp. 77–87.

Goulden, C. E., 1969, Temporal Changes in Diversity, in *Diversity and Stability in Ecological Systems,* G. M. Woodwell and H. H. Smith, eds., Brookhaven Symp. Biol. No. 22, U.S. Dept. Commerce, Springfield, Va., pp. 96–102.

Hutchinson, G. E., 1941, Limnological Studies in Connecticut. IV. The Mechanism of Intermediate Metabolism in Stratified Lakes, *Ecol. Monogr.* **11:**21–60.

Hutchinson, G. E., 1957, Concluding Remarks, in *Population Studies, Animal Ecology, and Demography,* Cold Spring Harbor Symp. Quant. Biol. No. 22, pp. 415–427.

Hutchinson, G. E., 1974, Aposematic Insects and the Masters of the Brussels Initials, *Am. Sci.* **62:**161–171.

Jansen, D. H., H. B. Juster, and I. E. Liener, 1976, Insecticidal Action of the Phytohemagglutinin in Black Beans, *Science* **192:**795–796.

Keating, K. I., 1976, *Algal Metabolite Influence on Bloom Sequence in Eutrophied Freshwater Ponds, EPA-600/13-76-081,* U.S. Environmental Protection Agency, Corvallis, Oregon, 148p.

King, C. E., 1964, Relative Abundance of Species and MacArthur's Model, *Ecology* **45:**716–727.

Linnaeus, C., 1758, *Systema Naturae, Regnum Animale,* 10th ed., 2 vols., Laurentii Salvii, Stockholm.

MacArthur, R. H., 1957, On the Relative Abundance of Bird Species, *Natl. Acad. Sci. (USA) Proc.* **43:**293–295.

MacArthur, R. H., 1958, Population Ecology of Some Warblers of Northeastern Coniferous Forests, *Ecology* **39:**599–619.

MacArthur, R. H., 1960, On the Relative Abundance of Species, *Am. Nat.* **94:**25–36.

MacArthur, R. H., and E. O. Wilson, 1967, *The Theory of Island Biogeography,* Princeton University Press, Princeton, N.J., 203p.

Odum, E. P., 1969, The Strategy of Ecosystem Development, *Science* **164:**262–270.

Patrick, R., 1949, A Proposed Biological Measure of Stream Conditions Based on a Survey of Conostoga Basin, Lancaster County, Pa., *Acad. Nat. Sci. Phila. Proc.* **101:**277–341.

Patrick, R., M. H. Hohn, and J. H. Wallace, 1954, A New Method for Determining the Pattern of the Diatom Flora, *Acad. Nat. Sci. Phila. Not. Nat.* **259:**1–12.

Preston, F. W., 1948, The Commonness, and Rarity, of Species, *Ecology* **29:**254–283.

Proctor, V. W., 1957, Studies of Algal Antibiosis Using Haematococcus and Clamydomonas, *Limnol. Oceanogr.* **2:**125–139.

Simpson, E. H., 1949, Measurement of Diversity, *Nature* **163:**688.

Wallace, A. R., and C. Darwin, 1858, On the Tendency of Species to Form Varieties; and on the Perpetuation of Varieties and Species by Natural Means of Selection, *Linnean Soc. (London) Proc.* **56:**45–62.

Whittaker, R. H., 1965, Dominance and Diversity in Land Plant Communities, *Science* **147:**250–260.

Whittaker, R. H., and S. A. Levin, eds., 1975, *Niche: Theory and Application,* Benchmark Papers in Ecology, vol. 3, Dowden, Hutchinson & Ross, Stroudsburg, Pa., 448p.

Williams, C. B., 1953, The Relative Abundance of Different Species in a Wild Animal Population, *J. Anim. Ecol.* **22:**14–31.

Williams, C. B., 1964, *Patterns in the Balance of Nature, and Related Problems of Quantitative Ecology,* Academic, New York, 324p.

Part I

DIVERSITY AND RELATED CONCEPTS

Editor's Comments
on Papers 1 and 2

In the field of ecology the term diversity is commonly used to describe the assemblage of species that interact . . . and form . . . a community. Species in the complex do not merely respond to a particular environment but create new conditions through their inter-actions with each other. For example, one species may modify an environmental factor such as light so that another species, or group of species, can live more successfully, or one species may be the food source for another or produce oxygen by photosynthesis, which is necessary for respiration of both. Through such interactions the community develops its identity and carries out its characteristic functions.

Diversity is a generalized term that refers to the structure of the community. In a sense, it expresses the genetic variability existing in the taxa that occur together and, therefore, the adaptive capacity of the assemblage. Thus, the measure of diversity is not merely a count of presence but rather it is a measure of the structural and functional interactions of the community.

When one considers the structure of communities of organisms, the first question that arises after one has the list of species in hand is why are there so many species with such different characteristics? Reasoning from our human experience, we might think that a single-species community structure might be more efficient. However, this is not the case. For example, at the herbivore level of an aquatic community we might find insects and fish of a variety of species, genera, and families feeding upon the plants. In another comparable community we might find protozoa, as well as insects and fish, serving as herbivores. Intuitively one would say that the gene pool present in these herbivore taxa was greater in the second case than in the first. Furthermore, the second set might consume a greater variety in size and taste preference. Its tolerance to natural products produced by various plants also might be broader; therefore, nutrient transfer from the primary production level might be more, not less, efficient (Freeland and Jansen, 1974).

In general we find that species seem to prefer a variety of food rather than a single species. Of course, a notable exception is parasitism. In the aquatic world, organisms from protozoa to fish generally prefer many species of diatoms as a food source. In contrast, blue-green algae and green algae such as *Cladophora glomerata* are the least preferred food sources. This may be due to lower food value and the presence of toxic chemicals in *Cladophora*. These data indicate that the characteristics of the prey as well as the preferences of the predator determine the efficiency of nutrient transfer in a diverse community.

One might consider diversity of organisms from a functional rather than a taxonomic viewpoint, although usually they go hand in hand. That is, in an environment that is favorable for many species, one will find an association composed of a large number of species each of which can utilize the environment in a somewhat different way and thus they can co-survive. MacArthur (1969) and others have pointed out that in such a case there may be a large invasion rate of species into the favorable habitat, with the result that more species will become established and will ultimately pack the area with the greatest number of species that can coexist by utilizing the resources in different ways. From this standpoint the number of species gives us the most information about the diversity of the environment. The most diverse environments exposed an adequate time to invasion would be characterized by many species with relatively small populations. The presence of disproportionately large populations might mean that the environment or the multidimensional habitats were not as diverse as they might be if there were more species with smaller populations, and therefore less redundancy. For example, in a riffle of a stream there is a range of size of rocks that will support the greatest number of species if they are distributed so as to produce a large variety of current patterns and protection against predators. Rocks that are too small roll or are shifted too much and are poor habitats. Rocks that are too large will produce a redundancy of habitats and will not support any more species, but may support more individuals of the same species. Thus, the diversity of organisms tells us about the diversity of the environment and vice versa.

9

In considering diversity of a community it is also relevant to know whether the species are mainly species with high population growth rates and high productivity ("r" selected species) or "K" species, which are more efficient at utilizing resources and have lower rates of population growth and production (MacArthur and Wilson, 1967). The "r" species often have short life cycles and may use resources less efficiently than "K" selected species. Most communities have a variety of "r" and "K" species. Areas in which most of the ecological factors are highly variable tend to have mostly "r" selected species, and species replacement is correlated with the shifting environment. Typically the existing species are not eliminated by unfavorable environmental conditions, but are greatly reduced in population size or occur in life-cycle stages that are dormant, cryptic, or are not collectable.

One may also think of diversity of species in the terms of their ability to disperse and invade communities (Diamond, 1975). Very stable "S" species are found only in species-rich communities and represent the extreme of "K" selection in MacArthur and Wilson's (1967) terminology. Then follows a series—A–B–C–D tramps—and finally supertramps that are widely distributed, do well in harsh environments, but can maintain themselves in species-rich communities.

Any discussion of diversity in ecosystems will find itself intertwined with the concept of stability (see Botkin and Sobel, 1975; Levin, 1970). In the last thirty years the terms diversity and stability have been the subjects of many scientific papers in response to questions such as: Why are there so many species? (Hutchinson, 1958); to findings by Patrick (1949) that in similar ecological habitats the numbers of species were similar and remained similar over time if severe perturbation did not occur; and to the Odums' (1962; 1957) concepts of community homeostasis derived from study of the energy flow through communities and the relationship between community structure and function. It has often been proposed that a direct relationship exists between diversity and stability, and research on this relationship reached a peak with the 1969 Brookhaven Symposium on *Diversity and Stability in Ecological Systems* (Woodwell and Smith, eds.).

Stability has many definitions but the two that pertain to our use are, first, a quality of endurance without alteration and, second, the ability to return to the original form after alteration. It is this double meaning that has contributed to the different interpretations of the meaning of stability in considering community structure.

Stable communities composed of a few species are found in harsh environments such as the *Spartina* marshes of the east coast of the United States. Niering (personal communication) has found by boring these marshes that the same species have been dominant over hundreds of years. There are many reasons why this is true. For example, the concentration of salt in the water and its high variability and the very high evapotranspiration rates in summer limit the number of species that can establish themselves in this environment. A second consideration is the characteristics of the *Spartina* plants. The species are perennials with well-branched rhizomes. In *Spartina alterniflora,* the most common species, the rhizomes are 4–7 mm thick. These rhizomes and roots form a tough mat and thus reduce the invasion of other species. These various factors would tend to mitigate invasion and, coupled with the harsh environment, produce stability of the species over time. This is an example of a highly productive community, in terms of carbon fixed, that is simple and stable.

A similar type of a relatively stable simple community has been observed in polluted streams. In this condition the nutrients may be adequate to support a diversified species community, but the pollutant contains a toxic substance in concentrations that only a few species can tolerate. These grow well and dominate the stream over time. An example is the dominant growth of *Stigeoclonium lubricum* in Lititz Run (Lancaster County, Pennsylvania). The stream receives a variety of toxic materials, particularly heavy metals. The absence of predators, which are killed by the toxic substances, allows the species to develop large standing crops. As long as the pollutant is released this simplified and productive (in terms of ^{14}C fixation) community is stable. Under natural conditions in this drainage basin other species are common and *Stigeoclonium lubricum* does not develop large populations.

These are examples of what might be classed as communities that live in harsh environments where stability is not correlated with diversity. There are, of course, other examples where stability and diversity in the community are highly correlated. Tropical rain forests and coral reefs are often cited as examples.

The other definition of stability—that is, the ability to return to the original state after displacement—is exemplified by communities found in natural streams. These communities are composed of many species forming interlocking chains or nets of nutrient transfer. Typically the species are "r" selection species, although a few may be "K" selection types. The communities are in a quasi-equilibrium state with significant inputs and outputs. The stream is a multi-dimensional resource area, with the resources fluctuating in an unpredictable manner. The species with short turnover time rapidly adjust

11

to the variable environment. Usually the number of species performing various functions remains fairly similar, although the kinds of species vary greatly in similar ecological habitats at the same time or in the same habitat during the same season over time (Patrick et al., 1969). Furthermore, these stream communities—because of the large available species pool, rapid invasion rates, and inputs and outputs—can recover rather quickly from severe perturbation; whereas smaller perturbations, such as increased nutrients, produce a readjustment in population sizes of the existing species rather than a shift in species.

McNaughton (1977) similarly found in his studies of the grassland of the Serengeti-Mara in Tanzania and Kenya that areas with large numbers of species adjust to the perturbation of chemical fertilization of the vegetation by shifts in the population sizes rather than shifts in species. Such communities were more stable than those with fewer species.

It is the natural oscillations of the multidimensional environment that help to maintain the large number of species, which in turn promotes functional stability.

From these comments it is evident that when one talks about species diversity one may be concerned with taxonomic diversity, functional diversity in the community, diversity in autecology of the species, or in reproductive strategy. In each case the primary datum is a count of species and individuals, and depending upon our knowledge of the taxa these counts may represent any relevant biological or ecological feature of the community.

It is interesting to note that the formation of diverse communities, whether they are bird or diatom communities, involves the same factors—that is, size of species pool, invasion rate, size and diversity of the area to be invaded (MacArthur and Wilson, 1963). Furthermore, the maintenance of a diverse community is dependent on density-independent factors, density-dependent factors, and predator pressure. The relative importance of these forces depends on the type of community being studied. A community in or near equilibrium is usually more dependent on density-dependent factors and predator pressure for its maintenance, whereas a quasi-equilibrium community (with substantial inputs and outputs as compared with storage) often is more dependent on the type of oscillation of density-independent factors and on predator pressure for its maintenance. Papers 1 and 2 illustrate some of the forces or conditions that operate to produce and maintain a diversified community (see also Harper, 1969; Hutchinson, 1957; Kohn, 1971; and Pianka, 1966). Subsequent sections focus on diversity in different types of communities—aquatic, terrestrial, and laboratory—and consider different groups of organisms.

REFERENCES

Botkin, D. B., and M. J. Sobel, 1975, Stability and Time Varying Ecosystems, *Am. Nat.* **109:**625–646.

Diamond, J. M., 1975, Assembly of Species Communities, in *Ecology and Evolution of Communities,* Belknap Press, Harvard University, Cambridge, Mass., pp. 342–445.

Freeland, W. J., and D. H. Jansen, 1974, Strategies in Herbivory by Mammals: The Role of Plant Secondary Compounds, *Am. Nat.* **108:**269–290.

Harper, J. L., 1969, The Role of Predation in Vegetational Diversity, in *Diversity and Stability in Ecological Systems,* G. M. Woodwell and H. H. Smith, eds., Brookhaven Symp. Biol. No. 22, U.S. Dept. Commerce, Springfield, Va., pp. 48–61.

Hutchinson, G. E., 1957, Concluding Remarks, in *Population Studies, Animal Ecology, and Demography,* Cold Spring Harbor Symp. Quant. Biol. No. 22, pp. 415–427.

Hutchinson, G. E., 1958, Homage to Santa Roaslia, or Why Are There so Many Kinds of Animals? *Am. Nat.* **93:**145–159.

Kohn, A. J., 1971, Diversity, Utilization of Resources, and Adaptive Radiation in Shallow-water Marine Invertebrates of Tropical Oceanic Islands, *Limnol. Oceanogr.* **16:**332–348.

Larson, R. A., in press, The Potential for Biological Controls of *Cladophora glomerata.* Part 2: Chemical Composition of *Cladophora glomerata* Cells. *(Environ. Prot. Agency Rep.)*

Levin, S. A., 1970, Community Equilibria and Stability, an Extension of the Competitive Exclusion Principle, *Am. Nat.* **104:**413–423.

MacArthur, R. H., 1969, Species Packing and What Competition Minimizes, *Natl. Acad. Sci. (USA) Proc.* **64:**1369–1371.

MacArthur, R. H., and E. O. Wilson, 1963, An Equilibrium Theory of Insular Zoogeography, *Evolution* **17:**373–387.

MacArthur, R. H., and E. O. Wilson, 1967, *The Theory of Island Biogeography,* Princeton University Press, Princeton, N.J., 203p.

McNaughton, S. J., 1977, Diversity and Stability of Ecological Communities: A Comment on the Role of Empiricism in Ecology, *Am. Nat.* **111:**515–525.

Odum, E. P., 1962, Relationships Between Structure and Function in Ecology, *Japanese J. Ecol.* **12:**108–118.

Odum, H. T., 1957, Trophic Structure and Productivity, Silver Springs, Florida, *Limnol. Oceanogr.* **2:**85–97.

Patrick, R., 1949, A Proposed Biological Measure of Stream Conditions Based on a Survey of Conestoga Basin, Lancaster County, Pennsylvania, *Acad. Nat. Sci. Phila. Proc.* **101:**277–341.

Patrick, R., B. Crum, and J. Coles, 1969, Temperature and Manganese as Determining Factors in the Presence of Diatom or Blue-green Algal Floras in Streams, *Natl. Acad. Sci. (USA) Proc.* **64:**472–478.

Pianka, E. R., 1966, Latitudinal Gradients in Species Diversity: A Review of Concepts, *Am. Nat.* **100:**33–46.

Woodwell, G. M., and H. H. Smith, eds., 1969, *Diversity and Stability in Ecological Systems,* Brookhaven Symp. Biol. No. 22, U.S. Dept. Commerce, Springfield, Va., 264p.

Reprinted from *Cambridge Philos. Soc. Biol. Rev.* **40**:510-533 (1965) by permission
of Cambridge University Press

PATTERNS OF SPECIES DIVERSITY

By ROBERT H. MacARTHUR

Department of Biology, Princeton University

(*Received 8 February* 1965)

CONTENTS

I. INTRODUCTION

Patterns of species diversity exist. Usually there is an increasing gradient of numbers of species from the poles toward the equator. There is also a general pattern of remote or small islands having fewer species than larger islands near the source of colonization. Finally, within a fairly small region there are what might be called within- and between-habitat patterns although the distinction is vague. The within-habitat pattern is the predictable number of species which co-exist in spite of regular overlap in their place of feeding, or which have made other adjustments to co-exist. But the total number of species present in a geographic area usually greatly exceeds the number within a component habitat, because different species are likely to occupy different habitats; this is the between-habitat pattern of diversity. If the patterns were wholly fortuitous and due to accidents of history, their explanation would be a challenge to geologists but not to ecologists. The very regularity of some of the patterns for large taxonomic groups suggests, however, that they have been laid down according to some fairly simple principles, and it is the purpose of this article to review the facts and some of the explanations which have been proposed. The reader may refer to Goodall (1962) for a very complete bibliography of the subject.

An analogy is unusually enlightening as a means of clarifying the relevant features of the problem. If we were to ask 'Why does one library contain more books than another?' there would be two categories of answer. If both libraries were known to be full, the answer would have to do with the total length of the shelves and the average distance between centres of books. If, on the other hand, the libraries were not yet full, a wholly different answer would be appropriate, expressed in terms of the ages of the libraries and their rates of acquisition and elimination of books. The same categories of answer are appropriate for the species-diversity question: if the areas being compared are not saturated with species, an historical answer involving rates of speciation and length of time available will be appropriate; if the areas are saturated with species then the answer must be expressed in terms of the size of the niche space (the biological equivalent of the length of the book shelves) and the limiting similarity of co-existing species (the equivalent of the distance between books). There is of course no necessity for different aspects of the diversity to have the same type of explanation. Indeed, we shall see that there is strong evidence that species reach saturation within a habitat rather quickly even though the total number of species in all habitats combined may still be increasing. Thus it is possible that the pattern of the total number of species may be best given an historical answer, while the within-habitat patterns of diversity may have an equilibrium explanation.

II. MEASURES OF SPECIES DIVERSITY

The simplest measure of species diversity is a count of the number of species. This applies to resident species and not those which are present by accident. Thus an ornithologist making a census of the breeding birds of a fifteen-acre forest will normally ignore some sea bird seen flying over once during a storm. He counts only those species which give evidence of breeding. The typical result of a breeding-bird census is a virtually accurate list of all species present with a slightly less accurate count of the number of pairs of each. In what follows we shall make considerable use of the data provided by such censuses. But the problem is much more difficult for the botanist counting tree species in the same forest, for he cannot so easily eliminate from his count an established tree which has no business there but was derived from a seed blown in many years ago from a more suitable adjacent habitat. By 'no business there' we mean that the species cannot maintain itself within the forest but must rely on immigration for its continued existence. G. H. Orians (personal communication) has even suggested that only the one most shade-tolerant tree species would persist in a large forest of uniform topography and soil, and that all others are in the same sense temporary or accidental. Hence the problem may be severe. However, trees, like birds, can be counted and identified relatively easily so that many censuses are available, even if they must be interpreted with care. Both the tree and bird counts can be complete: every tree or breeding-bird species in the fifteen-acre forest will be included. Many other species counts are really samples: the soil arthropods collected in a funnel or the fish from a net do not usually include members of every species present. Some of the rare species are likely to be missing. However, after repeated sampling, a relatively complete list of species can usually be assembled, even if no accurate estimate of individuals per species is available.

Thus, species counts or fairly complete samples (with or without data on the abundances of the species) are usually available. These quite simple species counts are adequate for studying some of the patterns of species diversity.

Various authors have used refined measures of species diversity designed to overcome the two principal drawbacks of using these species counts. The potential drawbacks of species counts are that they fail to take account of species abundance and

15

that they depend upon sample size. A census with ninety-nine individuals of one species and one of a second has the same number of species (two) as does one with fifty of one and fifty of the second. Yet most people intuitively feel that the second census should be assigned a greater species diversity. Two essentially different ways have been used to overcome this difficulty. First, it is possible to fit a variety of statistical distributions to the distribution of relative abundance of species. These may have parameters which can be used as diversity indices. Thus Fisher, Corbett & Williams (1943) guessed that a sample of species from an area should have a negative binomial distribution of abundance. They went on to approximate this by a log-series. These distributions fit the data for moth species caught in a light trap reasonably

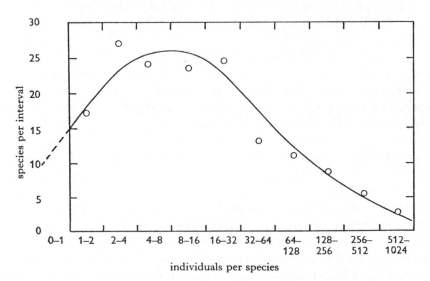

Fig. 1. The abundances of diatom species on a slide are plotted by Preston's method. The numbers of species with abundances 1–2, 2–4, 4–8, etc., are plotted as points and a truncated normal distribution is fitted to the points. (From Patrick, 1954.)

well and a parameter of the log-series is an appropriate measure of species diversity (see Williams (1964) for further results). Similarly, Preston (1948) showed that in a variety of censuses the number N of species of abundance x could be approximated by the lognormal distribution which, for convenience, Preston wrote in the form

$$y = y_0 \exp \{-aR^2\} = y_0 \exp \left\{ -\frac{R^2}{2\sigma^2} \right\}, \tag{1}$$

where R was the number of 'octaves' right or left of the mode, and each 'octave' is double the previous one. Thus the first octaves are 1–2 individuals, 2–4 individuals, 4–8 individuals, 8–16 individuals, etc., and the number of species, y, with abundances within these octaves are given by equation (1). Preston plots one point for each octave, giving the number of species whose abundances fall within the octave. He then fits a truncated normal distribution to the points. (Actually, since 2, 4, 8, 16, etc., are boundary abundances, belonging to two octaves, the species with these abundances

16

are half assigned to each appropriate octave.) Figure 1 shows a diatom count made by
Patrick (1954) plotted by Preston's technique. (A glass slide was suspended in fresh
water according to methods described by Patrick for a fixed time and then the diatoms
which had settled and reproduced were identified and counted. Since freshwater
diatoms are not free-living planktonic species, these slides represent very convenient
communities.) On the assumption that a doubling of the sample size will approximately double the number of each species present, and add a few new species represented by one or two individuals, we see that the shape of Preston's curve is virtually
(but not exactly) independent of sample size and that with a doubling of sample size
it simply moves one unit to the right, unveiling its left tail by one further unit. Thus,
for instance, each species which formerly had between 2 and 4 individuals now has
between 4 and 8; all which formerly had 4–8 now have 8–16, and so on. The rare
species will, however, only approximately double. This new curve is nearly the old
one moved one unit right. Preston reasoned that an infinitely large sample would
reveal all the species under the whole curve, even those parts now left of the origin
and hence currently undiscovered. This number of species, N, which is $y_0\,\sigma\sqrt{(2\pi)}$, is
a reasonable measure of species diversity. It also, as we have seen, takes species
abundance into account and is nearly independent of sample size. Both Preston
(1960, 1962) and Patrick (1963) have made very good use of this estimate in
investigating species diversity.

A similar parameter to estimate species diversity was suggested by many plant
ecologists (see Gleason, 1922; Goodall, 1952) and independently by Odum, Cantlon &
Kornicker (1960) who kept track of the number of species as the sample size increases.
They found that if they plotted cumulative numbers of species against the logarithm
of the number of individuals they got a nearly straight line passing through the
origin. The slope of this line, measuring how fast species accumulate as the number of
individuals increases, has many of the properties required of a species diversity
indicator.

A second way to take account of the abundances of the species is independent of any
hypothetical distribution of relative abundance. We ask the question: 'How difficult
would it be to predict correctly the species of the next individual collected?' This is a
problem which communication engineers have had to face, because they are interested
in the difficulty of predicting correctly the name of the next letter which might appear
in a message. When successive letters are chosen independently, the formula

$$H = -\sum_{i=1}^{26} p_i \log_e p_i$$

measures the uncertainty of the next letter, where p_i is the probability of the ith
letter of the alphabet. Similarly, if successive individuals in our census are independent of previous ones,

$$H = -\sum_{i=1}^{N} p_i \log_e p_i$$

is the appropriate measure of the uncertainty of the specific diversity of the next
individual in our census. Here N is the number of species in the count and p_i is the

proportion of the total number of individuals which belong to the ith species. Notice that if all N species are equally common, each is a proportion $1/N$ of the total. Thus the measure

$$-\sum_{i=1}^{N} p_i \log_e p_i \quad \text{takes on the value} \quad -N\left(\frac{1}{N}\log_e \frac{1}{N}\right).$$

This equals log N, so the measure of equally common species is simply the logarithm of the number of equally common species, meaning that E equally common species would have the same diversity as the N unequally common species in our census.

Returning to the example of a census with 99 individuals of one species and 1 of a second, we calculate

$$H = -p_i \log_e p_1 - p_2 \log_e p_2 = -0.99 \log 0.99 - 0.01 \log_e 0.01 =$$
$$0.0099 + 0.0461 = 0.0560.$$

For a census of fifty individuals of each of the two species we would get

$$H = -0.5 \log_e 0.5 - 0.5 \log_e 0.5 = -\log \tfrac{1}{2} = \log_e 2 = 0.693.$$

To convert these back to 'equally common species', we take $e^{0.0560} = 1.057$ for the first census and $e^{0.693} = 2.000$ for the second. These numbers, 1.057 and 2, accord much more closely with our intuition of how diverse the areas actually are, but it does not necessarily follow that E is more simple to predict from an examination of the habitat than the simple species count will be. This measure of species diversity was first used by Margalef (1957), who actually used the related measure

$$1/M \log M!/(M_1! \; M_2! \ldots M_2!)$$

in which M_1 is the actual abundance of the ith species and there are M individuals in all. By Stirling's theorem this is virtually identical to the other formula. Lloyd & Ghelardi (1964) calculated H and then gave tables relating it to what would be expected if the N species had abundances proportional to the lengths of the segments of a stick broken at $N-1$ randomly chosen points. This measure, H, is less dependent on sample size than is the species count, for the new species added as the sample size increases will be relatively rare and will make a small contribution to H.

The main virtue of the measurement H, or E, as an index of diversity is that it can be used equally well to measure species diversity from species abundances or to measure habitat diversity from knowledge of abundances of the components of the habitat. We shall see below that this permits us to find which components of the environment control species diversity.

In what follows we shall use Preston's measure of species diversity or the information-theoretic one, H, most often, but we shall not hesitate to use counts of species where their abundances are unknown, or where the actual number of species seems most appropriate.

We can make further interesting use of the information-theory diversity formula to obtain a measure of the faunal difference between two regions. We ask 'What multiple is the total fauna of the average of the simple censuses?' Suppose, for instance, each census had 8 equally common species. If they had no species in common the combined census would have 16 equally common species so $\frac{16}{8} = 2$ would be a measure of the difference—the largest this value could be. If they had no difference, but represented the same species, the total would have 8 equally common species and $\frac{8}{8} = 1$ would be the measure of difference—the minimum value the difference could take. Suppose, now, that the species in the censuses were not equally common; the information theory formula lets us convert the censuses to 'equivalent number of equally common species',

$$\exp\left\{\sum_{i=1}^{N} p_i \log_e p_i\right\} = E = e^H.$$

We take H_1 for the first census, H_2 for the second, and H_T for the total (in which p_i is the unweighted average, $(p_i(1)+p_i(2))/2$, of the proportions in the two censuses), and we use $\bar{H}$ for the unweighted mean of H_1 and H_2 and we then use $\exp\{H_T - \bar{H}\}$ as the measure of difference. This reduces to 2 and 1, respectively, for the simple cases described above, and can be applied to any censuses. It can be shown that this is approximately

$$\exp\left\{0.693 - 0.3(\pm 0.04)\left[\frac{O_1}{T_1} + \frac{O_2}{T_2}\right]\right\},$$

where O_1 is the numbers of individuals in census 1 of species also found in census 2; O_2, similarly is the number of individuals in census 2 of species also in census 1. T_1 and T_2 are the total numbers of individuals in the two censuses. This is a simpler formula and is very accurate when the censuses are rather different; for similar censuses, the correct formula must be used. For other measures of difference, see Greig Smith (1957) and Whittaker, (1960).

III. THE WITHIN-HABITAT COMPONENT OF SPECIES DIVERSITY

(1) *Facts*

If we examine any small, apparently homogenous, area we are likely to find that the number of species depends upon the structure of the habitat. Thus a few acres of forest support more breeding-bird species than do a few acres of field. Even if we census a field large enough to contain as many bird pairs (of all species combined), the forest has many more species. Since this aspect of species diversity is best understood for birds, we shall begin with the bird data. The experimental procedure is this: some measures of habitat complexity are guessed; to see which, if any, of these measures is responsible for the local bird species diversity it is sufficient to see, for a variety of bird censuses in a variety of different habitats, which habitat diversity measure is closely correlated with the bird species diversity. More precisely, a multiple regression of bird species diversity is calculated against all of the measures of habitat diversity which might be supposed to regulate the diversity of birds. This project has been carried out (MacArthur & MacArthur, 1961; MacArthur, 1964) and it was guessed that the species diversity of the plants and the 'profile' of the foliage into horizontal layers were likely features of the habitat controlling the birds. The success of the predictions justified the guess. Plant species diversity was measured with the same formula, $-\sum_i p_i \log_e p_i$, as was bird species diversity, but for the plants p_i referred to the leaf area of the ith species of plant divided by the total leaf area of all species. The 'foliage height diversity' was calculated as follows. For each height above the ground a density of foliage was estimated by optical means. A profile was constructed of height against density. By trial and error it was decided to subdivide the foliage into three layers corresponding to herbaceous ground cover (usually the part of the profile below 2 ft. in height), bushes and young trees (from 2 to 25 ft. in the profile) and canopy (usually over 25 ft. from the ground). The proportions of the total foliage area which lay in these three layers were called p_1, p_2 and p_3, and the same formula

$$- \sum_{i=1}^{3} p_i \log_e p_i$$

was used to compute the foliage height diversity. Remarkably enough, the bird species diversity was quite accurately predicted from the foliage height diversity

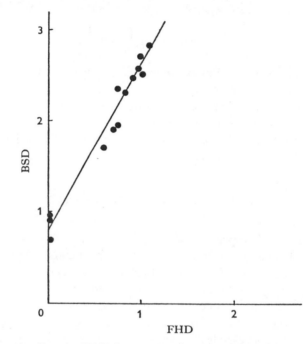

Fig. 2. Bird species diversity (BSD) is plotted against foliage height diversity (FHD) for bird censuses of a variety of habitats, each large enough to hold twenty-five pairs of all species combined.

alone (see Fig. 2) and knowledge of the plant species diversity did not enable an improved prediction to be made. The main point, however, is that there is a simple structural component to within-habitat species diversity. Furthermore, and this is more important, the prediction of Fig. 2 holds for such a wide geographic area (at least over much of the United States) that there appears to be a fairly uniform level of saturation of species. In other words, accidents of local history appear to have little or no effect on the within-habitat bird species diversity. (The between-habitat effect is another story, as we shall see later, although these facts, too, reinforce the present conclusion.)

For lizards, there are data of a similar nature, gathered by E. R. Pianka (personal communication), who found tentatively that the number of lizard species in flat, semi-desert areas of western U.S.A. was proportional to various correlated structural features of the environment. For instance, the relation between the number of lizard species in nine desert areas and the average volume of major shrub in cubic feet was related as in Fig. 3. Pianka's interesting study showed that most lizards, like the birds, appear to have reached a local saturation, so that the difference between habitats is more closely related to structure than to the time available for colonisation and rate of speciation. (He did find two species, however, which were unable to withstand the

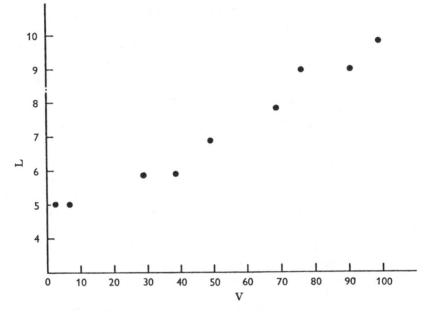

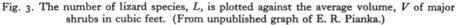

Fig. 3. The number of lizard species, L, is plotted against the average volume, V of major shrubs in cubic feet. (From unpublished graph of E. R. Pianka.)

colder climates of more northern deserts.) The closest parallel to this work, for plants, is that of Whittaker (1960 and earlier) who has shown that tree species occur, on gradients of temperature and moisture, only in specified places. The exact prediction of species diversity is, however, much more difficult for the trees. Whittaker recognizes the distinction between within- and between-habitat diversities and calls them alpha and beta diversities respectively.

For most defoliating insect species, a single tree constitutes a 'habitat'. Southwood (1961) showed, in apparent contradiction to these results, that the numbers of species of insects inhabiting various British tree species is proportional to the cumulative abundance of the various tree species throughout Quaternary history as measured by the total frequency of remains in the pollen record. These data suggest that insects inhabiting British trees have not had sufficient time to saturate their habitats. Thus, the historical answer seems appropriate; alternatively, we might suppose that a single tree constitutes more than one habitat and that Southwood is not dealing with within-habitat diversity.

Another aspect of within-habitat diversity is susceptible to fairly direct observation: the limiting similarity of co-existing species. Lack (1947) and Vaurie (1951) drew attention to the way in which the morphologies of two species diverge in the geographic area where their ranges overlap. Brown & Wilson (1956) reviewed many such cases, calling the phenomenon 'character displacement', and Hutchinson (1959) pointed out that there is a fairly constant relation, in the region of overlap, between the lengths of appropriate feeding appendages of the larger and smaller of the two species. This size ratio of larger to smaller is an indicator of how similar coexisting species can be; Hutchinson suggested 1·3 to be a typical value of the ratio. Klopfer & MacArthur (1961), in an attempt to show that the ratio is reduced in the tropics, meaning that species can be more similar there, measured some birds of species which appeared to be co-existing and found much lower values of this ratio. As several authors have pointed out in reply, there is no reason to think these species were character displacement pairs, so that the comparison was not legitimate. There was nothing wrong with the idea of Klopfer & MacArthur, however, and Schoener is currently examining the whole phenomenon critically.

(2) *Theory*

In Section IV it will be shown that the number of species within a habitat reaches saturation much sooner than the total number of species in a large area, since the latter continue to grow by increasing the species difference between habitats (i.e. the number of species which are found in some habitats and not others) even when each habitat is saturated. It is the purpose of this section, however, to show why there should be a limiting similarity of co-existing species, and hence a species saturation limit, and to relate this to the life-histories of the species involved.

We consider, first, the limiting similarity of resource-limited species, and defer until later those species which are limited by predators or other factors, singly or in combination. The exposition follows MacArthur & Levins (1964) and Levins & MacArthur (unpublished). A species limited by resources will normally be able to maintain its population whenever the density of the resources exceeds some threshold. Furthermore, the value of this threshold may be virtually independent of the size of the harvesting population. The rate of depletion of the resource will of course increase as the harvesters become more numerous, but at any instant of time it is the density of the resource and not of the harvesters which determines the yield to each harvesting animal. In symbols, where x is the population of a species and r_i, $(i = 1,2,\ldots,n)$, are the densities of the n resources,

$$\frac{dx}{dt} = \left(\sum_{i=1}^{n} a_i r_i - T \right) x,$$

where the a's are constants and the T is the threshold, so that x increases only when

$$\sum_{i=1}^{n} a_i r_i > T.$$

There are also equations for resource renewals, but they will not concern us here. For each resource, r_i, there is a value $R_i = T/a_i$ which indicates how much of resource i alone is needed to support species x. Different species, or different phenotypes within a species, will in general have different values, of R_i, because they will differ in their ability to locate, capture and produce offspring from the units of resource i. In fact, the values of R_i may be expected to vary as in the left-hand graphs of Fig. 4, in which

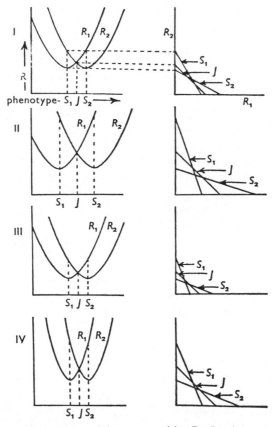

Fig. 4. In the left-hand graphs the minimum quantities R_1, R_2 of two resources to support a population are plotted against the population phenotypes. These graphs are in the category of 'raw data'. The right-hand graphs show how the data can be analysed to predict whether the jack-of-all-trades phenotype, J, will outcompete the specialist phenotypes, S_1 and S_2, in combination. The graphs assume resource-limited populations and also that specialists tend to reduce their favoured resource, while jacks-of-all-trades reduce whichever resource is commoner. For further details, see text.

for convenience two resources are considered and the values of R_1 and R_2 are plotted against a continuum of phenotype values. There will be a phenotype, s_1, which can maintain its population at a level of resource 1 which is lower than would be acceptable for any other phenotype. This is the specialist on resource 1. Similarly, there is a specialist, s_2, for resource 2. We can also distinguish a jack-of-all-trades, J, which is equally adept at using both resources. The curves in the figure can have a more or less arbitrary shape except that they first fall and then rise. Graph I gives a sample of two R curves such that the two specialists, in combination, are ousted by the jack-of-all-trades and kept from re-invading. This is seen from the right hand of the graphs I, which shows that the curve

$$a_1(\mathcal{J})R_1(\mathcal{J}) + a_2(\mathcal{J})R_2(\mathcal{J}) = T(\mathcal{J})$$

lies inside the intersection of the two curves of the specialists. In other words, when the resources are at the level R_1, R_2, determined by the intersection of

$$a_1(s_1)R_1(s_1) + a_2(s_1)R_2(s_1) = T(s_1) \quad \text{and} \quad a_2(s_2)R_2(s_2) + a(s_2)R_2(s_2) = T(s_2),$$

which is the lowest level to which the two specialists can reduce the resources, then $a_1(\mathcal{J})R_1 + a_2(\mathcal{J})R_2 > T(\mathcal{J})$ so that the jack-of-all-trades can still increase and thereby reduce the resources to an even lower level. Furthermore, the jack-of-all-trades, if he always decreases the commoner resource, will maintain the resources about midway along the $\mathcal{J}$ line, so that neither specialist can enter. Thus, no more than two species can persist on two resources and, if the resources are sufficiently similar, only one can, so there is a limiting similarity to co-existing species. (MacArthur & Levins (1964) give another reason why the jack-of-all-trades can prevent the specialists from re-entering.) The further biological significance of the figure comes from the comparisons of graphs II, III and IV with I. Graph II is identical to I except that the R curves are shifted farther apart (i.e. the resources are more different, requiring the specialist phenotypes to be more different). But the right-hand graph II shows that now the jack-of-all-trades line ($a_1(\mathcal{J})R_1(\mathcal{J}) + a_2(\mathcal{J})R_2(\mathcal{J}) = T(\mathcal{J})$, which is marked $\mathcal{J}$ in the figure) lies outside of the intersection of the specialist lines, so that the specialists in combination can oust the jack-of-all-trades and keep it from re-invading. We say the specialists are competitively superior. Graphs III show that the specialists also become competitively superior if the thresholds, $R = T/a$ are uniformly reduced, for the R curves in graph III are identical in shape and separation to those in graph I, but nearer to the phenotype axis. For birds, at least, lowering the clutch size may be expected to reduce the R curves by causing fewer mouths for the parents to feed and thus allowing the parents to bring up their brood with a lower resource density. Similarly, graphs IV show that with the same separation and threshold values as graph I, phenotypes s_1 and s_2 become competitively superior if the phenotypes become more specialized (as shown by narrower R curves). Mr Egbert Leigh has pointed out (personal communication) that increase in specialization and lowering of the specialists' resource thresholds are likely to go hand-in-hand, combining their advantages to the specialist. Both are presumably related to climatic variability and other aspects of the physical environment, as well as to hunting strategy. In fact, an increase in productivity, by reducing the time spent searching for food, will favour an increase in specialization.

Two other cases of interest are not shown in the figure. If one of the R curves is raised higher from the phenotype axis than the other, representing unequally harvestable resources, the advantage shifts away from the jack-of-all-trades. More important, however, is the situation which arises when the resources alternate temporally, so that resource 1 is present for some time in the absence of resource 2, and then resource 2 (but not 1) is present for some time. This situation clearly requires new equations and new right-hand graphs for Fig. 4, but it is clear that the advantage of the jack-of-all-trades is strengthened, since the specialists can only be as common as their respective resources will allow when these resources are minimal.

In summary, there is a limiting similarity in the species of co-existing resource-limited species and it is increased by increasing the specialization of the species, by reducing the resource thresholds (e.g. by reducing the number of young in a clutch for which the parents care), by increasing the productivity, by increasing the inequality in resource availability, and by reducing the temporal variability of the resources. Hutchinson (1965) has independently derived some of these results and applied them to the co-existence of African birds of prey.

Predator-limited species are much less simple to classify. A clever predator, such as man, can keep almost any number of prey species in a stable co-existence by maintaining each at a low population level but not so low that it becomes extinct. In fact, the most common danger is in the other direction: a species which is too common is in more danger both from diseases (a predator) and from the large ordinary predators than is a rare one (R. C. Lewontin, personal communication). Hence there is an advantage in predator-limited species being somewhat rare; yet if the individual species are rare, there can be many of them. As Lewontin has pointed out, this could conceivably be a reason why there are more tree species in a tropical habitat: each must be rare to avoid epidemics of tropical tree diseases, though why epidemics should be more prevalent in the tropics is still a problem.

If, however, the different species of prey are so similar that the predators consume them in the proportion in which they occur, then that prey species which is least tolerant of predation will be eliminated, as Leigh (1965) has pointed out to me. Hence there is also a limiting similarity to predator-limited species.

The analogy with the capacity of a library for books shows that there is another aspect of within-habitat diversity: the part analogous to the length of the shelves. Thus even though we know how similar co-existing phenotypes can be, we can only predict the within-habitat diversity if we also know what range of phenotypes are well adapted for some particular resource. Alternatively, since we know how similar the resources of co-existing species can be, we only need additional knowledge of the range of resources. It is the latter alternative which is easier to follow. The usable range of resources is not just the range of existing resources; rather it is the range of those existing resources which are maintained in sufficient quantity to support a species. Hence the more productive the habitat, the greater the fraction of existing resources which will be present in sufficient quantity. If, for instance, resources are normally distributed along an x-co-ordinate, with productivity determined by a constant, P, and with the actual range of resources controlled by the variance σ^2, then the usable range of resource will be the range along the x-axis, such that the distribution, $P \exp\{-x^2/2\sigma^2\}$, exceeds some constant, c. Solving $Pe^{-x^2/2\sigma^2} > c$ we get $x < \sigma \sqrt{(2 \log_e P/c)}$ so that the usable range would be proportional to the standard deviation, σ, and proportional to the square root of the logarithm of the productivity. Of course there is no reason to think that the resources will be exactly normally distributed, so this is just an indication of how the usable range might be related to the productivity and the actual range. T. Schoener has exploited this view in an unpublished manuscript. It is dangerous, however, to assume that increases in productivity should always be accompanied by increases in diversity. In fact, where productivity is increased there is often a correlated decrease in resource variety and greater inequality of existing resources. Both of these tendencies reduce species diversity. For instance, highly productive polluted rivers show reduced species diversity (Patrick, 1954). Hence, it is only where increased productivity is unaccompanied by a reduced spectrum of resources that we have any right to expect increased diversities.

The theory of within-habitat diversity is now reasonably complete. The number of species expected is the usable range of resources divided by the limiting similarity of resources which can be used by co-existing species. Both of these are qualitatively described above, and in summary the number of species within a habitat can be expected to increase with productivity (sometimes), with the structural complexity of the habitat, the lack of seasonality in resources, the degree of specialization and with reduced family size. Of these, only the structural complexity and to some extent the productivity are likely to vary over a small geographic area; hence the local variation in within-habitat diversity is expected to be explainable in structural terms (with a smaller and somewhat ambiguous productivity term included). Actually, productivity is often correlated with structural complexity, so a structural prediction may be quite accurate. This is the theoretical justification for the facts of within-habitat diversity given in the previous section.

It is interesting that every one of the features which increases the within-habitat diversity is found to a greater degree in the tropics. Hence there is no shortage of potential causes for a tropical increase in species diversity, and there is no reason to expect this increase to be controlled by any single potential cause. Most causes should be reflected in reduced character displacement in the tropics. However, as we shall see in later sections, much of the tropical increase in species diversity is of the between-habitat rather than within-habitat type.

IV. A THEORY OF WITHIN- AND BETWEEN-HABITAT DIVERSITY

We begin with two premises.

Premise 1. There is a limit to the similarity of co-existing species.

Premise 2. No species voluntarily restricts its habitat, but only does so when forced to, by competition or lack of suitable conditions. (We shall say two species which are too similar to co-exist belong to the same phenotype category.)

From these two premises it is simple to deduce a pattern of within- and between-habitat diversity of species. For concreteness, consider an island being colonized by an ever-increasing sample of the mainland species. The first species will, by the premises, occupy all appropriate habitats (a much wider selection of habitats than it occupied on the mainland where it was restricted by competitors). In all probability, the next species will be sufficiently different that it can co-exist with the first in all suitable habitats. The third species is likely to be sufficiently different from both that it too can occupy all suitable habitats, but the probability of it being too similar to one of the previous ones to co-exist is greater than it was for the second species. Eventually a species will arrive which is so similar to an existing species that both cannot co-exist. Then, since both species persist on the mainland, each will be likely to find a habitat on the island in which it is superior, and the first habitat selection will have occurred. As the fauna of the island becomes quite large, it will contain a representative of each phenotype category and no further increase in the number of species per habitat will take place. Each new species will, instead, cause a further subdivision of the habitats. In summary, during the initial stages of colonization the diversity of species will be wholly within-habitat; as the fauna increases there will be more and more between-habitat diversity, and eventually all new diversity will be between-habitat. This same distinction should hold in the comparison of any rich fauna with a poor fauna occupying an environment with the same amount of structure. Before an account of the empirical evidence, an indication will be given of how this theory can be made quantitative.

Picture the possible phenotypes as a continuous space, one co-ordinate for each measured variable of the phenotype (say a line or a plane) and subdivide this into a mosaic of equal-sized cells corresponding to the phenotype categories. Let the incoming species fall, like raindrops, in a Poisson distribution on the phenotype plane. Then, if there are, on the average, m species per phenotype category, the expected proportion of phenotype categories with 0, 1, 2, 3, ... species will, owing to chance, be

$$e^{-m}, \ me^{-m}, \ \frac{m^2}{2!}e^{-m}, \ \frac{m^3}{3!}e^{-m}....$$

Each habitat will be expected to contain representatives of all phenotype categories except those which have no representatives on the island. In other words, the proportion of 'vacant niches' will fall according to e^{-m} and the number of species in a given habitat will grow proportionally to the formula $1 - e^{-m}$. The fraction me^{-m} of phenotype categories with a single species contains those species which have no habitat selection imposed by competition. Hence, the fraction $(me^{-m}/1 - e^{-m})$ of species within a habitat should show no competitive habitat selection and be present on other suitable parts of the island. As the species present on the island have more time together, they can make certain evolutionary adjustments, and any slight morphological change which would enable a species to move into a vacant phenotype category, or even into a less thoroughly exploited one, might be favoured. Eventually, by this process, all phenotype categories would acquire equal, or at least less unequal, numbers of species. We can use the popular term 'disharmonic' to describe the initial, unadjusted, community, and can call the other community which has equally exploited phenotype categories 'harmonic'. Real communities presumably fall somewhere in between these extremes.

The actual data relating to this are impressive in their confirmation of the theory. Crowell (1961) confirmed the earlier impressions of Lack (1942) and others by measuring the habitats of Bermuda bird species, comparing these with mainland measurements for the same species and showing that the island birds had larger habitats than their mainland counterparts. Crowell's technique requires that the island species be also found on the mainland. Patrick (1949) has shown the same phenomenon in diatoms.

We have used a method (MacArthur & Recher, unpublished) applicable even when the island species are different from the mainland ones. Basically, the method consisted in measuring in a standard way a difference between habitats and a difference between the censuses of the birds breeding in the habitats. The difference formula of Section II was used for both the difference in bird species and the difference in profile of vegetation density between habitats. With these, one can see whether the

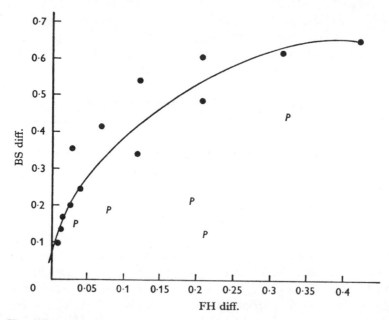

Fig. 5. The difference in bird species composition (BS Diff.) between census areas differing in foliage profile is plotted against foliage profile difference (FH Diff.), showing that in Puerto Rico (*P* points) the same difference in habitat causes much less change in bird species than on the mainland of U.S.A. (● points).

island bird species change less with habitat changes than do mainland species. Figure 5, from the data of Recher and the author, show very clearly that the birds of Puerto Rico change much less as habitats change than do the birds of the eastern U.S.A. Similarly, Whittaker (1960) showed that a richer flora was associated with reduced habitat selection by the plant species.

V. ISLAND SPECIES DIVERSITIES

There are two types of possible explanation for the impoverishment of island faunas and floras compared to the adjacent mainland, and both explanations are probably appropriate for different groups of organisms. The explanation most people give is that islands are impoverished because there has not been enough time: the full quota of mainland species has never had a chance to colonize. More recently Preston

(1962) and MacArthur & Wilson (1963) independently published theories accounting for an equilibrium in which the number of new immigrant species is balanced by the extinction of the rarer ones. The need for such a theory is apparent when one considers that Krakatau, which had at least all vertebrate and higher plant life destroyed by the volcanic explosion in 1883, had reconstitued a bird and insect fauna (of most orders) nearly equal to that of other islands of comparable size and remoteness within fifty years after the explosion (Dammermann, 1948). Clearly, then, birds and many kinds of insects approach equilibrium very quickly; mammals, on the other hand, are much slower and many remote islands may still be unsaturated with mammals, although the success of introductions is no proof, for the immigration rate is enormously higher when man introduces forms and an increase in the equilibrium fauna would be expected.

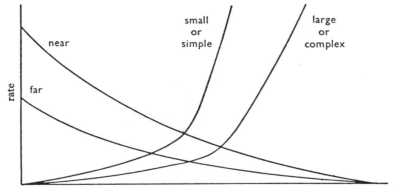

Fig. 6. The rates of immigration (falling curves) of new species, not on the island, and of extinction (rising curves) of species on the island are plotted for all numbers of species on an island. Where they intersect, the extinction balances the immigration. This equilibrium is shown to be greater for large, complex or near islands. See text for further discussion. (From MacArthur & Wilson, 1963.)

The theory of MacArthur & Wilson (1963) is summarized most conveniently in Fig. 6. It is permissible to ask what would be the rate of extinction of species and the rate of immigration of new species (not already on the island) for any number of species on the island. Thus, if the island has no species on it, all immigrants are new species and there are no species to become extinct. This is the left end of the curves in Fig. 6. As the number on the island grows, there are more species to become extinct, so this curve rises; there are also fewer new ones among the immigrants, so this curve falls, dropping to zero when the island has the full mainland fauna. Actually, the number of species will only grow until the number of new, immigrant species is balanced by the number of species becoming extinct. As Fig. 6 shows, this equilibrium number will be lower on small or uniform islands where extinction of additional species is likely (the rare ones with small niches are the likely candidates for extinction). It will also be lower on remote islands with lower rates of immigration from the sources of colonisation. The multiple regression analyses of Hamilton, Barth & Rubinoff (1964) and Watson (1965) show clearly that there is a habitat diversity component to the number of species on an island. This may be reflected in the extinction curves by the rather pronounced upward bend (see Fig. 6). This takes place where the number of species on the island roughly corresponds to the capacity of the island and is assumed to be proportional to the variety of habitats; species in excess find more difficulty in surviving and thus the extinction curves become steep. When this approximate number of species has been reached, the number of species varies but little with changes in the immigration rate, and instead depends mostly upon habitat diversity. This approach to understanding island diversities has the merit of explaining qualitatively both the distance effect and the area-habitat effect (Fig. 7); statistical consequences of the theory can also be used to distinguish between saturated and unsaturated island systems (MacArthur & Wilson, 1963). The exact shape of the curves is not simply predictable, however, so quantitative *apriori* pre-

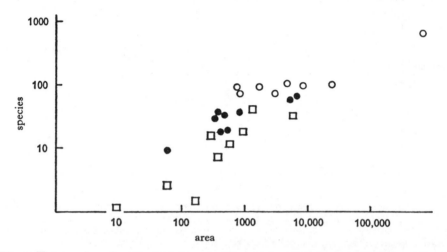

Fig. 7. The number of resident bird species is plotted against the area in square miles of Pacific islands in the Moluccas, Melanesia and Oceania. Islands less than 500 miles from the presumed source of colonization (New Guinea) are represented by circles, islands farther than 2000 miles by squares. Islands of intermediate distance are represented by dots. Part, but not all, of this apparent distance effect is due to the lesser elevation of the far islands. (From MacArthur & Wilson, 1963.)

dictions of island species diversity cannot be made. We turn to Preston's (1962) account for a better understanding of the extinction rates, and an alternate account ofthe whole theory in the case where the islands are of equal distance from the source ofcolonization.

Preston showed with considerable ingenuity how the lognormal distribution of species abundances can be used to describe species-area curves and other aspects of species diversity. (Preston's papers are too long to summarize here in their entirety.) As was shown in Section II, if a larger sample is taken from the same lognormal distribution, it essentially unveils more of the left-hand tail without altering the shape. When a biologist takes a larger sample, however, he is seldom sampling from the same distribution as in the small sample, for the large sample usually comes from a larger and hence more diverse area. Hence the unveiled curve has a slightly different shape from the original one, and we shall now show in what way it should be expected to differ. Some species will be adapted to many of the component habitats of the large area and these, the commonest, will be very abundant in the large sample—as abundant in fact as they would have been had the area sampled been uniform. Other species are adapted for only some of the component habitats, and these will not increase proportionally to the area sampled. Hence, the right-hand tail will move to the right faster than the mode, and the curve will become flatter—that is, it will be characterized by a larger value of the logarithmic standard deviation, σ. Preston obtains, by a circuitous route, an empirical relation between σ and the number of species, but it should be noted that, in his data, larger numbers of species arose from progressively larger areas sampled and not from richer faunas. In fact, as pointed out by Patrick (1963) and E. MacArthur (unpublished dissertation for Smith College, Northampton, Mass.), the value of σ decreases with the larger numbers of species in rich faunas or floras. In any case, Preston's relation between σ and the number of species reduced the number of independent constants in equation (1) to two, equivalent to the number of species, N, in the 'universe' (i.e. the number of species in the whole, untruncated curve) and the number of individuals in the sample. If we were to census each area until its complete fauna were counted, Preston assumes that its distribution would reveal the complete lognormal distribution (equation (1)), including the left tail to the point at which the rarest species has m individuals. (Empirically m is not far from 1; it is because of extinction that this is the tail of the distribution, so that almost no species are as rare as this.) This gives a relation, for these complete samples, which Preston calls 'isolates', between the remaining two independent constants, so that the relation between the number of species, N, in the isolate and the total number of individuals in the isolate is prescribed. Using the empirical relationships between constants, Preston obtained the equation $N = 2\cdot07(I/m)^{0\cdot262}$, where I is the number of individuals in the isolate. If we let p be the density of individuals and A the area of the isolate ($PA = I$), this becomes

$$N = 2\cdot07(P/m)^{0\cdot262}A^{0\cdot262}, \tag{2}$$

which is the most accurate species-area curve yet devised. It shows, as Preston noted that the number of species, N, grows, very roughly, with the fourth root of the area. This applies, of course, only to the complete samples, or isolates. Island faunas are such isolates, because extinction prevents many species from having abundance 1–2, and equation (2) describes island species diversities with considerable accuracy on the assumption that m is equal for all islands.

Both Preston's and MacArthur & Wilson's theories account for an equilibrium diversity on islands, but there is no reason to think all species on all islands have reached this equilibrium. In fact, we showed (MacArthur & Wilson, 1963) that the approach to equilibrium should be asymptotic, so virtually no island will have achieved complete equilibrium.

Dr Ruth Patrick has carried out most interesting experiments with diatoms which duplicate the whole process of island colonization. Each empty glass microscope slide which she suspends in the water is an island to be colonized, since she has pointed out that these diatom species only reproduce while attached and the floating ones are strictly immigrants. Bearing this in mind, she suspended small and large slides in uniform currents of low and high flow rates carrying immigrant diatoms. As expected, the small slides, like small islands, ended up with fewer species. More surprising was the spectacular demonstration of competition. The small slides with low flow rates initially had colonists of many of the available species; as the colonies grew, a number of these species became extinct by being crowded off the slides. Whether there would be further reductions in diversity as time went on is not yet clear. The small slides with high flow maintained many species. In another experiment, Dr Patrick has put slides of equal area in currents of equal flow rate and nutrient content. In one, however, the rate of immigration of diatoms is reduced by a filtering of some of the water. Preliminary counts show that the slide with reduced immigration has fewer species, suggesting that immigration is balancing extinction, as in the theory of Fig. 6.

VI. TOTAL DIVERSITY AND TROPICAL DIVERSITY

In Section IV it was shown that individual habitats tend to become saturated quite soon, as a total fauna becomes richer, and that the extra species—the excess of the total diversity over the within-habitat diversity—tend to spill over into other habitats in which they are better able to compete. Thus the total diversity of a taxonomic group over a large area of several habitats and the within-habitat diversity appear to be independently controlled. Whittaker (1960) recognized the importance of total diversity and called it gamma diversity. Most of the authors who have written on species diversity have been primarily interested in discovering why there are so many species (total diversity) in the tropics; this problem has been excellently reviewed by Fischer (1960), to which paper the reader may refer for abundant documentation of the total-diversity gradient which accompanies latitude changes. Simpson (1964) and especially Terent'ev (1963) have given fairly complete descriptions of mammal and other diversities, and Patrick (1961) has discussed the similarity in species diversity in river systems, providing new examples. Here we shall review selected parts of the old data, new data and the relevant explanations which have been suggested.

Preston (1960) plotted bird species counts for temperate and tropical areas on log-log co-ordinates, obtaining essentially Fig. 8 of this article. If the vertical distance between these curves were uniform, it would mean that tropical counts—over no matter what magnitude of area—were a uniform multiple of temperate counts from the same size area. This would mean that within-habitat diversity was solely responsible for the extra diversity of the tropics. If, on the other hand, the tropical curve rose faster it would mean that the tropics show more between habitat diversity. The data from Preston's graph are not adequate for discriminating between these possibilities. In fact, the data more recently collected suggest, tentatively, that for birds at least the two curves should almost meet on the left, that is, that very small tropical areas have only a few more species than temperate areas of the same size. For other organisms this is less certain. If this is confirmed, it will follow that a large part of the extra tropical diversity is 'between-habitat'. These considerations prompt two questions:

(1) Is the tropical increase in within-habitat diversity accounted for by the factors in the theory of Section III on within-habitat diversity, or has it some other—perhaps historical—explanation? (2) Is the greater between-habitat diversity due to a greater topographic diversity, or has it an historical explanation?

For birds, question one can already be answered. The within-habitat diversity on the relatively impoverished island of Puerto Rico is about as great as the within-habitat diversity of the temperate habitats and is in fact nearly equal to that of the very much richer tropical mainland. The way in which Puerto Rico is impoverished is, as was shown in Fig. 5, that the between-habitat component is so small. Now, all of the

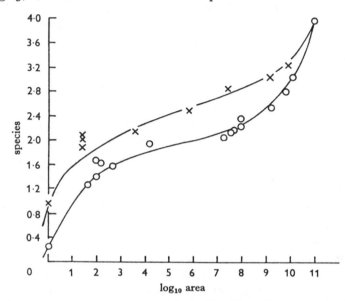

Fig. 8. Species-area curves for temperate birds in the New World (lower curve) and neotropical birds (upper curve). *N* is the number of species, *A* is the area in acres. (After Preston, 1960.)

factors which were shown theoretically to effect within-habitat diversity are just as tropical on Puerto Rico as on the tropical mainland, although their histories are very different. If the island were not at least partially satud, ratethe between-habitat component would not be so much larger on the mainland.

Question two is as yet unanswered. It is perfectly plausible—perhaps even probable—that this does have an historical answer. Perhaps more species have evolved which are adapted to tropical conditions and, by the process described in Section IV, the extra diversity is revealed by greater between-habitat differences. In fact, this historical explanation seems the only one that accounts for the astonishing species diversity in ancient lakes (Brooks 1950), which at least in some cases have no more structural diversity than the more impoverished recent lakes. Furthermore, Richards (1952) pointed out that the richest floras are the Mediterranean-type floras of unglaciated S.W. Australia and S. Africa, in contrast to the Mediterranean vegetation of the north temperate zone, which, although unglaciated, has had a more disturbed history.

33

There is a theoretical upper limit to total diversity set, not by limiting similarity as in within-habitat diversity, but by the abundance, or scarcity, of the species. For the total number of individuals hardly grows as the number of species increases, so the more species, the rarer each becomes. There obviously has to be at least one pair of each bi-sexual species, so the number of species cannot possibly rise further than half the number of individuals. In fact, of course, it cannot rise nearly that far, for the existence of many species is precarious when the abundance falls into the hundreds. More accurately, one must measure the stability, perhaps as Leigh (1965) has done but with other factors included, and compare this in temperate regions and tropics. It is conceivable that owing to fewer climatic hazards, species can be rarer in the tropics without running great danger of extinction. However, whether a limit has been reached or whether the number of species in both tropics and temperate regions is still increasing is hard to determine from present knowledge.

Various other theories of tropical total diversity have been proposed, and Connell & Orias (1964) have summarized the reasonable possibilities in one large scheme in which the roles of many mechanisms for increasing species diversity are interrelated. An increased diversity in species of lower trophic levels is certainly a potential cause of increased diversity among the species which consume them, as Hutchinson (1959) and Odum *et al.* (1960) have emphasized. However, this applies mostly to specialist species such as monophagous insects. It has been suggested, by analogy with human beings who have more occupations where populations are dense, that more species can persist where productivity is high. It is true that there often, but not always, seem to be more species where productivity is high. But the reason seems to be that given in Section III, where it was pointed out that only in areas of high productivity can a marginal niche support a species. Human occupations become diverse in order to exchange services effectively; such exchange of services is unknown among most organisms, so one should not be carried away by the analogy of species with human occupations.

In summary, although the total diversity—the total number of species of some phylogenetic group in a fairly wide geographic area composed of several habitats— has been the subject of a great deal of speculation and data tabulation, it is the aspect of species diversity which is least well understood.

No review of tropical species diversity could be complete without some mention of the groups which do not increase in diversity toward the tropics. Thorson (1957) pointed out that the *infauna* living in the material on the ocean bottom does not increase in diversity toward the tropics, although the *epifauna* living on the bottom does. Patrick (1964) found no increase either in the diatom diversity, the protozoan diversity or in the diversity of the freshwater insects when she sampled the upper Amazon in the same fashion in which she had sampled rivers of the eastern U.S.A. On the other hand, the fish fauna in the Amazon proper was richer. The explanation of these cases is obscure at present, but any general theory must include these exceptions.

VII. SUMMARY

1. Species diversity is most simply measured by counting species. More complicated measures, which take into account the relative abundance of the species, have been derived from information theory or from parameters of statistical distributions fitted to the census data. The information theory formulae can also be used to measure habitat diversity and differences between communities or habitats. In this way, changes in the pattern of species diversity can be compared with changes in the environment.

2. Small or remote islands and islands with uniform topography have fewer species than large or complex islands or islands nearer the source of colonization. For birds and some orders of insects it appears that the rate of colonization of new species is virtually balanced by the rate of extinction, so that the number of species has reached equilibrium. For other organisms, such as mammals, and for all organisms on the most remote islands, this equilibrium has probably not been reached and further increases in the fauna may be expected. The comparison of impoverished island faunas with the mainland faunas whence they were derived shows the effect of relaxed competition.

3. Local variations in the species diversity of small uniform habitats can usually be predicted in terms of the structure and productivity of the habitat. Habitats of similar structure on islands and mainland often have similar species diversities; the impoverishment of the island is reflected in the fact that different habitats on the island have nearly the same species, while different habitats on the mainland have more different species. This is interpreted as evidence that uniform habitats are nearly saturated with species and that new species usually colonize by occupying different habitats from present species.

4. The theory of competition and the facts of character displacement indicate that there is a limiting similarity to species which co-exist within a habitat. Species more similar than this limiting value must occupy different habitats. According to the theory, this limiting value should be less where productivity is high, where family size is low and where the seasons are relatively uniform. It should also be less for pursuing hunters than for species which search for stationary prey.

5. Total species diversities, from areas composed of many types of habitat, are usually, but not always, much greater in the tropics than in temperate regions. This is accomplished by a finer subdivision of habitats (habitat selection) more than by a marked increase in diversity within habitats. This total diversity may still be increasing and may have not reached saturation.

I have discussed this problem profitably with biologists too numerous to name here. I am especially grateful to J. Connell, G. E. Hutchinson, P. Klopfer, E. Leigh, R. Levins, R. Lewontin, E. W. MacArthur, G. Orians, E. Orias, R. Patrick, E. Pianka, T. Schoener and G. Watson who gave kindly let me see data or theories which are still unpublished. I also wish to thank M. Cody, J. MacArthur and H. Recher who helped me to collect some of the data presented here for the first time. My researches have been generously supported by the National Science Foundation and the American Academy of Arts and Sciences.

VIII. REFERENCES

BROOKS, J. L. (1950). Speciation in ancient lakes. *Q. Rev. Biol.* **25**, 30–176.

BROWN, W. L. & WILSON, E. O. (1956). Character displacement. *Syst. Zool.* **5**, 49–64.

CONNELL, J. H. & ORIAS, E. (1964). The ecological regulation of species diversity. *Am. Nat.* **98**, 399–414.

CROWELL, K. (1961). The effects of reduced competition in birds. *Proc. natn. Acad. Sci. U S.A.* **47**, 240–3.

DAMMERMAN, K. W. (1948). The fauna of Krakatau, 1883–1933. *Verh. K. Akad. Wet (Nat.)*, **44**, 1–594.

FISHER, R. A., CORBETT, A. S. & WILLIAMS, C. B. (1943). The relation between the number of species and the number of individuals in a random sample of an animal population. *J. Anim. Ecol.* **12**, 42–58

FISCHER, A. G. (1960). Latitudinal variations in organic diversity. *Evolution*, **14**, 64–81.

GLEASON, H. A. (1922). On the relation between species and area. *Ecology*, **3**, 158–62.

GOODALL, D. W. (1952). Quantitative aspects of plant distribution. *Biol. Rev.* **27**, 194–245.

GOODALL, D. W. (1962). Bibliography of statistical plant ecology. *Excerpt bot.* B, Bd 4, 253–322.

GREIG-SMITH, P. (1957). *Quantitative plant ecology.* London.

HAMILTON, T. H., BARTH, R. H. & RUBINOFF, I. (1965). The environmental control of insular variation in bird species abundance. *Proc. natn. Acad. Sci. U.S.A.* **52**, 132–40.

HUTCHINSON, G. E. (1959). Homage to Santa Rosalia, or, Why are there so many kinds of animals? *Am. Nat.* **93**, 145–59.

HUTCHINSON, G. E. (1965). *The ecological theater and the evolutionary play.* New Haven.

KLOPFER, P. H. & MACARTHUR, R. H. (1961). On the causes of tropical species diversity: niche overlap. *Am. Nat.* **95**, 223–6.

LACK, D. (1942). Ecological features of the bird faunas of British small islands. *J. Anim. Ecol.* **11**, 9–36.

LACK, D. (1947). *Darwin's finches.* Cambridge.

LEIGH, E. (1965). On a relation between the productivity, biomass, stability and diversity of a community. *Proc. natn. Acad. Sci. U.S.A.*

LLOYD, M. & GHELARDI, R. J. (1964). A table for calculating the equitability component of species diversity. *J. Anim. Ecol.* **33**, 217–26.

MACARTHUR, R. (1964). Environmental factors affecting bird species diversity. *Am. Nat.* **98**, 387–97.

MACARTHUR, R. & LEVINS, R. (1964). Competition, habitat selection and character displacement in a patchy environment. *Proc. natn. Acad. Sci.,* U.S.A. **51**, 1207–10.

MACARTHUR, R. & MACARTHUR, J. (1961). On bird species diversity. *Ecology,* **42**, 594–8.

MACARTHUR, R. & WILSON, E. O. (1963). An equilibrium theory of insular zoogeography. *Evolution,* **17**, 373–87.

MARGALEF, R. (1957). La teoria de la inforacion enecologia. *Memorias de la real academia de ciencias y artes (Barcelona),* **33**, 373–449.

ODUM, H. T., CANTLON, J. E. & KORNICKER, L. S. (1960). An organizational hierarchy postulate for the interpretation of species individual distributions, species entropy, ecosystem evolution and the meaning of a species variety index. *Ecology,* **41**, 395–9.

PATRICK, R. (1949). A proposed biological measure of stream conditions. *Proc. Acad. nat. Sci. Philad.* **101**, 277–341.

PATRICK, R. (1954). A new method for determining the pattern of the diatom flora. *Notulae Natural of the Academy of Natural Sciences of Philadelphia,* no. 259, 1–12.

PATRICK, R. (1961). A study of the numbers and kinds of species found in rivers in eastern United States. *Proc. Acad. nat. Sci. Philad.* **113**, 215–58.

PATRICK, R. (1963). The structure of diatom communities under varying ecological conditions. *Ann. N.Y. Acad. Sci., Wash.,* **108**(2), 353–8.

PATRICK, R. (1964). A discussion of the results of the Catherwood Expedition to the Peruvian headwaters of the Amazon. *Verh. int. Verein. theor. angew. Limnol.* **15**, 1084–90.

PRESTON, F. W. (1948). The commonness, and rarity, of species. *Ecology,* **29**, 254–83.

PRESTON, F. W. (1960). Time and space and the variation of species. *Ecology,* **41**, 785–90.

PRESTON, F. W. (1962). The canonical distribution of commonness and rarity. *Ecology,* **43**, 185–215. 410–32.

RICHARDS, P. W. (1952). *The tropical rain forest.* Cambridge.

SIMPSON, G. G. (1964). Species density of North American recent mammals. *Syst. Zool.* **13**, 57–73.

SOUTHWOOD, T. R. E. (1961). The numbers of species of insect associated with various trees. *J. Anim. Ecol.* **30**, 1–8.

Терентьев П. В. (Terent'ev P. V.) (1963). Опыт применения анализа вариансы к
    качественному богатству фауны наземных позвоночных СССР. *Вестник Ленинградского
    университета* No. 21, *Серия биологии выпуск* 4. *Vest. Leningradsk Univ.*, *Ser. Biol.*, **18**, 19–26.
Thorson, G. (1957). Bottom communities. In *Treatise on marine ecology and paleoecology* (ed. Ladd).
    Mem. geol. Soc. Am. **67**, 461–534.
Vaurie, C. (1951). Adaptive differences between two sympatric species of nuthatches (*Sitta*). *Proc.
    Xth int. Ornithol. Congr. Uppsala*: pp. 163–6.
Watson, G. (1964). Ecology and evolution of passernine birds on the islands of the Aegean Sea.
    Ph.D. dissertation, Yale University, New Haven.
Whittaker, R. H. (1960). Vegetation of the Siskiyou Mountains, Oregon and California. *Ecol. Monogr.*
    30, 279–338.
Williams, C. B. (1964). *Patterns in the balance of nature*. London.

ERRATA

Page 529, line 13 from the bottom of the page should read "If the island
    was not at least partially saturated, the between-habitat . . ."
Page 531, line 5 from the bottom of the page, the word "gave" should
    be deleted.

2

THE LACUSTRINE MICROCOSM RECONSIDERED

G. E. Hutchinson

The great intellectual fascination of limnology lies in the comparative study of a great number of systems, each having some resemblance to the others and also many differences. Such a point of view presupposes that each lake can in fact be treated as at least a partly isolated system.

Today[1] I want to begin by considering two rather different approaches implicit in such treatment, partly in the work of Birge and Juday during the time when they were making Lake Mendota famous throughout the scientific world, and partly in the earlier work of S. A. Forbes, from whom my title is of course derived.

Birge's mature point of view is expressed in his concept of the heat budget,[2] which, though derived from ideas of Forel and others, represented a highly original and important contribution because it first called attention to the lake as a natural system with an input and an output. This point of view has

1. Text of address given after the transfer of keys of the Laboratory of Limnology from the National Science Foundation to the Board of Regents of the University of Wisconsin, at Wisconsin Center Auditorium, Madison, May 8, 1964. The author on this occasion represented the International Association for Pure and Applied Limnology.

2. E. A. Birge, The heat budgets of American and European lakes, *Trans. Wis. Acad. Sci. Arts Lett., 18:*166–213, 1915.

tended to underlie most of what has been done in lake chemistry and in the study of primary productivity during the past three or four decades. Such a way of thinking, in which the lake is considered, in the jargon of the moment, as a black box, has been called elsewhere[3] the *holological* approach. It has been extremely fertile, but, since water is transparent, the black box is too restrictive an analogy. The time has perhaps come for further development of the antithetical *merological* approach, in which we discourse on the parts of the system and try to build up the whole from them. This is what Forbes was trying to do in his classical lecture on "The Lake as a Microcosm."[4]

It is desirable to think for a moment about certain scale effects characterizing the lacustrine microcosm when viewed by a human observer. If we suppose that an organism reproduces about once every week in the warmer half of the year and on an average about once every month in the cooler half, it will have about thirty generations a year. This corresponds in time to about a millennium of human generations, and considerably longer for those of forest trees. In the case of the latter, we should expect in thirty generations some secular climatic change to be apparent. We should not expect in a tree the seeds or resting stages to remain viable while thirty generations passed, and in the larger animals no such stages exist. The year of a cladoceran or a chrysomonad, in both of which groups rapid reproduction may alternate with the formation of resting stages, is thus in some ways comparable to a large segment of postglacial time, though in other ways the comparison either to several millennia, or to a year in the life of a human being or tree, is definitely misleading. Another peculiar scale effect is that, in passing from the surface to the bottom of a stratified lake in summer, we can easily traverse in 10 to 20 m. a range of physical

3. G. E. Hutchinson, Food, time and culture, *Trans. New York Acad. Sci.*, ser. II, 5:152–54, 1943.

4. S. A. Forbes, The lake as a microcosm, *Bull. Scient. Assoc. Peoria,* 1887:77–87, reprinted, *Illinois Nat. Hist. Surv.*, 15:537–50, 1925.

and chemical conditions as great or greater than would be encountered in climbing up a hundred times that vertical range on a mountain.

I would also emphasize how fantastically complicated the lacustrine microcosm is likely to be. There is probably no almost complete list of species of animals and plants available for any lake, but it would seem likely from the several hundred species of diatoms and of insects[5] known from certain lakes that a species list of the order of a thousand entries may be not unusual. This probably means that in the course of a season at least a thousand somewhat different ecological niches may for a time be recognizable. Most of this diversity is associated with the shallow marginal waters in which the bottom can form a solid substratum for attached aquatic plants.

Simpler situations in a lake are probably provided by the plankton, though it soon appears that they are not particularly simple and that we cannot regard the plankton, excluding the rest of the community, as an entirely satisfactory entity. We begin with the variously named and deductively respectable principle that two co-occurring organisms cannot form equilibrium populations in the same niche. In the phytoplankton we immediately meet the paradoxical situation of an enormously complicated association of phototrophic species all living together under conditions that do not seem to permit much niche specialization.

It is possible that the permanent and apparently almost monospecific *Anacystis* blooms recorded[6] under some conditions

5. N. Foged (On the diatom flora of some Funen lakes, *Folia Limnol. Scand.*, no. 6, 75 pp., 1954) lists up to 260 different diatom taxa from a single lake. L. Brundin (Chironomiden und andere Bodentiere der südschwedischen Urgebirgsseen, *Inst. Freshwater Res.*, Drottningholm, rep. 30, 914 pp., 1949) finds up to 140 species of insects of the family Tendipedidae in a single lake.

6. S. V. Ganapati, The ecology of a temple tank containing a permanent bloom of Microcystis aeruginosa (Kutz) Henfr. I., *Bombay Nat. Hist. Soc.*, 42:65–77, 1940; G. E. Hutchinson, The paradox of the plankton, *Amer. Nat.*, 98:132–46, 1961.

in tropical waters, notably temple tanks of South India, may represent a monospecific equilibrium of the kind to be expected from theory. Much more often, what I have elsewhere termed the paradox of the plankton intrudes itself. It is to be noticed that the paradox of a multispecific phototrophic phytoplankton only arises if we assume a closed system, providing a single niche, with enough time to permit the achievement of equilibrium. In general in a lake, we do not have a single niche system that is closed. The epilimnion if reasonably turbulent may approach a single niche system, but introductions from the littoral benthos are always possible. Moreover, there is no rule about the speed at which competitive exclusion excludes. As Hardin[7] has pointed out, in the theory all that is needed is an axiom which states that no two natural objects, or classes of objects, are ever exactly alike. What then happens is that under constant conditions one class, or population of reproducing objects, finally displaces the others. If the conditions are continually changing, the favored species might also change. This is what usually seems to be happening, but it must not be forgotten that a multispecific system never in equilibrium would be expected to suffer continual random extinctions and, if not quite closed, random reintroductions also, and should therefore drift in specific composition, probably more than is indicated by paleolimnological data.

It is possible that, in a lake, random extinction is primarily a danger for the rarer species which are never likely to be observed. In a square basin 1,000 m. across and 1 m. deep we should have 10^6 organisms so common that one occurred per cubic meter, 10^9 of those occurring a thousand times more often at a rate of one per liter, and 10^{12} of those with one individual in the average cubic centimeter. If we are considering ordinary phytoplankton organisms, the first organism would be far too

7. G. Hardin, The competitive exclusion principle, *Science, 131*:1292–97, 1960.

rare ever to find by ordinary techniques even though the population before us numbered a million.

I am now inclined to think that a large part of the diversity of the phytoplankton is in fact due to a failure ever to attain equilibrium so that the direction of competition is continually reversed by environmental changes, as suggested many years ago, moderated in two ways which insure that competitive exclusion does not continually and irreversibly remove bits of the association. The first moderating influence is the speed at which exclusion occurs. In spite of the ultimate validity of Hardin's axiom of inequality, Riley[8] feels that a sort of asymptotic approach to more or less equal adaptation is not unexpected in the phytoplankton. If we suppose two species S_1 and S_2, such that in niche N_1, S_1 displaces S_2, and in niche N_2, S_2 displaces S_1. Seasonal environmental changes now occur, so that at first only N_1 and then N_2 are available. If competition went fast compared with the rate of environmental change, S_1 would be eliminated and would not be available for a new cycle, but if the two species were almost equally efficient over a wide range of environmental variables, competitive exclusion would be a slow process. Both species then might oscillate in varying numbers, but persist almost indefinitely.

The second way of moderating the tendency to random extinction is the provision of resting stages, so that if S_1 is eliminated completely as an active competitor in the plankton, when N_1 gives place to N_2, later next season when the reverse change occurs, resting stages of S_1 can recolonize the environment now again providing niche N_1. In practice any plankter that really disappears and reappears rather than becoming alternatively rare or very common must have some such stages. Annual macrophytes and many small animals also have such stages as seeds, eggs, pupae, and the like. Perennials, moreover,

8. G. A. Riley, in *Marine Biology I, Proc. First Internat. Interdisciplinary Conf. on Marine Biol.*, Amer. Inst. Biol. Sci., Washington, D.C., 1963 (see pp. 69–70).

may hibernate in ways that take the individual out of competition. In the diatoms in many of which resting zygospores are still unknown, it is possibly relatively unmodified littoral or shallow-water benthic individuals that are involved in tiding the planktonic populations over competitively unfavorable conditions. In *Melosira*, Lund's beautiful work[9] shows how a relatively heavy diatom rests on the bottom for very long periods in a more or less unassimilative form; here there is doubtless some special physiological adaptation, so we are halfway between a species invading the plankton casually with a continuous littoral population and the condition in which morphologically specialized resting stages or cysts are produced. One of the most remarkable results of several recent paleolimnological studies, notably Nygaard[10] on Store Gribsø and our own work on Lago di Monterosi,[11] is the fantastic variety of chrysophycean cysts recognizable in the sediments, at least, of rather soft-water lakes. Resting stages of all sorts are of course particularly prone to occur in freshwater organisms, where they were doubtless developed primarily to promote survival under extreme physical conditions, notably desiccation and freezing. Once developed, they would however clearly be of great value in obviating extinction when conditions changed in favor of a competitor. It is therefore peculiar that Lund finds that planktonic desmids tend to lose such stages.

The diversity of the phytoplankton is clearly of primary importance in producing that of the zooplankton. Given the diversity of the phytoplankton, and some degree of food specificity in the animal forms, no striking paradoxical situation need

9. J. W. G. Lund, The seasonal cycle of the plankton diatom, *Melosira italica* (Ehr.) Kütz. subsp. *subarctica* o. Müll., *J. Ecol.*, *42*:151–79, 1954. Further observations on the seasonal cycle of *Melosira italica* (Ehr.) Kütz subsp. *subarctica* o. Müll., *J. Ecol.*, *43*:90–102, 1955. See also his Baldi lecture, *Verh. int. Ver. Limnol.*, *15*:37–56, 1964.

10. G. Nygaard, in K. Berg and I. C. Petersen, Studies on the humic acid Lake Gribsø., *Fol. Limnol. Scand.*, no. 8 (1956), 273 pp.

11. Elaine Leventhal, ms., to appear in a series of papers on this locality.

arise. Moreover, it is clear from all the available work on the seasonal succession of closely allied forms, such as the species of *Daphnia*, from Birge's[12] early studies on Mendota up to the very beautiful and elaborate investigations of Dr. J. L. Brooks and of Dr. Donald W. Tappa at Yale, shortly to be published, that the same sorts of seasonal phenomena that damp competition in plants also occur in animals.

MacArthur and Levins[13] have recently pointed out that two rather different extreme types of diversity between closely allied sympatric species (i.e. members of a genus or subfamily) are possible.

The two species may be specialized in such a way that they eat slightly different food, but hunt it over the same area. In this case morphological specializations, of which the simplest is a size difference, are to be expected. Probable examples, such as the hairy and downy woodpeckers, easily come to mind.

If the two species eat the same sorts of diversified food, they are likely to differ in the proportions in which they encounter it, and to specialize in habitat preferences without much morphological specialization becoming necessarily involved in feeding activity. MacArthur's own work on the American warblers provides a striking example. The existence of these two general situations has long been known, but MacArthur and Levins provide a good abstract theory of the phenomenon.

Over the whole vertical column in a stratified lake, even if only 10 m. deep, habitat differences are available in summer at least as great as over the range from the bottom and to the top of a mountain several hundred times that vertical range. In the turbulent epilimnion it is in general hard to develop habitat preferences and, within any layer in which free movement is habitual, size differences may be expected as the simplest special-

12. E. A. Birge, Plankton studies on Lake Mendota: *II*, The Crustacea of the plankton from July 1894, to December 1896, *Trans. Wis. Acad. Arts Sci. Lett.*, 1898.
13. R. H. MacArthur and Richard Levins, *Proc. Nat. Acad. Sci.*, June 1964.

ization increasing diversity, as is the case with Copepoda. In view of the extreme vertical variation when we leave a turbulent, freely mixed layer, the antithetic habitat difference type of specialization in the plankton is likely to be rather different from what is found terrestrially, involving fairly complete adaptation to very divergent physical factors rather than habitat preferences, though, in two species living together with vertical migration over partly overlapping ranges, we have the lacustrine analogue of birds feeding in different parts of the same tree. The rapid production of a number of generations per year permits a kind of seasonal succession in rotifers and Cladocera, though to a less extent in Copepoda, that is comparable to that in the phytoplankton. Considerable possibility of avoiding competitive exclusion is thus achieved by slow competition between species that have slightly different optima, and so succeed each other in time. Here the production of resting stages is of the greatest importance. That they are produced at the time of maximum population fits reasonably into this scheme quite independently of adaptation to unfavorable physical situations.

This succession in time may be coupled with size differences and habitat differences, probably producing, in for instance the genus *Polyarthra*[14] where five or six species can be sympatric but not always strictly synchronic, a very pronounced niche specificity.

An interesting question arises, namely to what extent sympatric species of a given taxon, say genus or family, not merely have different ecologies, but also have ecologies that, though different, are closer than any would be likely to be to that of a sympatric nonmember of the taxon picked at random.

If we compare a desert assemblage with a limnoplanktonic one, it is obvious that, if the first organism captured in the

14. Carlin, Die Planktonrotatorien des Motalaström: zur Taxonomie und Ökologie der Planktonrotatorien, *Meddel. Lunds Univ. Limnol. Inst.*, no. 5 (1943), 255 pp.

desert is a beetle, the probability that the next one of another species is also another beetle is higher than that it is a rotifer, and vice versa. It is however rather surprising to find in Carlin's data four unallied perennial rotifers, including the microphagous sedimenters *Keratella*, *Notholca*, and *Conochilus* and the selective predator *Asplanchna*, all reacting similarly to an unidentified difference, possibly involving an earlier decline in the late summer bloom of *Oscillatoria*, that distinguished 1940 from the other years of his study.

The relatively small development that has been possible in the study of the interrelationships of the plankton since Forbes wrote in 1887 and Birge in 1898, and of which some examples have just been given, has been due to a very large amount of work both in field observations, in very meticulous taxonomy, and in ecological theory. In the other parts of the lacustrine community the problems are more difficult though their solutions would have great fascination, as will be apparent from a single example. It we examine the lasion, "Aufwuchs" or fouling community of fresh waters, we find a variety of filamentous algae and diatoms with an associated fauna ordinarily of small motile forms. The biomass of the animals is doubtless ordinarily much less than that of plants, and on a surface near the bottom of the euphotic zone organisms will tend to be scarce. There may be a few sponges and bryozoans but they are not conspicuously important. In the sea, the parallel community, though largely algal in the tidal range, consists at most levels of an astonishing mass and variety of sessile animals—sponges, mollusks such as *Mytilus*, numerous hydroids, Bryozoa, and tunicates. The difference is presumably due to the lack of pelagic larvae other than copepodan nauplii in fresh water. The only exceptions are a very few mollusks of far from worldwide distribution, notably *Dreissena* and to a less extent *Corbicula*, the larval colonies of phylactolaematous Bryozoa, hardly ever noted in open water, perhaps a few transitory planula larvae (*Cordylophora*, *Limnocnida*, *Craspedacusta*), and the free-swimming larvae of trematodes which show odd diversity in behavior when allied species are compared, but which presumably do not enter into competition with other plankters. This is a very poor showing compared to the dozen phyla likely to be found in any series of marine neritic plankton samples. This difference has received various explanations, the most reasonable, essentially due to Needham,[15] probably being the difficulty that a small larval animal in freshwater, feeding on plant cells low in sodium and chloride, would have in acquiring enough salt before it could develop salt-absorbing organs. In a certain sense the adult animals of the marine littoral benthos, not all sessile nor all microphagous, are the resting stages removed from competition at least in the open-water plankton. We see in this type of relation a rather large-scale example of the sort of interaction which fascinated Forbes. It is hard, in reading Forbes on "The Lake as a Microcosm," to prevent the mind drifting back to the tangled bank of the last chapter of Darwin's *Origin of Species*,[16] and from there it is permissible, at least

15. J. Needham, On the penetration of marine organisms into freshwater, *Biol. Zbl.*, *50*:504–09, 1930.

16. The concluding paragraphs of Forbes' essay and of the *Origin of Species* may be profitably compared:

"Have these facts and ideas, derived from a study of our aquatic microcosm, any general application on a higher plane? We have here an example of the triumphant beneficence of the laws of life applied to conditions seemingly the most unfavorable possible for any mutually helpful adjustment. In this lake, where competitions are fierce and continuous beyond any parallel in the worst period of human history; where they take hold, not on goods of life merely, but always upon life itself; where mercy and charity and sympathy and magnanimity and all the virtues are utterly unknown; where robbery and murder and the deadly tyranny of strength over weakness are the unvarying rule; where what we call wrong-doing is always triumphant, and what we call goodness would be immediately fatal to its possessor—even here, out of these hard conditions, an order has been evolved which is the best conceivable without a total change in the conditions themselves; an equilibrium has been reached and is steadily maintained that actually accomplishes for all the parties involved the greatest good which the circumstances will at all permit. In a system where life is the universal good, but the destruction of life the well-nigh universal

in 1964, to look back still farther, remembering what Jane Austen said about Shakespeare, to that other "bank where the wild thyme blows."[17] Both Forbes and Darwin realize struggle but see that it has produced harmony. Today perhaps we can see just a little more. The harmony clearly involves great

---

occupation, an order has spontaneously arisen which constantly tends to maintain life at the highest limit—a limit far higher, in fact, with respect to both quality and quantity, than would be possible in the absence of this destructive conflict. Is there not, in this reflection, solid ground for a belief in the final beneficence of the laws of organic nature? If the system of life is such that a harmonious balance of conflicting interests has been reached where every element is either hostile or indifferent to every other, may we not trust much to the outcome where, as in human affairs, the spontaneous adjustments of nature are aided by intelligent effort, by sympathy, and by self-sacrifice?" (Forbes)

"It is interesting to contemplate a tangled bank, clothed with many plants of many kinds, with birds singing on the bushes, with various insects flitting about, and with worms crawling through the damp earth, and to reflect that these elaborately constructed forms, so different from each other, and dependent upon each other in so complex a manner, have all been produced by laws acting around us. These laws, taken in the largest sense, being Growth with Reproduction; Inheritance which is almost implied by reproduction; Variability from the indirect and direct action of the conditions of life, and from use and disuse: a Ratio of Increase so high as to lead to a Struggle for Life, and as a consequence to Natural Selection, entailing Divergence of Character and the Extinction of less-improved forms. Thus, from the war of nature, from famine and death, the most exalted object which we are capable of conceiving, namely, the production of the higher animals, directly follows. There is grandeur in this view of life, with its several powers, having been originally breathed by the Creator into a few forms or into one; and that, whilst this planet has gone cycling on according to the fixed law of gravity, from so simple a beginning endless forms most beautiful and most wonderful have been, and are being evolved." (Darwin)

17. I know a bank where the wild thyme blows,
Where oxlips and the nodding violet grows;
Quite over-canopied with luscious woodbine
With sweet muskroses, and with eglantine.
There sleeps Titania sometime of the night,
Lull'd in these flowers with dances and delight;
And there the snake throws her enamell'd skin,
Weed wide enough to wrap a fairy in.
A Midsummer Night's Dream
II.1.249–56.

diversity, and we now know, in the entire range from subatomic particles to human artifacts, that every level is surprisingly diverse. We cannot say whether this is a significant property of the universe; without the model of a less diverse universe, a legitimate but fortunately unrealized alternative, we cannot understand the problem. We can, however, feel the possibility of something important here, appreciate the diversity, and learn to treat it properly.[18]

18. The evening program ended with a performance of Mozart's Quartet in G Minor for Piano and Strings, K. 478.

Part II

DIVERSITY IN AQUATIC COMMUNITIES

Editor's Comments
on Papers 3 Through 8

Aquatic communities are characterized by the presence of water as the media through which the environment interacts with biological organisms. Because of the physical-chemical characteristics of water the aquatic medium can be very mobile, as well as reduce the impacts of violent change in temperature. Nevertheless, aquatic organisms invade and live in almost all aquatic habitats, even including the rocky margins of oceans where waves smash against the land constantly, in shifting sands and muds of river deltas, and in the deep oceans. As we would expect, these communities vary widely in their diversity. Unfortunately, in aquatic communities, as in terrestrial communities as well, it is very difficult to obtain a complete count of all species living in the habitat. This is especially true for microorganisms. The consequence is that we tend to use the taxonomic variety of collectable groups, such as fish or diatoms, to represent the community diversity. The papers chosen to represent aquatic systems illustrate these points.

First, in Paper 3, Sanders discusses the diversity of marine benthic communities. Data on bivalves and polychaetes were obtained in marine environments off of India and North and South America. Sanders stresses the twin role of physical control and time in governing biological adaptation, evolution, and the development of diversity. Where physiological stress is low, biologically accommodated communities have evolved, given sufficient time. As the physical gradient of stresses increases, the biologically accommodated community shifts toward a physically accommodated community. The number of species also declines on this gradient. Sanders's paper is also useful because he discusses the problem of methodology in evaluating the data on diversity—different methods may produce quite different counts of species and individuals.

Paper 4 presents data gathered and analyzed by the editor for stream communities. These freshwater systems provide a further example of the general pattern in marine benthic communities, but they illustrate special conditions of establishment and maintenance under constantly flowing water, change in sediment load, and change in depth and flooding. Stream communities frequently form a close connection with terrestrial or semiterrestrial communities on the stream margin. Stream diversity reflects all of these conditions.

In both of the above papers it is clear that diversity is not linearly related to an environmental factor but rather is a response to a complex of conditions characterizing the aquatic habitat. In Paper 5, Lubchenko shows how a generalized herbivore, the snail *Littorina littorea,* may increase or decrease algal diversity. Algae form the food of this snail, and herbivore food preference, as well as the competitive ability of the plants, governs the impact of the herbivores on diversity of the plant species.

Paine (Paper 6) develops this general point further from a comparison of rock intertidal communities across a latitudinal gradient. Latitude is not the factor controlling diversity here; rather it is the presence of predators that creates diversity. Communities with higher production support more complex foodwebs and more predators and thus tend to have higher diversity. The factors producing diversity vary in their importance in different communities and an ultimate understanding of the underlying causal processes can only be arrived at by study of local situations.

The observation that tropical diversity is greater than that in temperate areas has structured much other research. For example, Spight (Paper 7) compared gastropod communities in Costa Rica and

Washington. He showed that in those animals living on rocky shores, tropical species used fewer habitats and, more importantly, used niches that did not occur in temperate environments. Nevertheless, diversity and density varied more among adjacent habitats on any one beach than among latitudes.

Paper 8 continues the concepts developed in aquatic communities but from the perspective of fish and sea urchins in coral reef communities. Ogden and Lobel stress the coevolutionary relationships, also given prominent attention by Lubchenco, between herbivores and coral reef algae. The complex of adaptations on both plant and herbivore sides of the relationship govern the rate and type of feeding.

These papers emphasize the variation in diversity one may expect in natural environments caused by the interaction between species and their environment. Where there is sufficient information, variation in diversity is used to measure the degree of perturbation caused by man (Grassle and Sanders, 1973) and is particularly useful in monitoring the effect of pollution in aquatic ecosystems (Cairns et al., 1972; Gaufin, 1972; Patrick, 1972, 1977).

REFERENCES

Cairns, J., Jr., K. L. Dickson, and G. Lanza, 1972, Rapid Biological Monitoring Systems for Determining the Aquatic Community Structure in Receiving Streams, in *Biological Methods for the Assessment of Water Quality*, J. Cairns, Jr. and K. L. Dickson, eds., Am. Soc. Testing Mater. Tech. Publ. 528, American Society for Testing and Materials, Philadelphia, pp. 148–163.

Gaufin, A. R., 1972, Use of Aquatic Invertebrates in the Asessment of Water Quality, in *Biological Methods for the Assessment of Water Quality*, J. Cairns, Jr. and K. L. Dickson, eds., Am. Soc. Testing Mater. Tech. Publ. 528, American Society for Testing and Materials, Philadelphia, pp. 96–116.

Grassle, J. F., and H. L. Sanders, 1973, Life Histories and the Role of Disturbance, *Deep-Sea Res.* **20:**643–659.

Patrick, R., 1972, The Use of Algae, Especially Diatoms, in the Assessment of Water Quality, in *Biological Methods for the Assessment of Water Quality*, J. Cairns, Jr. and K. L. Dickson, eds., Am. Soc. Testing Mater. Tech. Publ. 528, American Society for Testing and Materials, Philadelphia, pp. 76–95.

Patrick, R., 1977, The Importance of Monitoring Change, in *Biological Monitoring of Water and Effluent Quality*, J. Cairns, Jr., K. L. Dickson, and G. F. Westlake, eds., Am. Soc. Testing Mater. Tech. Publ. 607, American Society for Testing and Materials, Philadelphia, pp. 157–190.

3

MARINE BENTHIC DIVERSITY: A COMPARATIVE STUDY*

HOWARD L. SANDERS

Woods Hole Oceanographic Institution, Woods Hole, Massachusetts 02543

INTRODUCTION

One of the major features of animal communities is their diversity, that is, the number of species present and their numerical composition. It has long been recognized that tropical regions, by and large, support a more diverse fauna than do regions of higher latitude. In the aquatic medium it is also evident that the marine habitats contain a greater wealth of species than do brackish regions. The reasons why certain environments harbor many kinds of organisms while others support a very limited number of species are still unclear. Various theories based on time (Fischer, 1960; Simpson, 1964), climatic stability (Klopfer, 1959; Fischer, 1960; Dunbar, 1960), spatial heterogeneity (Simpson, 1964), competition (Dobzhansky, 1950; Williams, 1964), predation (Paine, 1966), and productivity (Connell and Orias, 1964) have been proposed to explain these differences.

In the present paper, data collected from soft-bottom marine and estuarine environments of a number of differing regions will be used in a comparative study of within-habitat diversity. A new diversity measurement will be presented that is independent of sample size, and a hypothesis will be proposed to explain the observed patterns of diversity as well as to provide a framework for interpreting other diversity studies.

MATERIALS AND METHODS

The stations used in this study are as follows:

RH-14: Arabian Sea off Cochin, Kerala State, India (14 m).
RH-26: Bay of Bengal off Porto Novo, Madras State, India (20 m).
RH-28: Vellar River estuary, Porto Novo, Madras State, India (2 m).
RH-30: Bay of Bengal off Madras, India (15 m).
RH-33: Kakinada Bay, Andhra State, India (2 m).
RH-36: Bay of Bengal off Kakinada, Andhra State, India (37 m).
RH-41: Arabian Sea off Bombay, India (20 m).
RH-51: Indian Ocean off Hellville, Nossi Bé, Madagascar (18 m).
C#1: Outer continental shelf south of New England (40°27.2'N 70°47'W, 97 m).
S1.3: Upper continental slope south of New England (39°58.4'N 70°40.3'W, 300 m).
D#1: Upper continental slope south of New England (39°54.5'N 70°35'W, 487 m).
F#1: Lower continental slope south of New England (39°47'N 70°45'W, 1,500 m).

*Contribution No. 1959 from the Woods Hole Oceanographic Institution, Woods Hole, Massachusetts 02543.

G#1: Lower continental slope south of New England (39°42′N 70°39′W, 2,086 m).
GH#1: Abyssal rise south of New England (39°25.5′N 70°35′W, 2,500 m).
DR-12: Continental slope off northeast South America (07°09′S 34°25.5′W, 790 m).
DR-33: Continental slope off northeast South America (07°53.5′N 54°33.3′W, 535 m).
POC 1, 2, 3, 4: Pocasset River, Cape Cod, Massachusetts (0.5 m).
R Series: Buzzards Bay, Massachusetts (20 m).

All samples were collected with an Anchor dredge or a Higgins meio-benthic sled (the RH series of samples). The sediments were processed through a fine-meshed screen with 0.4 mm apertures, and the animals were carefully picked out and sorted in the laboratory. Sanders, Hessler, and Hampson (1965) gave details on the Anchor dredge and the methodology of processing.

THE RAREFACTION METHODOLOGY AND RESULTS

Since most diversity measurements are affected by sample size (see later discussion), it would be most useful to have a procedure which will allow one to compare directly samples of differing sizes. If this can be achieved, it may then be possible to perceive more clearly the factors influencing biological diversity. The rarefaction method, which permits each sample to generate a line, was developed to achieve this end. This methodology was applied to benthic marine samples collected from boreal estuary, boreal shallow marine, tropical estuary, tropical shallow marine, and deep-sea environments. The usual difficulty inherent in comparing samples of different sizes is that as sample size increases, individuals are added at a constant arithmetic rate but species accumulate at a decreasing logarithmic rate. The rarefaction method, instead, is dependent on the shape of the species abundance curve rather than the absolute number of specimens per sample. In all cases, the sediments were soft oozes and, therefore, comparable in regard to particle size.

The comparison was based on the polychaete-bivalve fraction of the samples rather than the entire fauna. Since these two groups comprise about 80% of the animals by number in most of the samples (Fig. 1), one can feel justified in generalizing from whatever results may be found. This study shows a systematic pattern of diversity that can be correlated with the variability of the physical environment.

With the method of diversity analysis developed for this study, samples with different numbers of specimens and from different regions of the world were compared directly. The procedure was to keep the percentage composition of the component species constant but reduce the sample size, that is, to artificially create the results that would have been obtained had smaller samples with the identical faunal composition been taken. Using this technique, the expected number of species present in populations of different sizes, that is, numbers of species per 10, 25, 50, 100, 200, . . ., 1,000, 2,000, etc., was determined.

In order to evaluate the validity of this method, one must understand how the values are obtained. The species are ranked by abundance, and

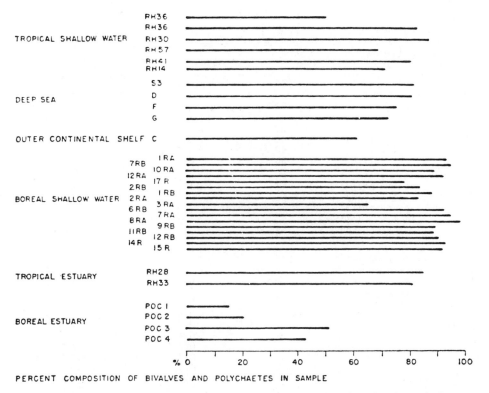

Fig. 1.—Percentage composition of the polychaete-bivalve fraction of the soft-bottom samples used in the analysis.

the percentage composition of each species and the cumulative percentage are plotted. In a hypothetical sample (Table 1) there are 1,000 individuals and 40 species. As an example, the number of species at the 25-individual level will be determined. The percentage composition is the same as in the original sample, but the number of individuals is reduced to 25. Since 25 specimens in this reduced sample represent 100% of the individuals present, then each individual specimen forms 4% of the sample. In the original sample, seven species each comprise 4% or more, and in total they compose 76% of the sample by number. Therefore, each of these seven species will be present in the reduced sample. This leaves a residue of 24% of the original sample comprising the remaining 33 species. Because none of these species forms more than 4% of the original sample, those species of this group that will appear in the reduced sample cannot be represented by more than one individual. Since one specimen comprises 4% of the reduced sample, therefore 24%/4% = 6 species; 7 + 6 = 13 species present per 25 individuals.

The determination of species per 100 individuals is as follows: (1) Since each individual represents 1% of the sample, then (2) 15 species in Table 1 each comprise $\geqq$ 1.0% of the fauna and cumulatively = 92.1% of the

TABLE 1

Hypothetical Sample with 1,000 Individuals and 40 Species

Rank of Species by Abundance	Number of Individuals	% of Sample	Cumulative of Sample %
1	365	36.5	36.5
2	112	11.2	47.7
3	81	8.1	55.8
4	61	6.1	61.9
5	55	5.5	67.4
6	46	4.6	72.0
7	40	4.0	76.0
8	38	3.8	79.8
9	29	2.9	82.7
10	23	2.3	85.0
11	21	2.1	87.1
12	15	1.5	88.6
13	13	1.3	89.9
14	12	1.2	91.1
15	10	1.0	92.1
16	8	0.8	92.9
17	7	0.7	93.6
18	7	0.7	94.3
19	6	0.6	94.9
20	6	0.6	95.5
21	5	0.5	96.0
22	5	0.5	96.5
23	5	0.5	97.0
24	4	0.4	97.4
25	3	0.3	97.7
26	3	0.3	98.0
27	3	0.3	98.3
28–33	2 each	0.2 each	99.3
34–40	1 each	0.1 each	100.0
Total Number	1,000		

sample. (3) The residue = 7.9% of the sample; 7.9%/1.0% = 7.9 species. (4) 15 + 7.9 = 22.9 species per 100 individuals.

For species per 200 individuals: (1) Each individual forms 0.5% of the sample, and (2) 23 species each represent $\geq$ 0.5% of the fauna and cumulatively = 97.0% of the sample. (3) The residue = 3.0%; 3.0%/0.5% = 6.0 species. (4) 23 + 6.0 = 29.0 species per 200 individuals.

Using this technique, we have made, in Figure 2, arithmetic plots of the number of species at different population levels up to the total number of individuals for samples from high latitude, low latitude, shallow water, deep sea, estuarine, and marine regions. The curvilinear nature of the lines is due to the fact that individuals are being added at a constant rate but the progressively rarer species are added at a continuously decreasing rate. The circles or the termination of the lines in Figure 2 give the actual number of individuals and species present in the samples. The curves themselves give the interpolated number of species at the different population levels.

What is significant is that each environment seems to have its own characteristic rate of species increment. Lowest diversity, that is, the fewest number of species per unit number of individuals, is found in the boreal

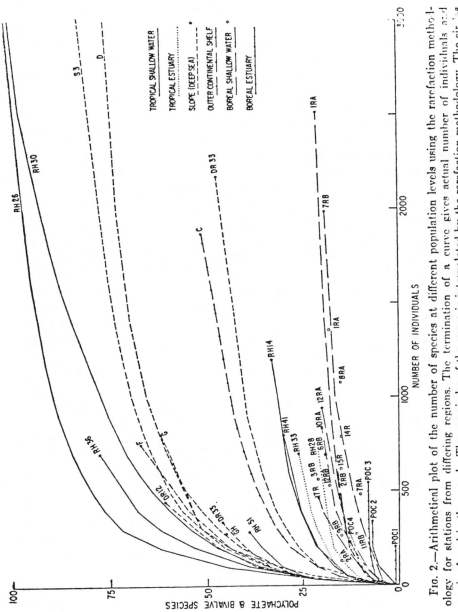

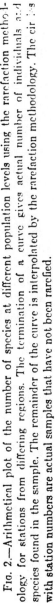

FIG. 2.—Arithmetical plot of the number of species at different population levels using the rarefaction methodology for stations from differing regions. The termination of a curve gives actual number of individuals and species found in the sample. The remainder of the curve is interpolated by the rarefaction methodology. The circles with station numbers are actual samples that have not been rarefied.

estuary as represented by the Pocasset River, Cape Cod, Massachusetts. The station highest up the estuary, POC 1 in Figure 2, with a mean sediment salinity of 7‰ and a range of 9.5‰ per tidal cycle, has the lowest diversity within the series. POC 2, next highest up the estuary, with a mean sediment salinity of 17‰ and with a 3‰ salinity range per tidal cycle, has the next lowest diversity. The diversity increases at POC 3, with a mean salinity of 20.7 and a tidal variation of 1.4‰. Still higher diversity is present at POC 4, where the mean salinity rises to 22.9‰ and the tidal variation is 1.8‰ (for details on faunal distribution in the Pocasset estuary and its relationship to salinity, see Sanders, Mangelsdorf, and Hampson, 1965). Besides the low and variable sediment salinities in the Pocasset River, there are, as well, large seasonal temperature changes.

Somewhat higher diversity values occur in tropical estuaries. These are represented by stations RH-28, the Vellar River estuary at Porto Novo, Madras State, India, and RH-33, at the mouth of the Godavari estuary at Kakinada, Andhra State, India. During the periods of heavy rainfall of October and November, the salinities in these shallow bodies of water are reduced to zero. Yet, in the dry season of May and June, the salinities are more than 34‰ (Jacob and Rangarajan, 1959). The probable reason for the greater diversity values in low- as compared with high-latitude estuaries is that it is easier to tolerate reduced salinities at high temperatures than at low temperatures (Panikkar, 1940). As a result, more marine forms are able to invade estuaries in the tropics than in higher latitudes.

Among the boreal shallow marine samples (the R-station series) diversity is modest. These samples were taken from Buzzards Bay, Massachusetts, in 20 m of water at all seasons of the year (Sanders, 1960). Here the annual temperature change is more than 23°C, with winter temperatures often less than −1.0°C and summer temperatures of more than 22°C. Such appreciable changes in annual temperature are as large as that found in any marine region of the world. This pronounced seasonal temperature variation probably accounts for the low diversity values. Note that within the R series of samples in Figure 2, the actual samples with lower densities, instead of being widely scattered throughout the graph, are clustered about the interpolated curves derived from the larger samples, thus verifying the validity of our methodology. This same clustering demonstrates that diversity values are sample-size independent when derived by the rarefaction technique. At the outer edge of the continental shelf, station C, the amplitude of temperature change has been reduced to 10°C and the faunal diversity has increased (Sanders, et al., 1965).

The most diverse values are found among the tropical shallow marine samples, although there is appreciable spread in the position of the curves. The highest values are from the three Bay of Bengal samples: RH-26, off Porto Novo, Madras State, India, in 20 m of water; RH-36, off Kakinada, Andhra State, India, in 37 m depth; and RH-30, off the city of Madras, in 15 m depth. All of these stations are too deep to be affected by the freshening of the surface water during the monsoons (LaFond, 1958;

Murty, 1958; Ramamurthy, 1953). Station RH-51, from a depth of 18 m off Nossi Bé, Madagascar, gives an intermediate value. Lowest values are from two stations in the Arabian Sea, RH-14 in 14 m off Cochin, India, and RH-41, in 20 m depth off Bombay, India. The probable cause for these modest diversity values is the low-oxygen minimum layer found throughout the northern Arabian Sea at the 100 to 200 meter depth. During the south-west monsoons, this low-oxygen water is pushed onto the continental shelf off India, creating a severe stress condition for the bottom fauna which is probably reflected in the reduced number of species present. Banse (1959) found, at almost precisely the site and depth of our Cochin station, oxygen values of only 5% saturation during the southwest monsoon, and Carruthers, Gogate, Naidu, and Laevastu (1959) obtained similar low-oxygen values at a location of equivalent depth near our Bombay station.

Our deepwater diversity curves, derived from stations Sl.3, D#1, F#1, and G#1 from the continental slope, station GH#1 on the abyssal rise (all south of New England), and stations DR-12 and DR-33 from the continental slope off northeastern South America, with but a single exception, are confined to a narrow sector of the graph. The physical factors in this environment are rigidly constant, with low temperatures, high salinity (see Sanders, et al., 1965), and high oxygen values. The single exception to our deepwater diversity pattern, DR-33, is due to the aggregation of a single polychaete species which forms more than 85% of the sample. If this species is arbitrarily excluded, the residual diversity, DR-33', is similar to that found in other deep-sea samples. Thus the deep-sea benthic fauna appears to possess a relatively high diversity of the same general order as that present in tropical shallow seas.

It should be clearly pointed out here that this method of measuring diversity is valid when the fauna is randomly or evenly distributed but not aggregated. Even in cases of aggregation, it may be possible to uncover inherent diversity (DR-33') by using this methodology. On the other hand, in samples with little aggregation, the diversity is only slightly increased by eliminating the most abundant species.

Applying confidence limits to the curves for certain of the environments is meaningless. In cases where the number of samples included is small, the confidence limits will be broad. The Bay of Bengal series, the pair of stations from the shallow depths of the Arabian Sea off India, and the two tropical estuarine stations suffer from this weakness.

The best that can be done is to represent the ranges in Figure 2 as bands of values (Fig. 3). The number of samples is given at each of the rarefied sample sizes from 100 individuals and larger. Environments and sample sizes with single samples, the Pocasset series, and aberrant station DR-33 are excluded. The clear separation of the environmental bands strongly indicate that these diversity differences are real.

Limitations of Methodology

The rarefaction method for measuring diversity must be used with dis-

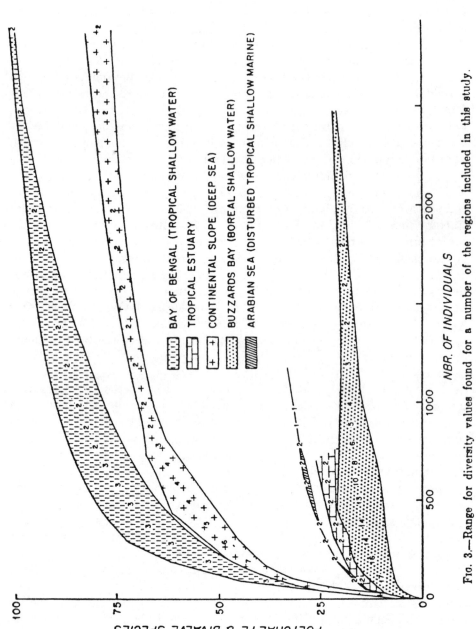

Fig. 3.—Range for diversity values found for a number of the regions included in this study.

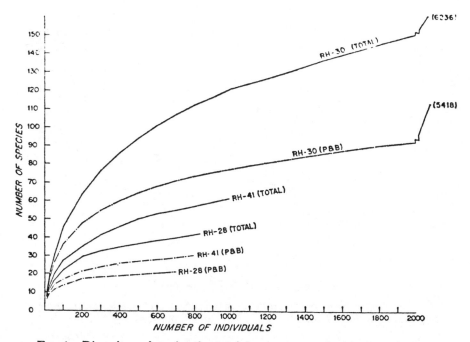

Fig. 4.—Diversity values for the total fauna compared with the polychaete-bivalve component of the same samples. Numbers in parentheses for station RH-30 are the actual numbers of specimens in the total sample and in the polychaete-bivalve fraction of that sample.

crimination to be meaningful. Such a technique is valid only when the same groups of organisms are compared and contrasted. With the inclusion of additional groups, the diversity values for a given faunal density increase. This phenomenon can be clearly observed with the few representative samples used in Figure 4. In each case, the diversity for the total fauna is decidedly higher than that of the polychaete-bivalve component of the sample.

Another requisite is that all the habitats sampled be similar; that is, the comparison must be made among a within-habitat (MacArthur, 1965) series of environments (in the present situation, the soft estuarine and marine oozes). Differing habitats from the same geographic region have differing diversity values (between-habitat comparison [MacArthur, 1965]). Thus the sand bottom fauna in Buzzards Bay is more diverse than the mud bottom fauna (Fig. 5). (Probably the fauna of stable sand bottoms will always be inherently more diverse because of the greater variety of microhabitats.)

In order to have the data comparable, it is necessary that the sampling procedures such as the type of gear used, the methodology utilized in processing the sample, and the screen size employed in washing the samples should be approximately similar.

Finally, this method does not specify which species taken from the residue will be present, and it can be used only to interpolate, not to extrapolate.

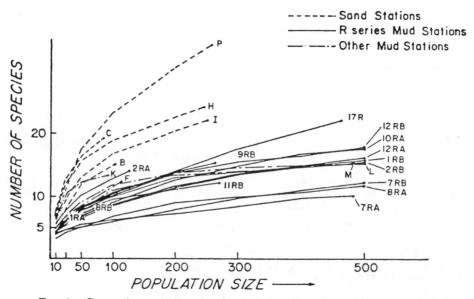

FIG. 5.—Comparison of diversity between mud- and sand-bottom samples from Buzzards Bay, Massachusetts, up to population sizes of 500 individuals.

THE PHYSICALLY CONTROLLED AND THE BIOLOGICALLY ACCOMMODATED COMMUNITIES

The interpretation of the curves in Figure 2 and the more general analyses of the total fauna might best be understood by describing two contrasting types of communities, both of which are abstractions. One can be called the "physically controlled community." In environments harboring this kind of community, the physical conditions fluctuate widely and the animals are exposed to severe physiological stress.

In the physically controlled community the adaptations are primarily to the physical environment. Examples of such communities are those found in hypersaline bays, high arctic terrestrial environments, and deserts. The physically controlled communities are always eurytopic and are characterized by a small number of species. A similar paucity of species occurs in environments of recent past history, such as most freshwater lakes.

The other extreme condition might be called the "biologically accommodated community." These communities are present where physical conditions are rather constant and uniform for long periods of time. Because of the historic constancy of the physical environment, physical conditions are not critical in controlling the success or failure of the species. With time, biological stress (intense competition, nonequilibrium conditions in prey-predator relationships, simple food web, etc.) is gradually mediated through biological interactions resulting in the evolution of biological accommodation. The resulting stable, complex, and buffered assemblages are always characterized by a large number of stenotopic species. The deep-sea regions, tropical shallow-water marine regions, and tropical rain forests best represent such conditions.

There is no such thing as a "pure" physically controlled or biologically accommodated community. All communities are the result of both their physical and biological components and are therefore somewhat intermediate between these extreme types. What determines the structure of any community is the relative proportions of these two parts.

In predominantly physically controlled communities there can be no close coupling of a species to its environment, as would be the case in the predominantly biologically accommodated communities. This is due to the variations in the amplitude of environmental factors; that is, there is no precise reproducibility from year to year. For example, one year the temperature may be slightly higher, so that one species is favored regarding breeding, which results in the "year-class" phenomenon, that is, a tremendous increase in the number of the new year class. (The year-class phenomenon probably is a characteristic feature of the predominantly physically controlled communities.) The next year, the temperature may be slightly lower, so that the same species is adversely affected, resulting in an unsuccessful breeding. At the same time, another species is favored in its breeding by the reduced temperature. The same would be true for the effects of other environmental variables, such as salinity and oxygen, on growth, breeding, metabolism, etc. Therefore, animal species of the physically controlled communities must adapt to a broad spectrum of physical fluctuations which does not allow the biological interrelationships to develop very far. (In some intertidal environments, the prey may be biologically controlled while its predator is physically controlled [Connell, 1961]).

From the concepts summarized above, it is possible to present the stability-time hypothesis in Figure 6. Where physiological stresses have been historically low, biologically accommodated communities have evolved. As the gradient of physiological stress increases, resulting from increasing physical fluctuations or by increasingly unfavorable physical conditions regardless of fluctuations, the nature of the community gradually changes

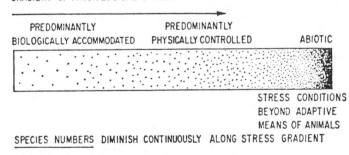

FIG. 6.—Bar graph representation of the stability-time hypothesis.

from a predominantly biologically accommodated to a predominantly physically controlled community. Finally, when the stress conditions become greater than the adaptive abilities of the organisms, an abiotic condition is reached. The number of species present diminishes continuously along the stress gradient.

When the stability-time hypothesis is applied to Figure 2, the closer the curves approach the ordinate, as shown with the shallow tropical marine and deep-sea samples, the nearer they approximate the biologically accommodated community. They describe assemblages in which there are large numbers of species per unit number of individuals (high diversity). In these environments, physical conditions are constant and have remained constant for a long period of time. The closer the curves approach the abscissa, as in the cases of the boreal estuary and boreal shallow marine samples, the greater are the physiological stress conditions imposed by the physical environment. In these assemblages there is a small number of species per unit number of individuals (low diversity). Here the physical conditions are highly variable and approach the idealized physically controlled community.

Thus each environment in Figure 2 appears to have its own unique family of curves. Such lack of randomness implies biological organization, with the nature of the organization or structure differing in different environments. Such organization is determined by the degree of stability of the physical environment and the past history of the physical environment, that is, to what degree an animal association is physically controlled and to what degree it is nonphysically regulated or biologically accommodated.

TIME

It requires appreciable time to evolve a highly diverse fauna, and the time component of our stability-time hypothesis is perhaps best illustrated with lakes. Most lakes are of a relatively transitory nature, or of recent geologic origin. It has been 10,000 years or less since the last glaciation, and the aquatic fauna from such recently glaciated regions shows limited diversification. However, there are a few ancient lakes—for example, the rift-valley lakes of Africa and Lake Baikal in Russian Siberia.

Lake Baikal was formed either about 30 million years ago in the middle Tertiary or at the end of the Tertiary and early Quaternary periods about one million years ago. (For references, see Kohzov, 1963.) This lake, in common with other ancient lakes, is characterized by a highly diverse fauna. One of the most diverse faunal elements are the gammarid amphipods, represented by 239 endemic and one nonendemic species.

To appreciate the full significance of such diversity, in all of what was glaciated North America there are no more than 28 species of gammarid amphipods (Bousfield, 1958). Certain of these crustaceans are confined to streams near the sea, can tolerate brackish water, and have recently evolved from closely related marine forms. Others are restricted to cave streams and springs. A few are limited to ponds. Still others occur in sloughs and temporary bodies of water. Only seven gammarid species are confined to lakes and rivers, and it is these few amphipods, distributed throughout vast areas of North America, which are the ecologic equivalents of the 240 species present in ancient Lake Baikal. (It should be mentioned that the diversity effects of time on a physically fluctuating environment of constant magnitude and periodicity remain unanswered.)

Lake Baikal has further implication to the concepts proposed in this paper. Two distinct and essentially separate faunas exist there. One element, broadly distributed through much of Siberia, is confined to the shal-

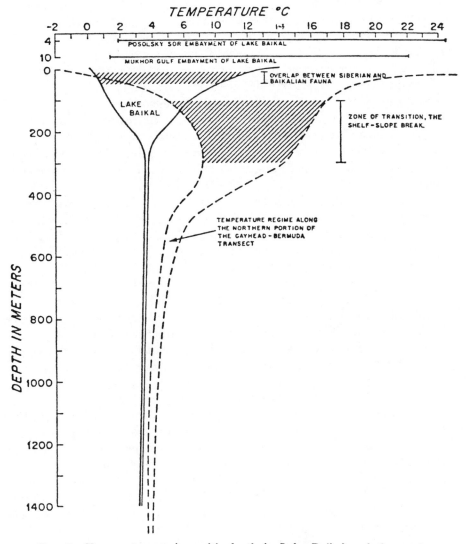

FIG. 7.—Temperature regime with depth in Lake Baikal and the northern part of the Gayhead–Bermuda transect.

low gulfs or bays and is not found deeper than 20 m in the lake. This group is represented by 137 species of nonendemic, free-living, benthic or pelagobenthic macroinvertebrates, 54% of which are insects. The other, an entirely endemic Baikalian fauna of 580 species of which are only 4% are insects, avoids the shallowed depths and embayments, and representative species are found down to 1,620 m, the maximum depth. (Lake Baikal is the deepest freshwater lake in the world.) The ecotone or region of overlap between these two faunas is very narrow (15 to 50 m depth), and but 22 species occur there (Fig. 7).

The Siberian eurytopic component exists in an environment of highly variable seasonal temperatures (see Fig. 7, Posolosky Sor and Mukhor Gulf) and low, fluctuating oxygen content (Kohzov, 1963). The deeper-dwelling Baikalian element, on the other hand, lives under physical conditions that are hardly varying. Such stable conditions, with time, have allowed the evolution of this highly diverse endemic stenotopic fauna, while the less diverse, nonendemic, eurytopic Siberian fauna remains confined to the shallow and physically more variable parts of the lake.

Similar rigidly stable physical conditions are encountered in the greater depths of Great Slave Lake in the Northwest Territories of Canada (Rawson, 1953). Yet, within the depth range from 200 to 600 m, only four species of macrofaunal benthic invertebrates were collected from this large "postglacial" lake, again demonstrating the significance of the time component of diversity.

These findings from Lake Baikal are entirely analogous to the conditions occurring in the boreal region of the Gayhead-Bermuda transect. The continental shelf, particularly in the shoaler depths, harbors an impoverished fauna. The continental slope, however, supports a benthic fauna of high diversity. The region of very rapid and pronounced faunal change occurs somewhere between the depth range of 100 to 300 m and is the most marked zoogeographical boundary encountered along the transect. Not only are there specific and generic differences, but in some groups these changes are of familial and even ordinal significance. Conceivably, the changeover from the diverse stenotopic deep-sea fauna, which is physiologically adapted to constant temperature conditions (as well as other constant environmental factors), to that of the relatively depauperate eurytopic boreal littoral fauna, which exists under a varying temperature regime (other environmental factors tend also to fluctuate here), occurs at that depth where seasonal changes in temperature become large (see Fig. 7).

The same interpretation might be applied to the faunal changes in Lake Baikal, although the zone of rapid transition from the eurytopic assemblages of limited diversity to the highly diverse fauna of deeper water takes place at shallower depths (Fig. 7). Note that the range of seasonal temperature change in the transitional zone is approximately the same in both regions (Fig. 7).

Both limnetic Lake Baikal and the marine area of study south of New England are in boreal regions dominated by a continental climate.

In both situations, the pronounced seasonal changes in temperature are imposed on the shallow-water fauna while the benthos of greater depths are insulated from these changes (Fig. 7). With time, similar patterns of diverse stenotopic faunal assemblages have evolved in the physically stable deeper waters while the highly unstable shallow waters continue to support a rather impoverished fauna. Thus these two unrelated freshwater and marine faunas, molded by similar physical forces and time, have evolved and diversified in a parallel and analogous manner.

Such an interpretation is not at variance with the stability-time hypothesis as shown diagramatically in Figure 6. A continuous diminution in stress conditions certainly takes place from shallow to deep depths, but over a spatially restricted portion of this gradient the rate of change is very great. The region of abrupt change represents the transitional zone from the predominantly physically controlled to the predominantly biologically accommodated community.

Fischer (1960) concluded that the greater diversity in the tropics occurs not only because of the greater stability and longer history of that environment but also because the temperature is nearer the midpoint of the temperature range that protoplasm can endure. High temperature, per se, does not play a critical role in promoting diversity, for the highly diverse assemblages of the deep-sea and the endemic fauna of Lake Baikal evolved in relatively low-temperature environments. The critical factors appear to be time and environmental stability.

EXTRAPOLATIONS FROM THE STABILITY-TIME HYPOTHESIS

Our proposed hypothesis has another use. It allows us to predict. Hutchins (1947) pointed out that in the Northern Hemisphere a much greater seasonal change in water temperature takes place along the western edges of oceans at temperate latitudes than along the eastern edges. Such temperature conditions result from the prevailing west-to-east wind patterns in the middle latitudes. Thus the coastal boreal regions of eastern United States and parts of eastern Asia are dominated by a continental climate of high summer temperature and low winter temperature, while the outer European coasts and the western coast of North America are dominated by a maritime climate of appreciably less seasonal temperature change.

On this basis we can predict that two distinct types of boreal shallow-water communities exist. One can be termed the "continental climate boreal community" and would be exemplified by our Buzzards Bay series of samples. This community will be characterized by low faunal diversity. Furthermore, many, if not most, of the infaunal species will cease to grow and become inactive during the cold winter months (personal data and observations). The other boreal marine shallow-water community can be called the "maritime climate boreal community," characterized by greater faunal diversity and without the complete cessation of growth among the infauna during the winter months.

The model also suggests that the great upwelling regions of the oceans, such as the areas off southwest Africa and the Peruvian and Chilean coasts of South America, will typically show low benthic diversity. The abundant organic matter depletes the available oxygen as it sinks, so that the bottom water contains little or no oxygen. The stress condition resulting from the reduced available oxygen will be reflected in low diversities and, with further oxygen depletion, in low faunal densities or ultimately in abiotic conditions. Indeed, data by Gallardo (1963) from the upwelling areas off northern Chile provide impressive evidence for this interpretation. He found that the oxygen content of the bottom water was less than 5% saturated and that the sediments were reduced at water depths from 50 to 400 m. The benthic samples yielded few individuals, averaging only 6 to 7 per cubic meter.

SPATIAL VARIATION AND TEMPORAL VARIATION

Let us now consider a possible mechanism that would give the few though often numerically abundant species in predominantly physically unstable environments and the many species in the physically stable environments. We must consider two types of variation: *temporal variation*, which has already been discussed in some detail, and *spatial variation*, or habitat diversity.

When temporal variation is large (physiological stress conditions in physically unstable environments), it masks the effects of spatial variation (i.e., the wide range of habitats utilized by lemmings in the high latitudes of North America; see also MacArthur [1965] on within- and between-habitat avian diversities below). When temporal variation is small (minimal stress conditions in physically stable environments), the effects of spatial variations are realized, resulting in the progressive division of species with time. Thus with spatial variation occurring within the distributional range of a species, different selective forces will be acting on the species in different parts or habitats of its range. Initially, this process may result in the formation of separate subspecies and, if the gene flow is sufficiently attenuated, of separate species.

MacArthur (1965) pointed out that in a new environment (this may be comparable to environments of large temporal variation or high physiological stress), such as the initial stages in the colonization of an island by birds, few species are present but are distributed through a number of habitats. With time (this may be comparable to environments of decreasing temporal variation or reduced physiological stress) the number of species increases, but this enrichment is reflected in the *between-habitat* or β diversity of Whittaker (1965) (increase in the total number of species for all habitats) rather than the *within-habitat* or α diversity of Whittaker (1965) (number of species in a specific habitat remains constant).

A concept somewhat similar to MacArthur's constancy of within-habitat diversity is the earlier "parallel community hypothesis" postulated by

70

Thorson (1952) for the marine infauna. He contended that while there is a continuous gradient of species diversity from the arctic to the tropics for the epifauna, the number of infaunal species remains approximately the same. The findings in the present investigation, which is clearly a study of within-habitat diversity, give diametrically opposite results. Samples from historically stable environments of long duration and low physiological stress give high within-habitat diversity values, while samples from historically recent and/or variable environments yield low within-habitat diversity values.

Thorson, at the time he suggested his concept (1952), had only a very limited amount of data on the tropical marine infauna. At least one tropical locality upon which this interpretation is based, the Persian Gulf, with very high salinities and temperatures, represents a stress environment. Our own findings in the present study show that diversity can be quite variable in the tropics. Regions of low stress, such as the Bay of Bengal (RH-26, RH-30, and RH-36), support a very diverse infauna. Conversely, tropical areas of high stress, as exemplified by the shallow-water samples from the west coast of India (RII-14, RII-41), give much reduced values. Thorson (1966) recently found very high tropical infaunal diversities at shelf depths off the west coast of Thailand and he now feels that the applicability of the parallel community hypothesis to tropical environments should be carefully scrutinized.

There still appears to be an underlying difference between avian populations and benthic infaunal invertebrates. Among birds, with time and a physically stable environment, an increase in species occurs. This species enrichment takes place entirely as between-habitat diversity, while within-habitat diversity remains unchanged. With our infaunal benthic organisms, on the other hand, species enrichment is reflected both in the between- and the within-habitat diversities.

A lucid genetic interpretation for the relationship of environmental stability to diversity has been given by Grassle (1967). He pointed out that populations present in physically stressed and unpredictable environments show broad adaptations to these conditions by maintaining a high degree of genetic variability. Thus, even though the stress may be expressed in a variety of ways, a portion of the polymorphic population will probably survive. These genetically flexible species are opportunistic and cosmopolitan, and they have little tendency to speciate.

The price paid for this variability is "the genetic load or loss of fitness relative to the maximum in a more uniform environment." In stable environments "the expression of deleterious genes outweighs the advantages obtained from maintaining genetic flexibility." Therefore, in stable and predictable environments, such genetic variability will be selected against.

Diversity differences found in the present study between stable and unstable environments can be interpreted on the basis of genetic variability. The flexibility needed for survival in an unstable environment necessitates a larger utilization of the environment by each species. Thus diversity and genetic flexibility would be inversely related. (For a comprehensive discussion of the genetic basis of benthic diversity, see Grassle [1967].)

THE DEFINITIONS AND MEANINGS OF DIVERSITY

We might pause here and ask what we precisely mean by the word "diversity." It is apparent from looking through the literature that there are two definitions. This has resulted in some confusion.

One kind of diversity is the *numerical percentage composition* of the various species present in the sample. The more the constituent species are represented by equal numbers of individuals, the more diverse is the fauna. The less numerically equal the species are, the less diverse the sample is or, conversely, the greater is the dominance in the sample. This is a measure of how equally or unequally the species divide the sample, and the number of species involved is immaterial. Diversity measurements of this kind include the MacArthur "Broken Stick" model (1957), the Preston lognormal distribution (1948), and the Simpson index (1949). Such diversity might be designated after Whittaker (1965) as *dominance diversity*.

The other kind of diversity is determined by the *number of species*. The more species in a sample or the more species present in a species list for a given environment, the greater the diversity. Measurements of this sort are the α values of Fisher, Corbet, and Williams (1943), Margalef's d values (1957), the methodologies of Gleason (1922) and of Hessler and Sanders (1967), and the rarefaction technique used in the present paper. Such diversity can be designated after Whittaker (1965) as *species diversity*.

Since the number of species present and the relative dominance or lack of dominance in a sample are both measures of diversity, one might assume that they must be highly correlated with one another. Thus, a large number of species per unit number of individuals reflects low dominance; alternatively, a small number of species indicates high dominance. In Table 2 we will test this assumption.

Eighteen of the stations used in Figure 2 are included in the analysis. Each station is represented by a single sample, except station R, which is a composite of 15 samples taken from the same locality in Buzzards Bay. (The Pocasset series of samples are excluded because of the very few species present.)

The samples are ranked by dominance diversity from highest (lowest dominance) to lowest (highest dominance) diversity. These values are determined by plotting the percentage composition of the species along the ordinate and ranking the species by abundance along the abscissa (Fig. 8). The resultant cumulative frequency curve is used as a measure of dominance diversity.

Maximum diversification occurs when all the species in a sample are represented by exactly the same number of individuals, and the cumulative frequency curve, in this case, is a diagonal straight line that can be described by the formula $x = y$. Such a straight line forms the base line (Fig.

TABLE 2

MEASUREMENT OF THE CORRELATION BETWEEN SPECIES DIVERSITY AND DOMINANCE
DIVERSITY AT DIFFERENT FAUNAL DIVERSITY LEVELS USING THE PEARSON
PRODUCT MOMENT CORRELATION FOR 18 STATIONS INCLUDED IN THIS STUDY

Sample Size	N	r Value	Critical r Value at 5% Level	Critical r Value at 1% Level
Spp./10 ind.	18	.974	.456	.575
Spp./25 ind.	18	.948	.456	.575
Spp./50 ind.	18	.894	.456	.575
Spp./100 ind.	18	.766	.456	.575
Spp./200 ind.	18	.642	.456	.575
Spp./300 ind.	16	.636	.482	.606
Spp./400 ind.	15	.649	.479	.623
Spp./500 ind.	14	.622	.514	.641
Spp./600 ind.	14	.623	.514	.641
Spp./700 ind.	13	.748	.532	.661
Spp./800 ind.	9	.832	.632	.765
Spp./900 ind.	8	.769	.666	.798
Spp./1,000 ind.	8	.819	.666	.798
Spp./1,200 ind.	8	.817	.666	.798
Spp./1,400 ind.	7	.832	.707	.834
Spp./1,600 ind.	7	.833	.707	.834
Spp./1,800 ind.	7	.832	.707	.834
Spp./2,000 ind.	6	.801	.754	.874
Spp./2,500 ind.	4	.572	.878	.959
Spp./3,000 ind.	4	.519	.878	.959

8). What is measured is the deviations in percentage composition of a given sample from a hypothetical sample containing the same number of equally abundant species, that is, the degree of departure from the base line. The greater the departure, the greater the dominance and, conversely, the smaller the diversity. (See also Sanders, 1963, pp. 87 and 88.)

The stations are also ranked by species diversity from highest (most species per unit number of individuals) to lowest (least species per unit number of individuals) diversity at various population levels from 10 to 3,000 individuals. The dominance-diversity ranking is compared to the species-diversity ranking at each of the derived sample-size levels, using the rarefaction method.

This relationship is measured using the Pearson product moment correlation. The findings are given in Table 2. The square of the correlation coefficient, r, gives the approximate correlation, that is, the variance accounted for by the correlation.

This analysis reveals that species diversity and dominance diversity are indeed correlated at the 5% significance level, except with the largest samples. Using the 1% significance level, a consistent correlation exists only with the smaller-size samples (10 to 400 individuals). At intermediate sample sizes (500 to 1,800 individuals) the relationship is marginal, with the correlation values fluctuating around the critical correlation value. Among the largest sample sizes (2,000 to 3,000 individuals), species diversity and dominance diversity are not significantly correlated.

Thus a correlation, although not a particularly intimate one, exists between species diversity and dominance diversity. This conclusion is in agreement with the recent suggestion of Whittaker (1965) that the rela-

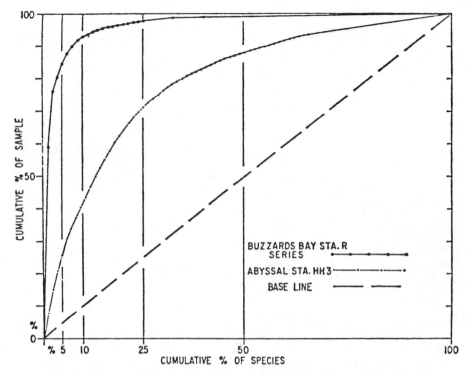

FIG. 8.—Degree of dominance of a sample related to numerical percentage composition of the included species plotted cumulatively. For explanation of figure, see text.

tionship between these two diversity measurements is weak. The degree of correlation appears to be sample-size dependent. A strong correlation is found with small sample sizes, a weaker correlation occurs among intermediate sample sizes, and either a weak or no significant relationship exists among the largest sample sizes.

Only at the smallest sample size does this correlation account for most of the variance (about 95% at the 10-individual level, about 90% per 25 individuals, and about 80% per 50 individuals). At such small population sizes only the most abundant species would normally be present, and one might expect a high level of correlation between dominance and species diversities. As sample size becomes larger, the less common species begin to appear and the percentage of variance accounted for by the correlation diminishes.

What, then, determines this relationship? As mentioned earlier, dominance diversity is independent of the number of species present. However, all species-diversity indexes are affected not only by the number of species but also by how a sample is divided among these species (percentage composition). There is always a dominance-diversity component in all species-diversity measurements because, while the height of the curve is determined by the number of species present, the shape of the curve is set

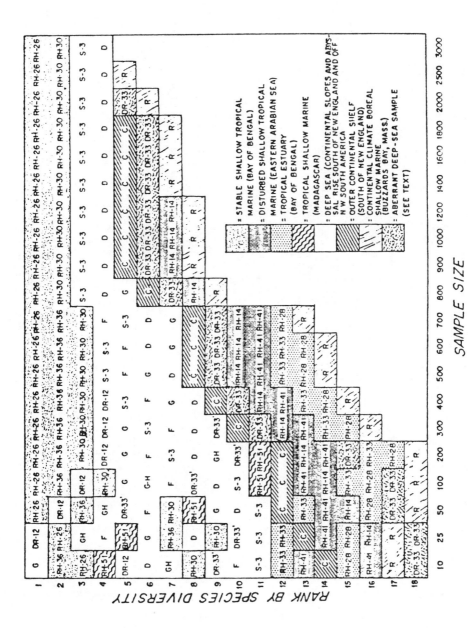

FIG. 9.—Ranking of stations by species diversity at different sample sizes.

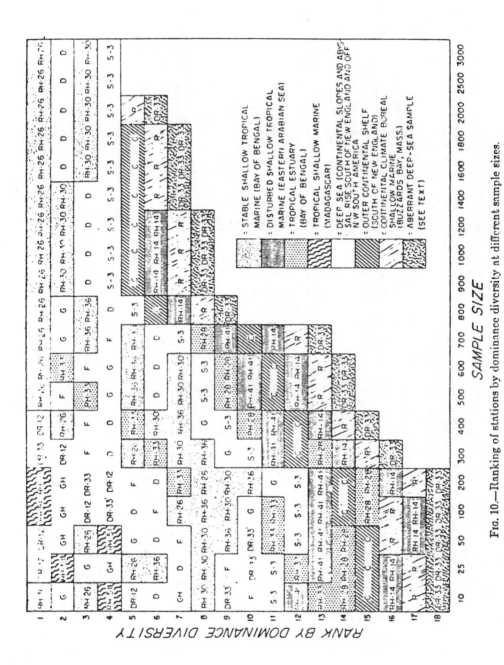

Fig. 10.—Ranking of stations by dominance diversity at different sample sizes.

by dominance. Probably it is the dominance dependency of species-diversity measurements that is primarily responsible for the percentage of variance accounted for by the correlation. Such an interpretation is consistent with the observation that the percentage of variance is very high with smallest sample sizes when only the numerically dominant species are present. Of critical importance, then, is the ecological significance of these two measures of diversity.

If these stations are ranked by both dominance diversity and species diversity, much better clustering of stations by environments is obtained within the species-diversity series (Figs. 9 and 10). Except at low densities, the species-diversity pattern remains stable by environment over the spectrum of population sizes. The stations with the highest diversity values are from the shallow marine depths of the Bay of Bengal (stations RH-26, RH-36, and RH-30). Then, except for aberrant station DR-33, there is a block of samples which includes all the deep-sea stations. They, in turn, are followed by the single shallow marine sample (RH-51) from Madagascar and the single station from the outer continental shelf (C). Next comes the previously mentioned atypical deep-sea station DR-33, the two stations from the stress shallow waters of the Arabian Sea off India (RH-14 and RH-41), the two tropical estuarine samples from India (RH-28 and RH-33), and, finally, the R series of samples from Buzzards Bay, Massachusetts.

Within the dominance-diversity series (Fig. 10), no such clear-cut groupings are present. Stations from a single environment often are widely separated. The Bay of Bengal stations (RH-26, RH-36, and RH-30) with the highest species-diversity values often show wide variability in ranking, both within a specific population size and among differing population sizes. The deep-sea stations do not form a solid block, but have stations from other environments interspersed among them. At those population sizes where the Madagascar station (RH-51) achieves first ranking in the dominance-diversity series, it does no better than eleventh rank using species-diversity criteria. The two tropical estuary stations, RH-33 and RH-28, have appreciably differing dominance-diversity values, yet their species-diversity values are almost identical. Station RH-33 gives intermediate to high dominance-diversity but low species-diversity values. The disturbed tropical stations, RH-14 and RH-41, show reasonable agreement in ranking when both diversity measurements are compared. The values are always low; yet the two stations are usually contiguous by species-diversity ranking and are always separated by dominance-diversity analyses. Outer continental shelf station C usually shows higher species- than dominance-diversity ranking. The R series of samples from Buzzards Bay give low values by both diversity methods. It occupies the lowest species-diversity rank over most of the sample-size spectrum and the next to lowest rank by dominance-diversity criteria. Aberrant deep-sea station DR-33 has last ranking in the dominance-diversity series, but its diversity values increase with sample size using species-diversity ranking. (The pronounced effect of the single numerically dominant species, forming 84.43% of this sample, is gradually mediated as the population size increases and the inherent species diversity begins to emerge. This effect is so overwhelming by dominance-diversity standards that station DR-33 is restricted to lowest ranking throughout the entire range of population sizes.) The only good agreement found between species- and dominance-diversity rankings occurs at the smallest population sizes, where only the most common species would be present.

In brief, then, the stations in the species-diversity series clearly and sharply sort themselves by environment and generally follow the gradient from the biologically accommodated to the physically controlled environmental situations. Within the dominance-diversity series, no such clear-cut groupings are present. Stations from a single environment often are widely separated. Further, there is often little agreement in the position of a station in one series as compared with the other.

These findings can only be interpreted to mean that the high level of agreement between environment and species diversity indicates that such a measure is a conservative and, therefore, ecologically powerful tool. On the other hand, the much poorer fit with dominance diversity suggests that this type of diversity is more variable in its relationship to the physical environment.

THE RELATIONSHIP OF OTHER DIVERSITY CONCEPTS TO THE STABILITY-TIME HYPOTHESIS

Pianka (1966) has summarized the various hypotheses advanced to explain the causes of latitudinal diversity gradients. He was able to separate them into six more or less distinct groupings, although most of the hypotheses contain components of more than one grouping. Presented in their most elemental form, they are:

a) The time theory (Simpson, 1964).—All communities tend to diversify with time. Older communities, therefore, are more diverse than younger communities.

b) The theory of spatial heterogeneity (Simpson, 1964).—The more heterogeneous and complex the physical (topographic) environment, the more complex and diverse its flora and fauna become.

c) The competition theory (Dobzhansky, 1950; Williams, 1964).—Natural selection in higher latitudes is controlled by the physical environment, while in low latitudes biological competition becomes paramount.

d) The predation hypothesis (Paine, 1966).—There are more predators in the tropics who intensively crop the prey populations. As a result, competition among prey species is reduced, allowing more prey species to coexist.

e) The theory of climatic stability (Klopfer, 1959; Fischer, 1960; Dunbar, 1960; Connell and Orias, 1964).—Because of the greater constancy of resources, environments with stable climates have more species than environments of variable or erratic climates.

f) The productivity theory (Connell and Orias, 1964).—All other things being equal, the greater the productivity, the greater the diversity.

To fit better into the context of the present paper, two of these theories are rephrased as follows:

b) The *theory of climatic stability* is generalized to the *theory of environmental stability*. The more stable the environmental parameters—such as temperature, salinity, oxygen—the more species present.

c) The *competition theory* is altered to state that in environments of high physiological stress, selection is largely controlled by the physical variables, but in historically low stress environments, natural selection results in biologically accommodated communities derived from past biological interactions and competition.

How do the data from the marine benthic samples presented earlier in this paper and the derived *stability-time hypothesis* fit these theories? The *time theory* and the *theory of environmental stability* are most directly applicable to the *stability-time hypothesis*. They, in turn, determine the expression of the *competition theory* and the *theory of spatial heterogeneity*. Biologically accommodated communities resulting from past biological interactions (including competition) are realized in physically stable environments of long temporal continuity. Similarly, the potentials of spatial heterogeneity can be achieved only under these same environmental conditions.

Neither the *predation theory* nor the *productivity theory* can readily be explained by the *stability-time hypothesis*. The *predation theory* was recently postulated by Paine (1966) for rocky intertidal marine organisms, although he feels it may have wider application. In the intertidal environment, the epibenthic animals experience alternating periods to exposure and immersion. Therefore, these organisms are subjected to desiccation, high salinity imposed by evaporation, exposure to freshwater rain, and air temperatures that are often significantly higher or lower than the seawater temperature. Thus, all rocky intertidal assemblages, independent of latitude, especially at the higher intertidal levels (see Connell, 1961), must be considered predominantly physically regulated communities, and the adaptations are primarily to the physical environment and the biological interactions are poorly developed.

Conceivably, the *productivity theory*, which says that the more food produced, the greater the diversity, may have some validity. Yet this effect is readily masked by numerous environmental variables (Hessler and Sanders, 1967). High productivity itself, from the sheer amount of organic matter produced, can create severe stress conditions and low diversity. For example, in some upwelling areas, high production is responsible for low oxygen content of the water on and just above the ocean floor. Similarly, the highly productive eutrophic lakes often have bottom water devoid of oxygen. In contrast, the high diversity values for the deep-sea benthos, shown by Hessler and Sanders (1967) and in the present paper, come from regions of low productivity.

COMPARISON USING CERTAIN OTHER FAUNAL INDEXES

Numerous indexes have been formulated to measure diversity. Odum, Cantlon, and Kornicker (1960) pointed out that in all types of presentation, logarithmic functions are involved. They recognize four categories. Three have pertinence to our paper:

1. *Cumulative species versus logarithm of abundance.*—This was exemplified by Gleason (1922), Fisher et al. (1943), and Margalef (1957). Such an index is obtained by determining the rate of species increase as additional samplings are made from the same population.

2. *Number of species of particular abundance versus logarithm of abundance.*—This method was formulated by Preston (1948) and is based on the premise that, if the presence of all species found in a given habitat can be revealed, the abundance distribution would follow a lognormal curve. Since such a complete revelation is usually impossible, the resulting curve is truncated at its rarer end. However, the shape of the curve allows one to approximate the total number of species in the habitat, including those as yet undiscovered.

3. *Abundance versus logarithm of rank.*—This type of index was proposed by MacArthur (1957). The observed abundances are compared with theoretical abundances derived from a model containing contiguous, non-overlapping niches.

One of the fundamental drawbacks of most diversity indices is that they are sample-size or density dependent. Hairston and Byers (1954), in an analysis of cumulative samples of soil arthropods from a singel habitat by both the logarithmic series of Fisher, et al. (1943) and the lognormal distribution of Preston (1948), found that the results depended on the size of the total sample. Margalef (1957) pointed out that his diversity measurement, in common with other diversity indexes, increases with enlarged samples. A similar finding was reported by Williams (1964) for the Simpson diversity index (1949). Hairston, in a later paper (1959), demonstrated that the MacArthur model (1957) is also density dependent. From the analysis of his own data, he interpreted this phenomenon of increased diversity with increased sample size to mean that rare species are clumped. With repeated samples, there will be a greater likelihood of obtaining a new rare species than of obtaining a member of a rare species already collected. He concluded "that an increase in heterogeneity with an increase in sample size lies in the spatial distribution of the species concerned, and the inverse relationship between clumping and abundance."

In our study, single samples were collected from comparable sediment environments. Both small samples and large samples from the same environment (Fig. 2, boreal shallow marine and deep-sea) fall along the same diversity curve. If we apply the rarefaction method to a series of 15 samples of greatly differing sizes (35 to 2,514 individuals) which were carefully selected for sediment homogeneity and taken during the course of a 2-year period from a single locality in Buzzards Bay (Sanders, 1960), we find in the semilogarithmic plot in Fig. 11 no tendency for smaller samples to be

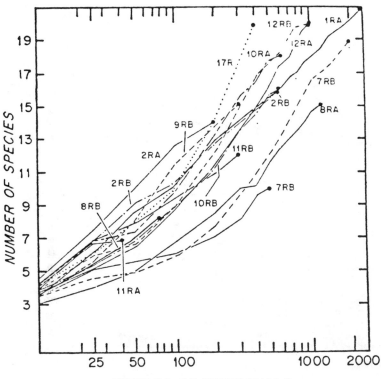

FIG. 11.—Rarefaction curves for the 15 station R samples from Buzzards Bay, Massachusetts.

less diverse than larger ones—that is, smaller-sized samples do not show a tendency to rise more steeply than larger-sized samples. Thus, both under the conditions of our sampling program and by using the rarefaction method of measuring diversity, no increase in heterogeneity takes place with increasing sample size. After all, what could be more homogeneous than a series of different-sized subsamples, each with the same percentage composition as the original samples? This, in essence, is what the rarefaction method does.

Now that we have demonstrated the effectiveness of the proposed rarefaction methodology in obtaining constancy of diversity at all population sizes in our samples, can similar stability be achieved when other diversity formulas are applied to the identical data whose internal homogeneity has been demonstrated over the entire range of sample size? We will attempt to answer this question by applying a number of diversity measurements to the data presented in Figure 2.

With the Preston truncated lognormal distribution analysis, the numbers of species with abundances of 1 to 2, 2 to 4, 4 to 8, etc., are plotted as points. The resulting curve is assumed to approximate a lognormal distribution. Such an estimate is essentially independent of sample size because a dou-

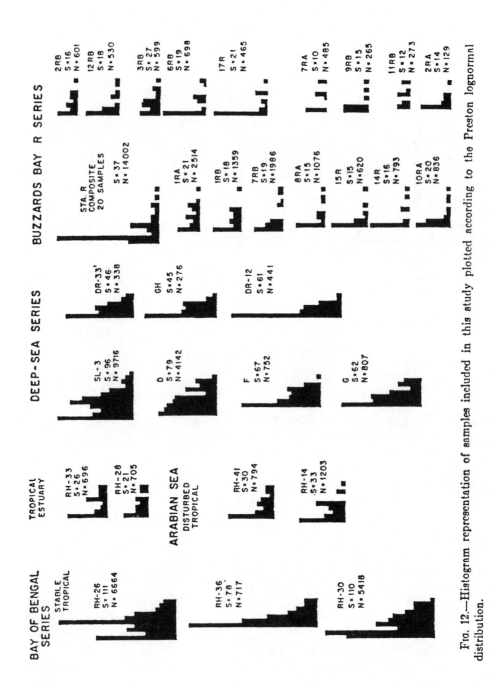

Fig. 12.—Histogram representation of samples included in this study plotted according to the Preston lognormal distribution.

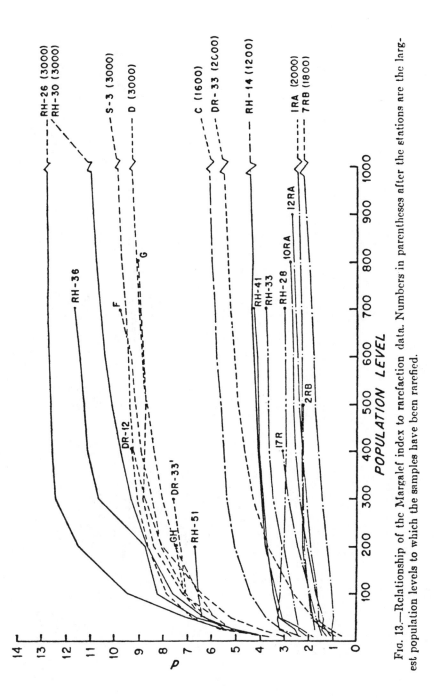

FIG. 13.—Relationship of the Margalef index to rarefaction data. Numbers in parentheses after the stations are the largest population levels to which the samples have been rarefied.

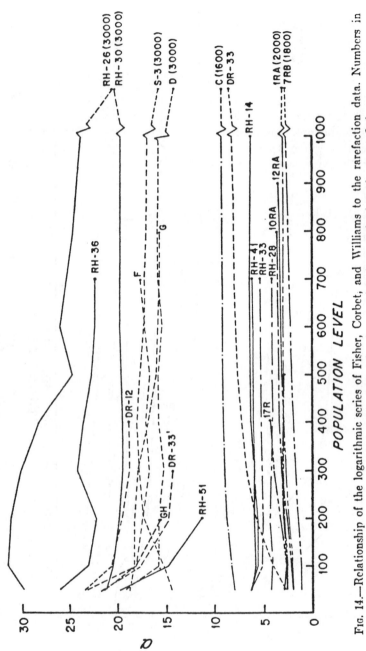

Fig. 14.—Relationship of the logarithmic series of Fisher, Corbet, and Williams to the rarefaction data. Numbers in brackets after the stations are the largest population levels to which the samples have been rarefied.

bling of individuals simply displaces the curve one unit to the right and adds a new unit to the left or rare end of the curve. When the entire suite of species is finally revealed, a lognormal rather than a truncated lognormal curve describes the situation.

Plotting our data by this method in Figure 12 gives histograms that are difficult to interpret. For example, the rarefaction curves for the two disturbed tropical shallow marine samples, RH-14 and RH-41, are essentially identical. The histograms derived for these same stations by using the Preston methodology are totally unlike (Fig. 11), and it would take a great amount of ingenuity to fit the histogram of station RH-14 to a truncated lognormal distribution. While the rarefaction curves for the two tropical shallow marine samples, RH-26 and RH-30, are somewhat alike, the Preston histograms are very different. The histogram for RH-26 also does not remotely fit a truncated lognormal pattern.

No attempt will be made to consider each of the histograms in Figure 12. It seems evident from the examples already chosen that the truncated lognormal distribution pattern cannot convincingly be made to fit these samples. Extrapolations of the total number of expected species made from such fitted truncated lognormal curves by adding to the left tail and converting them into normal distribution curves give results that are unrealistic. By such analyses, samples taken from the same environment and displaying similar rarefaction curves often have normal distribution curves containing appreciably differing numbers of species.

Using Margalef's (1957) index (which is essentially the same as Gleason's [1922] formulation), $d = (S - 1)/\ln N$, where S = number of species, N = number of individuals, and d = index of diversity, we find (Fig. 13) that d in all samples is initially low. As sample size becomes larger, the d value rapidly increases. At high densities (large sample sizes), the rate of increase gradually diminishes. This effect is most pronounced in samples with high diversities—the shallow waters of the Bay of Bengal and the deep sea. In high-stress environments—the Buzzards Bay series, tropical estuaries, and the disturbed tropical shallow marine—this effect is less pronounced. These data indicate that diversity indexes of the Margalef and Gleason types are influenced by sample size, even when such samples are internally homogeneous.

When the same samples are plotted using the α values of Fisher et al. (1943), $S = \alpha \ln (N/\alpha + 1)$, almost opposite results are obtained (Fig. 14). In the more diverse samples, the α values are highest at low densities, rapidly decrease as density increases, and then more slowly decrease until an approximate equilibrium is reached. With low-diversity samples, the tendency for higher α values at low density is either absent, poorly developed, or weakly opposed. When internally homogeneous samples are used, this diversity index also is not independent of sample size. In comparison with the d values, the α indexes tend to stabilize at lower faunal densities.

Application of the Simpson diversity index (1949), $C = (y/N)^2$, where

y = number of individuals in species N = total number of individuals, and C = measure of concentration of dominance, to these data does show an increase in diversity with sample size. The rate of change is decidedly less than when a or d indexes are used. This does not mean that the Simpson index is necessarily a more valid diversity measurement. The formula for calculating the diversity index, as shown by Williams (1964), greatly exaggerates the contributions of the few abundant species, while the influence of the many species with few individuals is insignificant. Thus there can be little increase in the diversity index as additional rare species are added. This is the explanation for the relatively low rate of diversity change with increasing sample size.

The degree of exaggeration by this methodology can be demonstrated by comparing the Simpson index per 100 individuals with the number of species for 100 individuals as determined by the rarefaction method for a number of the samples used in our study (Fig. 15). The maximum value by the rarefaction method is 4.93 times greater than the minimum value. Yet the maximum-to-minimum ratio, using the Simpson index, is 16.5:1.00.

We also compared the species numbers using the rarefaction method with the MacArthur model (1957):

$$\frac{N}{S} = \sum_{i=1}^{r} \frac{1}{(S - i + 1)},$$

where N = number of individuals, S = number of species, i = interval between successively ranked species and rarest, and r = rank in rareness. What we measured was the deviation in percentage composition of our actual and rarefied samples from the expected percentage composition derived from the MacArthur model for the same number of species. Like Hairston (1959) and others, we found the common species to be more common and the rare species to be rarer than expected from the model.

As shown in Table 3, there was a strong tendency within each environ-

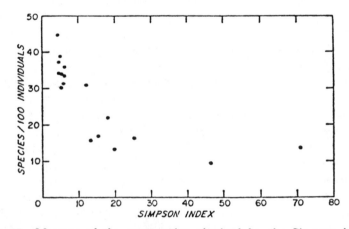

FIG. 15.—Measure of the exaggeration obtained by the Simpson index as compared with rarefaction methodology at the 100-individual level.

TABLE 3

PERCENTAGE DEVIATION OF THE RAREFACTION DATA FROM THE MACARTHUR MODEL
FOR THE ENTIRE SAMPLE AND AT THE 100-INDIVIDUAL
LEVEL FOR CERTAIN OF THE STATIONS

Station	Species Number	Number of Individuals	Deviation for Total Sample	Deviation/100 Individuals	Type of Environment
RH-26...	111	6,664	64.58	35.77	Shallow tropical marine
RH-30...	110	5,418	76.28	26.62	Shallow tropical marine
RH-36...	78	717	55.42	33.82	Shallow tropical marine
RH-51...	39	285	17.90	9.04	Shallow tropical marine
RH-14...	33	1,203	98.43	64.86	Stress shallow tropical marine
RH-41...	30	794	72.27	41.21	Stress shallow tropical marine
RH-28...	21	705	66.39	46.93	Tropical estuary
RH-33...	26	696	47.79	21.43	Tropical estuary
S1-3.....	96	9,716	89.32	49.23	Deep sea
D.......	79	4,142	63.12	18.51	Deep sea
DR-33*..	47	2,171	156.76	124.28	Deep sea
G.......	62	807	50.59	19.77	Deep sea
F........	67	752	49.75	32.78	Deep sea
DR-12...	61	449	34.63	15.84	Deep sea
DR-33'...	46	338	36.87	29.07	Deep sea
GH......	45	276	21.67	17.14	Deep sea
C........	51	1,861	89.69	56.55	Outer continental shelf
1RA.....	21	2,514	119.61	97.27	Shallow boreal marine
7RB.....	19	1,976	109.05	44.41	Shallow boreal marine
12RA....	20	959	104.27	64.79	Shallow boreal marine
10RA....	20	836	103.35	63.70	Shallow boreal marine
2RB.....	16	465	98.17	83.20	Shallow boreal marine
17R......	21	465	93.23	70.67	Shallow boreal marine

* The gross difference between this sample and the other deep-sea samples is caused by the pronounced aggregation of a single species (for further comments, see text).

ment for progressively smaller samples to show better agreement (less deviation) with the model. The rarefied samples containing 100 individuals always gave a better fit than the actual samples from which they were derived. Finally, when a large number of rarefied samples of different sample sizes from a common sample were compared (Fig. 16), the larger the sample size (except at very low numbers), the greater the departure from the model.

Thus even in homogeneous environments, the MacArthur model is markedly density dependent. For a given faunal density in Figure 16 and Table 3, samples from high-stress environments showed greater deviations from the model than samples from low-stress environments. Such effects are readily masked by sample size. In this regard, it is more than a coincidence that one of the few studies giving a good fit to the MacArthur model (Kohn, 1959) was based on small samples from a stable, low-stress environment. In all fairness to MacArthur, it should be pointed out that he has recently (1966) disavowed the validity of this index.

Lloyd and Ghelardi (1964) proposed the equitability concept as a measure of how a sample is apportioned among its constituent species. Because

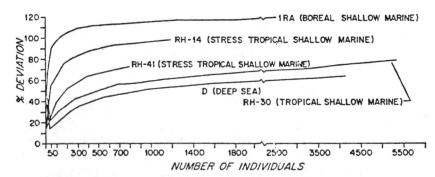

FIG. 16.—Deviation from the MacArthur model using the rarefaction method for certain stations of this study.

numerical equality is never achieved in practice, they suggested "equita-bility" rather than "evenness" as a more realistic standard. The Shannon-Wiener information function:

$$H(s) = - \sum_{r=1}^{s} p_r \log_2 p_r,$$

where s = total number of species and p_r = observed proportion of indi-viduals that belong to the rth species ($r = 1, 2, \ldots, s$), provides the basis for this measure when combined with some theoretical distribution of abun-dances, in their case, the MacArthur model shown above.

From this model, given an s, a hypothetical diversity function can be calculated:

$$M(s) = - \sum_{r=1}^{s} \pi_r \log_2 \pi_r,$$

where π_r is the theoretical proportion of individuals in the rth species ranked in order of increasing abundance from $i = 1$ to s. By setting $M(s') = H(s)$, a calculation for the number of hypothetical "equitably distrib-uted" species, s', is obtained. Equitability (ϵ) is the ratio of the "equitably distributed" species (s') to the actual number of observed species (s), $\epsilon = s'/s$. Higher values mean greater equitability; lower values, less equitable apportionment within the sample.

In Figure 17 we have plotted the equitability value (ϵ) for a range of population sizes for a number of our samples. Since in the rarefaction procedure the percentage composition of the original sample is unaltered, we should expect the equitability value to remain constant throughout the spectrum of population sizes. Figure 17 clearly shows that this is not the case. In every sample there is a continuous decrease in the equitability value with increasing sample size. This decrease is most pronounced at the smaller population sizes, particularly in samples from physically un-stable environments. Thereafter, there is a more gradual reduction in the magnitude of decrease with increasing sample size. At larger population

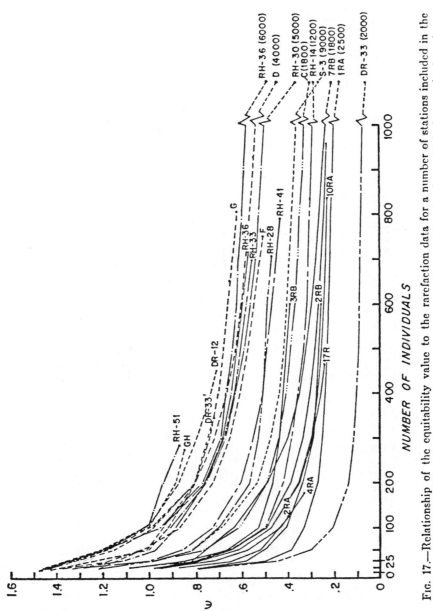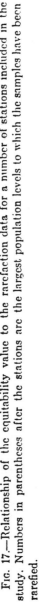

Fig. 17.—Relationship of the equitability value to the rarefaction data for a number of stations included in the study. Numbers in parentheses after the stations are the largest population levels to which the samples have been rarefied.

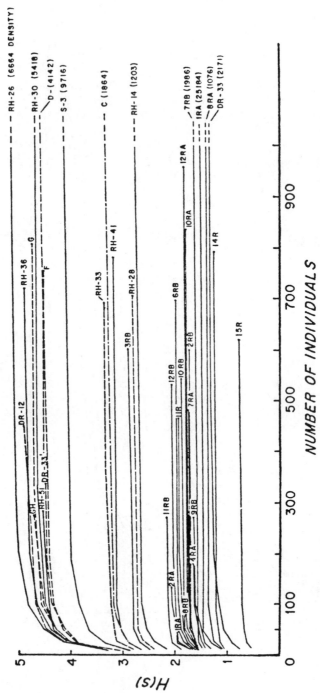

FIG. 18.—Relationship of the information function to the rarefaction data. Numbers in parentheses after the stations are the actual numbers of individuals obtained at those stations.

levels, the magnitude of decrease remains larger for samples from physically stable environments (shallow tropical marine and the deep sea).

Thus equitability is a measurement that is markedly sample-size dependent. This is not surprising when we remember that it is intimately related to the MacArthur model, which we have already shown to be highly sensitive to sample-size differences.

As a final diversity index, we will consider the Shannon-Wiener information function $H(s)$ itself. This function has the attribute of being influenced by both the number of species present and how evenly or unevenly the individuals are distributed among the constituent species. In other words, $H(s)$ is sensitive to both species and dominance diversities.

When the information function is plotted against the rarefaction data (Fig. 18), it very rapidly reaches a stable value and remains essentially constant over a broad spectrum of population sizes. Such stability is achieved at population sizes of about 200 individuals for the high-stress environments (boreal shallow marine, tropical estuary, and disturbed tropical shallow marine) and about 400 individuals for low-stress environments (tropical shallow marine and the deep sea).

Therefore, unlike the other diversity indexes tested, the information function is relatively sample-size independent, and samples of differing sizes, except at lowest faunal densities, can be directly compared.

In summary, when our series of samples were reduced to lower homogeneous population sizes by using the rarefaction method and then compared with various proposed diversity indexes, we found that most of these indexes were decidedly affected by sample size. On the other hand, the Shannon-Wiener information function, except when applied to small-sized samples, possesses the critical characteristic of a useful diversity index, that of being relatively sample-size independent.

SUMMARY

In this paper a methodology is presented for measuring diversity based on rarefaction of actual samples. By the use of this technique, a within-habitat analysis was made of the bivalve and polychaete components of soft-bottom marine faunas which differed in latitude, depth, temperature, and salinity. The resulting diversity values were highly correlated with the physical stability and past history of these environments. A stability-time hypothesis was invoked to fit these findings, and, with this hypothesis, predictions were made about the diversities present in certain other environments as yet unstudied. The two types of diversity, based on numerical percentage composition and on number of species, were compared and shown to be poorly correlated with each other. Our data indicated that species number is the more valid diversity measurement. The rarefaction methodology was compared with a number of diversity indexes using identical data. Many of these indexes were markedly influenced by sample size. Good agreement was found between the rarefaction methodology and the Shannon-Wiener information function.

ACKNOWLEDGMENTS

The ideas presented in this paper have been discussed with numerous individuals. I particularly would like to thank R. R. Hessler, J. H. Connell, and L. B. Slobodkin for their comments and criticisms. M. Rosenfeld and D. W. Spencer generously devoted many hours to both the statistical aspects of this paper and the application of appropriate computer programs. A General Electric 225 computer was used in the analyses.

Support for the acquisition of the various data used came from various sources. The tropical shallow water and estuarine samples from the Indian Ocean were collected as a result of support by the National Science Foundation as a part of the U.S. Program in Biology, International Indian Ocean Expedition. The boreal shallow-water samples from Buzzards Bay, Massachusetts, were collected during the period from 1956 to 1958 under grant NSF G-4812. Support for the collection of deep-sea samples was obtained under grants NSF G-15638, GB-3269, and GB-563. Grant NSF GB-563 also provided support for the collections of boreal estuarine samples from the Pocasset River, Massachusetts.

LITERATURE CITED

Banse, K. 1959. On upwelling and bottom-trawling off the southwest coast of India. J. Marine Biol. Ass. India 1:33–49.

Bousfield, E. L. 1958. Fresh-water amphipod crustaceans of glaciated North America. Can. Field-Natur. 72:55–113.

Carruthers, J. N., S. S. Gogate, J. R. Naidu, and T. Laevastu. 1959. Shoreward upslope of the layer of oxygen minimum off Bombay: Its influence on marine biology, especially fisheries. Nature 183:1084–1087.

Connell, J. H. 1961. Effects of competition, predation by *Thais lapillus,* and other factors on natural populations of the barnacle *Balanus balanoides.* Ecol. Monogr. 31:61–104.

Connell, J. H., and E. Orias. 1964. The ecological regulation of species diversity. Amer. Natur. 98:399–414.

Dobzhansky, T. 1950. Evolution in the tropics. Amer. Sci. 38:209–221.

Dunbar, M. J. 1960. The evolution of stability in marine environments. Natural selection at the level of the ecosystem. Amer. Natur. 94:129–136.

Fischer, A. G. 1960. Latitudinal variations in organic diversity. Evolution 14:64–81.

Fisher, R. A., A. S. Corbet, and C. B. Williams. 1943. The relation between the number of species and the number of individuals in a random sample of an animal population. J. Anim. Ecol. 12:42–58.

Gallardo, A. 1963. Notas sobre la densidad de la fauna bentonica en el sublitoral del norte de Chile. Gayana Zool. 10:3–15.

Gleason, H. A. 1922. On the relation between species and area. Ecology 3:158–162.

Grassle, J. F. 1967. Influence of environmental variation on species diversity in benthic communities on the continental shelf and slope. Unpublished Ph.D. dissertation. Duke Univ., Durham, N.C.

Hairston, N. G. 1959. Species abundance and community organization. Ecology 40:404–416.

Hairston, N. G., and G. W. Byers. 1954. The soil arthropods of a field in southern Michigan. A study in community ecology. Contrib. Lab. Vertebrate Biol. Univ. Michigan 64:1–37.

Hessler, R. R., and H. L. Sanders. 1967. Faunal diversity in the deep-sea. Deep-Sea Res. 14:65–78.

Hutchins, L. W. 1947. The basis for temperature zonation in geographical distribution. Ecol. Monogr. 17:325–335.

Jacob, J., and K. Rangarajan. 1959. Seasonal cycles of hydrological events in the Vellar estuary. First All-India Congr. Zool., Proc., Part 2, Scientific Papers, p. 329–350.

Klopfer, P. H. 1959. Environmental determinants of faunal diversity. Amer. Natur. 93:337–342.

Kohn, A. J. 1959. The ecology of *Conus* in Hawaii. Ecol. Monogr. 29:47–90.

Kohzov, M. 1963. Lake Baikal and its life. Monogr. Biol. 11. 352 p.

LaFond, E. C. 1958. Seasonal cycle of the sea surface temperatures and salinities along the east coast of India. Andhra Univ. Mem. Oceanogr. 2:12–21.

Lloyd, M., and R. J. Ghelardi. 1964. A table for calculating the "Equitability" component of species diversity. J. Anim. Ecol. 33:217–225.

MacArthur, R. H. 1957. On the relative abundance of bird species. Nat. Acad. Sci. Proc. 43:293–295.

———. 1965. Patterns of species diversity. Biol. Rev. 40:510–533.

———. 1966. Note on Mrs. Pielou's comments. Ecology 47:1074.

Margalef, R. 1957. La teoria de la informacion en ecologia. Memorias de la real academia de ciencias y artes (Barcelona) 33:373–449.

Murty, C. B. 1958. On the temperature and salinity structures of the Bay of Bengal. Current Sci. 27:249.

Odum, H. T., J. E. Cantlon, and L. S. Kornicker. 1960. An organizational hierarchy postulate for the interpretation of species individual distributions, species entropy, ecosystem evolution and the meaning of a species variety index. Ecology 41:395–399.

Paine, R. T. 1966. Food web complexity and species diversity. Amer. Natur. 100:65–75.

Panikkar, N. K. 1940. Influence of temperature on osmotic behavior of some crustacea and its bearing on problems of animal distribution. Nature 146:366–367.

Pianka, E. R. 1966. Latitudinal gradients in species diversity: A review of concepts. Amer. Natur. 100:33–46.

Preston, F. W. 1948. The commonness and rarity of species. Ecology 29:254–283.

Ramamurthy, S. 1953. Hydrobiological studies in Madras coastal waters. J. Madras Univ., Series B. 23:148–163.

Rawson, D. S. 1953. The bottom fauna of Great Slave Lake. J. Fisheries Res. Board Can. 10:486–520.

Sanders, H. L. 1960. Benthic studies in Buzzards Bay. III. The structure of the soft-bottom community. Limnol. Oceanogr. 5:138–153.

———. 1963. Components of ecosystems, p. 86–91. *In* Gordon A. Riley [ed.] Marine Biology I. First Int. Interdisciplinary Conf., Proc. Port City Press, Baltimore.

Sanders, H. L., R. R. Hessler, and G. R. Hampson. 1965. An introduction to the study of the deep-sea benthic faunal assemblages along the Gay Head–Bermuda transect. Deep-Sea Res. 12:845–867.

Sanders, H. L., P. C. Mangelsdorf, Jr., and G. R. Hampson. 1965. Salinity and faunal distribution in the Pocasset River, Massachusetts. Limnol. Oceanogr. 10 (Suppl.):R216–R228.

93

Simpson, E. H. 1949. Measurement of diversity. Nature 163:688.

Simpson, G. G. 1964. Species density of North American recent mammals. Syst. Zool. 13:57–73.

Thorson, G. 1952. Zur jetzigen Lage der marinen Bodentier-Ökologie: Verhandlungen Deut. Zool. Ges. Wilhelmshaven 1952, p. 276–327.

———. 1966. Some factors influencing the recruitment and establishment of marine benthic communities. Neth. J. Sea Res. 3:267–293.

Whittaker, R. H. 1965. Dominance and diversity in land plant communities. Science 147:250–260.

Williams, C. B. 1964. Patterns in the balance of nature. Academic, New York. 324 p.

4

STREAM COMMUNITIES

R. Patrick

This chapter will give a broad review of the species communities in fresh-water streams, and of the factors affecting the structure of these communities. As study objects, stream communities have a major advantage over vertebrate communities (such as those discussed in Chapters 10–14) in that they lend themselves to experimental manipulation. For example, because of the small sizes and high population densities (in individuals per square meter) of stream diatoms, it is feasible to create realistic "islands" on glass slides less than a few inches in diameter. The rate of colonization of these islands can be varied simply by changing the flow rate of stream water over the slides. Because of the short life cycles of diatoms, one can relatively rapidly perform detailed analyses of how environmental variables such as light, temperature, trace metal concentrations, and quality of the substrate affect the growth rates and relative competitive abilities of the species. Similar experiments on vertebrates usually require modifying many acres or square miles of habitat and waiting months or years for the results, because vertebrates are relatively large, live at low densities, and have long life cycles.

I shall begin by summarizing some distinctive characteristics of the natural history of stream communities, and the patterns of species diversity in these communities. Next I shall describe what factors bound the niches and affect the relative competitive abilities of stream species, and how predation affects community structure. Finally, species succession in stream communities will be briefly discussed.

Natural History of Stream Communities

In this section I point out three distinctive features of stream communities: the sizes, life cycles, and trophic relations of their species.

Size

Fresh-water stream communities are composed of a wide variety of both microscopic and macroscopic species. The important herbivores and carnivores not only exhibit a wide range of size but also belong to many different phyla, classes, and orders of animals (e.g., Protozoa, Annelida, Mollusca, Crustacea, Insecta, and Vertebrata). The primary producers are mainly microscopic algae, although macrophytes or large plants, such as Spermatophytes, Pteridophytes, or Bryophytes, may be common in some areas.

Most of the species in stream ecosystems are small compared with their terrestrial counterparts, and many are small compared with their marine counterparts.

For example, macroscopic plants are the primary producers in virtually all terrestrial communities, and also in the intertidal and rocky-shore zones of estuaries and oceans (e.g., the macroscopic algae *Fucus* and *Ascophyllum*). Even those river organisms that are relatively large, such as the porpoises of the Amazon River, are still generally smaller than their marine counterparts. Since stream communities are evolutionally ancient, why are their species skewed towards small size? The answer may be partly related to the total available area of habitat, because fish, for example, tend to be bigger in larger rivers than in small rivers. Perhaps the number of individuals of a large species that could be maintained in a stream system is sufficiently low that such populations and species would have short survival times to extinction. A further contributing factor may be the mechanical constraints of narrow channels, and the expenditure of energy necessary to hold a fixed position in swift currents.

Life Cycles

Compared with the species of terrestrial communities and some deep-lake and marine communities, stream organisms have short life cycles. In many species the life cycle only lasts one day, in others a few weeks or months. In only a few groups of stream organisms, especially fish and molluscs, do we find species whose cycles last several years. Yet even the longest such cycles are short compared with those of terrestrial communities, where many predators live for decades and many trees for centuries.

These short generation times mean that stream organisms typically have very high reproductive rates. For example, asexual reproduction of diatoms may occur daily, the blackfly *Simulium vitatum* often has three generations per year, and many other aquatic insects and several invertebrates of other classes have more than one generation per year. The offspring of each reproductive bout are numerous, relatively small in size, and rarely receive parental care (cf. Connell, Chapter 16). Clearly, stream species, like multivoltine butterflies (Shapiro, Chapter 7) and annual weeds (Schaffer and Gadgil, Chapter 6), have developed *r*- rather than *K*-strategies.

Food Webs

We routinely find four or five stages of energy transfer in stream communities. At any given place and time the species at a given trophic level belong to many taxonomic groups and have very diverse ecological requirements. For instance, at the base of food webs are primary producers and detritivores. The former are mainly algae of many major groups (Chlorophyta, Euglenophyta, Myxophyta, Bacillariophyta, and Chrysophyta), and the latter consist of bacteria, fungi, and sometimes invertebrates. The bacteria and fungi degrade detritus metabolically into much simpler chemical radicals that are utilized by algae and some invertebrates as a direct food source. In contrast, the invertebrate detritivores, such as certain worms and insects, usually break detritus particles into sizes acceptable to other detritivore species. R. Vannote (personal communication), for example, found that leaf particles in the feces of tipulid larvae are in the size range of food particles selected by mayflies (*Ephemerella*). The general importance of particle size as a criterion of food preference has been stressed for plankton communities by Brooks and Dodson (1965) and duplicates similar findings for vertebrate communities (Hespenheide, Chapter 7, Figures 1, 2, and 4; Brown, Chapter 13, Figures 6 and 7; Diamond, Chapter 14, Figures 30 and 31). As a further instance of digestion by one species generating food for other species, algae that pass through the guts of a crustacean may reproduce more effectively than uneaten algae and may thereby increase in food value to their consumers (Porter, 1973).

While algae constitute the main plant food of stream ecosystems, they also serve other functions. Although macroscopic algae and other plants are eaten by many stream organisms, it is often unclear whether this is for the food value of the plants themselves or of their epiphytes. The main significance of rooted aquatic plants (Spermatophyta, Pteridophyta, and Bryophyta) seems to be in furnishing shelter and protection from predators to various animals (cf. Connell, Chapter 16). Rooted or floating plants also serve as substrates on which other sessile organisms live, and thus increase the diversity of available habitats.

The approximate number of species performing a given trophic function remains fairly constant over time. For example, in the Savannah River communities we studied, carnivores comprise 40–50% of all species of fish, 10–30% of insect species, and 3% of invertebrate species other than insects. These figures are only approximate, since little definitive information on diets of invertebrates and fish is available. An important factor making it difficult to classify species by food habits is that different developmental stages of certain species have different food preferences. Thus, caddisfly larvae are detritus feeders for several months, then switch to feeding primarily on diatoms. The reason for the switch seems to be that detritus and diatoms contain very different nutrients and that caddisfly larvae need different amounts of nutrients at different stages of development (R. Vannote, personal communication). Without gut analysis and feeding experiments one cannot always be confident in deciding whether a certain consumer species at a particular stage of its life cycle is a carnivore, a herbivore feeding exclusively on plants, an omnivore, or a detritivore.

Species Diversity

Patterns of Species Diversity

Natural stream communities often support a very large number of species, as first noted by Thienemann (1920, 1939) and subsequently confirmed for stream communities by Patrick (1949, 1953, 1972), Tarzwell and Gaufin (1953), and Patrick, Cairns, and Roback (1967). The species belong to many different systematic groups, and each genus or family present may be represented by several to many species.

When the number of species in any

given aquatic community is carefully examined (Patrick, 1949), this number is found to remain constant with time as long as no exogenous perturbations occur. This is illustrated by Table 1, which shows that the number of species in each of four different groups (algae, protozoa, insects, and fish) surveyed in the Savannah River was relatively similar from year to year. We also find little difference among species numbers in different sites or streams with structurally similar habitats. This similarity is exemplified by comparison of the four stations in the Savannah River in a given year (Table 1), or by comparison of the 12 stations in nine streams listed in Table 2. These nine streams were selected for their wide geographic range: the Savannah River, North Anna, Rock Creek, and Potomac drain to the Atlantic; the Escambia River and North Fork of the Holston drain via the Mississippi River to the Gulf of Mexico; and the Ottawa drains into the St. Lawrence. Despite this geographic diversity, the species numbers in each of four systematic groups rarely deviate by more than 33% from the mean value for all rivers studied. When we separate hard-water rivers from soft-water rivers, the deviations from the means for each type of river are even less. Since these nine streams were censused by different groups of scientists, Table 2 provides compelling evidence that well-collected, similar-sized areas of different streams support similar numbers of species. This constancy of α-diversity or species packing level among similar stream communities parallels Cody's (Chapter 10) findings for terrestrial bird communities.

Although different streams share similar *total numbers* of species, the number of *species shared* between stations in different streams is low. Table 3, which is based on surveys of the same stations in these nine different streams, shows that more than half of the species in each of the four broad systematic groups were found at only one station, and that vanishingly few species were found at seven or more stations. This finding does not

Table 1. Numbers of species at four stations in the Savannah River.

Organisms	Stations															
	1	3	5	6	1	3	5	6	1	3	5	6	1	3	5	6
	Year: 1955				Year: 1956				Year: 1960a				Year: 1960b			
Algae	98	89	103	120	98	97	97	84	96	77	90	72	75	90	103	99
Protozoa	42	52	48	55	41	38	37	51	53	54	67	58	55	60	62	67
Insects	44	41	54	58	46	47	54	46	33	35	37	26	26	34	35	28
Fish	35	23	30	25	24	30	31	29	32	30	36	33	40	33	37	40

The table gives the number of species in each of four broad groups of organisms, at each of four collecting stations in the Savannah River at each of four times. Low flow prevailed during the 1955 and second 1960 (1960b) collection, high flow during the 1956 and first 1960 (1960a) collection. Silt loads were high in 1960 because of dredging. Note that the species number in each category is relatively constant in space and time.

Table 2. Total number of taxa in a local area in each of nine streams

Organisms	Soft-water Rivers						Hard-water Rivers								
	Escambia	Savannah 54	Savannah 55	North Anna	White Clay	Flint	N. Fork Holston	Rock Creek	Ottawa 55	Ottawa 56	Potomac 56	Potomac 57	Mean All Rivers	Mean Soft Rivers	Mean Hard Rivers
Algae	77	105	101	98	73	79	63	65	76	58	105	103	84	89	78
Protozoa	38	61	40	58	56	51		86		48	85	68	59	51	72
Insects	29	58	51	61	57		83	48	59	61	89	99	63	51	73
Fish	39	19	35	21	20	13	21	24	18	28	18	29	24	25	23

The number of taxa or species in each of four broad groups of organisms is given for a local collection at each of 12 stations in nine streams. The streams belong to three different drainage systems (Atlantic, Gulf of Mexico, St. Lawrence). Nevertheless, the number of species in a given category is relatively constant, especially if comparisons are confined to different hard-water or different soft-water streams.

imply that equally marked differences would be found if entire stream systems were collected and compared; the meaning is, instead, that if one collects small areas of different streams at particular times, the chance of finding the same species in such collections is low. In our Savannah River study (Table 1) there were few shared species between collections at the same station at different times but more shared species among different stations collected at the same time.

Table 3. Distribution of taxa in rivers cited in Table 2

Organisms	Total no. of taxa	Number of taxa and number of rivers in which they occur								
		1	2	3	4	5	6	7	8	9
Algae	354	197 (55.7%)	61 (17.2%)	38 (10.7%)	25 (7.1%)	16 (4.5%)	8 (2.3%)	6 (1.7%)		3 (0.9%)
Protozoa	299	188 (62.9%)	40 (13.4%)	32 (10.7%)	23 (7.7%)	8 (2.7%)	5 (1.7%)	1 (0.1%)	2 (0.7%)	
Insects	283	209 (73.9%)	31 (10.9%)	24 (8.5%)	9 (3.2%)	6 (2.1%)	3 (1.1%)	1 (0.1%)		
Fish	132	75 (56.8%)	22 (16.7%)	22 (16.7%)	9 (6.8%)		3 (2.3%)			

For collections from the twelve stations belonging to nine rivers listed in Table 2, and for each of four broad groups of organisms, the table gives the number of taxa or species found in 1, 2, 3, 4, 5, 6, 7, 8, or all 9 rivers. Column 2 gives the total number of taxa in each category; numbers in parentheses refer to percentages of total number of taxa in each category. Protozoa and insecta were studied in only eight rivers. Most species are confined to one or a few collections, indicating high species-turnover in space and time.

Thus, the relative constancy of α-diversity with geography or time conceals a high species turnover with geography or time, analogous to high β-diversity. Some of the reaons why different species occur at different places or times will be explored later. In the next section we consider how these high species diversities are established.

Experiments on
Stream Community "Islands"

Support for the MacArthur and Wilson (1967) equilibrium theory of island species diversities has come both from experimental tests (e.g., Simberloff & Wilson, 1969) and observational tests. Stream communities lend themselves much more readily to experimental study than do the mangrove-tree islands studied so fruitfully by Simberloff and Wilson. To create artificial islands in streams, I mounted small bits of glass on pedicels of glass slides, placed the slides in plastic containers through which water could flow, placed the containers in streams, and counted the numbers of individuals and species of diatoms that developed populations on the slides (Patrick, 1967, 1968). As long as one counts more than 8000 individuals, 95% or more of the individuals on each of a set of replicate slides belong to species encountered on the other slides of the set. Thus, one has very similar communities with which to experiment on the various slides. This system can be used to study the effects of island area, invasion rate, and species pool size on the species diversity of diatom communities.

To test the effect of island area on species diversity, I compared slides with areas of 9, 36, 144, or 625 square millimeters. Larger slides were found to support more species: e.g., 28–32 and 44–47 diatom species on slides with areas of 144 and 625 mm^2, respectively.

To test the effect of invasion rate, I let water from White Clay Creek flow over slides of constant area at either 1.5 or 650 liters per hour. As illustrated in Figure 1, reduction in flow rate caused a large decrease in the numbers of species forming the community, and also caused an increased variance in the population sizes of the species. The same two findings emerge from comparison of two very different flow rates in Darby Creek (Table 4, first four lines).

The effect of species pool size was studied in two series of experiments. In the first series I compared slides of various areas suspended in two Pennsylvania streams with diatom communities of very different sizes: Ridley Creek and Roxborough Spring Creek, which support about 250 diatom species and fewer than 100 diatom species, respectively, at any point in time. More than 160 species grew on the 36-mm^2 slides suspended in Ridley Creek, whereas only 14–29 species grew on slides of the same area in Roxborough Spring Creek. In the second series, I compared two species-rich continental streams (Hunting Creek in Maryland and Darby Creek in Pennsylvania on the North American mainland) with three species-poor insular streams (Canaries River on St. Lucia Island, and Layou River and Check Hall River on Dominica Island, in the West Indies). As summar-

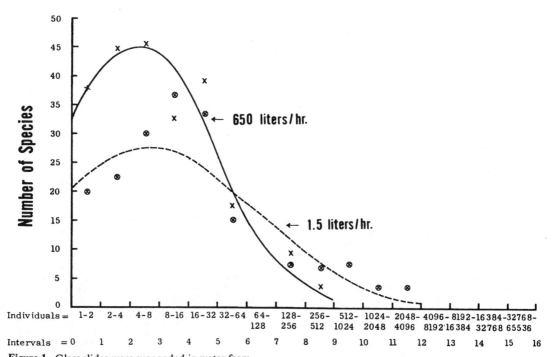

Figure 1 Glass slides were suspended in water from White Clay Creek, and the number of individuals of each diatom species that attached was counted. The ordinate is the number of species represented by the number of individuals given by the abscissa. The abscissa is a logarithmic scale, divided into intervals spanning a two-fold range of number of individuals. Stream water flowed over the slides either at 650 liters/hr (points x; the natural rate of stream flow), or at 1.5 liters/hr of new water recycled to give a total flow rate of 650 liters/hr (points ⊗). The curves are lognormal distributions fitted through the experimental points. Note that reduction in flow rate lowers the height of the mode and the total number of species represented, but increases the variance of abundance and the number of abscissa intervals covered by experimental points.

ized in Table 4, slides suspended in the continental streams supported 79–129 species, whereas slides of the same area in the insular streams supported only 46–61 species. The slides from insular streams also exhibited greater variation in population sizes than did the continental slides. Thus, lower species number, whether it results from lower invasion rate (Figure 1 and first four lines of Table 4)

Table 4.   Species-abundance relations of diatom communities

Streams	(1) Height of mode	(2) σ^2	(3) Observed number of species	(4) Theoretical number of species	(5) Intervals covered by curve
Continental streams					
Darby Creek (Pa.), Sept.–Oct. 1964					
550–600 liters/hr.	22.5	6.9	129	148	9
1.5 liters/hr.	13.9	12.6	100	124	12
Darby Creek, Oct.–Nov. 1964					
550–600 liters/hr.	22.4	6.2	123	140	9
1.5 liters/hr.	15.3	12.0	97	133	15
Hunting Creek (Maryland)	12.0	9.1	79	92	14
Insular streams					
Canaries River (St. Lucia Island)	8.4	9.3	61	64	10
Layou River (Dominica Island)	5.3	26.0	49	67	14
Check Hall River (Dominica Island)	5.2	21.6	46	60	14

Glass slides of constant area were suspended in water from two continental (North American) and three insular (West Indian) streams, and the number of individuals of each diatom species that grew on the slides was counted. The species-abundance relation was then plotted as in Figure 1. The first column gives the number of species with modal abundance (i.e., ordinate values of the maxima of curves such as Figure 1); the second column, a measure of the width or variance of the distribution; the third column, the observed number of species; the fourth column, the theoretical total number of species calculated from a lognormal distribution curve fitted to the observed species-abundance values; and the last column, the number of abscissa intervals (each representing a two-fold range of number of individuals) covered by the observed species-abundance values. Two successive experiments were carried out in Darby Creek; each experiment used two different flow rates of stream water over the slides. Note that more species grow on slides in species-rich continental streams than in species-poor insular streams (compare rows 1–5 with rows 6–8 in columns 3 and 4); that number of attached species decreases with decreasing flow rate (compare rows 2 and 4 with rows 1 and 3 in columns 3 and 4); and that decrease in species number resulting either from reduced species pool or reduced flow rate decreases the height of the mode (column 1) but increases the variance (column 2) of the species-abundance relation.

or from lower species pool (insular vs. continental streams of Table 4), is associated with greater variation in population size.

These experiments show that species diversity of island communities increases with "island area," and is maintained by invasion. The invasion rate is proportional to the product of water flow rate times species pool size. In nature, invasion results from a combination of passive downstream drift and active upstream movement of stream organisms. As pointed out by Waters (1962), downstream

drift of aquatic insects is a common phenomenon and is maximal at dusk and at dawn. In Darby Creek and White Clay Creek we noted that downstream drift of diatoms is much greater after sunset and before sunrise than during the daytime. The reason for the temporal pattern may be that at low light intensity aquatic insects forage on algae on the tops of rocks and dislodge the algae, which then drift. Active upstream movement of aquatic organisms has been noted by many authors. For example, many species of fish migrate upstream to spawn. To deposit eggs, aquatic insects may fly either upstream or downstream but usually fly upstream (Hynes, 1970).

Niche Limits and Competition

We have seen that brief, local sampling efforts show stream communities to be rich in species, but that there is much turnover of species in space and time. The spatial and temporal niche limits of each species are determined by physical factors, competition, and predation. In 'his section we shall discuss physical factors (especially current, substrate texture, light, temperature, nutrients, and trace metals) and competition, while predation is considered in the following section.

Effects of physical factors and of competition on niche limits are often linked. That is, a species may be confined to a region in time or space where a certain physical factor falls within a certain range of values, not because a value outside this range is directly lethal to the species but because at such a value the species is outcompeted by other species. For instance, trace metal concentrations and temperature greatly influence the relative abilities of algal species to outcompete each other. The species of stream communities have generally developed different strategies and different preferred physical conditions, in order to coexist and to reduce niche overlap and competition for resources. Thus, the same rock in a stream may support several species of caddisflies eating detritus and algae. However, these species eat different amounts and types of food at different stages of development, and at any moment only one species is eating large amounts of algae. As a result, similar amounts of algae are being cropped from the rock throughout the year, but different species are the principal grazers at different times (R. Vannote, personal communication). This competitive spacing of species is often reinforced by interspecific aggression. For example, species of Trichoptera occupy areas with certain preferred current rates and attack other species that try to occupy the areas (Scott, 1958; Edington, 1965). Similarly, crayfish may divide stream habitats into riffle and pool territories when two species are present, instead of one species occupying the whole area (Bovbjerg, personal communication).

Among the factors determining where various species of invertebrates and fish live, current is one of the most important (Hustedt, 1938; Ruttner, 1953; Hynes, 1970). The various organisms coexisting in stream communities prefer different kinds of current structure. Species that are known as rheophils and prefer rapidly flowing water include stoneflies, mayflies, and some filter feeders such as blackfly larvae and caddisflies. In tropical streams many species of fish have special adaptations, such as suckers, to permit them to live in rapid currents. Other species, such as tubificid worms, chironomid larvae, and many Crustacea, prefer pools or slower currents.

Substrate structure and roughness provide further habitat axes along which species segregate. Soft-textured habitats are preferred by certain burrowing species such as burrowing mayflies, oligochaete worms (especially the Tubificidae), and some chironomid larvae such as *Chironomus plumosus*. Certain diatom species such as *Nitzschia palea* and *Navicula cryptocephala* prefer to grow on soft sediments. In contrast, other species (including many mayflies, stoneflies, caddisflies, and diatoms such as *Synedra rumpens* and the stalked diatoms *Gomphonema parvulum, G. intricatum,* and *Cymbella lanceolata*) prefer very hard-textured substrates, regardless of the current speed. When habitat roughness is reduced by siltation, as often happens in channelizations, species diversity is greatly reduced (Patrick and Vannote, in U.S. Government, 1973). This homogenization of the habitat by siltation reduces not only substrate diversity but also diversity of light patterns and of current patterns.

It is well known that different species of algae grow better at different light intensities. Thus, algal species can coexist even when one is shaded by the other. Species also differ in their temperature tolerances and preferences. For instance, a community that is dominated by blue-green algae above 34°C is dominated by diatoms at lower temperatures.

Differential effects of nutrients on species may also occur. For example, certain algae, including some blue-green algae and dinoflagellates, can live on much lower levels of nutrients such as nitrogen and silica than can *Fragilaria crotonensis* and *Gomphonema parvulum*. Thus, the latter two species are outcompeted by the former types of algae when concentrations

of these nutrients are low.[1] In bird communities this finding is paralleled by the suspected ability of so-called high-S bird species to survive at lower nutrient levels than so-called supertramps. Such a gradient of nutrient requirement lends itself to displacement of the species with high requirements by species with low requirements, through overexploitation (Diamond, Chapter 14, especially Figures 18, 49, and 50). However, competitive effects due to nutrient limitation or to heterotoxins (toxic substances that are produced by one species and that affect populations of other species) are far less pronounced in stream communities than in lakes and other aquatic communities (cf. Levins, Chapter 1, Figures 3 and 4), and are probably less pronounced than in terrestrial communities (cf. examples among terrestrial plants discussed by J. Connell, Chapter 16). The reason is that nutrients are always being supplied to streams from the watershed, and that the flow of water continually removes autotoxins from the species producing them and dilutes autotoxins and heterotoxins. Little local recycling occurs in streams, and the effects of upstream-to-downstream recycling are masked by nutrient renewal from watershed runoff or from groundwater.

In recent experiments on algal communities we have found that variations in concentrations of trace metals may have

---

[1] This need not mean that the outcompeted species is completely absent from the community. More often, it either becomes so rare as to escape collection, or passes into a different stage such as an egg, spore, or (in the case of insects) pupa.

profound effects on the kinds of species present and on their population sizes. For example, blue-green algae such as *Schizothrix calcicola* and *Microcoleus vaginatus* can live on much lower concentrations of manganese (Mn^{++}) than are needed to support diatom communities (Patrick, Crum, and Coles, 1969). When Mn^{++} concentrations exceed 40 micrograms per liter ($\mu g/l$), diatoms outcompete blue-green algae and maintain much larger populations. At Mn^{++} concentrations below about $15\mu g/1$, blue-green algae outcompete diatoms. The reason may be related to lipid metabolism, since Mn^{++} is known to be required for lipid metabolism in the alga *Euglena gracilis* (Constantopoulos, 1970). Because diatoms metabolize and store lipids, it is reasonable that other species that do not store such large amounts of lipid should be able to outcompete diatoms when Mn^{++} is present in very low concentrations.

The effects of vanadium are opposite to those of manganese. At $3-4\,\mu g/1$ vanadium, diatoms increase in biomass and dominate the community, whereas at $4000\,\mu g/1$ blue-green algae replace diatoms as dominant species and attain high biomass. Similarly, blue-green algae are significantly stimulated by addition of $6\,\mu g/1$ nickel and become dominant at $36-40\,\mu g/1$ nickel (R. Patrick, unpublished observation). In some cases the effect of an element depends on its form. Thus, selenium in the form of selenite at 1 mg/l or 10 mg/l stimulates diatom growth, whereas selenate at 1 mg/l inhibits diatom growth but promotes blue-green algae and some unicellular green algae.

Effects of Predation

Although some predator species are relatively indiscriminate in their choice of food, other predators select for certain prey and against other prey. For instance, among ciliate Protozoa some species eat only diatoms, some prefer bacteria, and others that are generalists feed on a combination of diatoms and bacteria (Gizella and Gellert, 1958). Certain insect predators not only prefer diatoms as food but select particular diatom species, such as *Rhoicosphenia curvata*. Roop showed that the snail *Physa heterostropha* discriminates against the diatoms *Cocconeis placentula* and *Achnanthes lanceolata* in favor of other species (Patrick, 1970). Thus, if these two diatoms are allowed to continue to grow while snail predation reduces populations of other diatom species, selection by the predator results in a lowering of diatom species diversity. Similarly, it is well known that aquatic insects and other invertebrate predators greatly prefer diatoms to blue-green algae as prey. Thus, predation pressure allows populations of blue-green to multiply.

Depending on the degree of predator pressure and whether or not it is selective, predation can affect not only the biomass (standing crop) and species diversity but also the whole structure of a community. If herbivore grazing pressure is intense enough to control an aggressive species of primary producer that would otherwise become dominant, the result is increased plant diversity. This in turn supports a greater variety of herbivores, which in turn increases the diversity of carnivores

(Hairston et al., 1968). For example, the selective predation of large Crustacea by fish not only determines what species of Crustacea dominate the zooplankton, but also affects the composition of the phytoplankton, by permitting growth of species that had formerly been cropped by the larger zooplankton (Brooks and Dodson, 1965).

Connell (Chapter 16) has emphasized that the relative sizes of predator and prey often determine the intensity of predation, so that prey may be unable to escape predators until the prey somehow succeed in growing to a certain size. The prey population structure may then come to consist of widely spaced size classes, each reflecting a transient escape from predation. In stream communities, however, many species exhibit little variance in size in different stages of the life cycle. Because of the short life spans of most stream species, predators may eliminate many life stages of a prey species, including adults, and quickly reduce a large population of a prey species. Since the prey has no uneaten size classes left, the prey must either reinvade the community or else depend on a few surviving individuals to restore the population. Predation may have less marked effects on species in which the various age classes are of different sizes, since some size classes survive predation and continue to reproduce.

Connell also points out that prey species that can live in precarious or harsh environments are often able to develop larger populations than would be possible in benign areas where predators are com-

mon. In stream communities this phenomenon is exemplified by blackfly larvae, a prey species highly sought by predators. Except for a short time after the larvae hatch, their populations are small in most areas of natural streams. They become numerous only in rock crevices that are inaccessible to predators or in very fast water where the predators cannot survive.

Although competition affects the relative population sizes of species with similar ecological requirements and may reduce the number of species that can coexist, competition does not affect the structure of a whole food web or community as acutely as does predation. The surprising and far-reaching ways in which changes at one trophic level may affect other levels have been discussed at length by Levins (Chapter 1). This transmission of effects through the community is illustrated by the two cases discussed above, involving herbivore grazing pressure and fish predation on Crustacea. Effects of trace metals on stream communities provide another clear example. If changes in trace metal concentrations cause a shift among primary producers towards species less desirable as prey to herbivores (e.g., a shift from diatoms to blue-green algae, as caused by high vanadium and nickel concentrations or low manganese concentrations), the metals greatly affect the kinds and abundances of herbivores, hence also species at higher trophic levels. In the opposite direction along the trophic pyramid, the metals also shift the balance between primary producers and detritivores and change the levels of nutrients. Thus, the concentration of trace metals may alter the entire structure of a stream community.

Predictability and Succession of Communities

The values of some of the important parameters discussed in the section on Niche Limits and Competition, which control the existence and diversity of species in streams, vary unpredictably in space and time. For example, the identities and amounts of available nutrients vary as a function of the watershed use and runoff pattern at any moment. Storms and the resulting variable flows that scour the substrate of a stream often greatly reduce the abundances of previously dominant species and thereby open the community to invasion from the large available species pool. In analogy to observations by Simberloff and Wilson (1969) on recolonization of fumigated mangrove islands by arthropods, the species that invade a scoured stream often differ from the species reduced by scouring and are more diverse (Patrick, 1963). The new immigrants often correspond to what Diamond (Chapter 14) calls "tramp" or "supertramp" species.

As pointed out by Blum (1956) and observed by us at the Academy of Natural Sciences, temporal changes in stream communities and in terrestrial plant communities are very different. On land there is a predictable succession of communities, because shade, detritus, and other factors set by the community present at any instant determine the species composition of the subsequent community. Thus, species changes in succession can be predicted from a matrix of transition probabilities among species, as developed by Horn in Chapter 9. In contrast, the flushing of nutrients through streams, and the flushing-out of waste products, prevent the species of the moment from determining their successors. Streams do exhibit an annual cycle of seasonal succession, however, owing to seasonal changes in temperature, light, and nutrient levels. Superimposed on this seasonal regularity is a large element of unpredictability, as discussed in the preceding paragraph. Since the species of stream communities are relatively small in size and have short life cycles and high reproductive rates, the communities are always very young.

Many of the perturbations that man causes in stream communities arise from minute changes in the concentrations of substances such as trace metals, often by a few micrograms per liter. By affecting the physiology, behavior, and resistance to disease of individual species as well as competitive and trophic relations between species, such changes lead to large changes in ecosystem structure and to the dominance of certain species. Therefore, in examining the factors that develop and maintain diversity in stream ecosystems, one must consider not only obvious factors such as predator pressures competition, but also effects of substances present in minute amounts.

The species composition of a stream community at any place and time is determined by many independent variables, the unpredictable and ever-changing parameters of a randomly fluctuating environment. As discussed by May (Chapter 4), the Central Limit Theorem leads to the expectation that the species-abundance relation of such a community is lognormal (Figure 1). Although the fluctuations in species composition are largely unpredictable, the combination of high invasion rates, diverse physical structure of streams, and continual supply of nutrients and other factors serves to perpetuate stream communities.

References

Blum, J. L. 1956. Application of the climax concept to algal communities of streams. *Ecology* 37:603–604.

Brooks, J. L., and S. I. Dodson. 1965. Predation, body size, and composition of plankton. *Science* 150:28–35.

Constantopoulos, G. 1970. Lipid metabolism of manganese-deficient algae. *Plant Physiol.* 45:76–80.

Edington, J. M. 1965. The effect of water flow on populations of net-spinning Trichoptera. *Mitt. Internat. Verein. Limnol.* 13:40–48.

Gizella, T., and J. Gellert. 1958. Über Diatomeen und Ciliaten aus dem Aufwuchs der Ufersteine am Ostufer der Halbinsel Tihany. *Ann. Inst. Biol. Hungar. Acad. Sci.* (Tihany) 25:240–250.

Hairston, N. G., J. D. Allen, R. K. Colwell, D. J. Futuyma, J. Howell, M. D. Lubin, J. Mathias, and J. H. Vandermeer. 1968. The relationship between species diversity and stability: an experimental approach with protozoa and bacteria. *Ecology* 49:1091–1101.

Hustedt, F. 1938. Systematische and ökologische Untersuchungen über die Diatomeen-Flora von Java, Bali and Sumatra. *Arch. Hydrobiol.* Vol. 15 (Suppl.), pp. 131–177, 187–295, 393–506.

Hynes, H. B. N. 1970. *The Ecology of Running Water.* University of Toronto Press, Toronto.

MacArthur, R., and E. O. Wilson. 1967. *The Theory of Island Biogeography.* Princeton University Press, Princeton.

Patrick, R. 1949. A proposed biological measure of stream conditions based on a survey of Conestoga Basin, Lancaster County, Pennsylvania. *Proc. Acad. Nat. Sci. Philadelphia* 101:277–341.

Patrick, R. 1953. Biological phases of stream pollution. *Proc. Pennsylvania Acad. Sci.* 27:33–36.

Patrick, R. 1963. The structure of diatom communities under varying ecological conditions. *In: Conference on the Problems of Environmental Control of the Morphology of Fossil and Recent Protobionta. Trans. New York Acad. Sci.* 108:359–365.

Patrick, R. 1967. The effect of invasion rate, species pool, and size of area on the structure of the diatom community. *Proc. Nat. Acad. Sci. U.S.A.* 58:1335–1342.

Patrick, R. 1968. The structure of diatom communities in similar ecological conditions. *Amer. Natur.* 102:173–183.

Patrick, R. 1970. Benthic stream communities. *Amer. Scientist* 58:546–549.

Patrick, R. 1972. Benthic communities in streams. *Trans. Conn. Acad. Arts and Sci.* 44:272–284.

Patrick, R., J. Cairns, Jr., and S. S. Roback. 1967. An ecosystematic study of the fauna and flora of the Savannah River. *Proc. Acad. Nat. Sci. Philadelphia* 118:109–407.

Patrick, R., B. Crum, and J. Coles. 1969. Temperature and manganese as determining factors in the presence of diatom or blue-green algal floras in streams. *Proc. Nat. Acad. Sci. U.S.A.* 64:472–478.

Porter, K. G. 1973. Selective grazing and differential digestion of algae by zooplankton. *Nature* 244:179–180.

Ruttner, F. 1953. *Fundamentals of Limnology.* University of Toronto Press, Toronto.

Scott, D. 1958. Ecological studies on the Trichoptera of the River Dean, Cheshire. *Arch. Hydrobiol.* 54:340–392.

Simberloff, D. S., and E. O. Wilson, 1969. Experimental zoogeography of islands: the colonization of empty islands. *Ecology* 50:278–295.

Tarzwell, C. M., and A. R. Gaufin. 1953. Some important biological effects of pollution often disregarded in stream surveys. Proc. 8th Industr. Waste Conf., *Purdue University Engineering Bulletin* 295–316.

Thienemann, A. 1920. Untersuchungen über die Beziehungen zwischen dem Sauerstoffgehalt des Wassers und der Zusammensetzung der Fauna in norddeutschen Seen. *Arch. Hydrobiol.* 12:1–65.

Thienemann, A. 1939. Grundzüge einer allgemeinen Okologie. *Arch. Hydrobiol.* 35:267–285.

U. S. Government. 1973. Report on channel modifications submitted to Council on Environmental Quality. No. 4111-00015. *U. S. Government Printing Office,* Washington, D.C.

Waters, T. F. 1962. Diurnal periodicity in the drift of stream invertebrates. *Ecology* 43:316–320.

ERRATUM

Page 457, column 2, line 3 from the bottom of the page should read "such as predator pressures and competition, . . ."

5

Reprinted from *Am. Nat.* **112**:23–39 (1978) by permission of The University of Chicago Press

PLANT SPECIES DIVERSITY IN A MARINE INTERTIDAL COMMUNITY: IMPORTANCE OF HERBIVORE FOOD PREFERENCE AND ALGAL COMPETITIVE ABILITIES

JANE LUBCHENCO*

Biological Laboratories, Harvard University, Cambridge, Massachusetts 02138

Since Hutchinson (1959) drew attention to the question "Why are there so many kinds of animals?" investigation of the causes of species diversity has proven to be a fertile area of ecological endeavor. A major current emphasis is on mechanisms creating and/or maintaining diversity (Connell 1971, 1975; Dayton 1971, 1975; Dayton et al. 1974; Jackson and Buss 1975; Janzen 1970; MacArthur 1972; B. Menge and Sutherland 1976; Paine 1966, 1971, 1974; Pianka 1967, 1969; Ricklefs 1973). One such mechanism, the predation hypothesis, suggests that predators, by keeping the abundance of their prey in check, prevent competitive exclusion and thus permit or maintain a higher species richness than would occur in their absence (Paine 1966, 1971).

Experimental removals or additions of aquatic carnivores have resulted in changes (decreases or increases, respectively) in the local species diversity of lower trophic levels over ecological time (Paine 1966, 1971, 1974; Hall et al. 1970; B. Menge 1976). In contrast, the effect of herbivores on local species diversity patterns is confusing, in part because few experimental studies have been done. In some instances, herbivores appear to increase plant diversity (Harper 1969; Paine and Vadas 1969), decrease plant diversity (Harper 1969), or both (Harper 1969; Paine and Vadas 1969; Vadas 1968). The key to understanding such variable results may reside in understanding consumer prey preferences and competitive abilities of the food species. A number of authors have suggested that only when a consumer (predator or herbivore) preferentially feeds on the competitively dominant prey can the consumer increase diversity (Hall et al. 1970; Harper 1969; MacArthur 1972; Paine 1971; Patrick 1970; Van Valen 1974). In this paper I present results of an experimental evaluation of the effect of generalized herbivores on plant diversity in a rocky intertidal community. In this system, knowledge of (1) food preferences of the herbivores, (2) competitive relationships between the plants, and (3) how these relationships change according to physical regimes in microhabitats permits an analysis of the importance of the relationship between herbivore food preference and competitive ability of the plants.

---

* Present address: Department of Zoology, Oregon State University, Corvallis, Oregon 97331.

TABLE 1

FOOD PREFERENCES OF *Littorina littorea**

Preference Ranking	Chlorophyceae (Greens)	Phaeophyceae (Browns)	Rhodophyceae (Reds)
High	Cladophora	Ectocarpus-Pylaiella	Ceramium
	Enteromorpha	Elachistea	Porphyra
	Monostroma	Petalonia	...
	Spongomorpha	Scytosiphon	...
	Ulva	...	...
	Ulothrix-Urospora	...	...
Medium	Rhizoclonium	Dictyosiphon	Asparagopsis
	...	...	Cystoclonium
	...	...	Dumontia
	...	...	Halosaccion
	...	...	Phycodrys
	...	...	Polysiphonia lanosa
	...	...	P. flexicaulis
Low	Chaetomorpha	Agarum	Ahnfeltia
	Codium	Ascophyllum	Chondrus
	...	Chorda	Euthora
	...	Chordaria	Gigartina
	...	Desmarestia	Polyides
	...	Fucus	Rhodymenia
	...	Laminaria	...
	...	Ralfsia	...
	...	Saccorhiza	...

* Preferences were determined by laboratory two-way choice experiments. Only large individuals of any algal species were used. A group of 20–40 snails was placed in the middle of the bottom of a filled 20-gal aquarium (standing new seawater) and surrounded by equal amounts of two species of algae, with the same species on opposing sides. The probability of any snail's contacting species 1 was equal to that of its contacting species 2. These periwinkles did not appear to detect food at a distance, but relied on tactile-chemical methods once plants were contacted. Once a snail contacted a piece of alga it would either move away or remain there and feed. The numbers of snails on the two species of algae were compared using χ^2 after 30–90 min. All large algae had had micro- and macroscopic epiphytic algae removed from them. Results were usually clear-cut and are arranged here in three preference categories. Most experiments were repeated at least once, rotating positions of algae and using a new group of snails. Further details and discussion of these experiments will be presented in a later paper.

HERBIVORE FOOD PREFERENCES

Along the rocky shores of New England, the most abundant and important herbivore in the mid and low intertidal zones is the periwinkle snail *Littorina littorea* (J. Menge 1975). Often attaining a length of 2–3 cm, this snail forages primarily when under water or during cool, humid low tides. It is a generalist with respect to both size and species of food, consuming most local species of microscopic and macroscopic algae (J. Menge 1975). Laboratory choice experiments indicate that *L. littorea* has strong food preferences (table 1). In general, the preferred algae are primarily ephemeral small and tender species (like the green *Enteromorpha* spp.), which appear to lack either structural or chemical means of deterring herbivorous snails. Algae in the lowest preference category are either never eaten by *L. littorea* or are eaten only if no other food has been available for a considerable length of time. These plants (like the perennial red *Chondrus crispus* [Irish moss]) are all tough compared to those in the high category. More detailed information on these preferences and how they correlate to potential antiherbivore mechanisms of the algae will be published later. The snails and many of the algae occur both in tide pools and on emergent substrata, i.e., rock exposed to air at low tide. Since different mechanisms and relationships exist in these two different habitats, the effects of the herbivores on the algae will be considered separately for each habitat.

EFFECT OF *Littorina littorea* ON COMPOSITION
OF TIDE POOL ALGAE

Normally there is considerable variation in the algal composition of tide pools in the upper half of the rocky intertidal region ($+5.9-+12.0$ ft or $+1.8-+3.7$ m). Comparable variation exists for European pools where a classification scheme of tide pools based solely on the dominant type of algae present has been suggested (Gustavsson 1972). Algal composition of high tide pools (hereafter called pools) in New England ranges from the extremes of almost pure stands of the opportunistic green alga *Enteromorpha intestinalis* or of the perennial red alga *Chondrus crispus* with a variety of intermediate situations (i.e., pools inhabited by many different types and species of algae). *Littorina littorea* appears to colonize these pools primarily by settlement from the plankton as newly metamorphosed snails (≤ 0.2 cm long). Experimental manipulations and subsequent monitoring of snail density (described below) indicate that immigration and emigration of adult *L. littorea* ($\geq 1.2-1.5$ cm long) are rare despite the pools being inundated approximately every 6 h. There is wide variation in *L. littorea* density between but not within pools. In any single pool, the snail density remains relatively constant over time. To examine the role of *L. littorea* in controlling the macroscopic algal composition of these pools, periwinkle densities were experimentally altered. In September 1973, three pools of similar height, salinity, size, depth, and exposure to light at the Marine Science Institute, Nahant, Massachusetts, were selected which subjectively appeared to represent the two extremes of the continuum in types of algae present. One pool was dominated by an almost pure stand of *Entero-morpha* (97% cover, see initial point in fig. 1*B*) and had a low density of *L. littorea* (four per m^2). The other two pools were dominated by *Chondrus* (85% and 40% cover; see initial points in fig. 1*A* and 1*C*, respectively) and had high densities of *L. littorea* (233 and 267/m^2).

Because *Enteromorpha* is one of *L. littorea*'s preferred food species (table 1) and *Chondrus* is not eaten by the snails, I hypothesized that the observed correlation between littorine abundance and algal composition was causal. It appeared that intense snail grazing may be eliminating ephemeral algae such as *Enteromorpha* and allowing inedible *Chondrus* to persist. To test whether *L. littorea* was responsible for the algal differences between pools, I removed all *L. littorea* from one *Chondrus* pool, added them to the *Enteromorpha* pool, and left the second *Chondrus* pool undisturbed as a control. These experiments were

112

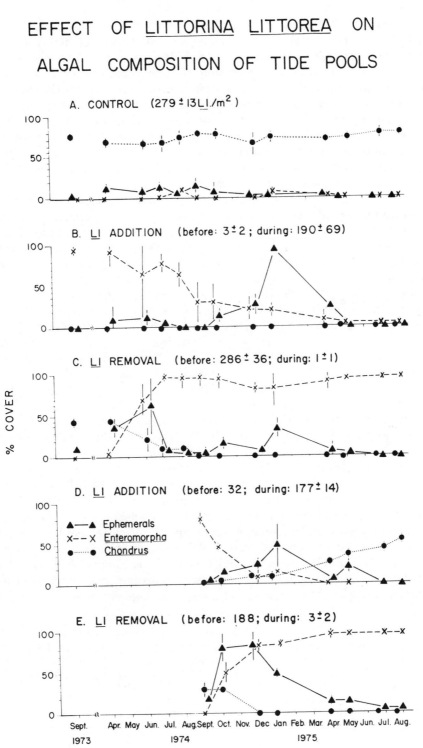

EFFECT OF LITTORINA LITTOREA ON ALGAL COMPOSITION OF TIDE POOLS

initiated in April 1974 after the samples shown for that date had been taken (fig. 1A–C). The percentage of cover of algae and density of herbivores were monitored throughout the following $1\frac{1}{2}$ yr. The few species of encrusting algae (*Ralfsia, Hildenbrandia,* and encrusting corallines) are not included here because of sampling and/or field identification problems. Their distribution in the various pools appears uniform. Percentage of cover of upright algae was estimated by observing what species was under each of 100 dots on a 0.25-m^2, $\frac{1}{2}$-inch Plexiglas quadrat placed over the area. The coordinates of the dots were obtained using a table of random numbers. As many quadrats as could be fit onto the bottom of each pool were sampled. Three quadrats fit in pools A, D, and E and four quadrats in pools B and C.

In the control pool, *Chondrus* abundance remained high throughout this monitoring period (fig. 1A); *Enteromorpha* and other ephemeral algae were present but never abundant. In *Chondrus* pools, the periwinkles feed on microscopic plants and sporelings and germlings of many ephemeral algae that settle on *Chondrus.* In the *L. littorea* addition pool (formerly *Enteromorpha* dominated, fig. 1B), *Enteromorpha* gradually declined in abundance to $< 5\%$ cover by April 1975. Snails could be seen actively ingesting *Enteromorpha* throughout this time period. Comparison of this experimental pool to the control (compare fig. 1B with 1A) supports the hypothesis that *L. littorea* is the cause of this decline in *Enteromorpha.* Note that ephemeral algal species (e.g., *Ectocarpus confervoides, Petalonia fascia,* and *Scytosiphon lomentaria*) became seasonally abundant even in the presence of *L. littorea.* Periwinkles are less active during the winter (J. Menge 1975), and as a result, ephemeral algae can temporarily swamp them (e.g., in January 1975). From January to April, periwinkles became increasingly active and eliminated nearly all edible algae from the pool. (Inedible algae include *Chondrus* and crusts.) No *Chondrus* has yet appeared in this pool. However, my observations in the low intertidal suggest that *Chondrus* recruits slowly (J. Menge 1975). I therefore predict that this alga will eventually settle and become abundant in the *L. littorea* addition pool. In the *L. littorea* removal pool (formerly with 40% cover of *Chondrus,* fig. 1C), *Enteromorpha* and several seasonal, ephemeral species immediately settled or grew from microscopic sporelings or germlings and became abundant. These include *Cladophora* sp., *Rhizoclonium tortuosum, Spongomorpha lanosa, Ulva lactuca* (all green algae), *Chordaria flagelliformis, Petalonia, Scytosiphon* (browns), *Ceramium* spp., *Dumontia incrassata* (reds), and filamentous diatoms. In spite of the presence of these species, *Enteromorpha* quickly became

FIG. 1.—Effect of *Littorina littorea* on algal composition of high tide pools at Nahant, Massachusetts. Means $\pm95\%$ confidence intervals of angularly transformed (Sokal and Rohlf 1969) percentage of cover data are indicated. All pools are 1–2 m² in surface area and 10–15 cm deep. Three to four permanent quadrats (0.25 m²) were sampled per pool. The mean density of *L. littorea* (1974) is indicated after each caption. "Before" percentage of cover and density data were taken in September and April for *A, B,* and *C* and in September for *D* and *E*. Removals or additions were begun immediately after April (*A, B, C*) or September (*D, E*) sampling. See legend in *D. Chondrus* is deemed "present" only when upright thalli occur.

the most abundant alga in this pool. Although individuals of *Enteromorpha* are ephemeral, the species appears able continually to monopolize space by reproducing and recruiting throughout the year. Hence, individuals of *Enteromorpha* initially colonizing this pool are probably not still present a year later, but have been replaced by other individuals. I have often observed *Enteromorpha* in tide pools releasing swarmers (spores or gametes) during low tide which may increase the probability that offspring will recruit near their parents (e.g., Dayton 1973).

Careful examination of the *L. littorea* removal pool revealed that the disappearance of upright thalli of *Chondrus* was not simply the result of its being hidden by the canopy of *Enteromorpha*. *Enteromorpha* settled on *Chondrus* and on primary substratum and appears to have outcompeted the long-lived *Chondrus*. Following settlement of *Enteromorpha* on *Chondrus*, the thalli (upright portions) of the latter became bleached and then disappeared. However the encrusting holdfasts of *Chondrus* remain.

A second set of *L. littorea* addition and removal experiments were initiated in September 1974 to determine the effect of seasonal differences on these results, since many ephemeral algal species in tide pools are different in spring-summer and fall-winter (J. Menge 1975). These experiments demonstrated that upon removal of *L. littorea* (fig. 1E), ephemerals (primarily the brown algae *Ectocarpus*, *Scytosiphon*, and *Petalonia*) are initially more abundant. However, *Enteromorpha* eventually prevails, as in the removal experiments initiated in April. Notes (but no data) taken on other pools lacking *L. littorea* indicated that *Enteromorpha* had continually covered about 80%–100% of the pool substratum for at least 3 yr.

In the second set of experiments, addition of *L. littorea* resulted in an immediate increase in ephemeral algal abundance (fig. 1D). These algae were apparently able to settle because the *Enteromorpha* abundance had been reduced by grazers. However, both the ephemerals and *Enteromorpha* were eventually eaten. Contrary to the first addition experiment, encrusting holdfasts of *Chondrus* were present in this pool. Upright *Chondrus* thalli began to appear after grazers removed the ephemerals and *Enteromorpha*. Thus *Chondrus* increased in abundance in pool D via vegetative growth whereas it had not yet appeared in pool B, probably because of slow recruitment. Excepting these variations in *Chondrus* abundance, the outcome of both sets of experiments was the same. Removal of *L. littorea* resulted in a near pure stand of *Enteromorpha* (fig. 1C and E), while addition of this snail resulted in elimination of *Enteromorpha* and, in one case, eventual dominance by *Chondrus* (fig. 1D).

These experiments suggest that in tide pools *Enteromorpha* is the dominant competitor for space. However, because it and most of the ephemeral seasonal algae are preferred species of *L. littorea* (table 1), their abundance decreases when this grazer is common and active. *Chondrus* persists in tide pools where *L. littorea* is dense because it is not eaten. In such pools, the periwinkles feed on microscopic plants, sporelings, and germlings of many ephemeral algae that

RELATIONSHIPS BETWEEN ALGAL COVER, CRABS, AND NEWLY SETTLED L. LITTOREA IN TIDE POOLS

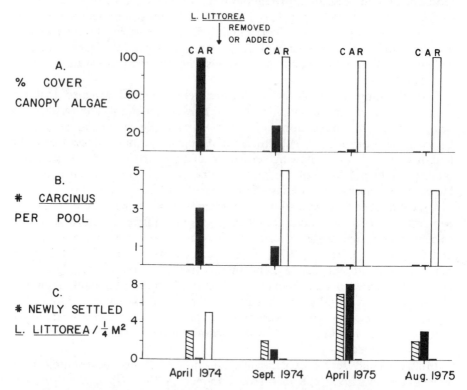

FIG. 2.—Changes in canopy algae, abundance of green crabs, and abundance of newly settled *Littorina littorea* following periwinkle manipulations in tide pools. *L. littorea* settles primarily during the spring and early summer. C (control, cross-hatched bars), A (*L. littorea* addition, solid bars), and R (*L. littorea* removal, blank bars) pools are the same as in figure 1A, B, and C, respectively. The other two experimental pools show similar changes. Canopy algae ($\geq$ 10 cm tall) is primarily *Enteromorpha* but may also include up to 5% total of *Scytosiphon* and/or *Rhizoclonium*. Percentage of canopy algae (X) and no. *Carcinus*/pool (Y) are positively correlated (linear regression: $Y = -0.02 + 0.04X$, $r^2 = .96$). No. *Carcinus*/pool (X) and no. newly settled *L. littorea* are negatively correlated ($Y = 5.69 - 1.59X$, $r^2 = .73$ for spring data).

settle on *Chondrus*. Thus *L. littorea* exerts a controlling influence on the algal composition of these pools.

In both of the *L. littorea* removal experiments (pools C and E), the green crab *Carcinus maenas* became abundant after a canopy of *Enteromorpha* was established (fig. 2A and B). Thereafter, very few tiny, newly metamorphosed *L. littorea* (0.2–0.3 cm long) were counted in the pools, in contrast to an abundance of them in the *L. littorea* addition and control pools (fig. 2C). Examination of numerous nonexperimental pools confirms that *Enteromorpha*-

dominated pools harbor many *Carcinus*, while *Chondrus*-dominated pools lack this crab. A dense canopy probably provides protection for this crab from sea gull (*Larus argentatus* and *L. marinus*) predators. Laboratory experiments demonstrate *Carcinus* readily preys upon small (but not medium or large) *L. littorea*. Thus, once *L. littorea* is absent from a pool long enough for *Enteromorpha* to become abundant, *Carcinus* may invade and prevent young *L. littorea* from recruiting from the plankton into the pools. Such a mechanism would explain the continued existence of low periwinkle density pools filled with highly desirable food. Hence, gulls probably indirectly affect the type of algae in pools. The pools may represent two alternative stable nodes (Lewontin 1969): (1) pools dominated by *Chondrus* because a dense contingent of herbivores continually removes superior competitors, and (2) pools dominated by the competitively superior *Enteromorpha* because predators prevent herbivores from being established. In this situation such alternative stable points may exist because the pools are essentially islands for which immigration and emigration are limited.

EFFECT OF *L. littorea* ON DIVERSITY OF ALGAE

When Competitive Dominant Is Preferred

The relationship between *Littorina littorea* density and the diversity of algae in tide pools can be seen in figure 3*A* and *B*. These data are from September 1974 after *L. littorea* grazing and competition between algal species have eliminated heavy spring and early summer recruitment of many short-lived algal species (Lubchenco and Menge 1978). Because the food preferences of *L. littorea* are known (table 1), and because this herbivore has been demonstrated to have a controlling effect on algae in tide pools (fig. 1), the source of between-pool variations in algal compositions and diversity can be interpreted as follows: When *L. littorea* is absent or rare, *Enteromorpha* outcompetes other algal species in pools, reducing the diversity. When *L. littorea* is present in intermediate densities, the abundance of *Enteromorpha* and various ephemeral algal species is reduced, competitive exclusion is prevented, and many algal species (ephemerals and perennials) coexist. At very high densities of *L. littorea*, all edible macroscopic algal species are consumed and prevented from appearing, leaving an almost pure stand of the inedible *Chondrus*. (New microscopic algae probably continue to settle and provide food for *L. littorea*.) Both the number of species (S) and H', an index of diversity based on both species number and the relative abundances, reveal the same unimodal relationship of macroscopic algal diversity to *L. littorea* density. A similar relationship was found for sea urchins grazing algae (Vadas 1968; Paine and Vadas 1969) and is suggested by qualitative results in certain terrestrial systems (Jones, cited in Harper 1969). These results support the theoretical predictions of Emlen (1973).

When Competitive Subordinate Is Preferred

Other workers have suggested that the effect of consumers on prey species diversity depends on the relationship between food preferences of the poten-

EFFECT OF <u>LITTORINA LITTOREA</u> ON THE DIVERSITY OF ALGAE

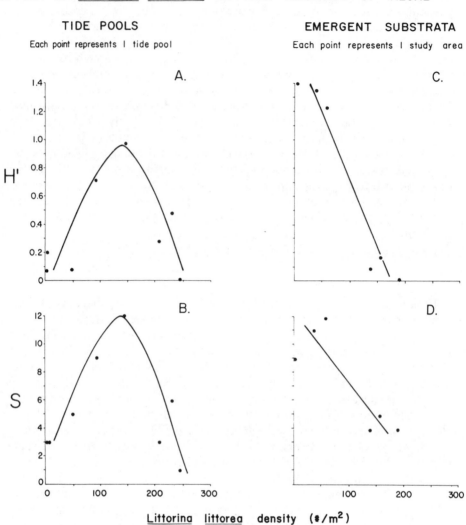

TIDE POOLS
Each point represents 1 tide pool

EMERGENT SUBSTRATA
Each point represents 1 study area

Littorina littorea density (#/m²)

FIG. 3.—Effect of *Littorina littorea* density on the diversity of algae in high tide pools (*A, B*) and on emergent substrata in the low intertidal zone (*C, D*). *S* = no. species, *H'* is an index of diversity, here based on the percentage of cover of each species. Each point in *A* and *B* is from four (0.25 m²) quadrats. Each point represents a different pool at Nahant, Massachusetts, September 1974. Each emergent substratum point was from 10 (0.25 m²) quadrats in the low zone at six different areas in Massachusetts and Maine, June and July 1974 (see J. Menge [1975] for descriptions of areas). Regression equations: (*A*) tide pool $H' = -0.0409 + 0.01250X - 0.00005X^2$, $r^2 = .65$; (*B*) tide pool $S = 1.64 + 0.1357X - 0.00056X^2$, $r^2 = .73$; (*C*) low emergent substratum $H' = 1.58 - 0.0089X$, $r^2 = .94$; (*D*) low $S = 11.58 - 0.0415X$, $r^2 = .72$. For mid-zone regressions (not illustrated), density of *L. obtusata* was converted to units of *L. littorea* density where 1 g wet weight *L. obtusata* is presumed to be equal to 1 g wet weight *L. littorea*. *X*, then, = "units of *L. littorea*," i.e., actual density of that snail plus presumably equivalent units of *L. obtusata*. Mid $H' = 1.65 - 0.004X$, $r^2 = .91$; mid $S = 8.03 - 0.017X$, $r^2 = .81$.

tially controlling consumer and competitive hierarchies of the food species (Paine 1969; Harper 1969; Patrick 1970; MacArthur 1972). The importance of this relationship can be seen when the effect of *L. littorea* on diversity of algae in tide pools is compared to its effect on diversity of algae on emergent substrata. In the New England rocky intertidal zone the competitive dominance of the most abundant tide pool plants is actually reversed when they interact on emergent substrata. Perennial brown algae (*Fucus vesiculosus, F. distichus,* and *Ascophyllum nodosum*) are competitively superior to other algae in the mid zone, while the perennial red alga *Chondrus* is competitively dominant in the low zone (J. Menge 1975; Lubchenco and Menge 1978). *Littorina littorea's* preferences remain the same in and out of tide pools. Consequently there is an inverse correlation between periwinkle abundance and algal species diversity on emergent substrata (fig. 3C and D). Specifically where *L. littorea* is scarce or absent (e.g., at areas exposed to wave action, or in experimental removals at protected sites) at least 14 ephemeral species coexist with *Fucus, Ascophyllum,* and *Chondrus.* (These include *Enteromorpha, Spongomorpha spinescens, S. arcta, Rhizoclonium tortuosum, Ulva lactuca* [greens]; *Chordaria flagelliformis, Dictyosiphon foenicularis, Ectocarpus* spp., *Elachistea fucicola, Petalonia fascia, Pylaiella littoralis, Scytosiphon lomentaria* [browns]; and *Ceramium* spp. and *Dumontia incrassata* [reds].) Although the ephemeral species are all eventually outcompeted by the perennials on primary space, the former can coexist with the latter by occupying patches of primary space cleared by disturbances or by settling and growing epiphytically upon the perennials. If no such refuge were possible, there would be no relationship between *L. littorea* density and algal diversity, and the herbivores would simply increase the rate at which the eventual dominance by competitively superior plants was attained. On emergent substrata in New England, when *L. littorea* is abundant, it preferentially eats the ephemerals, leaving the more unpalatable fucoids and *Chondrus,* thus decreasing diversity. *Littorina obtusata* has the same effect as its congener in the mid zone (J. Menge 1975). Similar results have been suggested by Patrick (1970) for freshwater snails grazing diatoms (based on unpublished laboratory data of K. Roop) and by Harper (1969) for Milton's data on sheep grazing pastures.

DISCUSSION

Space has been shown to be a primary limiting resource in many rocky intertidal communities (Connell 1961, 1971, 1972; Dayton 1971, 1975; Lubchenco and Menge 1978; B. Menge 1976; J. Menge 1975; Paine 1966, 1974). This resource is modified by physical conditions that may determine the outcome of competitive interactions. Thus in tide pools in New England, ephemeral algae like *Enteromorpha* are competitively superior while on emergent substrata the more hardy perennials dominate primary space. Spatial and temporal heterogeneity in physical conditions are undoubtedly important in maintaining the coexistence of these competing species.

The result that in some habitats an opportunistic species can continually outcompete perennials is worth emphasizing. Evidently in the absence of herbivores, ephemerals such as *Enteromorpha* can outcompete the perennial fucoids and Irish moss in tide pools but probably not on emergent substrata. In the latter habitat, ephemerals like *Enteromorpha* can temporarily "outcompete" perennials on newly opened primary substrata by growing faster (J. Menge 1975). However, these ephemerals are eventually replaced by perennial species (e.g., *Fucus* in the mid zone), which can recruit slowly and take over when the ephemerals die. The key difference between the ephemerals' performance in the two habitats appears to be that they can continually recruit and replace themselves in tide pools but not on emergent substrata. The brown alga *Postelsia* is another short-lived species which appears to outcompete later successional species, barnacles and mussels, by virtue of its ability to recruit offspring near parent plants (Dayton 1973). Thus in microhabitats where ephemerals outcompete perennials, the successional sequence does not progress except where herbivores remove the ephemerals.

Concomitant with the reversal of competitive dominance in different microhabitats is the alteration of effects of herbivores. The plants that seem best suited to cope with the physical rigors of the intertidal zone, the perennials, are also least attractive to periwinkle herbivores. Thus these herbivores have a negative or small effect on algal species diversity on emergent substrata. In tide pools, however, because the preferred algae are also competitively dominant, herbivores determine the algal composition and species diversity in this microhabitat. This shift in competitive dominance in different microhabitats and consequent alteration of effects of consumers may be widespread, but it needs additional documentation.

The effects of *Littorina littorea* on plant species diversity in New England are complex. The key to understanding these effects lies in knowledge of the food preferences of the herbivore. These results may typify the effects of many generalized consumers: When the competitively dominant species is preferred by the consumer, there is a unimodal relationship between prey diversity and consumer density, with the highest diversities at intermediate consumer densities. When the competitively inferior species are preferred, there is an inverse correlation between prey diversity and consumer density (e.g., fig. 3). Under the former conditions, effects of consumers counteract competitive dominance. In the latter situation, feeding reinforces effects of competitive dominance.

As indicated earlier, the general effects of herbivores on plant species diversity have not been clear. After reviewing a number of case studies of effects of terrestrial herbivores on plant species diversity, Harper (1969) concluded that the results were too variable and inconsistent to warrant any generalizations. In view of the results presented here, it is possible to reinterpret the studies cited by Harper. I believe knowledge of three critical factors is necessary to obtain the proper insight into these studies. These factors are (1) the relationship between herbivore preferences and competitive ability of the plants, (2) the length of time the experiment was monitored after the manipulation, and (3) the initial relative abundance of the herbivores. The importance of the second and third factors can be seen in the following example. In figure 3, if the initial density of *L. littorea* were $250/m^2$, its removal would result in an initial increase in species richness followed by a decrease as competitive interactions occurred. In contrast, if the initial density were only $100/m^2$, snail removal would result in a decrease in algal diversity. Thus the initial relative intensity of grazing or predation and the amount of time the experiment is monitored will determine what effect removal of the consumer has.

The importance of all three of the above factors is indicated by numerous studies. First, experimental manipulations of predators or herbivores that selected competitively superior food species conform to either the left half of the or the whole unimodal curves in figure 3A and B, depending on the range in abundance of consumers (Hall et al. 1970; Harper 1969; Paine 1966, 1971, 1974; Paine and Vadas 1969; Vadas 1968). Second, studies in which consumers preferred competitively inferior species comply with results in figure 3C and D (Harper 1969; Patrick 1970). Third, examples of overgrazing (i.e., the right half of fig. 3A and B curves) caused by a high density of consumers abound (Bartholomew 1970; Dayton 1971, 1975; Earle 1972; Harper 1969; Kitching and Ebling 1961, 1967; Leighton et al. 1966; Lodge 1948; Ogden et al. 1973; Paine and Vadas 1969; Randall 1965; Southward 1964; Vadas 1968). Taken together, these results suggest that generalized herbivores can have the same effect on species diversity as do generalized predators. This effect appears to depend primarily on the relationship between the consumer's food preferences and the competitive interactions of the food species.

The effects of periwinkles on algae presented here are the result of local manipulations of herbivores over ecological time. To what degree consumers affect broadscale biogeographic patterns of species diversity of lower trophic levels over evolutionary time remains to be seen. However, I believe the effects may be comparable. In the following example, regional differences in plant species diversity perhaps caused by the effects of an herbivore may be intermediate between local and broadscale patterns.

There is a striking difference in the low intertidal algal species diversity of New England rocky coasts and that of the rocky shores in the Bay of Fundy, which may be a function of the abundance of the herbivorous sea urchin *Strongylocentrotus droebachiensis*. Unlike *L. littorea*, this urchin readily eats *Chondrus*, and several lines of evidence suggest it has an effect on low-zone algal diversity comparable to the periwinkle's effect in tide pools.

In the low zone of New England, this urchin is rare and *Chondrus* outcompetes most other algae and dominates the zone (cols. 1, 2 in table 2; Lubchenco and Menge 1978). Changes in algal species diversity at different New England areas are caused by varying densities of periwinkles affecting algal epiphytes on *Chondrus* (fig. 3). Sea urchins are common in the low zone at rocky areas in the Bay of Fundy. Field experiments done with this urchin indicate it readily eats *Chondrus* and can prevent this plant from monopolizing space in the low zone (fig. 4; Lubchenco and Menge 1978). Thus sea urchins may prevent *Chondrus* from dominating the low zone in the Bay of Fundy and allow many other algae to coexist there (col. 3, table 2). This example would parallel the results of Dayton (1975) and Vadas (1968) for the Pacific northwest

TABLE 2

EFFECT OF SEA URCHINS ON ALGAL SPECIES DIVERSITY IN THE LOW
INTERTIDAL ZONE IN NEW ENGLAND AND THE BAY OF FUNDY

	NEW ENGLAND		BAY OF FUNDY	
	Chamberlain, Maine (1)	Canoe Beach, Cove, Nahant, Mass. (2)	Cape Forchu, Yarmouth, Nova Scotia (3)	Quoddy Head, Maine (4)
Herbivore densities:*				
Strongylocentrotus	0	0	4.2 ± 2.6	26.4 ± 13.8
Acmaea	0	$.1 \pm .2$	$.5 \pm .8$	21.2 ± 7.1
L. littorea	0	126.8 ± 60.0	0	0
Algal diversity and percentage of cover:				
No. species†	8	3	27	6
H'_e	1.20	.23	2.03	1.14
$\bar{X}\%$ cover canopy	0	0	78.6 ± 19.7	$.4 \pm .7$
$\bar{X}\%$ cover understory ..	80.6 ± 17.0	89.9 ± 10.2	$125.6 \pm 20.7‡$	2.6 ± 2.1
$\bar{X}\%$ cover Chondrus ...				
(= in understory) ...	74.5 ± 18.2	83.3 ± 11.4	14.6 ± 10.2	0

NOTE.—Data are from June–July 1975–1976.

* Densities are $\bar{X} \pm 95\%$ confidence intervals/0.25m^2 for 10 quadrats at each area; see Lubchenco and Menge (1978) or J. Menge (1975) for methods.

† Includes both canopy and understory species. In the low zone, L. littorea grazes and affects only epiphytic algae on Chondrus; sea urchins graze and affect Chondrus and most other understory and canopy species.

‡ Percentage of cover > 100% reflects the dense multilayer arrangement of the understory at Cape Forchu.

coast of North America. In New England, urchins occur only in the subtidal region; in the Bay of Fundy they are both subtidal and intertidal. It is not clear why urchins are more abundant in the low intertidal zone in the Bay of Fundy. Possible factors are (1) a lower level of gull predation, (2) less desiccation stress, and (3) less wave action in Fundy. When urchins are exceedingly abundant, e.g., at Quoddy Head, Maine (on the Bay of Fundy), algal species diversity and abundance are very low (col. 4, table 2). In such areas, only encrusting coralline algae, a few unpalatable species (Desmarestia and Agarum; Vadas 1968), and some recently settled ephemeral species are present. Thus there is evidently a unimodal relationship between low intertidal algal species diversity and sea urchin density similar to that in figure 3A and B (see also Vadas 1968). That this correlation is causal can be demonstrated only through experimental manipulations.

Throughout this paper I have emphasized the importance of preferential predation on the competitively dominant prey. However, it may be possible for nonselective predation to have comparable effects on prey species diversity. This could be accomplished with differential recruitment and/or growth rates of the prey such that with nonselective predation the competitively dominant prey are prevented from outcompeting other prey. For example, Chondrus may

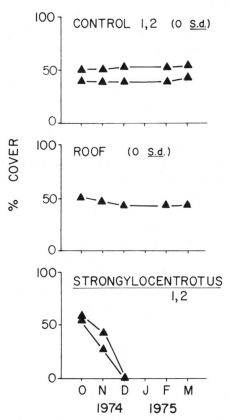

EFFECT OF STRONGYLOCENTROTUS
ON CHONDRUS

FIG. 4.—Effect of sea urchins on *Chondrus* in the low zone at Grindstone Neck, Maine, where urchins are normally absent. Two $10 \times 10 \times 4$-cm stainless steel mesh cages containing one (3.5 cm test diameter) and two (both 3.2 cm) urchins were placed over *Chondrus*. Two unmanipulated controls and one stainless steel mesh roof indicate the normal percentage of cover of *Chondrus* without urchins. In the urchin enclosures, urchins completely removed *Chondrus*, including the encrusting holdfast.

be prevented from dominating the low zone by urchins in the following manner. If urchins graze most plants present in proportion to their abundance (except for certain unpalatable algae), and many of these plants have higher recruitment and/or growth rates than does *Chondrus*, *Chondrus* could not take over. In other words, the critical factor is the effect the consumer has on the competitive dominant (here, prevention of competitive exclusion), not the means (selective or nonselective grazing) by which it is accomplished.

From the above discussion, it is obvious that not all herbivores have similar effects on plant species diversity. If the sea urchins are removing *Chondrus* and preventing it from outcompeting other low zone algae in the Bay of Fundy, they have a very different effect on the emergent substratum community than do the periwinkle snails that do not graze *Chondrus*, but only its epiphytes. The size, manner of feeding, degree of food specialization, and mobility of herbivores may be important determinants of their effects on vegetation. For example, spatial escapes of seeds and young trees from their specialized and less mobile herbivores may provide an important mechanism of maintaining tropical rain forest tree diversity (Connell 1971, 1975; Janzen 1970). Nonetheless, it appears that a critical determinant of the effect many generalized consumers have on their food resources is the relationship between consumer food preferences and food competitive ability.

SUMMARY

Field experiments demonstrate that the herbivorous marine snail *Littorina littorea* controls the abundance and type of algae in high intertidal tide pools in New England. Here the highest species diversity of algae occurs at intermediate *Littorina* densities. This unimodal relationship between algal species diversity and herbivore density occurs because the snail's preferred food is competitively dominant in tide pool habitats. Moderate grazing allows inferior algal species to persist and intense grazing eliminates most individuals and species. In contrast to pools, on emergent substrata where the preferred food is competitively inferior, this herbivore decreases algal diversity. Thus, the effect of this consumer on plant species diversity depends on the relationship between herbivore food preference and competitive abilities of the plants. These results may apply to most generalized consumers and provide a framework within which previously confusing results can be understood.

Thus predators or herbivores do not simply increase or decrease species diversity of their food, but can potentially do both. The precise effect a consumer has probably depends both on the relationship between its preferences and the food's competitive abilities and on the intensity of the grazing or predation pressure.

ACKNOWLEDGMENTS

I gratefully acknowledge P. K. Dayton, B. A. Menge, R. T. Paine, T. W. Schoener, F. E. Smith, J. R. Young, and an anonymous reviewer for discussions and comments on this manuscript. This paper is contribution no. 42 from the Marine Science Institute, Northeastern University, Nahant, Massachusetts, where facilities were kindly made available by N. W. Riser and M. P. Morse. This paper represents part of a Ph.D. thesis submitted to Harvard University. The research was supported in part by National Science Foundation grants to J. Lubchenco Menge (GA-40003) and to B. A. Menge (GA35617 and DES72-01578 A01).

LITERATURE CITED

Bartholomew, B. 1970. Bare zone between California shrub and grassland communities: the role of animals. Science 170:1210–1212.

Connell, J. H. 1961. The influence of interspecific competition and other factors on the distribution of the barnacle *Chthamalus stellatus*. Ecology 42:710–723.

———. 1971. On the role of natural enemies in preventing competitive exclusion in some marine animals and in rain forest trees. Pages 298–312 in P. J. den Boer and G. R. Gradwell, eds. Dynamics of populations. Proceedings of the Advanced Study Institute on Dynamics of Numbers in Populations, Oosterbeek, 1970. Centre for Agricultural Publishing and Documentation, Wageningen.

———. 1972. Community interactions on marine rocky intertidal shores. Annu. Rev. Ecol. Syst. 3:169–192.

———. 1975. Some mechanisms producing structure in natural communities: a model and evidence from field experiments. Pages 460–490 in M. L. Cody and J. Diamond, eds. Ecology and evolution of communities. Belknap, Cambridge, Mass.

Dayton, P. K. 1971. Competition, disturbance, and community organization: the provision and subsequent utilization of space in a rocky intertidal community. Ecol. Monogr. 41:351–389.

———. 1973. Dispersion, dispersal and persistence of the annual intertidal alga Postelsia palmaeformis Ruprect. Ecology 54:433–438.

———. 1975. Experimental evaluation of ecological dominance in a rocky intertidal algal community. Ecol. Monogr. 45:137–159.

Dayton, P. K., G. A. Robilliard, R. T. Paine, and L. B. Dayton. 1974. Biological accommodation in the benthic community at McMurdo Sound, Antarctica. Ecol. Monogr. 44:105–128.

Earle, S. A. 1972. The influence of herbivores on the marine plants of Great Lameshure Bay, with an annotated list of plants. Sci. Bull. Los Angeles County Natur. Hist. Mus. 14:17–44.

Emlen, J. M. 1973. Ecology: an evolutionary approach. Addison-Wesley, Reading, Mass. 493 pp.

Gustavsson, Ulla. 1972. A proposal for a classification of Marine rockpools on the Swedish West Coast. Bot. Marina 15:210–214.

Hall, D. J., W. E. Cooper, and E. E. Werner. 1970. An experimental approach to the production dynamics and structure of freshwater animal communities. Limnol. Oceanogr. 15:839–928.

Harper, J. L. 1969. The role of predation in vegetational diversity. Brookhaven Symp. Biol. no. 22, pp. 48–62.

Hutchinson, G. E. 1959. Homage to Santa Rosalia, or why are there so many kinds of animals? Amer. Natur. 93:145–159.

Jackson, J. B. C., and L. W. Buss. 1975. Allelopathy and spatial competition among coral reef invertebrates. Proc. Nat. Acad. Sci. USA 72:5160–5163.

Janzen, D. H. 1970. Herbivores and the number of tree species in tropical forests. Amer. Natur. 104:501–528.

Kitching, J. A., and F. J. Ebling. 1961. The ecology of Lough Ine. XI. The control of algae by Paracentrotus lividus (Echinoidea). J. Anim. Ecol. 30:373–383.

———. 1967. Ecological studies at Lough Ine. Advance. Ecol. Res. 4:198–291.

Leighton, D. L., L. G. Jones, and W. J. North. 1966. Ecological relationships between giant kelp and sea urchins in Southern California. Pages 141–153 in E. G. Young and J. L. McLachlan, eds. Proceedings of the Fifth International Seaweed Symposium. Pergamon, New York.

Lewontin, R. C. 1969. The meaning of stability. Brookhaven Symp. Biol. no. 22, pp. 13–23.

Lodge, S. M. 1948. Algal growth in the absence of Patella on an experimental strip of foreshore, Port St. Mary, Isle of Man. Proc. Trans. Liverpool Biol. Soc. 56:78–85.

Lubchenco, J., and B. A. Menge. 1978. Community development and persistence in a low rocky intertidal zone. Ecol. Monogr. (in press).

MacArthur, R. H. 1972. Geographical ecology. Harper & Row, New York. 269 pp.

Menge, B. A. 1976. Organization of the New England rocky intertidal community: role of predation, competition and environmental heterogeneity. Ecol. Monogr. 46:355–369.

Menge, B. A., and J. P. Sutherland. 1976. Species diversity gradients: synthesis of the roles of predation, competition and temporal heterogeneity. Amer. Natur. 110:351–369.

Menge, J. L. 1975. Effect of herbivores on community structure of the New England rocky intertidal region: distribution, abundance and diversity of algae. Ph.D. diss. Harvard University.

Ogden, J. C., R. A. Brown, and N. Salesky. 1973. Grazing by the echinoid *Diadema antillarum* Philippi: formation of halos around West Indian patch reefs. Science 182:715–717.

Paine, R. T. 1966. Food web complexity and species diversity. Amer. Natur. 100:65–75.

———. 1969. The *Pisaster-Tegula* interaction: prey patches, predator food preference and intertidal community structure. Ecology 50:950–961.

———. 1971. A short-term experimental investigation of resource partitioning in a New Zealand rocky intertidal habitat. Ecology 52:1096–1106.

———. 1974. Intertidal community structure: experimental studies on the relationship between a dominant competitor and its principal predator. Oecologia 15:93–120.

Paine, R. T., and R. L. Vadas. 1969. The effects of grazing by sea urchins, *Strongylocentrotus* spp., on benthic algal populations. Limnol. Oceanogr. 14:710–719.

Patrick, R. 1970. Benthic stream communities. Amer. Sci. 58:546–549.

Pianka, E. R. 1967. On lizard species diversity: North American flatland deserts. Ecology 48:333–351.

———. 1969. Habitat specificity, speciation, and species density in Australian desert lizards. Ecology 50:498–502.

Randall, J. E. 1965. Grazing effect of sea grasses by herbivorous reef fishes in the West Indies. Ecology 46:255–260.

Ricklefs, R. E. 1973. Ecology. Chiron, Newton, Mass. 861 pp.

Sokal, R. R., and F. J. Rohlf. 1969. Biometry. Freeman, San Francisco. 776 pp.

Southward, A. J. 1964. Limpet grazing and the control of vegetation on rocky shores. Pages 265–273 *in* D. J. Crisp, ed. Grazing in terrestrial and marine environments. Blackwell, Oxford.

Vadas, R. L. 1968. The ecology of *Agarum* and the kelp bed community. Ph.D. diss. University of Washington. 280 pp.

Van Valen, L. 1974. Predation and species diversity. J. Theoret. Biol. 44:19–21.

6

Reprinted from *Am. Nat.* **100**:65–75 (1966) by permission of The University of Chicago Press

FOOD WEB COMPLEXITY AND SPECIES DIVERSITY

ROBERT T. PAINE

Department of Zoology, University of Washington, Seattle, Washington

Though longitudinal or latitudinal gradients in species diversity tend to be well described in a zoogeographic sense, they also are poorly understood phenomena of major ecological interest. Their importance lies in the derived implication that biological processes may be fundamentally different in the tropics, typically the pinnacle of most gradients, than in temperate or arctic regions. The various hypotheses attempting to explain gradients have recently been reviewed by Fischer (1960), Simpson (1964), and Connell and Orias (1964), the latter authors additionally proposing a model which can account for the production and regulation of diversity in ecological systems. Understanding of the phenomenon suffers from both a specific lack of synecological data applied to particular, local situations and from the difficulty of inferring the underlying mechanism(s) solely from descriptions and comparisons of faunas on a zoogeographic scale. The positions taken in this paper are that an ultimate understanding of the underlying causal processes can only be arrived at by study of local situations, for instance the promising approach of MacArthur and MacArthur (1961), and that biological interactions such as those suggested by Hutchinson (1959) appear to constitute the most logical possibilities.

The hypothesis offered herein applies to local diversity patterns of rocky intertidal marine organisms, though it conceivably has wider applications. It may be stated as: "Local species diversity is directly related to the efficiency with which predators prevent the monopolization of the major environmental requisites by one species." The potential impact of this process is firmly based in ecological theory and practice. Gause (1934), Lack (1949), and Slobodkin (1961) among others have postulated that predation (or parasitism) is capable of preventing extinctions in competitive situations, and Slobodkin (1964) has demonstrated this experimentally. In the field, predation is known to ameliorate the intensity of competition for space by barnacles (Connell, 1961b), and, in the present study, predator removal has led to local extinctions of certain benthic invertebrates and algae. In addition, as a predictable extension of the hypothesis, the proportion of predatory species is known to be relatively greater in certain diverse situations. This is true for tropical vs. temperate fish faunas (Hiatt and Strasburg, 1960; Bakus, 1964), and is seen especially clearly in the comparison of shelf water zooplankton populations (81 species, 16% of which are carnivores) with those of the presumably less productive though more stable Sargasso Sea (268 species, 39% carnivores) (Grice and Hart, 1962).

127

In the discussion that follows no quantitative measures of local diversity are given, though they may be approximated by the number of species represented in Figs. 1 to 3. No distinctions have been drawn between species within certain food categories. Thus I have assumed that the probability of, say, a bivalve being eaten is proportional to its abundance, and that predators exercise no preference in their choice of any "bivalve" prey. This procedure simplifies the data presentation though it dodges the problem of taxonomic complexity. Wherever possible the data are presented as both number observed being eaten and their caloric equivalent. The latter is based on prey size recorded in the field and was converted by determining the caloric content of Mukkaw Bay material of the same or equivalent species. These caloric data will be given in greater detail elsewhere. The numbers in the food webs, unfortunately, cannot be related to rates of energy flow, although when viewed as calories they undoubtedly accurately suggest which pathways are emphasized.

Dr. Rudolf Stohler kindly identified the gastropod species. A. J. Kohn, J. H. Connell, C. E. King, and E. R. Pianka have provided invaluable criticism. The University of Washington, through the offices of the Organization for Tropical Studies, financed the trip to Costa Rica. The field work in Baja California, Mexico, and at Mukkaw Bay was supported by the National Science Foundation (GB-341).

THE STRUCTURE OF SELECTED FOOD WEBS

I have claimed that one of the more recognizable and workable units within the community nexus are subwebs, groups of organisms capped by a terminal carnivore and trophically interrelated in such a way that at higher levels there is little transfer of energy to co-occurring subwebs (Paine, 1963). In the marine rocky intertidal zone both the subwebs and their top carnivores appear to be particularly distinct, at least where macroscopic species are involved; and observations in the natural setting can be made on the quantity and composition of the component species' diets. Furthermore, the rocky intertidal zone is perhaps unique in that the major limiting factor of the majority of its primary consumers is living space, which can be directly observed, as the elegant studies on interspecific competition of Connell (1961a,b) have shown. The data given below were obtained by examining individual carnivores exposed by low tide, and recording prey, predator, their respective lengths, and any other relevant properties of the interaction.

A north temperate subweb

On rocky shores of the Pacific Coast of North America the community is dominated by a remarkably constant association of mussels, barnacles, and one starfish. Fig. 1 indicates the trophic relationships of this portion of the community as observed at Mukkaw Bay, near Neah Bay, Washington (ca. 49° N latitude). The data, presented as both numbers and total calories consumed by the two carnivorous species in the subweb, *Pisaster ochraceus*,

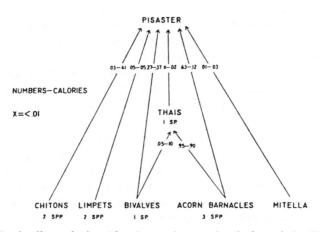

FIG. 1. The feeding relationships by numbers and calories of the *Pisaster* domi-nated subweb at Mukkaw Bay. *Pisaster*, N = 1049; *Thais*, N = 287. N is the num-ber of food items observed eaten by the predators. The specific composition of each predator's diet is given as a pair of fractions; numbers on the left, calories on the right.

a starfish, and *Thais emarginata*, a small muricid gastropod, include the observational period November, 1963, to November, 1964. The composition of this subweb is limited to organisms which are normally intertidal in dis-tribution and confined to a hard rock substrate. The diet of *Pisaster* is re-stricted in the sense that not all available local food types are eaten, al-though of six local starfish it is the most catholic in its tastes. Nu-merically its diet varies little from that reported by Feder (1959) for *Pisaster* observed along the central California coastline, especially since the gastropod *Tegula*, living on a softer bottom unsuitable to barnacles, has been omitted. *Thais* feeds primarily on the barnacle *Balanus glandula*, as also noted by Connell (1961b).

This food web (Fig. 1) appears to revolve on a barnacle economy with both major predators consuming them in quantity. However, note that on a nutritional (calorie) basis, barnacles are only about one-third as important to *Pisaster* as either *Mytilus californianus*, a bivalve, or the browsing chiton *Katherina tunicata*. Both these prey species dominate their respec-tive food categories. The ratio of carnivore species to total species is 0.18. If *Tegula* and an additional bivalve are included on the basis that they are the most important sources of nourishment in adjacent areas, the ratio becomes 0.15. This number agrees closely with a ratio of 0.14 based on *Pisaster*, plus all prey species eaten more than once, in Feder's (1959) general compilation.

A subtropical subweb

In the Northern Gulf of California (ca. 31° N.) a subweb analogous to the one just described exists. Its top carnivore is a starfish (*Heliaster kubiniji*), the next two trophic levels are dominated by carnivorous gastropods, and the main prey are herbivorous gastropods, bivalves, and barnacles. I have

collected there only in March or April of 1962–1964, but on both sides of
the Gulf at San Felipe, Puertecitos, and Puerta Penasco. The resultant
trophic arrangements (Fig. 2), though representative of springtime conditions and indicative of a much more stratified and complex community, are
basically similar to those at Mukkaw Bay. Numerically the major food item
in the diets of *Heliaster* and *Muricanthus nigritus* (a muricid gastropod),
the two top-ranking carnivores, is barnacles; the major portion of these
predators' nutrition is derived from other members of the community, primarily herbivorous mollusks. The increased trophic complexity presents
certain graphical problems. If increased trophic height is indicated by a
decreasing percentage of primary consumers in a species diet, *Acanthina
tuberculata* is the highest carnivore due to its specialization on *A. angelica*,
although it in turn is consumed by two other species. Because of this, and
ignoring the percentages, both *Heliaster* and *Muricanthus* have been placed
above *A. tuberculata*. Two species, *Hexaplex* and *Muricanthus* eventually
become too large to be eaten by *Heliaster*, and thus through growth join it
as top predators in the system. The taxonomically-difficult gastropod
family Columbellidae, including both herbivorous and carnivorous species
(Marcus and Marcus, 1962) have been placed in an intermediate position.

The Gulf of California situation is interesting on a number of counts. A
new trophic level which has no counterpart at Mukkaw Bay is apparent, in-
terposed between the top carnivore and the primary carnivore level. If
higher level predation contributes materially to the maintenance of diversity, these species will have an effect on the community composition
out of proportion to their abundance. In one of these species, *Muricanthus*,

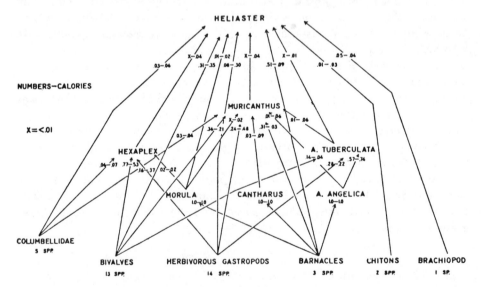

FIG. 2. The feeding relationships by numbers and calories of the *Heliaster*
dominated subweb in the northern Gulf of California. *Heliaster*, N = 2245; *Muricanthus*, N = 113; *Hexaplex*, N = 62; *A. tuberculata*, N = 14; *A. angelica*, N = 432;
Morula, N = 39; *Cantharus*, N = 8.

the larger members belong to a higher level than immature specimens (Paine, unpublished), a process tending to blur the food web but also potentially increasing diversity (Hutchinson, 1959). Finally, if predation operates to reduce competitive stresses, evidence for this reduction can be drawn by comparing the extent of niche diversification as a function of trophic level in a typical Eltonian pyramid. *Heliaster* consumes all other members of this subweb, and as such appears to have no major competitors of comparable status. The three large gastropods forming the subterminal level all may be distinguished by their major sources of nutrition: *Hexaplex*—bivalves (53%), *Muricanthus*—herbivorous gastropods (48%), and *A. tuberculata*—carnivorous gastropods (74%). No such obvious distinction characterizes the next level composed of three barnacle-feeding specialists which additionally share their resource with *Muricanthus* and *Heliaster*. Whether these species are more specialized (Klopfer and MacArthur, 1960) or whether they tolerate greater niche overlap (Klopfer and MacArthur, 1961) cannot be stated. The extent of niche diversification is subtle and trophic overlap is extensive.

The ratio of carnivore species to total species in Fig. 2 is 0.24 when the category Columbellidae is considered to be principally composed of one herbivorous (*Columbella*) and four carnivorous (*Pyrene, Anachis, Mitella*) species, based on the work of Marcus and Marcus (1962).

A tropical subweb

Results of five days of observation near Mate de Limon in the Golfo de Nocoya on the Pacific shore of Costa Rica (approx. 10° N.) are presented in Fig. 3. No secondary carnivore was present; rather the environmental resources were shared by two small muricid gastropods, *Acanthina brevidentata* and *Thais biserialis*. The fauna of this local area was relatively simple and completely dominated by a small mytilid and barnacles. The co-occupiers of the top level show relatively little trophic overlap despite the broad nutritional base of *Thais* which includes carrion and cannibalism. The relatively low number of feeding observations (187) precludes an accurate appraisal of the carnivore species to total web membership ratio.

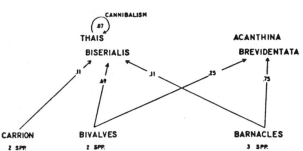

FIG. 3. The feeding relationship by numbers of a comparable food web in Costa Rica. *Thais*, N = 99; *Acanthina*, N = 80.

CHANGES RESULTING FROM THE REMOVAL OF THE TOP CARNIVORE

Since June, 1963, a "typical" piece of shoreline at Mukkaw Bay about eight meters long and two meters in vertical extent has been kept free of *Pisaster*. An adjacent control area has been allowed to pursue its natural course of events. Line transects across both areas have been taken irregularly and the number and density of resident macroinvertebrate and benthic algal species measured. The appearance of the control area has not altered. Adult *Mytilus californianus*, *Balanus cariosus*, and *Mitella polymerus* (a goose-necked barnacle) form a conspicuous band in the middle intertidal. The relatively stable position of the band is maintained by *Pisaster* predation (Paris, 1960; Paine, unpublished). At lower tidal levels the diversity increases abruptly and the macrofauna includes immature individuals of the above, *B. glandula* as scattered clumps, a few anemones of one species, two chiton species (browsers), two abundant limpets (browsers), four macroscopic benthic algae (*Porphyra*-an epiphyte, *Endocladia*, *Rhodomela*, and *Corallina*), and the sponge *Haliclona*, often browsed upon by *Anisodoris*, a nudibranch.

Following the removal of *Pisaster*, *B. glandula* set successfully throughout much of the area and by September had occupied from 60 to 80% of the available space. By the following June the *Balanus* themselves were being crowded out by small, rapidly growing *Mytilus* and *Mitella*. This process of successive replacement by more efficient occupiers of space is continuing, and eventually the experimental area will be dominated by *Mytilus*, its epifauna, and scattered clumps of adult *Mitella*. The benthic algae either have or are in the process of disappearing with the exception of the epiphyte, due to lack of appropriate space; the chitons and larger limpets have also emigrated, due to the absence of space and lack of appropriate food.

Despite the likelihood that many of these organisms are extremely long-lived and that these events have not reached an equilibrium, certain statements can be made. The removal of *Pisaster* has resulted in a pronounced *decrease* in diversity, as measured simply by counting species inhabiting this area, whether consumed by *Pisaster* or not, from a 15 to an eight-species system. The standing crop has been increased by this removal, and should continue to increase until the *Mytilus* achieve their maximum size. In general the area has become trophically simpler. With *Pisaster* artificially removed, the sponge-nudibranch food chain has been displaced, and the anemone population reduced in density. Neither of these carnivores nor the sponge is eaten by *Pisaster*, indicating that the number of food chains initiated on this limited space is strongly influenced by *Pisaster*, but by an indirect process. In contrast to Margalef's (1958) generalization about the tendency, with higher successional status towards "an ecosystem of more complex structure," these removal experiments demonstrate the opposite trend: in the absence of a complicating factor (predation), there is a "winner" in the competition for space, and the local system tends toward simplicity. Predation by this interpretation interrupts the successional process and, on a local basis, tends to increase local diversity.

No data are available on the microfaunal changes accompanying the gradual alteration of the substrate from a patchy algal mat to one comprised of the byssal threads of *Mytilus*.

INTERPRETATION

The differences in relative diversity of the subwebs diagrammed in Figs. 1-3 may be represented as Baja California (45 spp.) >> Mukkaw Bay (11 spp.) > Costa Rica (8 sp.), the number indicating the actual membership of the subwebs and not the number of local species. All three areas are characterized by systems in which one or two species are capable of monopolizing much of the space, a circumstance realized in nature only in Costa Rica. In the other two areas a top predator that derives its nourishment from other sources feeds in such a fashion that no space-consuming monopolies are formed. *Pisaster* and *Heliaster* eat masses of barnacles, and in so doing enhance the ability of other species to inhabit the area by keeping space open. When the top predator is artificially removed or naturally absent (i.e., predator removal area and Costa Rica, respectively), the systems converge toward simplicity. When space is available, other organisms settle or move in, and these, for instance chitons at Mukkaw Bay and herbivorous gastropods and pelecypods in Baja California, form the major portions of the predator's nutrition. Furthermore, *in situ* primary production is enhanced by the provision of space. This event makes the grazing moiety less dependent on the vagaries of phytoplankton production or distribution and lends stability to the association.

At the local level it appears that carnivorous gastropods which can penetrate only one barnacle at a time, although they might consume a few more per tidal interval, do not have the same effect as a starfish removing 20 to 60 barnacles simultaneously. Little compensation seems to be gained from snail density increases because snails do not clear large patches of space, and because the "husks" of barnacles remain after the animal portion has been consumed. In the predator removal area at Mukkaw Bay, the density of *Thais* increased 10- to 20-fold, with no apparent effect on diversity although the rate of *Mytilus* domination of the area was undoubtedly slowed. Clusters (density of 75-125/m^2) of *Thais* and *Acanthina* characterize certain rocks in Costa Rica, and diversity is still low. And, as a generality, wherever acorn barnacles or other space-utilizing forms potentially dominate the shore, diversity is reduced unless some predator can prevent the space monopoly. This occurs in Washington State where the shoreline, in the absence of *Pisaster*, is dominated by barnacles, a few mussels, and often two species of *Thais*. The same monopolistic tendencies characterize Connell's (1961a,b) study area in Scotland, the rocky intertidal of northern Japan (Hoshiai, 1960, 1961), and shell bags suitable for sponge settlement in North Carolina (Wells, Wells, and Gray, 1964).

Local diversity on intertidal rocky bottoms, then, appears directly related to predation intensity, though other potential factors are mentioned below. If one accepts the generalizations of Hedgpeth (1957) and Hall

(1964) that ambient temperature is the single most important factor influencing distribution or reproduction of marine invertebrates, then the potential role of climatic stability as measured by seasonal variations in water temperature can be examined. At Neah Bay the maximum range of annual values are 5.9 to 13.3 C (Rigg and Miller, 1949); in the northern Gulf of California, Roden and Groves (1959) recorded an annual range of 14.9 to 31.2 C; and in Costa Rica the maximum annual range is 26.1 to 31.7 C (Anon., 1952). Clearly the greatest benthic diversity, and one claimed by Parker (1963) on a regional basis to be among the most diverse known, is associated with the most variable (least stable) temperature regimen. Another influence on diversity could be exercised by environmental heterogeneity (Hutchinson, 1959). Subjectively, it appeared that both the Mukkaw Bay and Costa Rica stations were topographically more distorted than the northern Gulf localities. In any event, no topographic features were evident that could correlate with the pronounced differences in faunal diversity. Finally, Connell and Orias (1964) have developed a model for the organic enrichment of regions that depends to a great extent on the absolute amount of primary production and/or nutrient import, and hence energy flowing through the community web. Unfortunately, no productivity data are available for the two southern communities, and comparisons cannot yet be made.

PREDATION AND DIVERSITY GRADIENTS

To examine predation as a diversity-causing mechanism correlated with latitude, we must know why one environment contains higher order carnivores and why these are absent from others. These negative situations can be laid to three possibilities: (1) that through historical accident no higher carnivores have evolved in the region; (2) that the sample area cannot be occupied due to a particular combination of *local* hostile physiological effects; (3) that the system cannot support carnivores because the rate of energy transfer to a higher level is insufficient to sustain that higher level. The first possibility is unapproachable, the second will not apply on a geographic scale, and thus only the last would seem to have reality. Connell and Orias (1964) have based their hypothesis of the establishment and maintenance of diversity on varying rates of energy transfer, which are determined by various limiting factors and environmental stability. Without disagreeing with their model, two aspects of primary production deserve further consideration. The animal diversity of a given system will probably be higher if the production is apportioned more uniformly throughout the year rather than occurring as a single major bloom, because tendencies towards competitive displacement can be ameliorated by specialization on varying proportions of the resources (MacArthur and Levins, 1964). Both the predictability of production on a sustained annual basis and the causation of resource heterogeneity by predation will facilitate this mechanism. Thus, per production unit, greater stability of production should be correlated with greater diversity, other things being equal.

The realization of this potential, however, depends on more than simply the annual stability of carbon fixation. Rate of production and subsequent transfer to higher levels must also be important. Thus trophic structure of a community depends in part on the physical extent of the area (Darlington, 1957), or, in computer simulation models, on the amount of protoplasm in the system (Garfinkel and Sack, 1964). On the other hand, enriched aquatic environments often are characterized by decreased diversity. Williams (1964) has found that regions of high productivity are dominated by few diatom species. Less productive areas tended to have more species of equivalent rank, and hence a greater diversity. Obviously, the gross amount of energy fixed by itself is incapable of explaining diversity; and extrinsic factors probably are involved.

Given sufficient evolutionary time for increases in faunal complexity to occur, two independent mechanisms should work in a complementary fashion. When predation is capable of preventing resource monopolies, diversity should increase by positive feedback processes until some limit is reached. The argument of Fryer (1965) that predation facilitates speciation is germane here. The upper limit to local diversity, or, in the present context, the maximum number of species in a given subweb, is probably set by the combined stability and rate of primary production, which thus influences the number and variety of non-primary consumers in the subweb. Two aspects of predation must be evaluated before a generalized hypothesis based on predation effects can contribute to an understanding of differences in diversity between *any* comparable regions or faunistic groups. We must know if resource monopolies are actually less frequent in the diverse area than in comparable systems elsewhere, and, if so, why this is so. And we must learn something about the multiplicity of energy pathways in diverse systems, since predation-induced diversity could arise either from the presence of a variety of subwebs of equivalent rank, or from domination by one major one. The predation hypothesis readily predicts the apparent absence of monopolies in tropical (diverse) areas, a situation classically represented as "many species of reduced individual abundance." It also is in accord with the disproportionate increase in the number of carnivorous species that seems to accompany regional increases in animal diversity. In the present case in the two adequately sampled, structurally analagous, subwebs, general membership increases from 13 at Mukkaw Bay to 45 in the Gulf of California, a factor of 3.5, whereas the carnivore species increased from 2 to 11, a factor of 5.5.

SUMMARY

It is suggested that local animal species diversity is related to the number of predators in the system and their efficiency in preventing single species from monopolizing some important, limiting, requisite. In the marine rocky intertidal this requisite usually is space. Where predators capable of preventing monopolies are missing, or are experimentally removed, the systems become less diverse. On a local scale, no relationship between latitude (10° to 49° N.) and diversity was found. On a geographic scale, an increased stability of annual production may lead to an increased capacity for systems to support higher-level carnivores. Hence tropical, or other, ecosystems are more diverse, and are characterized by disproportionately more carnivores.

LITERATURE CITED

Anon. 1952. Surface water temperatures at tide stations. Pacific coast North and South America. Spec. Pub. No. 280: p. 1–59. U. S. Coast and Geodetic Survey.

Bakus, G. J. 1964. The effects of fish-grazing on invertebrate evolution in shallow tropical waters. Allan Hancock Found. Pub. 27: 1–29.

Connell, J. H. 1961a. Effect of competition, predation by *Thais lapillus*, and other factors on natural populations of the barnacle *Balanus balanoides*. Ecol. Monogr. 31: 61–104.

———. 1961b. The influence of interspecific competition and other factors on the distribution of the barnacle *Chthamalus stellatus*. Ecology 42: 710–723.

Connell, J. H., and E. Orias. 1964. The ecological regulation of species diversity. Amer. Natur. 98: 399–414.

Darlington, P. J. 1957. Zoogeography. Wiley, New York.

Feder, H. M. 1959. The food of the starfish, *Pisaster ochraceus*, along the California coast. Ecology 40: 721–724.

Fischer, A. G. 1960. Latitudinal variations in organic diversity. Evolution 14: 64–81.

Fryer, G. 1965. Predation and its effects on migration and speciation in African fishes: a comment. Proc. Zool. Soc. London 144: 301–310.

Garfinkel, D., and R. Sack. 1964. Digital computer simulation of an ecological system, based on a modified mass action law. Ecology 45: 502–507.

Gause, G. F. 1934. The struggle for existence. Williams and Wilkins Co., Baltimore.

Grice, G. D., and A. D. Hart. 1962. The abundance, seasonal occurrence, and distribution of the epizooplankton between New York and Bermuda. Ecol. Monogr. 32: 287–309.

Hall, C. A., Jr. 1964. Shallow-water marine climates and molluscan provinces. Ecology 45: 226–234.

Hedgpeth, J. W. 1957. Marine biogeography. Geol. Soc. Amer. Mem. 67, 1: 359–382.

Hiatt, R. W., and D. W. Strasburg. 1960. Ecological relationships of the fish fauna on coral reefs of the Marshall Islands. Ecol. Monogr. 30: 65–127.

Hoshiai, T. 1960. Synecological study on intertidal communities III. An analysis of interrelation among sedentary organisms on the artificially denuded rock surface. Bull. Marine Biol. Sta. Asamushi. 10: 49–56.

———. 1961. Synecological study on intertidal communities. IV. An ecological investigation on the zonation in Matsushima Bay concerning the so-called covering phenomenon. Bull. Marine Biol. Sta. Asamushi. 10: 203–211.

Hutchinson, G. E. 1959. Homage to Santa Rosalia or why are there so many kinds of animals? Amer. Natur. 93: 145–159.

Klopfer, P. H., and R. H. MacArthur. 1960. Niche size and faunal diversity. Amer. Natur. 94: 293–300.

————. 1961. On the causes of tropical species diversity: niche overlap. Amer. Natur. 95: 223–226.

Lack, D. 1949. The significance of ecological isolation, p. 299–308. In G. L. Jepsen, G. G. Simpson, and E. Mayr [eds.], Genetics, paleontology and evolution. Princeton Univ. Press, Princeton.

MacArthur, R., and R. Levins. 1964. Competition, habitat selection, and character displacement in a patchy environment. Proc. Nat. Acad. Sci. 51: 1207–1210.

MacArthur, R. H., and J. W. MacArthur. 1961. On bird species diversity. Ecology 42: 594–598.

Marcus, E., and E. Marcus. 1962. Studies on Columbellidae. Bol. Fac. Cienc. Letr. Univ. Sao Paulo 261: 335–402.

Margalef, R. 1958. Mode of evolution of species in relation to their place in ecological succession. XVth Int. Congr. Zool. Sect. 10, paper 17.

Paine, R. T. 1963. Trophic relationships of 8 sympatric predatory gastropods. Ecology 44: 63–73.

Paris, O. H. 1960. Some quantitative aspects of predation by muricid snails on mussels in Washington Sound. Veliger 2: 41–47.

Parker, R. H. 1963. Zoogeography and ecology of some macro-invertebrates, particularly mollusca in the Gulf of California and the continental slope off Mexico. Vidensk. Medd. Dansk. Natur. Foren., Copenh. 126: 1–178.

Rigg, G. B., and R. C. Miller. 1949. Intertidal plant and animal zonation in the vicinity of Neah Bay, Washington. Proc. Calif. Acad. Sci. 26: 323–351.

Roden, G. I., and G. W. Groves. 1959. Recent oceanographic investigations in the Gulf of California. J. Marine Res. 18: 10–35.

Simpson, G. G. 1964. Species density of North American recent mammals. Syst. Zool. 13: 57–73.

Slobodkin, L. B. 1961. Growth and regulation of Animal Populations. Holt, Rinehart, and Winston, New York.

————. 1964. Ecological populations of Hydrida. J. Anim. Ecol. 33 (Suppl.): 131–148.

Wells, H. W., M. J. Wells, and I. E. Gray. 1964. Ecology of sponges in Hatteras Harbor, North Carolina. Ecology 45: 752–767.

Williams, L. G. 1964. Possible relationships between plankton-diatom species numbers and water-quality estimates. Ecology 45: 809–823.

7

DIVERSITY OF SHALLOW-WATER GASTROPOD COMMUNITIES
ON TEMPERATE AND TROPICAL BEACHES

Tom M. Spight

Woodward-Clyde Consultants, 3 Embarcadero Center, Suite 700, San Francisco,
California 94111

Tropical communities generally include more species than temperate ones. Presumably, diversities reflect measurable environmental qualities, and diversity-environment relationships based on a few communities will lead to predictions for most communities. Studies to date suggest several major generalizations about diversity-environment relationships. However, most of these generalizations are based on limited work, much of which has been confined to bird communities. The objective of this paper is to determine whether five of these generalizations are applicable to tropical and temperate rocky shore gastropod communities.

GENERALIZATIONS

1. *Latitudinal diversity gradients (species per area) are steeper for large areas than for small ones.* Large tropical regions usually support many more species than do equally large temperate regions. Small tropical areas are also more diverse than small temperate ones; however, the difference is less pronounced than for the large areas. For example, Ecuador has seven times as many breeding land birds as does Vermont, but one finds only 2.5 times as many species in 5 acre (2 hectare) or 6 mile^2 (15.5 km^2) plots (MacArthur 1969).

2. *Adjacent sites in both temperate and tropical communities frequently differ in diversity.* Diversity frequently varies substantially from site to site within a small region. For example, in both Panama and Illinois, some patches have five to six times as many bird species as do other nearby patches (Karr 1971).

3. *Species numbers usually correspond to habitat complexity.* Adjacent habitats often differ greatly in physical qualities (such as soil type, water supply, slope, and topographic uniformity). These physical qualities will affect the suitability of a habitat for any given species. Therefore, one would expect to find significant relationships between diversity and physical qualities, at least when comparing adjacent communities that have similar evolutionary histories and climatic regimes. MacArthur (1965) has asserted that "local variations in the species diversity of small uniform habitats can usually be predicted in terms of the structure and productivity of the habitat" (p. 531). Diversities have been successfully related to structural features for a variety of organisms including

birds (Karr and Roth 1971), lizards (Pianka 1967), gastropods (Kohn 1967), and decapod crustaceans (Abele 1974).

4. *Tropical species are generally distributed more patchily than temperate ones.* When progressively larger areas are considered, species numbers rise relatively more rapidly in tropical communities than in temperate ones (generalization 1). Consequently, adjacent small areas have fewer species in common in the tropics than in the temperate zone. For example, seven times as many bird species are found in Ecuador as in Vermont, but only 2.5 times as many species are found on Ecuadoran plots. Therefore, there are the equivalent of 2.8 times as many unique plot compositions in Ecuador (in MacArthur's [1965] terms, the between-habitat component of diversity changes more than the within-habitat component does).

Where different species are found on adjacent spaces, individual species must also be distributed patchily. Tropical birds are distributed more patchily than temperate ones on a large scale; " many tropical species . . . may be absent at a considerable fraction of the localities offering a suitable habitat for them" (Diamond 1973). Temperate bird species do not show as many examples of range discontinuities (MacArthur 1972).

5. *Patchiness may correspond to more restricted distributions* (tropical species are more specialized than temperate ones). Klopfer (1959) suggested that tropical species make finer habitat discriminations than temperate ones do, and Karr and Roth (1971) provide supporting evidence. MacArthur (1965) has suggested that as competing species become more numerous and species become more specialized, each will tend to be replaced by its competitors in adjacent habitats and distributions will become progressively patchier.

QUALITY OF GENERALIZATIONS

Although each of these generalizations is supported by the available data, plausible alternatives for the last three can be identified: (1) comparable structures may correspond to substantially different diversities in latitudinally separated communities; (2) tropical species may be distributed no more patchily than temperate ones from equally diverse communities; and (3) patchiness may correspond to a lower predictability rather than to greater specialization. To evaluate these alternative hypotheses, I have examined the characteristics of rocky shore prosobranch gastropod assemblages on temperate (Washington State) and tropical (Costa Rica) beaches of the eastern Pacific Ocean.

Snails encounter similar habitats on both tropical and temperate beaches. Major structural habitats (rock surfaces, crevices, tide pools, undersides of stones, etc.) can be found at both temperate and tropical sites. These habitats are not highly modified by the development of complex plant communities; most herbivores obtain their food supplies from the plankton or from micro-algae attached to rock surfaces.

The rocky intertidal zone is also suitable for tropical-temperate comparisons because assemblages of various complexities can be found side by side at all latitudes. Risks of dehydration (Davies 1970) and predation (Vermeij 1972a), as well as quantity of food available (either time available for feeding on plankton [Barnes and Barnes 1968] or amount of surface microalgae [Sutherland 1970]) are all correlated with height on the shore. Stressful environments frequently contain fewer species than benign environments (Sanders 1968), and this is also the case in shore communities. The number of species decreases gradually as one moves from benign lower shore to the stressful upper shore on both hard and soft substrates (Hewatt 1937; Johnson 1970). Tidal excursions are nearly equal at the sites studied (Friday Harbor, Washington, height difference between mean higher high water and mean lower low water, 2.51 m, Puerto Culebra, Costa Rica, difference between mean higher high water and mean lower low water, 2.46 m [U.S. Department of Commerce 1969]). These sloping shores should provide comparable diversity gradients in both temperate and tropical communities. Since most species have fairly narrow vertical ranges on the shore, each height zone can be treated as a separate habitat with its own species pool.

Snails were collected from rocky intertidal surfaces at Playas del Coco, Costa Rica, and on two small reefs, Bird Rock and Turn Rock, on opposite sides of Shaw Island, Washington. The data from these collections were used to answer five specific questions: (1) Do small temperate and tropical areas (individual samples) have more similar species numbers than large temperate and tropical areas (beaches or regions)? (2) Does diversity change with shore height in the same way on tropical and temperate shores? (3) Do particular changes in physical structure correspond to similar diversity changes in both tropical and temperate assemblages? (4) Are tropical species distributed more patchily than temperate ones in equally diverse habitats? and (5) If tropical species are patchier, are their distributions more restricted, or are they patchier because representatives are less likely to be found in a patch of suitable habitat?

RESULTS

The Tropical Beach Assemblage

During low-tide periods 34 quadrat samples, each 0.1–6 m^2 (29 were 1–4 m^2), were taken from the rocky shores around Playas del Coco, Costa Rica (27 from February 8 to March 21, 1970 and seven from February 7 to 14, 1971). The quadrats yielded 6,970 individuals belonging to 79 species. Of these individuals, 78% belonged to 10 species: *Planaxis planicostatus* (1,217 individuals in three samples), *Nerita funiculata* (967 individuals in 15 samples), *Acanthina brevidentata* (640 individuals in 11 samples), *Littorina aspera* (573 individuals in three samples), *Anachis lentiginosa* (493 individuals in 14 samples), *Nerita scabricosta* (447 individuals in five samples), *Anachis costellata* (358 individuals in four samples), *Anachis rugulosa* (268 individuals in three samples), *Thais melones* (236 individuals in 20 samples), and *Anachis nigricans* (232 individuals in seven samples). The average sample included 9.5 species and 158 individuals

TABLE 1

CHARACTERISTICS OF GASTROPOD ASSEMBLAGES FOUND IN COSTA RICA AND
WASHINGTON STATE

Substrate	Samples (No.)	Mean No. of Species	SD	Mean Density per m^2	SD
Costa Rica:[a]					
Rock	11	6.54	2.58	85.36	78.14
Low rock[b]	6	7.83	.98	50.5	57.08
Cobble.............	17	11.53	6.53	181.99	292.33
Low cobble[b]	15	12.67	6.08	119.12	172.94
All samples[c]	31	9.52	5.53	157.74	245.52
Washington:[d]					
Rock	18	5.22	1.80	97.52	110.87
Low rock[b]	15	5.53	1.73	85.93	77.52
Cobble[e].............	6	5.83	2.14	81.00	75.91
All samples[c]	29	5.31	1.75	101.71	97.96

[a] Samples 33, 37, and 38 excluded from all totals (sample 33 is an 0.01 m^2 subsample of sample 35, and samples 37 and 38 are resamples of samples 11 and 8).
[b] Height classes A and B.
[c] Includes tidepool and gravel samples.
[d] Excludes limpets listed in table 2.
[e] All samples are low cobble.

(table 1). The number of species was not correlated with sample size. A complete species list and additional details of sampling procedures are given in Spight (1976a).

The Temperate Beach Assemblage

During low-tide periods 29 samples, 0.25–2.5 m^2, were taken from the rocky shores of Bird Rock and Turn Rock, Washington, July 26–27, 1968. The quadrats yielded 2,487 individuals belonging to 16 species. Of the total number of individuals, 91% belonged to six species: *Thais canaliculata* (805 individuals in 19 samples), *Littorina scutulata* (500 individuals in 20 samples), *Margarites pupillus* (357 individuals in 14 samples), *Thais lamellosa* (339 individuals in 25 samples), *Littorina sitkana* (152 individuals in 18 samples), and *Bittium eschrichtii* (121 individuals in seven samples). The average sample included 5.3 species and 102 individuals (table 1). A complete species list is given in Spight (1976).

Limpets (except one individual of *Acmaea mitra*) were not collected from the Washington quadrats. Since limpets are found on most quadrats and outnumber all other prosobranchs, some correction must be made before temperate data can be compared with tropical data. Dayton (1971) has sampled limpet populations on Turn Rock, and his densities and diversities (table 2) can be added to mine (table 1) to obtain totals for the temperate quadrats. A mean prosobranch density of 400 individuals/m^2 and a mean diversity of nine species per sample are obtained by this technique.

TABLE 2

Distribution and Density of the Limpets *Acmaea scutum, A. pelta, A. digitalis,*
and *A. paradigitalis* on Turn Rock, San Juan Islands, Washington

Height (Feet)	Species (No.)	Density per m²
<1	3	300
1–2	4	312
2–3	4	371
3–4	4	483
4–5	2	109
5–6	4	287
6+	[a]	[a]
Average	...	309

Source.—Dayton 1971.
[a] No observation.

Habitat Structure and Diversity

Diversity varies markedly from site to site on both beaches. As many as 25 and as few as two species were collected from Costa Rican quadrats. In Washington I found between two and nine species, omitting limpets. The limpets are found at all shore levels (table 2), and when they are added, one obtains a range of from four to six to 13 species per sample for the temperate site.

To determine how closely diversity reflects the characteristics of the sites sampled, I compared species counts with substrate types. Four substrate types were recognized: rock (variously dissected rock reefs), tide pools, cobble (quadrats including one or more movable stones, one liter or larger, resting on mud, gravel, or coral debris), and gravel. Most samples (28 tropical, 24 temperate, table 1) were from rock and cobble substrates, and only these two substrates will be considered in detail. Neither of these substrates had consistently more species than the other. From nine to 10 species are found on each substrate in Washington (table 1 plus limpets). The average low-shore cobble sample from Costa Rica contained 12.7 species, while the average rock sample contained only 7.8 species (table 1). However, the cobble samples were not consistently more diverse (range four to 25 species) than the rock samples (seven to nine species), and the means are significantly different only at the $P < .10$ level. Low-shore cobble may be a heterogeneous collection of substrates or a single substrate inhabited by a fauna that is much less predictable than that found on low-shore rock.

To evaluate relationships between shore height and diversity, the quadrats were grouped into six height classes (height above mean lower low water): A (< 0.46 m), B (0.46–0.92 m), C (0.92–1.23 m), D (1.23–1.54 m), E (1.54–1.85 m), and F (> 1.85 m). Low-shore quadrats yielded more species than upper-shore ones did at both sites (fig. 1). To quantify the shore-level effect, species numbers were regressed on shore heights. Overall regressions for both sites are significant ($P < .005$), and variations in height account for 27% (Washington) to 34% (Costa Rica) of the variations in number of species (table 3). Separate

142

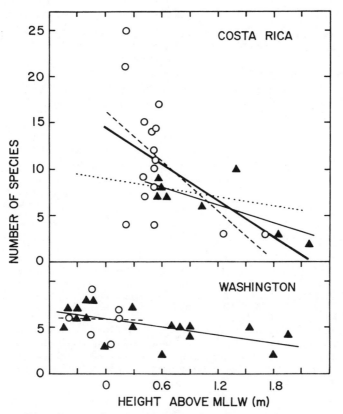

Fig. 1.—Diversity as a function of height above MLLW in Costa Rica (upper panel) and Washington (lower). Circles and dashed line, cobble quadrats; triangles and solid line, rock quadrats; heavy solid line, regression for all data. The rock line for Washington is repeated as a dotted line in the upper panel but has been elevated throughout the height range by three species to adjust for limpets.

TABLE 3

Relationships between Height on the Shore (X, FT) and No. Species in Sample for Samples Taken at Playas del Coco, Costa Rica, and on Bird and Turn Rocks, Washington

Habitat	Relationship	No.	r^2	MSE	$F(1, N-2)$	$\overline{X}$
Costa Rica:						
Rock	$Y = 10.04 - .96X$	8	.53	4.196	6.870	3.67
Cobble...	$Y = 16.12 - 2.69X$	16	.29	31.546	5.747	1.88
Total ..	$Y = 14.31 - 1.92X$	24	.34	23.487	11.549	2.48
Washington:						
Rock	$Y = 5.83 - .42X$	18	.36	2.198	9.074	1.44
Cobble...	$Y = 5.81 - .09X$	6	.00	5.704	.003	−.21
Total ..	$Y = 5.80 - .41X$	24	.27	2.648	8.072	1.03

Note.—Relationships do not include data for 1971 (exact heights were not measured) and data for tidepool and gravel samples.

143

TABLE 4

RELATIONSHIPS BETWEEN HEIGHT AND SUBSTRATE AND THE NO. OF SPECIES OR
INDIVIDUALS PER SAMPLE AND SPECIES POOL FOR GASTROPODS OF
WASHINGTON AND COSTA RICA

Substrate	Height Class	Samples (No.)	Average Species (No.)	Individuals (No.)	Species Pool
Samples from Costa Rica:					
Rock	F	3	3.0	127.0	4
	D	1	8.0	253.0	8
	B,C	7	7.6	44.9	18
Tidepool	D,E	3	9.0	285.7	17
Cobble	E	1	3.0	182.0	3
	D	1	3.0	1,125.0	3
	B	8	11.0	154.4	37
	A	7	14.3	78.6	50
Samples from Washington:					
Rock	F	1	4.0	429.0	4
	E	2	3.5	18.5	5
	B	5	4.2	102.6	6
	A	10	6.1	77.6	14
Tidepool	A	3	5.3	110.0	9
Gravel	A	2	4.5	189.0	6
Cobble	A	6	5.9	81.0	13

NOTE.—A species pool is the total no. of species found in samples from the height and substrate listed (height classes and substrates are defined in the text); totals for Washington do not include limpets, and totals for Costa Rica do not include data from samples 33, 37, and 38 (see table 1).

relationships for rock and cobble quadrats (table 3) are not significantly different at either site (Costa Rica, $F[1,20] = 1.77$; Washington, $F[1,20] = .08$ for test of difference between slopes).

Patchiness and diversity.—Relationships among latitude, diversity, and patchiness were obtained by plotting diversity (Y, the mean number of species per sample) against species pool size (X, the total number of species per habitat). Each height-substrate combination (table 4) represented by only one sample was omitted. Average diversities for the 11 remaining habitat types fall around a straight line:

$Y = 0.231X + 2.997$ (MSE $= 0.696$, $F[1,9] = 163.893$; fig. 2). For pools including from three to 50 species, variations in pool size account for 95% of the variations in number of species per sample ($r^2 = .95$). The data for the temperate site fall amid those for the tropical site (fig. 2).

Distributional patterns.—The distributional pattern of each species was described by assigning each sample to a habitat type and then listing all habitats in which the species was found. Only six habitat types were identified: low cobble (LS, height classes A and B), high cobble (HS), low rock (LR), high rock (HR), tide pool (TP, high at Playas del Coco and low at Friday Harbor), and gravel (GR, low, Friday Harbor only). By this method, one obtains 13 distributional patterns for the tropical species (left-hand entries of table 5),

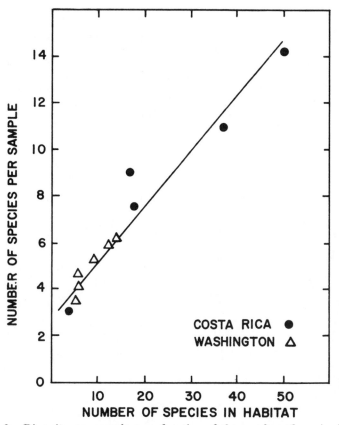

Fɪɢ. 2.—Diversity per sample as a function of the number of species in the species pool (total number of species found in all samples from habitat category).

and these reveal various kinds of habitat restrictions. The average tropical snail was found in 11.7% of the samples (314 occurrences of 79 species × 34 samples = 2,686 possible ones). By recognizing six habitats, the fidelity is greatly increased. The average species was found in 24.2% of the samples from habitats within its distributional range (for 1,300 possible occurrences within the included habitats).

The fidelities were examined to determine whether absence from a habitat type can be regarded as significant. The average tropical species was found in only about one sample of every four from its distributional pattern (e.g., fidelity = 24.2%, above), and therefore a habitat can be confidently ($P < .1$) excluded from the pattern if at least eight samples were taken from that habitat. This criterion is met in three instances: if a species was never found in (1) LS, (2) LR, or (3) in any upper-shore quadrat (HS + HR + TP). Only 17 tropical species were found on both LS and LR quadrats, and therefore all other low-shore species are substrate specialists. Only 19 tropical species were ever found on the upper shore (HS, HR, TP, table 5), and therefore all other species are

TABLE 5

Distributional Patterns of the Species Found at Playas del Coco, Costa Rica, and Friday Harbor, Washington

Habitats in Distributional Pattern	Costa Rican Species (No.)		Washington Species (No.)	
	With Listed Distribution	Found Only Once in Each Habitat	With Listed Distribution	Found Only Once in Each Habitat
LS	42	23	0	0
LR	7	4	1	1
HR	2	0	0	0
TP	2	1	0	0
GR	0	0	1	1
LS/LR	11	1	2	1
LS/TP	2	0	0	0
LR/HR	0	0	1	0
LR/TP	2	0	0	0
HS/TP	2	1	0	0
LS/LR/HR	0	0	1	0
LS/LR/TP	5	1	4	1
LS/LR/GR	0	0	1	0
LS/HS/TP	1	0	0	0
LS/TP/GR	0	0	1	0
LR/HR/TP	1	0	0	0
HS/HR/TP	1	0	0	0
LS/LR/HS/TP	1	0	0	0
LS/LR/HR/TP	0	0	1	0
LS/LR/TP/GR	0	0	1	0
LS/LR/HR/TP/GR	0	0	2	0
Total	79	31	16	4

Note.—Habitats: low cobble (LS), low rock (LR), high cobble (HS), high rock (HR), and gravel (GR). For example, on line 6, 11 Costa Rican species were found in samples from both LS and LR but no other habitats. One of these species was found only once in each of these habitats.

low-shore specialists. Most tropical species are indicated to have restricted distributions, and enough samples were taken from the excluded portions of the shore to establish that these restricted distributions are real.

The temperate shore species also have restricted distributions: the 16 species fall into 11 different patterns (table 5). By recognizing the six habitats, fidelity is increased from 33.0% (153 occurrences in 29 × 16 possible ones) to 44.8% (that is, each temperate species is present in nearly every other sample taken within its distributional range). Since the major habitat restrictions indicated in table 5 (absence from LS, or LR, or HR + TP + GR) are based on absence from at least six samples, they should be real ($P < .02$).

By utilizing all data, one obtains an exaggerated distributional range for each species because most species are found occasionally in habitats where they are never common (incidental occurrences—a habitat would be included if a tropical species had been found in as few as one of 15 samples). Fidelity is greatly decreased when incidental occurrences are frequent. To eliminate incidental occurrences, I constructed new distributional ranges, including a habitat only if the species was found in at least two samples from that habitat.

Only for rare species are the new distributional patterns substantially different from the initial ones. Most single occurrences belong to species that were never found more than once in any habitat type. Only 48 of the tropical occurrences (15.3%) were incidental (last col., table 6) but these included all 35 occurrences for 31 species (the rare species; no habitat in table 6). For the remaining species (the common species), only 4.7% of 279 occurrences (and four distributional patterns) were incidental (table 6). When these are eliminated, there are four fewer distributional patterns, and fidelity increases from 32.9% of 849 possible occurrences to 34.2% of 776 possible occurrences for the 48 common species. Thus, some species are found regularly in one or more habitats, and others are found occasionally in one or more habitats, but few species are found regularly in some habitats and occasionally in others.

For the temperate species, 15 of 153 occurrences (9.8%) were incidental (table 6) including all occurrences for four species and 5.1% of the 138 occurrences for the remaining species. When the incidental occurrences are eliminated, there are two fewer distributional patterns, and fidelity increases from 51.8% of 280 possible occurrences to 54.8% of 252 possible occurrences for the 12 common temperate species.

At both sites, resulting fidelities are high enough (by either criterion for including habitats) to assert that the indicated range restrictions are real. Average fidelity is increased substantially by eliminating incidental occurrences. However, most incidental occurrences belong to rare species, and their elimination does not significantly affect either the fidelities or the distributional patterns of the common species. Therefore, the distributional patterns obtained by either method (tables 5, 6) can be used to test the hypothesis that tropical species are more specialized than temperate ones (see Discussion).

On average, each temperate species was found on about 55% of the sample sites regarded as suitable, while a typical tropical species was found only about 34% as often as expected (totals, table 6). Fidelities would be low if snails of each species normally used only a few of the sites that are suitable for the species. However, fidelities would also be low if the investigator recognized only a few of the habitat distinctions made by the snails. Nearly every temperate species is found in a unique collection of habitats. If each tropical snail also has a unique distribution, then many more habitat parameters would have to be used to characterize tropical patterns than temperate ones. Therefore, the available data are not sufficient to compare fidelities at the two sites rigorously.

Distributions of generalists can be described while using a small number of habitat types, while those of specialists require finer habitat distinctions. If too few habitats have been recognized, then the most widely distributed species at both sites should have higher fidelities than sympatric specialists do. On

TABLE 6

Fidelity of Gastropod Species to Habitats When Only Habitats with Two Occurrences Are Included in a Species' Distribution Pattern

Habitats in Distributional Pattern	Samples (No.)	Species with Pattern (No.)	Observed Occurrences (No.)	Possible Occurrences (No.)	Fidelity (%)	Extraneous Occurrences (No.)
Costa Rica:						
LS	15	23	89	345	25.8	4
LR	11	5	13	55	23.6	3
HS	2	1	2	2	100.0	1
HR	3	3	8	9	88.9	2
TP	3	1	2	3	66.7	0
LS/LR	26	7	74	182	40.6	0
LS/TP	18	2	11	36	30.5	1
LR/TP	14	2	8	28	28.6	1
LS/LR/TP	29	4	59	116	50.9	1
No habitat	34	31	...	...	...	35
Total	34	79	266	776	34.3	48
Washington:						
LR	15	1	3	15	20.0	1
LR/HR	18	1	9	18	50.0	0
LR/LS	21	1	6	21	28.6	2
LR/TP	18	1	4	18	22.2	1
LR/LS/HR	24	3	54	72	75.0	3
LR/LS/GR	23	1	6	23	26.1	0
LS/TP/GR	11	1	7	11	63.6	0
LR/LS/TP	24	2	25	48	52.1	0
LR/LS/TP/GR	26	1	24	26	92.3	1
No habitat	29	4	...	...	...	7
Total	29	16	138	252	54.8	15

NOTE.—For example, on line 6, 15 samples were taken from LS and 11 from LR; seven species were found in at least two samples from each of these habitats. These seven species were found 74 times of a total possible $7 \times 26 = 182$ times, for a fidelity of 40.6%. None of these species was found in any habitat other than LS or LR (last column). The "no habitat" pattern includes species that were found only once in any one of the six habitat types (cols. 3 and 5 of table 5).

148

both beaches, species confined to a single low-shore habitat were less faithful and the low-shore species with the widest distributions were more faithful than most other species (table 6, compare fidelity with number of habitats at left). The species of the tropical high shore are highly faithful specialists and therefore provide an exception to this pattern (table 6, line 3). The habitat classification used is probably adequate for high-shore tropical species, is somewhat more suitable for temperate sites than tropical ones, and is inadequate for tropical low-shore species.

DENSITY

Individuals in some habitats regularly encounter many more snails than do ones in other habitats. Density varies greatly from place to place on both the tropical beach (one to 1,125 snails/m^2) and the temperate one (two to 429 snails/m^2; table 1). Including limpets (table 2), temperate snail densities are much higher than tropical ones (note that the tropical limpet fauna is as diverse as the temperate one, even though none of the six species of acmaeid limpets found at Playas del Coco is as common as any of the mid-level Washington limpets). Density varies more from habitat to habitat within either beach than within one habitat between latitudes (table 4).

In contrast to diversity, density varies markedly with latitude and is poorly correlated with height on the shore (table 3). Density would tend to be negatively correlated with diversity on the rocky shore because many species and few individuals are found on the lower shore and on the upper shore a few species generally form dense populations (*Nerita scabricosta, Littorina aspera, Planaxis planicostatus*). However, densities are poorly correlated with shore level at both sites (table 7). Density varies independently of number of species at both sites (Costa Rica, r^2 = .022; Washington, r^2 = .001). MacArthur et al. (1973) and Karr (1971) also failed to find consistent relationships between density and diversity.

DISCUSSION

Regional and Patch Diversity

Many more prosobranch gastropod species live in shallow tropical waters than in corresponding temperate waters. According to Keen (1971), 1,874 prosobranch species inhabit tropical west America, and according to Rice (1968), 362 prosobranch species can be found on the Washington and British Columbia coasts, or 5.2 tropical prosobranchs per temperate one. This regional tropical-temperate diversity ratio is of the same order as the 7 : 1 regional ratio found for birds (MacArthur 1969).

Individual rocky beach assemblages do not differ as much from each other as do regional faunas. I collected only four times as many snail species on the Costa Rican beach (79) as on the Washington beach (20). Furthermore, collecting tactics inflated the latitudinal difference. The Costa Rican samples were

149

TABLE 7

RELATIONSHIPS BETWEEN HEIGHT ON THE SHORE (X, ft) AND No. OF INDIVIDUALS IN
SAMPLE (Z) FOR SAMPLES TAKEN AT PLAYAS DEL COCO, COSTA RICA,
AND ON BIRD AND TURN ROCKS, WASHINGTON

Habitat	Relationship	No.	r^2	MSE	$F(1, N - 2)$	$\bar{X}$
Costa Rica:						
Rock..........	$Z = 31.82\ X - 41.21$	8	.57	4,025.8	7.803	3.67
Cobble	$Z = 109.00\ X - 12.70$	16	.22	74,318.1	4.004	1.88
Total	$Z = 35.53\ X + 65.41$	24	.06	62,440.5	1.480	2.48
Washington:						
Rock..........	$Z = 14.12\ X + 77.13$	18	.11	11,662.3	1.917	1.44
Cobble	$Z = 94.37\ X + 100.66$	6	.71	2,078.0	9.866	−.21
Total	$Z = 15.18\ X + 77.74$	24	.12	9,536.4	3.061	1.03

NOTE.—Relationships do not include data for 1971 (exact heights were not measured)
and data for tidepool and gravel samples.

deliberately allocated to include as many habitats as possible, and therefore
the species list should be relatively complete. Appropriately, after 3 mo of
daily visits to the beach, I had encountered only 10 species that were not found
in any of the 34 quadrat samples (*Conus dalli, C. chaldeus, Latirus medi-
americanus, L. hemphilli, Vasum caestus, Modulus disculus, Morum tuberculosum,
Hexaplex regius, Homalocantha oxyacantha,* and *Neorapana muricata*). Most
Washington samples were taken along three transects from two islands, and
all were taken within a 2-day period. These tactics should yield a relatively small
species list. Appropriately, additional species were found during other surveys
of these islands (*Haliotis kamtschatkana, Trichotropis cancellata, Fusitriton
oregonensis,* and *Calliostoma annulatum* [Illg 1963]). I also collected 30 species
on a 200 m² area less than a mile from Turn Rock (Shady Cove [Spight 1972]).
Although the Shady Cove collections were made over a 5-yr period, they
encompass a smaller range of habitats than are found on the two islands. When
Oregon and Costa Rican beaches were sampled with one technique, only 2.4
times as many species were found in Costa Rica (Miller [1974] found 63 prosobranch species on eight transects from three Costa Rican sites and 26 prosobranch species on six transects from six Oregon sites). The tropical-temperate
diversity difference for snails on individual rocky beaches is similar to that for
birds on 6-mile² land areas (MacArthur 1969). For both birds and snails, the
latitudinal diversity change is smaller on small areas than in the regions to
which the small areas belong.

On small patches of habitat, tropical diversities are even more similar to
temperate ones. The average tropical sample (to 6 m²) contained only about as
many species (9.5 species) as the average temperate sample (nine species including limpets; table 1). Quadrats are large enough to include the daily
wanderings of most snails (unpublished observations on temperate *Thais*).
Therefore, individual snails on both beaches will encounter about the same
number of different species each day. However, temperate samples contained

150

many more individuals (on average, 400, including limpets) than the tropical ones did (158; table 1).

The overall sample means of table 1 are probably not proportional to the true means for these beaches. Each beach is a complex assemblage of habitats, and both density and diversity vary from habitat to habitat (table 4). Most Costa Rican samples were from low-shore habitats, while species-poor high-shore areas form the greatest portion of the beach. If habitats had been sampled at random, the mean diversity for Costa Rica might have been lower than that for Washington. Miller (1974) actually obtained such a difference; he found on average only about four species per sample (1 m^2) from Costa Rica and about six per sample in Oregon. Since I made no inventory of habitat frequencies, the observed average diversity cannot be adjusted to obtain a true mean for either beach.

Structure and Diversity

If diversity were uniquely determined by habitat structure, then a given structural change would lead to the same diversity change at all latitudes and diversity analysis would be most effective if limited to a single latitude. Diversities of adjacent habitats frequently differ more than those of latitudinally separated habitats. The most diverse temperate habitat studied by Karr (1971) supports 5.9 times as many bird species as the least diverse; differences among habitats at a tropical site are actually somewhat less (5.3 times). In contrast, the average tropical habitat has only 2.5 times as many bird species as does a comparable temperate one (Karr 1971). Critical structural changes among adjacent habitats should also be much easier to isolate than ones between latitudinally separated sites. Long-distance comparison of "similar" temperate and tropical sites must deal with a whole host of latitude-correlated variables (such as temperature regime, climate, productivity, stability, and disturbance). The separate effects of these variables cannot be readily identified, and with such a long list key factors may be overlooked. If Karr's within- and between-habitat changes are typical, and if each structure has the same value at all latitudes, then latitudinal comparisons would be a relatively ineffective means for analyzing relationships between habitat structure and community organization.

Diversity varies from habitat to habitat on both temperate and tropical beaches (table 4, fig. 1). However, species are not distributed in the same way among the habitats on the two beaches. High-shore rock habitats are most diverse in the temperate community (four species in Costa Rica and four species plus four limpets in Washington). Low-shore cobble habitats are much more diverse in the tropical community (50 species) than in the temperate one (13 species). The same number of species (18) can be found on low-shore rock habitats at both beaches (species pool, table 4). If these habitat types are structurally comparable at the two latitudes (as they appear to be), then structure does not uniquely determine diversity, and global structure-diversity relationships will not be forthcoming.

151

The diversity associated with a given structure also changes from site to site for other types of organisms. As in this study, sites picked to be "physically comparable" usually support more species in the tropics (e.g., Karr 1971, although, in contrast to the snail community, parallel increases were observed in the various habitat types). Many more species may be found within one habitat type in some tropical regions than are found within the same habitat type in other tropical regions (Vermeij 1974; Karr 1976).

Factors other than habitat structure have also been shown to influence diversity. Many aspects of community structure cannot be explained without reference to historical or geographic factors which bear no direct relation to the physical habitat of the site. If adjacent patches of a temperate forest community were disturbed at different times in the past, they typically will support different successional stages (Loucks 1970). Each stage is an assemblage of unique composition and structure, and stages usually differ in diversity. All or most stages will occupy each piece of ground (e.g., a single physical habitat) at one time or another. Island biotas are the resultant of prevailing immigration and extinction rates, which are in turn correlated with island size and isolation (MacArthur and Wilson 1967; Simberloff 1974). Although habitat characteristics must influence community composition, attempts to demonstrate their role independent of area are equivocal (Hamilton et al. 1963; Johnson et al 1968; Johnson 1975; Johnson and Simberloff 1974). Similar rate models have accounted for species numbers in several continental habitats (Janzen 1968; Opler 1974; Culver et al. 1974; Vuilleumier 1970), including temperate and tropical areas as wholes (Terborgh 1973). Local stand composition often reflects the pattern of activity of predators (intertidal, Paine 1974; tropical rain forest, Janzen 1970; temperate lakes, Dodson 1974). In summary, most sites can support several different species assemblages. Relative abundances and species lists vary from year to year (Root 1973; Spight 1974). Not only is it impossible to predict the current species list from physical habitat features alone, but often one cannot predict the list accurately when given composition at some previous time. History and relationship to source populations are important habitat characteristics that cannot be measured at the site. Because these characteristics are unique to each site they will obscure habitat structure-diversity relationships common to latitudinally separated communities.

Predation intensity should affect how snails use the different rocky shore habitat types. When predation is intense, hiding places will be at a premium. Cobble areas generally have many hiding places, since stones always have undersides. In contrast, hiding places on rock reefs may be widely separated. Locations of suitable crevices are unpredictable, and crevices cannot be readily found by following simple clues such as slope. One would expect to find more snails on cobble habitats than on rock habitats when predation is more intense.

Predation intensity varies with both latitude and shore level in the rocky shore communities. Snail predators (including fish, crabs, octopods, and other snails) are more numerous and more diverse in Costa Rica than in Washington (personal observations). Most of these predators remain below the water level

152

or are active only when immersed. Consequently, predation is more intense on the lower shore than on the upper shore and more intense in Costa Rica than in Washington.

Faunal distributions correspond to differences in predation intensity. In Washington, snails at all tidal levels remain on the exposed rock surfaces for several days at a time. Most retreat to crevices during periods of long exposure to air or low food abundance but otherwise remain exposed. In Costa Rica, several species form dense populations on mid and upper shore (*Littorina aspera*, *L. modesta*, *Nerita scabricosta*, *Fossarus* sp.), and typical individuals remain on the exposed rock surface throughout the day. Upper-shore cobble areas also do not have substantially more species than upper-shore rock areas (table 4). On the lower shore, where predation is more intense, most individuals (including those of *Nerita funiculata*) retreat to hiding places during the day and most species are associated with cobble.

Adjacent habitats may contain as few as three to four species and as many as 50 in Costa Rica or as few as four and as many as 14 in Washington (table 4). These ranges substantially exceed those found within any one habitat type when one compares different latitudes (excepting low-shore cobble habitats). Judging from this fact alone, an individual investigating correlates of diversity should find studies of adjacent habitats at one latitude to be more productive than comparisons of temperate and tropical communities.

Observations at a single latitude may be deceptive. Snail distributions can be related to structural features of the habitats on both temperate and tropical beaches. However, habitat structure alone cannot account for the number of species one finds at a site, because structure-diversity relationships are different at the two latitudes. Predation intensity determines how useful the various habitat types will be to the snails on these beaches. In this and analogous situations, predictive diversity relationships must be tested in widely spaced communities to demonstrate that variables appropriate for most situations have indeed been identified.

Patchiness and Diversity

Tropical species are frequently distributed more patchily than temperate ones are. However, tropical communities also include more species. If many species are found in a community, but each sample contains only a few, then on average each species will be found in only a few of the samples—that is, it will be distributed patchily. When diversity per sample increases more slowly than diversity per community, patchiness will increase in proportion to diversity. As long as no other factors promote patchiness, patchiness and diversity are merely restatements of the same phenomenon.

In the rocky-shore communities, tropical species are distributed no more patchily than temperate ones when one compares habitats with the same number of species (fig. 2). Species packing (number per sample) corresponds to the size of the species pool and is otherwise the same at both latitudes. In consequence, one cannot explain high diversity by reference to finer habitat selection, because one cannot have many species without also having patchy distributions.

Habitat Structure and Distribution

Most organisms restrict their activities to favorable habitats, and therefore their patchiness can be equated in part with habitat selection. In particular, "intertidal invertebrates react to a wide range of environmental variables which will tend to restrict them to localized areas of the shore" (Meadows and Campbell 1972). The characteristics used to define shore habitats (height on shore and substrate) are significant habitat parameters for intertidal gastropods (Vermeij 1972a, 1972b). If species are distributed nonrandomly on the shore, then species distributions should correspond to these or other discernible environmental features.

Many more species live together on the tropical beach, and each of these species is distributed more patchily, even though the temperate and tropical beaches both have the same physical structure. Tropical species would be patchier if each had a more restricted distribution among the habitats on the beach (e.g., tropical species are more specialized). On the other hand, each tropical species could be both unspecialized and patchy if at any one time individuals use only a few of the available habitat patches. Measurable environmental features often do not account for all aspects of distribution because historical events modify faunas and ecologically similar species have different abilities to take advantage of particular events (Fager 1968; Levin 1974).

Typical tropical species have more restricted distributions than typical temperate ones (fig. 3). Fully 81% of the tropical species were either found in only one habitat or were rare (found in no more than one sample from each habitat), while on the temperate beach only 31% had such restricted distributions (table 5). The patchiness of the tropical snails is in part systematic; the average species is found in fewer of the available beach habitats than a temperate species would be (fig. 3).

Although many tropical species have relatively narrow distributions, some are habitat generalists; 15 species are regularly found in two or three habitats (table 5). Judging from Dayton's data (table 2), the four temperate limpets are found in at least two habitats (e.g., high and low), and therefore there are also 15 habitat generalists on the temperate beach.

There are more snails in Costa Rica because there are many niches with no temperate counterparts (the substrate specialists), rather than because several tropical species fill the role of each temperate one. Equally many habitat generalists are found at both sites, and at each site most of the individuals found on the rock reef belong to these species. Most of the tropical substrate specialists (46 of 64 species) are restricted to a single habitat: low cobble (table 6). No temperate species is confined to this habitat.

Latitudinal changes in the rocky-shore snail communities are similar to those observed for other taxonomic groups. Tropical bird and mammal communities

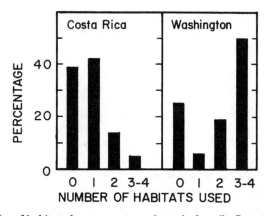

Fɪɢ. 3.—Use of habitats by temperate and tropical snails. In total, six habitats were identified. "No habitat" means that the species was found in only one sample from each of one or more habitat types.

also have a core of species with niches similar to those found in temperate communities. The core species are augmented by many specialists which have no counterparts in temperate areas (Orians 1969; Fleming 1973).

SUMMARY

Prosobranch gastropod assemblages on a tropical (Costa Rica) and a temperate (Washington) rocky shore were compared to evaluate how composition changes when there is a five-fold difference in regional species lists. Typical tropical beach quadrats (1–4 m²) contained no more species than did ones from the temperate beach; therefore, on the whole, tropical species are distributed more patchily than temperate ones. However, some shore habitats are more diverse than others at each latitude and temperate species are distributed as patchily as tropical ones when equally diverse habitats are compared. Locally, diversity varies with habitat structure (shore height and substrate). However, structure-diversity relationships are different at the two latitudes: High-shore habitats support more species in the temperate community, low-shore cobble habitats support many more in the tropical community, and low-shore rock habitats are equally diverse at both latitudes. Individuals are probably distributed among these habitats in response to prevailing predation rates (e.g., biotic interactions determine how useful physical structures are to the snails and consequently how they are used). Most tropical species use fewer habitats than temperate species do; only 18% were found on two or more of six identified habitat types. However, habitat generalists were equally numerous in Costa Rica and in Washington (15 species). Most tropical habitat specialists were confined to a single habitat: low-shore cobble. This tropical snail community is diverse because there are many niches with no temperate counterparts (substrate specialists), rather than because typical temperate niches have been divided more finely. Diversity and density vary more among adjacent habitats on any one beach than among latitudes within a particular habitat type.

ACKNOWLEDGMENTS

I wish to thank S. Jordan, M. MacGinitie, and S. Reichert for assisting in fieldwork in Washington and C. Birkeland, E. M. Birkeland, B. Patten, and R. Spight for supporting various aspects of the field program in Costa Rica. Fieldwork was greatly facilitated by assistance from various members of the Organization for Tropical Studies field office in San Jose, Costa Rica. G. J. Vermeij, J. R. Karr, J. M. Diamond, E. G. Leigh, and A. Miller provided valuable comments on the manuscript and A. Miller provided a copy of his thesis. I gratefully acknowledge support from OTS grant 69-34 and from NSF grant GB 6518 X to the University of Washington for fieldwork and from Woodward-Clyde Consultants for publication.

APPENDIX

1. Both G. Vermeij and A. Miller feel that the upper-shore fauna is under-represented in my Costa Rican samples. Miller collected nine species above 1.5 m in Oregon; I collected eight of these (*Tegula funebralis*, which is not found in the Friday Harbor area, was missing). Miller collected 11 species above 1.5 m in Costa Rica, but I collected only four of these above 1.85 m (level F). Of Miller's other upper-level species, *Planaxis* and *Acanthina* were present above 1.5 m (level E) in my samples but were abundant lower on the shore, *Purpura columellaris* (which I did not find) and *P. pansa* are conspicuous but uncommon at upper levels, and the remaining species (*Scurria mesoleuca*, *S. stipulata*, and *Tegula pellisserpentis*) were encountered at Playas del Coco but only as a few individuals at lower levels.

2. Several reviewers pointed out that birds are not mentioned. In some areas birds consistently eat large quantities of shore snails and appear to affect the snail populations (e.g., Feare 1966, 1970). A. Miller (personal communication) indicates that oyster catchers may have significant effects on Eastern Pacific limpet populations. Bird predation is sporadic on the Eastern Pacific shores I have examined and probably does not affect patterns of habitat use (Spight 1976b). The role of birds on tropical shores should be clarified.

LITERATURE CITED

Abele, L. G. 1974. Species diversity of decapod crustaceans in marine habitats. Ecology 55: 156–161.

Barnes, H., and M. Barnes. 1968. Egg numbers, metabolic efficiency of egg production and fecundity: local and regional variations in a number of common cirripedes. J. Exp. Marine Biol. Ecol. 2:135–153.

Culver, D., J. R. Holsinger, and R. Baroody. 1974. Toward a predictive cave biogeography: the Greenbrier Valley as a case study. Evolution 27:689–695.

Davies, P. S. 1970. Physiological ecology of *Patella* IV. Environmental and limpet body temperatures. J. Marine Biol. Assoc. U.K. 50:1069–1077.

Dayton, P. K. 1971. Competition, disturbance, and community organization: the provision and subsequent utilization of space in a rocky intertidal community. Ecol. Monogr. 41:351–389.

Diamond, J. M. 1973. Distributional ecology of New Guinea birds. Science 179:759–769.

Dodson, S. I. 1974. Zooplankton competition and predation: an experimental test of the size-efficiency hypothesis. Ecology 55:605–613.

Fager, E. W. 1968. The community of invertebrates in decaying oak wood. J. Anim. Ecol. 37:121–142.

Feare, C. J. 1966. The winter feeding of the purple sandpiper. Brit. Birds 59:165–179.

———. 1970. Aspects of the ecology of an exposed shore population of dogwhelks *Nucella lapillus* (L.). Oecologia 5:1–18.

Fleming, T. H. 1973. Numbers of mammal species in North and Central American forest communities. Ecology 54:555–563.

Hamilton, T. H., I. Rubinoff, R. H. Barth, Jr., and G. L. Bush. 1963. Species abundance: natural regulation of insular variation. Science 142:1575–1577.

Hewatt, W. G. 1937. Ecological studies on selected marine intertidal communities of Monterey Bay, California. Amer. Midland Natur. 18:161–206.

Illg, P. L. 1963. Cute and Turn Rock faunal survey. Zoology 533 report. Friday Harbor Laboratories, Friday Harbor, Wash.

Janzen, D. H. 1968. Host plants as islands in evolutionary and contemporary time. Amer. Natur. 102:592–595.

———. 1970. Herbivores and the number of tree species in tropical forests. Amer. Natur. 104:501–529.

Johnson, M. P., L. G. Mason, and P. H. Raven. 1968. Ecological parameters and plant species diversity. Amer. Natur. 102:297–306.

Johnson, M. P., and D. S. Simberloff. 1974. Environmental determinants of island species numbers in the British Isles. J. Biogeogr. 1:149–154.

Johnson, N. K. 1975. Controls of number of bird species on montane islands in the Great Basin. Evolution 29:545–567.

Johnson, R. G. 1970. Variations in diversity within benthic marine communities. Amer. Natur. 104:285–300.

Karr, J. R. 1971. Structure of avian communities in selected Panama and Illinois habitats. Ecol. Mongr. 41:207–233.

———. 1976. Within- and between-habitat avian diversity in African and neotropical lowland habitats. Ecol. Mongr. 46:457–481.

Karr, J. R., and R. R. Roth. 1971. Vegetation structure and avian diversity in several new world areas. Amer. Natur. 105:423–435.

Keen, A. M. 1971. Sea shells of Tropical West America. 2d. ed. Stanford University Press, Stanford, Calif. 1,064 pp.

Klopfer, P. H. 1959. Environmental determinants of faunal diversity. Amer. Natur. 93:337–342.

Kohn, A. J. 1967. Environmental complexity and species diversity in the gastropod genus *Conus* on Indo-West Pacific reef platforms. Amer. Natur. 101:251–259.

Levin, S. A. 1974. Dispersion and population interactions. Amer. Natur. 108:207–228.

Loucks, O. L. 1970. Evolution of diversity, efficiency, and community stability. Amer. Zool. 10:17–25.

MacArthur, R. H. 1965. Patterns of species diversity. Biol. Rev. 40:510–533.

———. 1969. Patterns of communities in the tropics. Biol. J. Linn. Soc. 1:19–30.

———. 1972. Geographical ecology. Harper & Row, New York. 269 pp.

MacArthur, R. H., J. MacArthur, D. MacArthur, and A. MacArthur. 1973. The effect of island area on population densities. Ecology 54:657–658.

MacArthur, R. H., and E. O. Wilson. 1967. The theory of island biogeography. Princeton University Press, Princeton, N.J. 203 pp.

Meadows, P. S., and J. I. Campbell. 1972. Habitat selection by aquatic invertebrates. Advance. Marine Biol. 10:271–382.

Miller, A. C. 1974. A comparison of gastropod species diversity and trophic structure in the rocky intertidal zone of the temperate and tropical West Americas. Ph.D. thesis. University of Oregon.

Opler, P. A. 1974. Oaks as evolutionary islands for leaf-mining insects. Amer. Sci. 62:67–73.

Orians, G. H. 1969. The number of bird species in some tropical forests. Ecology 50:783–801.

Paine, R. T. 1974. Intertidal community structure. Oecologia 15:93–120.

Pianka, E. R. 1967. On lizard species diversity: North American flatland deserts. Ecology 48:333–351.

Rice, T. C. 1968. A checklist of the marine gastropods from the Puget Sound Region: from the mouth of the Columbia River to the northern tip of Vancouver Island. Of Sea and Shore, Port Gamble, Wash. 169 pp.

Root, R. B. 1973. Organization of a plant-arthropod association in simple and diverse habitats: the fauna of Collards (*Brassica oleracea*). Ecol. Monogr. 43:95–124.

Sanders, H. L. 1968. Marine benthic diversity: a comparative study. Amer. Natur. 102: 243–282.

Simberloff, D. S. 1974. Equilibrium theory of island biogeography and ecology. Annu. Rev. Ecol. Syst. 5:161–182.

Spight, T. M. 1972. Patterns of change in adjacent populations of an intertidal snail. *Thais lamellosa*. Ph.D. thesis. University of Washington.

———. 1974. Sizes of populations of a marine snail. Ecology 55:712–729.

———. 1976a. Censuses of rocky shore prosobranchs from Washington and Costa Rica. Veliger 18:309–317.

———. 1976b. Colors and patterns of an intertidal snail, *Thais lamellosa*. Res. Pop. Ecol. 17:176–190.

Sutherland, J. P. 1970. Dynamics of high and low populations of the limpet, *Acmaea scabra* (Gould). Ecol. Monogr. 40:169–188.

Terborgh, J. 1973. On the notion of favorableness in plant ecology. Amer. Natur. 107: 481–501.

U.S. Department of Commerce, Coast and Geodetic Survey. 1969. Tide Tables 1970: West Coast North and South America. Government Printing Office, Washington, D.C. 224 pp.

Vermeij, G. J. 1972a. Intraspecific shore-level size gradients in intertidal molluscs. Ecology 53:693–700.

———. 1972b. Endemism and environment: some shore molluscs of the tropical Atlantic. Amer. Natur. 106:89–101.

———. 1974. Regional variations in tropical intertidal gastropod assemblages. J. Marine Res. 32:343–357.

Vuilleumier, F. 1970. Insular biogeography in continental regions. I. The northern Andes of South America. Amer. Natur. 104:373–388.

8

Copyright ©1978 by Dr. W. Junk B.V., Publishers
Reprinted from *Environ. Biol. Fishes* **3**:49–63 (1978)

THE ROLE OF HERBIVOROUS FISHES AND URCHINS IN CORAL REEF COMMUNITIES*

J. C. Ogden and P. S. Lobel

Keywords:
Algae, Co-evolution, Behavior, Ecology, Communities, Seagrass, Feeding selectivity, Predator-prey, Fish morphology, Herbivore

Synopsis

Herbivorous fishes and invertebrates are conspicious elements of coral reef communities where they predominate both in numbers and biomass. Herbivores and the coral reef algae on which they feed represent a co-evolved system of defense and counter-defense. Algal species have developed toxic, structural, spatial and temporal defense or escape mechanisms, while the herbivores employ strategies that involve anatomical, physiological and behavioral adaptations. Current research demonstrates that many reef fishes are highly selective in the algae they consume. Food selection in these fishes may be correlated with their morphological and digestive capabilities to rupture algal cell walls. Sea urchins select more in accordance with relative abundance, although certain algal species are clearly avoided.

The determinants of community structure on coral reefs have yet to be established but evidence indicates a strong influence by herbivores. Reef herbivores may reduce the abundance of certain competitively superior algae, thus allowing corals and cementing coralline algae to survive. We discuss how the foraging activities of tropical marine herbivores affect the distribution and abundance of algae and how these activities contribute to the development of coral reef structure and the fish assemblages which are intimately associated with reef structure.

Introduction

Coral reefs contain the most diverse assemblages of organisms in the marine world. About 15–25% of the

Received 25.6.1977   Accepted 5.8.1977

* This paper forms part of the proceedings of a mini-symposium convened at Cornell University, Ithaca, N.Y., 18–19 May 1976, entitled 'Patterns of Community Structure in Fishes' (G. S. Helfman, ed.).

fish species diversity and biomass on these reefs consists of herbivorous fishes in the families Scaridae, Acanthuridae, and Pomacentridae, in addition to other families which derive at least part of their energy from benthic plants (Bakus 1964). Adjoining seagrass beds, particularly conspicious in the Caribbean, have small resident species of herbivorous fishes and are heavily exploited by larger fishes from nearby reefs (Ogden & Zieman 1977). The tropical situation contrasts dramatically with temperate waters where herbivorous fishes are either absent or where plants are taken incidental to invertebrate feeding (Bakus 1964). Thus herbivorous fishes are largely confined to tropical regions where they play a major role in the transformation of energy fixed by benthic plants.

Also conspicuous in the tropical system are the herbivorous invertebrates, particularly the regular sea urchins. An extensive literature has developed around the relationships of algae and sea urchins (see Lawrence 1975). There is often difficulty in separating the effects of sea urchins from those of fishes; as the urchins are more easily manipulated, they have been much more intensively studied (e.g. Ogden 1976, Vadas, in press).

Within the coral reefs and seagrass beds of the tropics, the activities of herbivores tend to be concentrated in the top 20 m of water. This is the region of greatest development and highest productivity of benthic plants. Below these depths herbivorous fishes and invertebrates do occur, but are much less prominent.

The following is a brief review of the functional morphology, feeding behavior and ecology, and the influence of herbivores on plant distribution and community structure. Our purpose is to present what evidence is known, while recognizing that these data do not yet explain the whole story. As much as this is a review, we are also describing the direction of our own research.

General characteristics of herbivorous fishes

Much of the diversity among modern acanthopterygian fishes involves adaptations associated with feeding and with living on coral reefs. The teleost radiation began during the Jurassic period but these fishes did not become successful on reefs until acquiring the morphological advances which distinguished the acanthopterygians during the Cretaceous (Romer 1966). These were carnivorous fishes. Hobson (1974) has discussed the evolution of predaceous reef fishes. Herbivorous fishes appear to be one of the most recent trophic innovations attained during the second coral reef epoch beginning in the Eocene. Few taxa are herbivorous, but they often predominate on coral reefs (Randall 1963, Hobson 1974).

Superficially, herbivorous fishes differ from ancestral predaceous forms in the structure of the mouth. Predators generally possess gaping jaws with teeth suited for grasping prey (e.g. barracudas, eels, etc.). More specialized predators are adept at 'inhaling' prey by rapid expansion of the buccal cavity (e.g. anglerfish, groupers, trumpetfish, etc.). They do not necessarily have grasping teeth but may have reduced rasping teeth instead. In contrast, fishes which eat plants have short, blunt snouts with teeth side by side to form a cropping edge. In its most extreme form, represented by the scarids, the teeth have totally fused to form a beak with which these fishes bite into the inorganic substratum and obtain algal mat and endolithic algae.

Four morphological trends characterize the alimentary tract of herbivorous fishes (Lobel in prep.):
1) An elastic stomach and long intestine (e.g. damselfishes, some surgeonfishes).
2) A thick-walled stomach and long intestine (e.g. some surgeonfishes, some angelfishes).
3) A pharyngeal mill and long intestine but no stomach (e.g. scarids).
4) A pharyngeal mill and an elastic stomach with a long intestine (e.g. some cichlids).
The first three types are found among reef fishes. The fourth type, characteristic of cichlids, is restricted to freshwaters and will not be considered here.

Cellulase or comparable enzymes are lacking and microflora are not present in sufficient quantity to break down algal cell walls during the time food is in the gut of a fish (Phillips 1969). Stickney & Shumway (1974) found traces of microfloral cellulase in the guts of some estuarine fishes, but found no correlation between cellulase activity and the food habits of the fish. *Mugil cephalus* was the only herbivore of 17 fishes with cellulase activity. Odum (1970) has shown that flagellates occur in the intestine of *M. cephalus*, but the possible role of microorganisms in digestion has not been evaluated.

Studies conducted thus far are too limited to make generalizations concerning how fish herbivores utilize plant material. The possibility that bacteria and protozoans that can break down cellulose and other polysaccharides are present in the guts of herbivorous fishes has not been adequately explored and should not be dismissed yet.

Fishes may utilize a chemical mechanism to rupture cell walls. Moriarty (1973) and Bowen (1976) have demonstrated that extreme acidity in the stomach of freshwater cichlids (*Tilapia nilotica* and *T. mossambica*)* functions to lyse the cell walls of blue-green algae, detrital bacteria and some green algae. The possible functions of pH and alimentary morphology in herbivorous reef fishes in relationship to the foods they eat is currently under investigation (Lobel in prep.).

Herbivorous fishes are classed as 'browsers' or 'grazers' according to their ingestion of inorganic substratum (Hiatt & Strasburg 1960, Jones 1968). Grazers pick up large quantities of the inorganic substratum while feeding either by rasping or sucking. Browsers bite or tear benthic macro-algae and rarely ingesting any inorganic material. The thick-walled, gizzard-like stomach which *may* be capable of trituration (its function has not yet been demonstrated) and the grinding pharyngeal mill are common to most grazers. Fishes with thin-walled stomachs are generally browsers. Fishes with gizzard-like stomachs have shorter intestinal lengths than species with thinwalled stomachs, suggesting that food is better prepared for digestion in fishes possibly capable of trituration (Jones 1968). The abilities of different fishes to break down plants differ significantly but may be partially compensated for by the length of time the food remains in the gut. Where a fish lacks the necessary mechanics or pH for quick breakdown of plants, it may lengthen the time in which food is in the intestine. Such seems to be the strategy of fishes on a high fat or waxy diet (Patton & Benson 1975).

* = *Sarotherodon*.

In general, herbivorous fishes have a low assimilation efficiency and high ingestion rate with a retention time of a few hours. Maximum nutrient absorption apparently results from a large and continuous ingestion rate and short retention time. Assimilation efficiency is probably high for a few components of the diet but rather low when the entire bulk of ingested material is examined (Odum 1970).

Differences in morphology of the alimentary tract and a diversity of feeding mechanisms may reflect an ability by reef fishes to consume and utilize certain types of algae. These differences may serve as a basis from which to evaluate competitive interactions and the effect of fishes on the distribution and abundance of particular algae.

Feeding behavior and feeding ecology of herbivorous fishes

Herbivorous fishes are conspicuous and active during the day. As twilight progresses, they seek hiding places within the reef, often moving over relatively long distances. At night they are generally inactive.

Table 1. Habitat and foraging strategy of herbivorous fishes (see text for references and details).

Habitat	Foraging strategy	Fishes
Shallow reef	Single and mixed species aggregations	surgeonfishes, parrotfishes, angelfishes
	Territorial	damselfishes, small parrotfishes
	Home range	angelfishes, surgeonfishes
Deep reef	Home range	angelfishes, surgeonfishes
	Single and mixed species aggregations	parrotfishes
Sand	Wide roaming schools and aggregations	mullet, milkfish, surgeonfish
Intertidal at high tide	Single and mixed species aggregations	sea chubs, parrotfishes, surgeonfishes
	Territorial	surgeonfishes, damselfishes
Seagrass	Home range	parrotfishes, surgeonfishes

During the day, herbivorous fishes have essentially three foraging strategies: territorial defense, group foraging and individual home ranges. The strategy adopted appears to relate to habitat type as much as to taxonomic status (Table 1).

Territorial behavior

The most conspicuous holders of feeding territories on a reef are the damselfishes (Pomacentridae). These fishes have been the object of intensive behavioral study, but their ecology and relationships with algae are only just coming to be known.

The territories of pomacentrids are easily defined by a patch of algae often of characteristic color and consistency. The algae in pomacentrid territories are known for only a few cases (Foster 1972, Belk 1974, Brawley & Adey 1977, Ogden, unpubl.); no published studies have as yet confirmed that the algae present in territories are actually consumed by the resident, although we have observed such feeding. Stomach contents of territorial pomacentrids contain some invertebrates, but their presence in the territories and importance of these to the nutrition of the fish has not yet been shown.

Several studies (Vine 1974, Moran & Sale 1977, Ebersole, in press) have shown conclusively that the pomacentrid territory is primarily a feeding territory and secondarily a nest site. The activities of *Eupomacentrus planifrons* in the Caribbean can lead to the destruction of corals and to the subsequent formation of new territory space (Kaufman 1977). Settling plates placed within territories rapidly develop algal growths more characteristic of territories than outlying areas (Vine 1974) and there is some suggestion that the fish may take an active role in 'weeding' its territory (Foster 1972, Ogden, unpubl.). Removal of a damselfish will lead to territory expansion by contiguous territory holders, or to rapid reduction of the algal biomass within the territory by scarid and acanthurid foraging groups. The presence of high densities of territorial pomacentrids in certain reef areas may consequently affect reef productivity, coral growth, nitrogen fixation, and reef cementation (Vine 1974, Kaufman 1977, Brawley & Adey 1977).

Territoriality will be advantageous if the actions of a fish increase the yield of the guarded resource. We therefore do not expect nor do we find territorial herbivores in deep water (> 30m) where such defense would not significantly increase the standing crop of algae in the territory (Vine 1974). Standing crop and

161

territory size are inversely correlated (Syrop 1974). Since the rate of growth of algae decreases with depth, we would expect larger territories in deeper water. Vine (1974) found a sharp decline in algal growth on protected plates at depths greater than 12 m. At 20 m there was no significant difference between standing crops of algae on protected vs. exposed plates. This matches the lower distributional limit of territorial, herbivorous pomacentrids, but not of other herbivores such as pomacanthids in Hawaii (Gosline 1965). Standing crop on exposed vs. protected plates may give a measure of algal replacement rates and implies that territory size is not only dependent on the standing crop but also on the replacement or growth rate of the algae.

The minimum size of an herbivore's territory should approximate the bioenergetic requirements of the resident divided by the maximum sustainable yield of the territory minus that which is predictably lost to intruders. Fluctuations in territory size may reflect changes in benthic productivity, degree of intrusion, and metabolic requirements of the resident at various reproductive and growth stages. The maximum size of a territory approaches the limit of area that a resident can economically defend.

While pomacentrids are the most conspicuous reef-associated territory holders, scarids are also known to hold feeding territories. Ogden & Buckman (1973) and Buckman & Ogden (1973) showed that *Scarus croicensis* will hold territories that appear to serve dual functions of feeding and reproduction. These territories are defended against conspecifics and other benthic feeding fishes (Robertson et al. 1976). The food supply in this species consisted of filamentous algae, diatoms and flocculent material that covered the reef surface. The sites in Panama were close to large rivers which periodically brought in abundant sediments. Interestingly, in St. Croix, U.S. Virgin Islands, *S. croicensis* is less common and does not hold territories. This may be due to the more dispersed nature of the available food (Ogden, unpubl.).

Foraging groups

Foraging groups of variable size are characteristic of the families Scardiae and Acanthuridae in tropical waters. These groups may be homotypic or heterotypic. Where heterotypic the groups often consist of several herbivorous species as well as a few carnivores which presumably feed upon invertebrates and small fishes disturbed by the foraging group (Eibl-Eibes-feldt 1965, Hobson 1974, Ogden & Buckman 1973,

Robertson et al. 1976, Alevizon 1976). The groups probably serve a dual function. Aggregations provide some protection from predators while they also serve to increase individual feeding efficiency, especially in areas where territorial fishes control much of the surface available for feeding (Barlow 1974, Robertson et al. 1976, Morse 1977).

In the Pacific, aggregating *Acanthurus* spp. are the most devastating invaders of algal territories. In particular, *Acanthurus triostegus* forms the nucleus of many mixed species aggregations. Other herbivorous species join the aggregation and feed along with *A. triostegus*. Benthic foraging predators, such as labrids and aulostomids, also accompany the aggregation (Hobson 1974). The feeding aggregation swarms over the reef and descends onto lush algal patches, which are usually defended by pomacentrids. At first the territorial pomacentrid will vigorously attack the marauders, but after a few moments the resident waits listlessly until they are gone. The number of attacks by the resident fish per fish in an aggregation are not enough to drive the aggregation away (Barlow 1974). Other acanthurids are more solitary and graze over the reef in general.

A. triostegus is also unique in that it does not possess a muscular stomach as do many other acanthurids (Randall 1961, Jones 1968). Its alimentary morphology appears functionally similar to that of the herbivorous pomacentrids. Unlike the highly aggressive pomacentrids, *A. triostegus* is passive, a necessary prerequisite for group cohesion.

There is a possible correlation between the occurrence of aggregations and the density of territorial herbivores (Barlow 1974). The herbivorous *Acanthurus nigrofuscus* defends territories much like a pomacentrid. *A. triostegus* will form aggregations where *A. nigrofuscus* is abundant but is solitary where *A. nigrofuscus* is rare. Too much crucial evidence is missing, however, to identify the relationship as causal or merely incidental to some other factor such as food availability, threat from predation, or time of day of observation. Another territorial herbivore is *Acanthurus lineatus* (Nursall 1974). These fish are found just below the intertidal zone. They are most territorial during high tide and become less defensive with a change in habitat due to low tide. During high tide, *A. lineatus* are especially aggressive toward transient scarids which enter the reef flat in aggregations to graze at that time (Nursall 1974). Other tidal zone fishes display similar behavioral responses to changing water levels (Gibson 1969, Nursall 1974).

Heterotypic schools have been described as traveling together but separating into smaller single species groups while actually feeding. The smaller groups reunite when moving to a new location. A further breakdown by size was apparent with small *Scarus croicensis* separating from larger individuals of the same species while feeding, but joining together with larger *S. croicensis* for 'travel' (Earle 1972 and pers. comm.).

Depletion of food in one particular locality is probably not the reason for movement by an herbivore aggregation. The first few fishes may leave a location for a variety of reasons and other fish may respond to these departing movements and follow. Thus a mechanism involved in group cohesion may prevent an algal patch from being overgrazed past its regenerative potential. At least enough resource is expected to remain to provide the essential nutrition for the territory resident while the territory recovers. A similar phenomenon has been noted for bird flocks (Moynihan 1962, Morse 1970, 1977).

Home range

Many reef fishes are known to limit their movements to a specific reef area in which their feeding is concentrated (Reese 1973, 1977, Sale 1977). Although these fishes may defend their 'home' against potential food competitors, as is the case for some chaetodontids (Reese 1975), others do not, such as the pomacanthid *Centropyge* (Lobel ms.). An animal with a feeding home range may be territorial over some other limited resource such as a shelter or nesting site. Concerning food, a home range seems to differ from a feeding territory in at least three ways. The home range does not increase the productivity of the food resource and stronger defense would not likely do so either. Home ranges are not defended against all potential food competitors; the species attacked are usually of close taxonomic relationship and possibly competitors for shelter or spawning sites. Finally, home range behavior develops where food is not the limiting factor for the species population or where food resources are not widely utilized by other species.

Selectivity in feeding by herbivorous fishes

Major studies have been conducted on the trophic habits of tropical fish communities in the Pacific (Hiatt & Strasburg 1960, Hobson 1974) and the Caribbean (Randall 1967, Carr & Adams 1973). These have included examination of herbivorous fishes. Stomach-content data alone provide only limited information and have led to the notion that most herbivorous fishes are generalist feeders. Detailed field studies involving both stomach contents and availability are lacking; there have been few complete studies of feeding preferences in herbivorous fishes (Table 2).

The common Pacific surgeonfish, *Acanthurus triostegus,* appears to prefer the algae which it can most readily utilize. *A. triostegus* is able to digest the proteins of *Enteromorpha* and *Polysiphonia* and some carbohydrates of *Enteromorpha*, but is unable to utilize *Sargassum* (Pfeffer 1963). It is possible that the restricted release of reducing groups by *Sargassum* is due to its storage of polysaccharides as polymers of mannuronic acid rather than as glucose (Blinks 1951, Pfeffer 1963). Bile from the gall bladder is able to kill *Enteromorpha, Polysiphonia,* and *Sargassum*, but not the blue-green alga *Phormidium*. There was no evidence that the cell walls were digested, although the

Table 2. Plants preferred by fishes during choice experiments.

Family	Species	Habitat	Preferred plant (genus)	Total No. plants Tested	Authority
Acanthuridae	*Acanthurus triostegus*	Pacific reefs	*Polysiphonia & Enteromorpha*	29	Randall 1961
Scaridae	*Sparisoma radians*	Caribbean seagrass beds	*Thalassia*	9	Lobel & Ogden (in prep.)
Siganidae	*Siganus spinus*	Pacific reefs	*Enteromorpha*	62	Bryan 1975
	Siganus rostratus	Pacific reefs	*Enteromorpha*	56	Tsuda & Bryan 1973
Pomacanthidae	*Pomacanthus arcuatus*	Caribbean reefs	*Codium*	11	Earle 1972

163

cytoplasmic structure was disrupted and the cell contents clumped into a central mass (Pfeffer 1963). Randall (1961) demonstrated that blue-greens and a red alga, *Asparagopsis* (known for its ketone content) were not eaten by *A. triostegus*. The brown algae were only sparingly consumed. The preference for algal species seems to correlate with the ability of a fish to kill and digest the algae.

Although the evidence is weak, it appears that not all preferred algae are of equal value and if given an opportunity, a fish will feed on an alga in proportion to its food value. *A. triostegus* died after two months when fed only *Enteromorpha* sp., while fish fed only *Polysiphonia* sp. lived four months until sacrified (Pfeffer 1963). Pfeffer determined that *Polysiphonia* contained more protein than *Enteromorpha*. Unfortunately, it is not known whether three times the quantity of *Enteromorpha* will maintain fish as well as one unit of *Polysiphonia*. *Polysiphonia* appears to be a more important energy source and is eaten in greater amounts than *Enteromorpha* when both are offered in copious amounts (Tandall 1961). It may be significant that *Polysiphonia* is usually encrusted with epiphytes which may add nutrition while *Enteromorpha* is not (Pfeffer 1963). Both *Polysiphonia* and *Enteromorpha* are preferred food by *A. triostegus* over other algae when presented a choice in an aquarium (Randall 1961).

Not all algae consumed by fishes may be intentionally eaten. The presence of some organisms in the guts of fishes may be simply incidental. Study of *A. triostegus* in the Line Islands revealed that 40 species of algae were eaten, exceeding the number previously known from that region (Dawson, Alleem & Halstead 1955). Of all the algae the five most abundant in the stomach contents were *Pterocladia* sp., *Sphacelaria furcigera*, *Lyngbya majuscula* (and/or *L. aestuaria*), *Bryopsis pennata* and *Lophosiphonia* sp.. After reviewing this study and from his own work Randall (1961) suggested that many algae were consumed by the fish incidentally according to their abundance on the reefs. At least *Lyngbya* spp. were avoided by *A. triostegus* in aquaria. *Lyngbya* and *Asparagopsis* (also not eaten; see above) are two out of three algae which commonly reach a height of 5 cm or more in Hawaii (Randall 1961). The third species is a red alga, *Plocamium sandvicense*, which has not been tested for palatability to fish. It appears as if some of the material ingested by fish are types which would be avoided if possible. In Hawaii, Jones (1968) found only 40 out of the 160-odd algal genera available to browsing acanthurids in fish's stomachs. In the individual algal divi-

sions, 6 of 16 available browns, 9 of 27 greens, 15 of 97 reds and 2 of 24 blue-greens were eaten.

Siganids are the only other marine fish herbivores whose food habits, preferences and assimilation efficiencies are known, principally from the studies of Tsuda & Bryan (1973) and Bryan (1975). Their work was prompted by the yearly occurrence of nearly 13 million juvenile siganids appearing on the reef flats of Guam. Soon after the population explosion, the reef flats were defoliated, especially of preferred species of algae. Dead individuals of *Siganus spinus* were collected, 74% of which contained no food in their guts. The rest contained only a small amount of filamentous and non-calcareous fleshy algae. The initial swarm, however, consisted of two siganid species, *Siganus spinus* and *S. rostratus* = (*S. argenteus*). Tsuda & Bryan (1973) speculated that the larger *S. rostratus* was able to outcompete *S. spinus* for food. They qualitatively state that *S. rostratus* is more aggressive; however, no empirical evidence is presented that interference competition is occurring. It may be that *S. rostratus* is actually more efficient at assimilating the same and/or more species of algae and consequently attains larger size.

In preference trials, the siganids chose mostly the same algae. Both fishes chose the algae (1) *Enteromorpha* (2) *Feldmannia* and *Derbesia*, and (3) *Cladophoropsis membranacea* (Tsuda & Bryan 1973). Differences between the fishes appeared in the avoidance of a different alga by each. *Siganus spinus* did not eat *Clorodesmis fastigiata* while *Siganus rostrata* did. *Siganus rostrata* did not eat *Polysiphonia* while *Siganus spinus* did (Tsuda & Bryan 1973, Bryan 1975). Only 12 of 56 algal genera offered were always completely eaten by both siganids.

Tsuda & Bryan (1973) noticed that the algae, *Enteromorpha*, *Caulerpa*, *Boodlea*, and *Cladophoropsis* were the first to vanish from the reef flats with the appearance of the siganid swarms. Thus, it appears that algae which were preferentially chosen by fishes in the laboratory were also heavily grazed upon in the field. Both siganid species fed preferentially on filamentous algae but generally did not eat blue-greens or calcareous algae. This agrees with Randall's (1961) observation that blue-greens are among the most abundant algae on reefs and are generally not eaten by fishes.

The preference for a particular alga is not necessarily expressed in the composition of food items found in the guts (Bryan 1975). However, this may be a reflection of the differential digestibility of the items, resulting in the identifiability of the contents.

Thus, what is often identified as a bulk item in the stomach contents may not really indicate what is nutritionally important for growth and reproduction. For example, *Enteromorpha* is the most preferred food of *Siganus spinus* yet it has an intermediate 'importance value' based upon its presence in gut contents (Bryan 1975). Preferred foods are assumed to be so because they are most important to the consumer in bioenergetic terms. The general diet may just reflect what is available in the habitat where the fish is grazing. The test of the assumption that the most preferred food item would also yield the most nutritional value would be to actually maintain consumers on specific diet combinations.

Selectivity in feeding by sea urchins

The regular sea urchins are the other major group of marine herbivores whose diet and feeding ecology are reasonably well known. In contrast to the herbivorous fishes, the sea urchins are nocturnally active. Their specializations for dealing with plants do not involve the modifications in the morphology of the gut that are present in fishes. The complex jaw apparatus, the Aristotle's lantern, is connected to a short esophagus and then to a thin-walled gut. The jaw scrapes the substrate or bites pieces of plants and a food pellet is formed in the esophagus. Sea urchins are known to have a gut flora which enables the digestion of various sugars, starches, and agar, but they apparently have a very limited ability to digest cellulose (Lawrence 1975, Prim & Lawrence 1975). A caecum which may serve the function of culture of microflora is known in the gut of *Diadema* (Lewis 1964, Lawrence 1975).

The feeding habits and ecology of sea urchins of the Caribbean are summarized in Table 3. Sea urchins appear to be differentiated primarily upon the basis of habitat specialization and much less on the basis of food specialization. *Diadema antillarum* is a reef-dwelling urchin which has the capability to eat whole plants or to subsist upon mat and boring algae in highly overgrazed situations. It is quite mobile and appears in striking numbers in reef areas especially by night (Lewis 1964, Ogden et al. 1973). *Echinometra lucunter* lives in holes or burrows, rarely moving more than a few centimeters at night. A considerable portion of its diet consists of algae captured raptorially from the drift. The spatial separation necessary for this type of feeding is maintained by aggressive behavior (Grünbaum et al. in press). It may overlap in distribution with *Diadema antillarum* in some cases, but is generally confined to wave-washed terraces and reef tops (Abbott et al. 1974). *Tripneustes ventricosus* and *Lytechinus variegatus* are commonly associated with seagrass beds. *Tripneustes* feeds exclusively as a browser on seagrasses and associated macroalgae (Keller 1976, Ogden & Abbott unpubl.). *Lytechinus* is a more generalized feeder and can eat *Thalassia* and macroalgae as well as sand and detritus (Keller 1976).

Feeding preferences have been shown in some of these urchins. *Diadema antillarum* fed upon many algae species in approximate relationship to their abundance on reefs, but some species were avoided, no-

Table 3.   Habitat, foraging strategy and preferred food of common Caribbean Sea urchins.

Urchin	Habitat	Foraging strategy	Preferred plants* (genera)	Authority
Diadema antillarum	Reef to 10 m; occasionally found in aggregations in sandy areas	Mobile; generalist; capable of browsing, grazing hard carbonate surfaces, and sand feeding	*Herposiphonia, Jania, Thalassia*	Ogden, Abbott, & Abbott 1973; Ogden 1976
Echinometra lucunter	Reef to 10 m; most abundant in burrows in wave-washed reef and beachrock	Sedentary; movement restricted to burrow mouth; feeds on drift and attached plants	*Dictyota, Jania, Thalassia* (drift)	Abbott, Ogden & Abbott 1974
Lytechinus variegatus	Seagrass beds and sandy areas	Mobile; generalist; capable of browsing and sand feeding	– – –	– – –
Tripneustes ventricosus	Seagrass beds to 10 m; occasionally on reefs with heavy growth of macroscopic algae	Mobile; diet restricted to seagrasses and macro-algae	*Thalassia, Syringodium*	Abbott & Ogden (unpubl.)

*   Preferences based on percentage of algal species in stomach contents per percentage occurrence of the plant in the habitat.

tably *Penicillis* sp., *Sargassum* spp., *Turbinaria turbinata*, and *Laurencia obtusa* (Ogden et al. 1973, Ogden 1976). When preliminary field preference tests were performed an avoidance of *Laurencia, Sargassum* and *Halimeda* was noted (Ogden 1976). The factors involved in this selection are not known, but all of the avoided plants grow abundantly in the presence of urchins. *Penicillis* and *Halimeda* are heavily calcified, *Sargassum* is fleshy and stiff and contains tannins (Ogino 1962), and *Laurencia* is fleshy. Abbott & Ogden (unpubl.) analyzed the diet of *Echinometra lucunter* feeding on attached plants and plants captured from the drift. The urchin showed strong avoidance of *Dictyopteris, Sargassum*, and *Chondria*.

Vadas (1968, in press) has shown strong feeding preferences in the temperate sea urchins *Strongylocentrotus drobachiensis* and *S. franciscanus*. Both species prefer the brown alga *Nereocystis luetkeana* and appear to avoid *Agarum fimbriatum*. Feeding preferences did not correlate with caloric content but did correlate strongly with absorption efficiencies, the preferred species being absorbed with greater efficiency by the urchins. Growth and reproduction (gonad index) correlated with the preference rankings. The most preferred species gave the greatest growth rates and gonad indices. Vadas concludes that the urchin preferences lead to an optimization of fitness as measured by growth and reproduction.

Herbivore-plant relationships

Plant distributions

There is abundant evidence that plant distributions in coral reef areas are influenced by grazing and browsing fishes and invertebrates. Reef surfaces usually appear bare and often the tooth marks of fishes and invertebrates can be seen. On closer examination the surface is covered with a thin mat of filamentous algae and the carbonate surface may be extensively pitted with boring forms. A number of studies have used herbivore exclusion cages and in almost every case rapid growth of benthic algae within the cage occurs (Stephenson & Searles 1960, Randall 1961, Earle 1972, Mathiesson et al. 1975). Heavy growths of benthic algae occur where fishes have difficulty feeding, such as the tops of wave washed reefs or intertidal regions or in areas of the reef devoid of shelter. Seagrass beds where they contact reefs in the Caribbean are heavily grazed up to 10 meters from the reef and a characteristic 'halo zone' may develop (Randall

1965, Ogden, Brown & Salesy 1973). Exploitation of beds further from reefs is limited to small resident species capable of hiding in the seagrasses. Large predators such as barracuda and mackerel may be responsible for this pattern by restricting the movements of small fishes (Earle 1972, Ogden & Zieman 1977).

Adey et al. (1976) have described 'fleshy algal pavements' on the Atlantic sides of high islands such as Martinique in the Lesser Antilles in the Caribbean. Dense masses of fleshy algae (up to 4 kg wet wt. m^{-2}), including *Sargassum, Gracilaria* and many other species, are present from near the surface to 10 to 20 m deep. The algae grow on a pavement of crustose coralline algae covering an *Acropora* framework and may be near a normal live *Acropora palmata* reef with a dense population of grazers. The development of the pavements is open to speculation. A local disturbance such as a storm may have destroyed a section of the original *Acropora* reef. This would destroy habitats utilized by reef grazers, particularly the sea urchin *Diadema* and also herbivorous fishes. The elevated nutrient levels present in the area may have allowed relatively rapid and sustained growth of algae; the grazers could have been prevented from recolonizing by the heavy surge and deep wave base characteristic of the Atlantic coasts of the Antilles. The new equilibrium state attained is quite different from the original reef, and is structurally more similar to a seagrass bed than a coral reef. The fishes occupying the area tend to be those characteristic of seagrass beds (Adey et al. 1976). Fleshy algal pavements are exceptional developments, but they serve as an indication of the balance between herbivory and algal growth characteristic of tropical marine regions.

Community structure

Beginning with the studies of Paine (1966, 1969) and extended by Vadas (1968, 1977), Dayton (1972) and Menge (in press), consumers have been shown to exert a strong effect upon the structure of marine ecosystems. Invertebrate herbivores present in intermediate densities in intertidal and subtidal areas in the temperate zone consume the fastest growing and most abundant algal species which would otherwise out-compete many slower-growing species. The structure of the system, measured by species diversity, is enhanced by such predation (Paine & Vadas 1969, Menge 1977, Vadas 1977).

Ogden, Brown & Salesky (1973) removed over

7,000 *Diadema* from a small patch reef in the Virgin Islands. Dramatic changes occurred very rapidly. Within six months a dense cover of algae obscured the reef surface and even overgrew and killed some of the smaller coral colonies. The halo zone disappeared after six to eight months. Sammarco et al. (1974) compared the same cleared patch reef with nearby 'control' reefs and found not only increased biomass, but also dramatic shifts in dominance and equitability in the algae on the cleared reef. Ogden (unpubl.) has followed this reef for several years subsequent to clearing. The reef has not been recolonized by *Diadema*. The large seagrass-dwelling *Tripneustes ventricosus* has moved from surrounding seagrass beds and taken up residence within the leafy algae on the reef surface. Analogous to the fleshy algal pavements, the reef has become much like a seagrass bed in terms of the response of organisms to its physical structure. *Diadema* is apparently excluded because of its inability to hold itself in position in surge when the substrate is covered with plants. *Tripneustes*, with shorter spines, can easily hold onto the plants with its tube feet and become firmly anchored to the substrate.

Tsuda & Kami (1973) found that selective browsing by herbivorous fishes favors the dominance of blue-green algae. These algae are the persisting forms which are continuously available as food for animals. Apparently few fishes, however, are known to consume these. Within pomacentrid territories, selected species of algae are maintained which are attractive to the majority of herbivorous fishes. Thus it is advantageous for fishes such as acanthurids, siganids and scarids to develop behavioral strategies such as feeding aggregations aimed at overwhelming aggressive pomacentrids. Where reef surfaces are exposed to grazing, the blue-greens composing the turf (rarely exceeding 2 mm in height) and certain macroalgae as *Jania, Halimeda, Laurencia* and *Caulerpa* (exceeding 10 mm in height) persist (Earle 1972, Dahl 1973, Tsuda & Kami 1973, Belk 1975, Ogden 1976). These species are the types that should be surveyed for possible defensive strategies.

Herbivores have been shown to enhance the productivity of plants in terrestrial and planktonic systems (Mattson & Adea 1975, Dyer & Bokhan 1976, Porter 1976). This effect, occurring in a variety of ways, is the consequence of a long history of co-evolution of herbivore and plant. Herbivory may also enhance nitrogen cycling in plants and positively affect plant production (Mattson & Adey 1977, Owen & Wiegert 1976). The presence in damselfish territories of certain highly productive species of algae may have an overall effect on reef productivity (Brawley & Adey 1977). Parrotfishes feeding in seagrasses selectively remove the older, more heavily epiphytized leaves, possibly opening up areas for new growth (Ogden, Lobel & Clavijo, unpubl.). Acanthurids feeding selectively on plant epiphytes may enhance individual plant productivity. The heavy grazing characteristic of open reef surfaces leads to the development of a very productive algal mat and holds the plant community in a high-turnover, early successional state.

The action of grazing fishes and invertebrates also enhances the growth and settlement of corals (Stephenson & Searles 1960, Sammarco 1974, Birkeland 1977, Kaufman 1977). Grazing opens up solid substrates upon which corals settle and prevents overgrowth by algae until the colonies escape in size. Birkeland (1977) showed that fishes will graze around newly settled corals without touching or damaging the colony. Grazing may also promote the growth of coralline algae by keeping competitively superior fleshy algae from overgrowing the corallines.

There has been much confusion about whether herbivorous fishes actually eat living corals. Randall (1974) concludes corals are not intentionally consumed by scarids. It is possible that some of the larger scarids are able to bite into living corals because of their stronger jaws, however, corals have not been identified as a main item in the diet (Hiatt & Strasburg 1960, Randall 1967, Hobson 1974). Scarid grazing on carbonate substrates may be a significant factor in their erosion (Bardach 1961, Glynn 1973, Ogden 1977).

Plant-herbivore co-evolution

The diversity of tropical marine plants and the abundance of herbivores has undoubtedly provided a rich background for the evolutionary interplay of plant defenses and herbivore foraging strategies.

According to the scheme we have presented on the morphology of herbivorous fishes, certain algae appear more suitable as food than others. Fishes with only an elestic stomach may be dependent primarily upon their bite as a means of releasing algal cytoplasm. These fishes are expected to select those macro-algae with relatively large cell size. Fishes with a stomach capable of trituration are expected to be able to utilize a different and possibly wider range of algae. This would include smaller cell sizes that could

Table 4. Potential benthic plant defense mechanisms against fishes and urchins (after Vadas, in press).

Defense	Example(s)
Escape in time (highly productive, short-lived)	Small filamentous forms (e.g. *Polysiphonia, Herposiphonia*)
Escape in space (wave-washed areas, cracks and crevices, from herbivore habitat)	Intertidal algal turfs, cryptic reef species, seagrass beds and associated macroalgae
Structural defense (hard, fleshy or stiff, calcareous, lowered food quality, small cell size)	*Turbinaria turbinata, Penicillis, Halimeda*, crustose coral-lines
Chemical defense (unpalatable to toxic)	*Sargassum* (tannins), *Asparagopsis* (ketones), *Caulerpa* (caulerpicin, caulerpin), *Laurencia* (halogenated compounds), *Dictyopteris* (dictyopterene), blue-greens (?)

be ground with ingested sand particles in the stomach. Finally, fishes with a pharyngeal mill capable of complete grinding of all foods should have the greatest range of suitable food items. Thus, large-celled macroscopic algae appear most susceptible to all these herbivores, while small-celled algae are possibly least utilized by fishes. Other potential plant defenses (Table 4) include toxicity or chemical defense, cryptic or inaccessible growth, toughness, and lowered caloric value or decreased food quality due to calcification (Paine & Vadas 1969, Ogden 1976, Lobel in prep.).

Caloric value of algal foods has been examined in the temperate zone by Paine & Vadas (1969) and in the Caribbean (Ogden unpubl.). In neither case was the energy content of the food directly correlated with the stomach contents or preferences based on laboratory tests or field availability. Caloric value may not be the most appropriate measure of energy as some of the carbohydrates are bound in structures (e.g. cellulose) which are not equally accessible to all consumers. Some algae may have lowered their food value or desirability by developing a highly silicified or calcified structure, such as *Penicillis* spp. and *Halimeda* spp..

Chemical defenses may exist in benthic marine algae. Many such secondary metabolites found in terrestrial plants are known or presumed to be defensive in nature (Whittaker & Feeny 1971, Freeland & Janzen 1974, Cates & Orians 1975, Levin 1976). Freshwater algae, especially the blue-greens, are also known to be toxic (see Porter 1977). While few marine algae

are known to be toxic or distasteful to herbivorous fishes and invertebrates, many algae produce biologically active compounds which have anti-bacterial or fungicidal properties (Bhakuni & Silva 1974). It seems likely that more intensive study will reveal a rich fabric of chemically-mediated relationships between algae and marine herbivores.

Hashimoto & Fusetani (1972) screened 48 species of algae in the Amami and Tokunoshima Islands and found 38 species that contained toxins lethal to mice or hemolytic to rabbit blood cells. Abbott et al. (1974) and Abbott & Ogden (unpubl.) found that some of the algae avoided by the sea urchin *Echinometra lucunter* contained terpenoids, pungent lipids and other compounds which possibly had a defensive role. Doty & Santos (1970) described the toxic compounds caulerpin and caulerpicin present in the green alga *Caulerpa* and concentrated by an herbivorous saccoglossid gastropod. The compounds may be defensive both in the plant and the invertebrate herbivore. Similarly the toxin known to cause ciguatera in humans who eat certain large predaceous fishes is believed to concentrate biologically in food chains from an origin in blue-green algae or dinoflagellates (Banner 1976).

Ehrlich & Raven (1965) speculated that plant secondary compounds have led to the rich development of insect-plant specializations which are characteristic of the Lepidoptera. The strategies of herbivorous fishes appear to have been very different. In contrast to insects, foraging herbivorous fishes such as the scarids have the ability to move over a wide region while feeding, have an excellent visual sense, and probably have good learning ability. In many ways their strategies may approach those of the mammals discussed by Freeland & Janzen (1974). The guts of herbivorous fishes are long and highly differentiated in some cases. Since they are known to eat some toxic algae they probably can cope with low levels of at least some plant toxins. They have diversified their diets to include a variety of plants and may quickly learn to feed largely on familiar foods. We may expect, however, that temporal differences in plant availability will keep them experimenting with diet and thus remaining flexible. Preferences, where shown by diet/availability studies or direct preference tests, may demonstrate that marine herbivores prefer foods with small amounts of secondary compounds, or select parts of the plants known to be low in such compounds.

On the other hand, the pomacentrids have evolved a strategy of closely tending a patch of bottom and encouraging the growth of algal species which are suitable as food. They do not forage for suitable foods and we may expect them to be much more selective and less likely to sample different plant types than such foraging herbivores as the scarids and acanthurids.

Selectivity in the feeding of sea urchins appears to be a compromise between urchin preferences and availability. Sea urchins are mobile, but their movements are much more restricted than fishes. They are non-visual, but appear to be chemosensitive where this sense assists them in locating perferred foods. Vadas (1968, in press) found that *Strongylocentrotus drobachiensis* was strongly attracted to sites where fronds of *Nereocystis* had fallen to the bottom. This preferred food was unpredictably available and the urchins appeared to have a finely developed chemical sense to locate such patches of food. In contrast, *Tripneustes ventricosus*, which feeds on the superabundant resource of *Thalassia* beds, does not show chemosensitivity to preferred foods in tank or Y-maze tests (Ogden, unpubl.).

The tropical sea urchins are faced with a more diverse food supply than temperate urchins and are relatively incapable of long distance movement and rapid sampling of available foods. These urchins have become generalists in feeding and are capable of getting along on a very wide range of diets. *Diadema*, for example, is capable of feeding on macroscopic plants, algal mats, sand, and will even become carnivorous (Lewis 1964, Ogden et al. 1973). Some preferences are shown, but more dramatic is the tendency to avoid certain genera of algae which then become highly suspect of having strong anti-herbivore defenses.

Conclusions

The preliminary evidence which we have selected and presented here is suggestive of a number of lines of continuing research. Very little is known about the ecology of tropical algal seasonality, reproduction and recruitment, nor especially of the competitive interactions between plant species in reef communities. There appears to be a rich evolutionary interplay between herbivores and algae involving algal structure and chemistry, productivity, and growth habit as well as herbivore feeding strategies and digestive physiology. The herbivores examined in detail so far have shown at least some diet selectivity based on many of these parameters. Detailed studies of herbivore diet, selectivity and sequencing are lacking. Close examina-

tion of the structure, growth, and reproductive strategies of particular algal species in relation to the number of consumers of that species would be very revealing. Herbivores appear to be intimately involved in the community structure of coral reefs; this influence is particularly strong with respect to algal species diversity and the outcome of competitive interactions between algae and stony corals. Manipulations of density of invertebrate herbivores have proven to be very useful in studies of reef community structure. Fishes are less easily manipulated, but the opportunity exists for comparative studies in areas of differing population densities of fish herbivores. Close monitoring of the recruitment of juvenile herbivorous fishes could provide natural experiments in algal-herbivore ecology.

Acknowledgements

We thank I. Clavijo, N. Ogden, M. Robblee, and T. Taxis for assistance in various aspects of our work. Phillip S. Lobel also thanks K. Liem, J. Menge and R. Robertson for discussions and research support. We appreciate the constructive comments of the reviewers and Gene Helfman. Partial research support to P. S. Lobel was supplied by the Anderson Fund and Department of Fishes, Harvard University and the Pacific Equatorial Research Laboratory, Fanning Atoll (M. Vitousek, Director).

Portions of this work were supported by National Science Foundation Grant No. NSF-OCE 7601304 to John C. Ogden. Our study is a contribution to the International Decade of Ocean Exploration Seagrass Ecosystem Study program. This is Contribution Number 42 of the West Indies Laboratory, Fairleigh Dickinson University, St. Croix, U.S. Virgin Islands.

Questions and answers

Keast: Is the picture of these herbivorous fish much like that in the African ungulate situation, where there is a great amount of actual overlap with certain chosen, favored plants coming through very strongly?

Ogden: There is some beautiful work on the grazing African ungulates and there are some very tempting parallelisms with the Caribbean seagrass system. Caribbean seagrass beds consist of two dominant grasses and a variety of associated algae which are extensively grazed by a variety of vertebrate and invertebrate herbivores some of which are resident in the beds, some of which venture into the beds short distances from reefs, and others, such as the green turtle, which range very widely. Where we have looked at food preferences, herbivores often show definite preferences for particular components of the vegetation. We are also looking at the influence of herbivores on plant community structure experimentally. Unfortunately we as yet do not know enough to fit our preliminary results into the dynamics of the whole system. One interesting recent observation concerns grazing in seagrass beds by green turtles. Turtles create large patches in the beds by biting off the grasses just above the substrate. The patches persist for several months and appear to be favored areas for fish grazing during recovery.

Brussard: Do any of the algae species that seem to be resistant to generalized herbivore grazing have their own specialist herbivores, analogous to the butterflies in the terrestrial system?

Ogden: Our work has concentrated upon relatively large, mobile herbivores which are clearly generalist feeders. There are some specialist herbivores such as the saccoglossids which eat *Caulerpa* and which are known to concentrate the toxins of the alga, but these specialists are not prominent. Our preliminary data indicate that the generalist herbivores must mix their diets. Where we have conducted feeding experiments with fishes and sea urchins on single plant species, some produce significantly lowered growth rate and others are toxic.

Dale: One slide you had of *Laurencia* with the red tips on its branches made me think that possibly a fish might mistake that for a *Condylactis* anemone, and that might be a mimicry there. Is that conceivable?

Ogden: The picture I showed was quite magnified and the alga is much smaller than the anemone. However, your question brings up an interesting point. Where terrestrial plants are known to be defended by toxic chemicals, we know that these can occur in specific parts of the plant. It is not inconceivable that an alga could advertise its defended parts with striking coloration.

Earle: A special circumstance that may interest the authors exists in the Galapagos Islands and is the subject of research that I have underway at present. Herbivorous fishes, urchins, and in some areas, iguanas,

are conspicuous and abundant in warm surface waters. Volcanic rock substrate is devoid of plants or hosts fast-growing filamentous species or kinds that are not favored as food (including *Sargassum* and certain red algae). A thermocline with a 6° C temperature difference (17° vs 23°) occurs in depths ranging from 20 to 40 m. Below the thermocline, plants are far more abundant and fish herbivores are rare. Algae regarded as 'cold water species' flourish in the colder water, including species of *Desmarestia, Laminaria, Eisenia,* and numerous large fleshy Rhodophyta.

In the tropics, temperature is rarely as conspicuously influential in shaping plant and herbivore distribution as in this special case, but it is a factor that should be considered in concert with light, substrate, etc.

I would like to question the conclusion that 'large-celled macroscopic algae appear most susceptible to all these herbivores, while small-celled algae are possibly least utilized by fishes.' The circumstances are more complex than this generalization implies, so much so, that I think it should be re-evaluated. Species of *Halimeda* have large cells and are calcified except in the youngest stages, and are favored food of parrotfishes. The related genus *Avrainvillea*, not calcified, and species of *Penicillus* (some calcified, some not) curiously are seldom eaten. Filamentous algae — such as species of *Dasya* (red), *Giffordia* (brown), and *Cladophora* (green) tend to have smaller cells, but may also be more 'digestible'. They seem to be readily consumed by various fish herbivores. Some large-celled uncalcified green algae that appear not to be favored as food are species of *Valonia, Dictyosphaeria,* and *Microdictyon.* The question is an intriguing one and the point about cells large enough to be macerated and thus more readily digested is a good one. But I feel there are too little data about circumstances too variable to make valid generalizations at this time.

Lobel: My original suggestion [concerning cell size and utilization] was based upon the assumption that fishes are dependent upon the use of dentition, a muscular stomach, or a pharyngeal mill for release of plant cytoplasm from the confines of a cellulose cell wall. A fish should choose a plant based on relative fish-plant morphologies; this is consistent with existing data on the morphology and physiology of herbivorous reef fishes and has been the working hypothesis in my research. Where the observed results differ from the expected based purely on morphological grounds, we will be led, hopefully, to the discovery of new mechanisms by which fishes cope with plants as food or a reason why fish did not choose a supposedly optimum food (i.e. plant defense mechanisms). We said that cellulases have not yet been found in reef fishes and fish are, therefore, more dependent upon morphological than chemical mechanisms. I will briefly provide one example and one counter-example as points in discussion. These are the results of my current research efforts and will be presented in detail elsewhere (Lobel, in prep.).

First, the counter-example. According to my fish morphology-plant selectivity scheme, herbivorous damselfish (Pomacentridae) should feed preferentially off large celled algae. However, this is clearly *not* the case for the Caribbean *Eupomacentrus planifrons* which feeds principally upon epiphytic blue-green algae growing within its territory (probably *Oscillatoria* sp., mean volume in diet 55 ± 44%, N = 23, Lobel, unpubl., and D. R. Robertson, pers. comm.). Utilization of this algae by the damselfish is not explainable using the morphological relationship model. The blue-green algae are of very small relative cell size. This led me to examine other potential digestive strategies, such as the role of pH in the stomach. *E. planifrons* has one of the most acidic stomachs (pH 2.38 ± 0.54, N = 12) I have found among 27 species of Pacific and Caribbean herbivorous fishes. These results strongly suggest a unique alternative digestive mechanism similar to that of the cichlids we mentioned earlier.

Secondly, an example of morphologically based feeding selectivity. We have evaluated the basis for food selection by the Caribbean bucktooth parrotfish, *Sparisoma radians*. Under field conditions, *S. radians* frequently took bites of *Halimeda incrassata*, yet this alga composed only 2 to 3% of a fish's total food intake. In laboratory preference experiments, *H. incrassata* and *Penicillus pyriformis* were the *least* preferred food of nine plants tested. The basis for plant selection in this case was not a function of relative plant cell size (as the model predicted) but was instead correlated with how easily the fish could bite and grind its food. Scarids are unique among herbivorous reef fishes: they can thoroughly macerate food in a pharyngeal mill apparatus. *Halimeda* is heavily calcified and may be 'difficult' and energetically more costly for the fish to handle. This has been demonstrated by calculation of amount of food value per unit effort (incorporating measures of absorption efficiencies, plant toughness, etc.) and by measuring comparative growth of the scarid on single and mixed diets. The results show that while *Halimeda*

was often sampled in the field, it is not a significant nutritional resource for this scarid. However, this does not discount *Halimeda*'s potential value as a supplementary dietary item for essential vitamins and/or minerals.

Another example of algae with very large individual cells are *Caulerpa* spp., which possess a toxin known to affect humans; they are generally avoided by fishes. It appears that both *Halimeda* and *Caulerpa* would be prime food for herbivores were it not for their effective anti-herbivore adaptations. I should mention that neither alga is totally immune to being eaten; Earle is, of course, correct in that *Halimeda* is eaten by some parrotfishes, and *Caulerpa racemosa* is eaten by the Pacific angelfish, *Centropyge flavissimus* (pers. obs.). Thus, these various algal defense mechanisms may function to deter some but not necessarily all herbivores.

Admittedly, the basic model relating fish and plant morphologies has its limitations. Even so, I have found it to be an effective scheme in which to examine herbivorous fish and plant relationships.

References cited

Abbott, D. P., J. C. Ogden & I. A. Abbott (eds.). 1974. Studies on the activity pattern, behavior and food of the echinoid *Echinometra lucunter* (L.) on beachrock and algal reefs at St. Croix, U. S. Virgin Is. West Indies Lab. Spec. Publ. 4. 111 pp.

Adey, W., P. Adey, R. Burke & L. Kaufman, in press. The reef systems of Eastern Matinique. Atoll Res. Bull.

Alevizon, W. S. 1976. Mixed schooling and its possible significance in a tropical Western Atlantic parrotfish and surgeonfish. Copeia 1976: 796–798.

Bakus, G. 1964. The effects of fish grazing on invertebrate evolution in shallow tropical waters. Allan Hancock Found. Occ. Pap. No. 27. 29 pp.

Banner, A. H. 1976. Ciguatera: A disease from coral reef fish, pp. 177–214. In: O. A. Jones & R. Endean (ed.), Biology and Geology of Coral reefs, 3, Acad. Press, N. Y.

Bardach, J. E. 1961. Transport of calcareous fragments by reef fishes. Science 133: 98–99.

Barlow, G. W. 1974. Extraspecific imposition of social grouping among surgeonfishes (Pisces: Acanthuridae). J. Zool. Lond. 174: 333–340.

Belk, M. S. 1975. Habitat partitioning in tropical reef fishes *Pomacentrus lividus* and *P. albofasciatus.* Copeia 1975: 603–617.

Bhakuni, D. S. & M. Silva. 1974. Biodynamic substances from marine flora. Botanica Marina 17: 40–51.

Birkeland, C. 1977. The importance of rate of biomass accumulation in early successional stages of benthic communities to the survival of coral recruits. Proc. Third Internat. Coral Reef Symp. 1: 15–22.

Blinks, L. R. 1951. Physiology and biochemistry of algae, pp. 263–291. In: G. W. Smith (ed.), Manual of Phycology,

171

Cronica Botanica Co. Mass.

Bowen, S. H. 1976. Mechanism for digestion of detrial bacteria by the cichlid fish *Sarotherodon mossambicus* (Peters). Nature 260: 137–138.

Brawley, S. H. & W. H. Adey. 1977. Territorial behavior of threespot damselfish (*Eupomacentrus planifrons*) increases reef algal biomass and productivity. Env. Biol. Fish. 2: 45–51.

Bryan, F. G. 1975. Food habits, functional digestive morphology and assimilation efficiency of the rabbit fish, *Siganus spinus* (Pisces: Siganidae) on Guam. Pac. Sci. 29: 269–277.

Carr, W. E. S. & C. A. Adams. 1972. Food habits of juvenile marine fishes occupying seagrass beds in the estuarine zone near Crystal River, Florida. Trans. Am. Fish. Soc. 102: 511–540.

Cates, R. G. & G. H. Orians. 1975. Successional status and the palatability of plants to generalized herbivores. Ecology 56: 410–418.

Dahl, A. L. 1973. Benthic algal ecology in deep sea and sand habitat off Puerto Rico. Botanica Marina 16: 171–175.

Dawson, E. Y., A. A. Aleem & B. W. Halsted. 1955. Marine algae from Palmyra Island with special reference to the feeding habits and toxicology of reef fishes. Allan Hancock Found. Occ. Pap. No. 17: 1–38.

Dayton, P. K. 1971. Competition, disturbance and community organization: The provision and subsequent utilization of space in a rocky intertidal community. Ecol. Monogr. 41: 351–389.

Doty, M. S. & G. A. Santos. 1970. Transfer of toxic algal substances in marine food chains. Pac. Sci. 24: 351–354.

Dyer, I. M. & V. G. Bokhari. 1976. Plant-animal interactions: studies of the effects of grasshopper grazing on blue gramma grass. Ecology 57: 762–772.

Earle, S. A. 1972. The influence of herbivores on the marine plants of Great Lameshur Bay, pp. 17–44. In: B. B. Collette & S. A. Earle (ed.), Results of the Tektite Program: Ecology of Coral Reef Fishes. Nat. Hist. Mus. L. A. County, Sci. Bull. 14.

Ebersole, J. P., in press. The adaptive significance of interspecific territoriality in the reef fish *Eupomacentrus leucostictus*. Ecology.

Ehrlich, P. R. & P. H. Raven. 1965. Butterflies and plants: a study in coevolution. Evolution 18: 586–608.

Foster, M. S. 1972. The algal turf community in the nest of the ocean goldfish, *Hypsypops rubicunda*. Proc. 7th Intl. Seaweed Symp. Sect. I., Univ. of Tokyo Press: 55–60.

Freeland, W. J. & D. H. Janzen. 1974. Strategies in herbivory by mammals: the role of plant secondary compounds. Am. Nat. 108: 269–289.

Gibson, R. N. 1969. The biology and behavior of littoral fish. Oceangr. Mar. Biol. Ann. Rev. 7: 367–410.

Glynn, P. W. 1973. Aspects of the ecology of coral reefs in the Western Atlantic region, pp. 271–234. In: O. A. Jones & R. Endean (ed.), Biology and Geology of Coral Reefs, 2, Acad. Press, N. Y.

Gosline, W. A. 1965. Vertical zonation of inshore fishes in the upper layers of the Hawaiian Islands. Ecology 46: 823–831.

Grünbaum, H., G. Bergman, D. P. Abbott & J. C. Ogden, in press. Intraspecific agnostic behavior in the rock-boring sea urchin *Echinometra lucunter* (L.), Echinodermata: Echinoidea. Bull. Mar. Sci.

Hashimoto, V. & N. Fusetani. 1972. Screening of the toxic

algae on coral reefs. Proc. 7th Intl. Seaweed Symp. Section IV. Univ. of Tokyo Press: 569–572.

Hiatt, R. W. & D. W. Strasburg. 1960. Ecological relationships of the fish fauna on coral reefs of the Marshall Islands. Ecol. Mongr. 30: 65–127.

Hobson, E. S. 1974. Feeding relationships of teleostean fishes on coral reefs in Kona, Hawaii. Fish Bull. U.S. 72: 915–1031.

Jones, R. S. 1968. Ecological relationships in Hawaiian and Johnston Island Acanthuridae (surgeonfishes). Micronesica 4: 309–361.

Kaufman, L. 1977. The three spot damselfish: effects on benthic biota of coral reefs. Proc. Third Int. Coral Reef Symp. I: 559–564.

Keller, B. D. 1976. Sea urchin abundance patterns in seagrass meadows. The effects of predation and competitive interactions. Ph.D. thesis, Johns Hopkins University, 39 pp.

Lewis, J. B. 1964. Feeding and digestion in the tropical sea urchin *Diadema antillarum* Philippi. Can. J. Zool. 42: 549–557.

Lawrence, J. M. 1975. On the relationships between marine plants and sea urchins (Echinodermata: Echinoidea), pp. 213–286. In: H. Barnes (ed.), Oceanogr. Mar. Biol. Ann. Rev. 13.

Levin, D. A. 1976. The chemical defense of plants to pathogens and herbivores. Ann. Rev. Ecol. Syst. 7: 121–159.

Lobel, P. S., ms. Lunar periodicity, annual seasonality and reproduction behavior of an Hawaiian coral reef fish.

Mathieson, A. C., R. A. Fralick, R. Burns & W. Flashive. 1975. Phycological studies during Tektite II at St. John, U. S. V. I., pp. 77–103. In: S. A. Earle & R. J. Lavenberg (ed.), Results of the Tektite Program: Coral Reef Invertebrates and Plants. Nat. Hist. Mus. L. A. County, Sci. Bull. 20.

Mattson, W. J. & N. D. Addy. 1975. Phytophagous insects as regulators of forest primary production. Science 190: 515–522.

Menge, J., in press. Plant species diversity in a marine intertidal community. Am. Nat.

Montgomery, W. L. 1977. Diet and gut morphology in fishes with special reference to the monkeyface stickleback, *Cebidichthys violaceus* (Stichaeidae: Blennioidei). Copeia 1977: 178–182.

Moriarty, D. J. W. 1973. The physiology of digestion of blue-green algae in the cichlid fish, *Tilapia nilotica*. J. Zool. Lond. 171: 25–39.

Morse, D. H. 1970. Ecological aspects of some mixed species foraging flocks of birds. Ecol. Mongr. 40: 119–168.

Morse, D. H. 1977. Feeding behavior and predator avoidance in heterospecific groups. Bioscience 27: 332–339.

Moynihan, M. 1962. The organization and probable evolution of some mixed species flocks of neotropical birds. Smith. Misc. Coll. 143: 1–140.

Nursall. J. R. 1974. Some territorial behavioral attributes of the surgeonfish *Acanthurus lineatus* at Heron Island, Queensland. Copeia 1974: 950–959.

Odum, W. E. 1970. Utilization of the direct grazing and plant detritus food chains by the striped mullet *Mugil cephalus*, pp. 222–240. In: J. H. Steele (ed.), Marine Food Chains, Oliver and Boyd, Edinburg, G. B.

Ogden, J. C. 1976. Some aspects of herbivore-plant relationships on Caribbean reefs and seagrass beds. Aquatic Botany 2: 103–116.

Ogden, J. C., in press. Carbonate sediment production by parrotfishes and sea urchins on Caribbean reefs. In: S. Frost and M. Weiss (ed.), Caribbean Reef Systems: Holocene and Ancient. Amer. Assoc. Petrol. Geol. Spec. Paper 4.

Ogden, J. C., R. Brown & N. Salesky. 1973. Grazing by the echinoid *Diadema antillarum* Phillipi: Formation of halos around West Indian patch reefs. Science 182: 715–717.

Ogden, J. C., D. P. Abbott & I. A. Abbott (ed.). 1973. Studies on the activity and food of the echinoid *Diadema antillarum* Phillipi on a West Indian patch reef. West Indies Lab. Spec. Publ. 2, 96 pp.

Ogden, J. C. & J. C. Zieman. 1977. Ecological aspects of coral reef-seagrass bed contacts in the Caribbean. Proc. Third. Int. Coral Reef Symp., U. of Miami 1: 377–382.

Ogino, C. 1962. Tannins and vacuolar pigments, pp. 437–442. In: Lewin, R. A. (ed.), Physiology and Biochemistry of Algae. Academic Press, N. Y.

Owen, D. F. & R. G. Wiegert. 1976. Do consumers maximize plant fitness? Oikos 27: 488–492.

Paine, R. T. 1966. Food web complexity and species diversity. Am. Nat. 100: 65–78.

Paine, R. T. 1969. *Pisaster-Tegula* interaction: prey patches, predator food preference, and intertidal community structure. Ecology 50: 950–962.

Paine, R. T. & R. L. Vadas. 1969. The effects of grazing by sea urchins, *Strongylocentrotus* spp. on benthic algal populations. Limnol. Oceanogr. 14: 710–719.

Patton, J. S. & A. A. Benson. 1975. A comparative study of wax ester digestion in fish. Comp. Biochem. Physiol. 52B: 111–116.

Pfeffer, R. 1963. The digestion of algae by *Acanthurus sandvicensis*. Masters thesis, University of Hawaii. 47 pp.

Phillips, A. M. 1969. Nutrition, digestion and energy utilization, pp. 391–432. In: W. S. Hoar & D. J. Randall (ed.), Fish Physiology 1.

Porter, K. G. 1976. Enhancement of algal growth and productivity by grazing zooplankton. Science 192: 1332–1334.

Porter, K. G. 1977. The plant-animal interface in freshwater ecosystems. Am. Sci. 65: 159–170.

Prim, P. & J. M. Lawrence. 1975. Utilization of marine plants and their constituents by bacteria isolated from the gut of echinoids (Echinodermata). Mar. Biol. 33: 167–173.

Randall, J. E. 1961. A contribution to the biology of the convict surgeonfish of the Hawaiian Islands, *Acanthurus triostegus sandvicensis*. Pac. Sci. 15: 215–272.

Randall, J. E. 1963. An analysis of the fish populations of artificial and natural reefs in the Virgin Islands. Carib. J. Sci. 3: 1–16.

Randall, J. E. 1965. Grazing effects on seagrasses by herbivorous reef fishes in the West Indies. Ecology 46: 255–260.

Randall, J. E. 1967. Food habits of reef fishes of the West Indies. Stud. Trop. Oceanogr. 5: 665–847.

Randall, J. E. 1974. The effect of fishes on coral reefs. Proc. Second Int. Coral Reef Symp. 2: 159–166.

Reese, E. S. 1973. Duration of residence by coral reef fishes on 'home' reefs. Copeia 1973: 145–149.

Reese, E. S. 1975. A comparative field study of the social behavior and related ecology of reef fishes of the family Chaetodontidae. Z. Tierpsychol. 37: 37–61.

Robertson, D. R., H. P. A. Sweatman, E. A. Fletcher and M. G. Cleland. 1976. Schooling as a mechanism for circumventing the territoriality of competitors. Ecology 57: 1208–1220.

Romer, A. S. 1966. Vertebrate Paleontology. Univ. Chicago Press. 468 pp.

Sale, P. F. 1977. Maintenance of high diversity in coral reef fish communities. Am. Nat. 111: 337–359.

Sammarco, P. W. 1975. Grazing by *Diadema antillarum* Phillipi (Echinodermata: Echinoidea): density dependent effects of coral and algal community structure. Assoc. Is. Mar. Labs. Carib. 11th meeting (abstract).

Sammarco, P. W., J. S. Levinton & J. C. Ogden. 1974. Grazing and control of coral reef community structure by *Diadema antillarum* Phillipi: a preliminary study. J. Mar. Res. 32: 47–53.

Stephenson, W. & R. B. Searles. 1960. Experimental studies on the ecology of intertidal environments at Heron Island. Aust. J. Mar. Freshwater Res. 2: 241–267.

Stickney, R. R. & S. A. Shumway. 1974. Occurrence of cellulase activity in the stomachs of fishes. J. Fish. Biol. 6: 779–790.

Syrop, S. 1974. Three selected aspects of the territorial behavior of a pomacentrid fish, *Pomacentrus jenkinsi*. Masters thesis. University of Hawaii. 48 pp.

Tsuda, R. T. & P. G. Bryan. 1973. Food preferences of juvenile *Siganus rostratus* and *S. spinus* in Guam. Copeia 1973: 604–606.

Tsuda, R. T. & H. T. Kami. 1973. Algal succession on artificial reefs in a marine lagoon environment on Guam. J. Phycology 9: 260–264.

Vadas, R. L. 1968. The ecology of *Agarum* and the kelp bed community. Ph.D. thesis. University of Washington. 282 pp.

Vadas, R. L., in press. Preferential feeding: an optimization strategy in sea urchins. Ecol. Monogr.

Vine, P. F. 1974. Effects of algal grazing and aggressive behavior of the fishes *Pomacentrus lividus* and *Acanthurus sohal* on coral reef ecology. Mar. Biol. 24: 131–136.

Whittaker, R. H. & P. P. Feeny. 1971. Allelochemics: chemical interactions between species. Science 171: 757–770.

ERRATA

Page 58, column 2, line 13 from the bottom of the page, "disversified" should be "diversified."

Page 59, column 1, line 12 from the top of the page, "perferred" should be "preferred."

Part III

DIVERSITY IN TERRESTRIAL
COMMUNITIES

Editor's Comments
on Papers 9 Through 13

The papers in this section were chosen to acquaint the reader with the various types of diversity and factors influencing diversity in major groups of terrestrial organisms. Paper 9 is an important theoretical article by Whittaker, who has written more about diversity in terrestrial plant communities than anyone else. Reviewing the broad patterns in plant species diversity, Whittaker concluded that there is a trend of decreasing plant community diversity toward environments less stable, less favorable, and more extreme, but that these trends are strongly modified by those relationships involving plant strata and life forms and the effects of dominant species on their associates. Thus, the observations of terrestrial plants parallel those of the aquatic communities described in the previous section.

However, in addition Whittaker concluded that plant community diversity is an evolutionary product, with no ceiling or saturation level for diversity. He found that the chemical differentiation of higher plants virtually implies unlimited potential for the addition of different species with different interactions with predators and seedling soil requirements. In this sense, Whittaker concluded that plants and insects act similarly, while bird diversity does reach a ceiling, because the possibility of niche subdivision is ultimately limited for birds.

The relationships between plants and associated animals mentioned above have been the subject of numerous studies, ever since McArthur and McArthur (1961) demonstrated that bird species diversity is related to the structural diversity of vegetation. Murdoch and coworkers (Paper 10) extended this relationship to plants and sucking insects (Homoptera) in their analysis of an old-field community in Michigan. They found that species diversity as well as a measure of foliage complexity (foliage height diversity) were equally good correlates of the Homopteran species diversity.

Plant communities of forests and old fields provide a reasonably constant set of resources or structure to which animals may respond. Desert systems are quite different terrestrial systems with strongly pulsed patterns of production and resource availability. Studies of the diversity of desert organisms are similar to studies of diversity in extreme aquatic environments. Davidson (Paper 11) examined the patterns of diversity of a guild of desert insects—the seed-eating ant complex. These ants collect, store, and use seeds, which are produced intermittantly when plant growth conditions are favorable. Davidson found a close correlation between ant diversity and mean annual precipitation, which is, in turn, an index to plant production. These ant communities are structured on the basis of competition for food, which is influenced by worker body size and colony foraging strategies. These ants act very similarly to another group of key desert animals—the seed-eating rodents (Brown, 1973). Pianka (Paper 12) also focused on diversity of animals in desert environments. In this case, he compared desert lizard communities in three deserts, in Australia, Kalahari in Africa, and in North America. Pianka found that the correlates of lizard diversity differed widely between desert systems. There was no one desert pattern; rather, significance of food, place, and time niches depended upon the desert studied.

In a further variant in terrestrial analyses, Terborgh (Paper 13) examined the species diversity of birds across an altitudinal gradient in highland Peru. In this case, the investigator traced a single group of organisms across widely different communities, all located in a relatively restricted area. As the altitude increased the forest was reduced in canopy height and in plant species. Terborgh found that bird species diversity was highly correlated with the pattern of foliage. However, when he subdivided birds into functional groups (such as insectivores, frugivores, and nectarivores), he found that the subgroups responded to patterns other than the foliage habitat structure.

Perhaps the reader of this section will ponder the differences and similarities of the strategies developed or the conditions of the habitats, which enable the maintenance of diversity in these very different terrestrial communities.

REFERENCES

Brown, J. H., 1973, Species Diversity of Seed-eating Desert Rodents in Sand Dune Habitats, *Ecology* **54:**775–787.

McArthur, R. H., and J. W. McArthur, 1961, On Bird Species Diversity, *Ecology* **42:**594–598.

9

Reprinted from pages 178–195 of *Diversity and Stability in Ecological Systems,*
G. M. Woodwell and H. H. Smith, eds., Brookhaven Symp. Biol. No. 22, U.S. Dept.
Commerce, Springfield, Va., 1969, 264p.

Evolution of Diversity in Plant Communities

ROBERT H. WHITTAKER

Section of Ecology and Systematics, Cornell University, Ithaca, New York 14850

INTRODUCTION

By organization we refer, I think, to the means by which a functioning complexity is maintained through time. My concern is with the organization of plant communities and the convergent insights into that organization offered by research on diversity and gradient analysis. We need to consider the manner of organization of plant communities in place, as systems of populations occurring together in particular habitats; and in space, as patterns of populations in the spatial or vectorial patterns of landscapes.[cf.35]

VERTICAL ORGANIZATION

It is useful to begin with specific cases. G.M. Woodwell and I have been working for some years[111-12,117-19] with the oak-pine forest of the sandy plains of southern Long Island. It is a forest of modest size and modest diversity which we think is of all the more interest for its manageable dimensions. Figure 1 is a dominance-diversity or importance-value curve for the forest. Vascular plant species have been arranged by their productivity on the ordinate and by their sequence from most important to least important on the abscissa. The curve approaches lognormal form, but has wide dispersion for such.

Trees, shrubs, and herbs tend each to be concentrated in one of the three reaches, or slopes, of the curve. We may go back in the tradition of plant ecology for the idea of treating separately the different strata. Our justification may be that species of a particular stratal grouping or synusia are more nearly alike in response to environment and relation to resources, more directly in competition with one another as mature plants, than species of different strata. Division of the plants into synusiae is thus comparable to division of an animal community into guilds.[90] Importance-value curves by strata are shown in Figure 2. In the tree stratum the four major species are arranged in line, forming a geometric series, while the minor tree species, some of them extraneous to the community, form another slope. The shrubs, in the middle, are arranged along a line. In the herbs the three major species form a group or line, but there is also a sigmoid curve of minor herbs, some quite rare and some also extraneous to the community.

For interpretation of such curves, let us assume some correspondence between the share of a community's resources a species utilizes, the share of the community's niche space it occupies, and the share of the community's productivity it realizes. We may then treat relative importance values (relative productivity, especially) as ex-

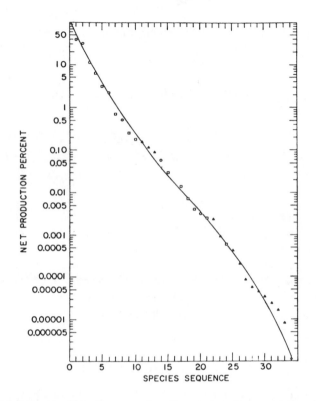

Figure 1. A dominance-diversity curve for all vascular plant species in a 0.5-ha sample in the Brookhaven forest.[112] Species' percentages of community above-ground net primary production (dry weight) are used as the importance values in a logarithmic scale on the ordinate, and species' numbers in the sequence from most productive to least productive on the abscissa. ○, Trees; ⊏, shrubs; △, herbs.

pressions of relative niche size, and ask how niche space may be divided to produce the distribution of productivity among species that we observe. Three hypotheses are familiar:

1. The random niche boundary hypothesis of MacArthur states that the boundaries of niches are located at random in niche hyperspace.[50,52,101] The distribution of sizes of the resulting hypervolumes corresponds in form to the distribution of lengths of segments of a line onto which points separating the segments have been cast at random.[113] A sigmoid curve of low slope results from plotting segments by size and size-sequence in the manner illustrated.

2. The niche pre-emption hypothesis states that the first or dominant species pre-empts, by its competitive success, some fraction of total niche space, that the second species may take a similar fraction of the remaining space, and the third species a similar fraction of space unoccupied by the first two species, and so on. A geometric series, such as first commented on by Motomura, results. [64,68,96] Allowing for sampling error and for random fluctuation of the ratio from one species to the next, just such arrangement of species along straight lines as shown for the Brookhaven trees and shrubs would be expected.

180

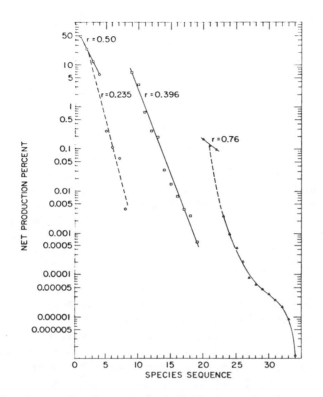

Figure 2. Dominance-diversity curves for individual vascular plant strata in the Brookhaven forest,[112] plotted as in Figure 1. The r values are slopes of the lines fitted to segments of the curve as geometric series. The three strata as wholes suggest a geometric series in this community; for the ratios of shrub to tree production, and of herb to shrub production, are about 0.08. ○, Trees; □, shrubs; △, herbs.

3. The lognormal hypothesis of Preston[84],[85] states that species importances are determined by a number of variables relatively independent in effect. Species importances will then approach a normal frequency distribution for which, however, a logarithmic scale of importance values is appropriate. Sigmoid curves of much steeper slope than those predicted by the random niche boundary hypothesis result.

The form of the random niche boundary distribution is approached by some samples from animal communities – notably in territorial birds but also in some other taxocenes, including Goulden's[26] microcrustaceans. When curves of a good number of animal samples are compared, however, they range from this form through lognormal distributions to geometric slopes.[100],[29],[30],[45],[18],[44] The random niche boundary form thus appears as a limiting case of minimum dominance or maximum equitability,[47] especially for some samples of closely related animals of more stable populations. Geometric slopes are often encountered in samples from plant communities – both some whole communities of low diversity[106] and particular strata not too rich in species, as in Figure 2. Samples from communities or strata that are rich in species – whether the tree stratum of a tropical rainforest,[81] or a microfilm of diatoms in a stream,[75],[76] or a soil arthropod sample[18],[29] – in general form lognormal distributions.[2],[106]

181

One notes a difference in manner of statement of two of these hypotheses, compared with the third. The random niche boundary and niche pre-emption hypotheses concern the way niches of competing species are determined; they concern species with contiguous niche hypervolumes. The lognormal distribution entails no such assumption. From this one might judge, first, that samples comprising small numbers of species in closely related niches should approach either the geometric series – if dominance is well developed among the organisms concerned – or the random niche boundary form – if regulatory processes affecting the populations act in restraint of dominance, i.e. toward reduced contrast of resource use and importance among the species. (One should allow also for various intermediate conditions.) One may judge, second, that samples including larger numbers of species, many of them with no effective competitive contact with one another, should approach lognormal distributions. Thus when the Brookhaven stratal samples, individually of geometric form, are combined into a community sample the latter approaches lognormal form. One judges further that gamma samples (combining into one sample species from a number of communities in an area) should be of lognormal form.

It is thus possible to interpret the forms of importance-value curves. First thoughts suggested that one type of curve – one of these three or the Fisher[20,114-16] logarithmic series – should fit all samples; and none does so. A second thought may be that there is interest and significance in the manner in which these curves express different manners of competition and organization of species in communities. Certain qualifying third thoughts may also be in order. Vandermeer and MacArthur[101] have shown that the (a) and (b) hypotheses (of non-overlapping or overlapping niches) lead to curves which are of quite similar form when applied to small samples. Cohen[7] has shown that two hypotheses or models different from the random niche boundary model imply curves of the same form. Still other interpretations are possible. (It may be considered that the flat-sigmoid curves observed for territorial bird communities are a modified lognormal distribution in which the manner of population regulation has much reduced contrast of importance values of species, and hence the dispersion of the lognormal distribution. Alteration of the constant a in the lognormal expression* $S_r = S_0 e^{-(aR)^2}$ can produce reasonable fits to bird community samples – as well as, in the opposite direction, such curves of high dispersion as shown in Figure 1. Or, it may be considered that importance values tend to approach geometric series within groups of more closely competing species – strata, synusiae, or guilds – within a community. Combining these groups, of different geometric slopes, can produce importance-value curves ranging in form from flat-sigmoid to the steep-sigmoid of Figure 1.)

The forms of dominance and diversity curves are thus underdetermined by our theoretical interpretations. If we think of an importance-value curve as a sentence about the organization of a community, the individual words – our measurements of productivity, etc. – seem unambiguous. But the sentence is in a language of which we have no assured knowledge of the syntax, and a number of different rules of grammar implying different meanings from the relations of the words to one another may be possible. One hopes for means of further research into the syntax.

---

*S_r is the number of species in an octave R octaves distant from the modal octave, which contains S_0 species.

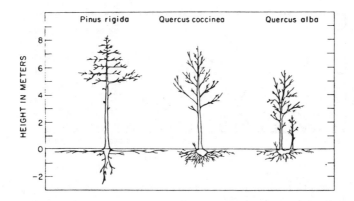

Figure 3. Branch and root profiles of the three dominant tree species in the Brookhaven forest.[112]

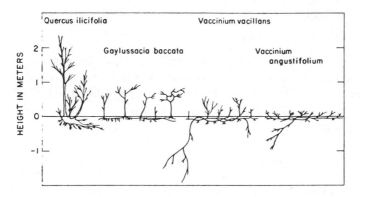

Figure 4. Branch and root profiles of the four major shrub species in the Brookhaven forest.[112]

For the present I prefer interpretations based on the niche and niche hyperspace as constructs, postulates by which we relate and interpret a range of phenomena in this area. These are animal ecologists' concepts for the most part; one thinks of the authors of the niche[23,17] and niche hyperspace[36,39] concepts and most applications as concerned primarily with animals. One may question, rather than assuming, application to plants.

Figure 3 shows profiles of the three major trees in the Brookhaven forest. The species are different in ways that are significant in relation to nutrient circulation. *Quercus coccinea* has oblique branches that conduct rainwater in toward the stem, and the rainwater with leached nutrients flows down the stem and into the hemisphere of soil densely occupied by the roots of the tree. The horizontal branches of *Pinus rigida* spread rainwater more widely, but wide-spreading roots collect water and nutrients from a still broader area, while the pine also has a taproot to ground water. The third species, *Quercus alba*, differs to some extent from both and in this community is smaller than, and fits into fire succession before, *Q. coccinea*. Figure 4 shows the four major shrub species of the forest; and again they differ in profile and root relationships, and also in foliage height and successional position.[88] It seems clear

183

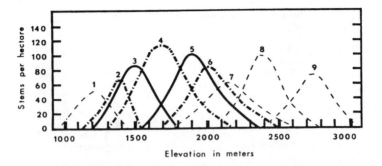

Figure 5. Distribution of oaks and other broadleaf trees along the elevation gradient in the Santa Catalina Mountains, Arizona. Species shown with solid lines are evergreen oaks of the black oak subgenus (*Erythrobalanus*), those with dot-and-dash lines are evergreen oaks of the white oak subgenus (*Lepidobalanus*), and those with broken lines are other broadleaf tree species. Species shown: (1) *Vauquelinia californica*, (2) *Quercus oblongifolia*, (3) *Q. emoryi*, (4) *Q. arizonica*, (5) *Q. hypoleucoides*, (6) *Q. rugosa*, (7) *Q. gambelli*, (8) *Acer grandidentatum*, and (9) *Acer glabrum*.

that, if one constructs a niche hyperspace in which plant height, root depth, and successional position are among the axes, the seven major woody species of the community will occupy clearly different parts of that space. Such observations[cf. 1,109] may contribute evidence that our postulates about niches and niche hyperspace for plants are not ungrammatical.

HORIZONTAL PATTERNING

The two oaks which dominate the Brookhaven forest raise a further question. Many oak forests and woodlands from Long Island to California are dominated by two oaks, one of the black and one of the white subgenus. The meaning of these subgenera in relation to niche differences is quite unknown; but one may ask whether the oaks form co-adapted pairs, evolved into stable and obligate niche accommodation to one another, and consequently distributed in close association with one another. In the Santa Catalina Mountains of Arizona a number of oak species are distributed along the elevation gradient (Figure 5). The black and white oaks are not paired but seem staggered, syncopated as it were, along the gradient in apparent avoidance of close distributional association. We infer that, even though these oaks may be niche differentiated, they tend to disperse their populations along the gradient in such a way that competition is reduced also by habitat difference.

When a larger number of species are observed along an environmental gradient, their population distributions together take the forms revealed by gradient analysis and shown in Figure 6. These vascular plant populations are partial competitors; but habitat differentiation reduces intensity of competition in their population centers, while niche differentiation permits co-occurrence in stands and broadly overlapping population distributions, forming community continua or coenoclines. These population continua are both for tree species along the topographic moisture gradient in mountains, in a maritime climate above and a continental climate below. They differ conspicuously in degree of floristic change along the gradient, in beta diversity

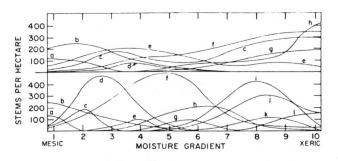

Figure 6. Population curves for tree species along topographic moisture gradients in mountains.[107] Above, Siskiyou Mountains, Oregon, 460 to 760 m elevation, with moderately low beta diversity; below, Santa Catalina Mountains, Arizona, 1830 to 2140 m elevation, with high beta diversity.

as we can measure this with similarity comparisons of the extremes, or with half-change units.[109,107]

It is on this basis that I would interpret the observations of MacArthur on diversities of tropical bird communities. Bird communities respond to the structure of vegetation[51,53,55-57] and evolve toward a relative saturation level in their alpha diversities, toward a ceiling on the number of bird species sharing the niche space and resources of a given structural kind of vegetation. The occurrence of such relative saturation in bird communities is supported by Recher's[87] Australian data and Cody's[5,6] work on grassland birds, as well as MacArthur's[54] studies. Through evolutionary time, however, additional bird species are able to fit themselves in along environmental gradients by habitat differentiation. By this means tropical forest bird communities have evolved to higher beta diversities, and consequently greater richness on the gamma level of combined samples representing landscapes and faunas of geographic regions.

The variation of communities in a landscape may be interpreted by means of a number of environmental gradients and corresponding community continua. The latter form together a community pattern which is also a complex population continuum.[104,107] Species evolve toward scattering of their population centers in the pattern; but, though the species thus behave "individualistically," there is a degree of orderliness or predictableness to the community composition to be expected in a given kind of habitat (at a given successional stage). The community pattern is an expression of the manner in which interacting species populations are organized in relation to one another by competition and differentiation along the spatial gradients of environment.

THE EVOLUTIONARY DANCE

There is a metaphor I like to use in relating results from gradient analysis and dominance and diversity studies. Consider first a hyperspace of the environmental factors in relation to which species evolve. Some of these factors involve positions, resources, and interactions within the community. These are niche factors, and the abstract space defined by these factors as axes is the niche hyperspace. Some of the factors involve gradients of physical environment, topography and soil character-

185

istics. These are habitat factors, and an abstract space defined by these factors as axes we may call a habitat hyperspace. There are, furthermore, successional gradients to which species respond, and these gradients are not simply either niche or habitat gradients, though having some of the characteristics of both. Consider now a compound hyperspace defined by all these axes – of habitat, niche, and succession – in relation with which the species of a given area evolve.

One might think of the compound hyperspace as an n-dimensional dance floor on which occur, as a kind of evolutionary dance, those adaptive maneuvers with which we are concerned. The couples move on the dance floor in such ways that each has a dancing space of its own with minimum competition from other couples. They move to disperse themselves in the space available. In moving on the dance floor to avoid one another the dancers will move in both niche and habitat dimensions at the same time. A species can move by either combination of one and two steps in relation to niche and habitat gradients, or one and a half of each. Such possibilities are illustrated by Schoener's[95] observation of wider niche difference on small islands, which restrict habitat difference, and Grant's[27] observation of three congeneric pairs of birds – one pair with strong habitat differentiation and small bill-length difference, one pair with little habitat differentiation but large bill-length difference, and a third pair with an intermediate degree of habitat overlap and bill-length difference. As additional dancers enter the floor, such maneuvers make space for them, with reduction of the dance areas of the remaining couples. There may come a time, however, when a part of the floor becomes so crowded that the rate at which new dancers enter is equaled by the rate of departure of couples discouraged or crowded off the floor.

It might be possible to observe parts of the dance floor in which niche relations are the principal direction of interaction, and other parts in which habitat relations are the principal direction of interaction. In the former case we see the dancers dispersing themselves in niche hyperspace, spacing themselves by somewhat different dance rules in different areas, rules that lead to the different patterns of niche differentiation, of dominance and alpha diversity, and of species interactions of which some are stability-increasing, that we observe. In the latter case we see the dancers dispersing themselves in habitat hyperspace, forming the patterns of species habitat differentiation, the landscape patterns of communities as complex population continua, and the degrees of beta differentiation within these patterns, that have been observed in gradient analysis. When we think of both aspects at once we are concerned with gamma diversities, the flux of association and disassociation of species in successional and evolutionary time, and the enormous richness of the living world in species remaining from Simpson's[98] three billion years of community evolution.

The salient organizing principles in these phenomena seem to be – not neglecting other community processes – competitive interplay and adaptive diversification. It is from competition, and the selective effects of competition on evolutionary maneuver of species populations, that niche structure of communities results. It is relative "success" in competition for niche space and resources that is expressed in dominance and in importance-value curves. It is from evolution toward different ways, with reduced competition, of exploiting the resources offered by different choices and combinations of other species, that the complexity of food webs is produced and some of the relative stability of animal populations may result. It is from dispersion along habitat gradients of species with overlapping niches that the population patterning of communities in the field results. This kind of organization of communities in place and of community patterns in space – based on diversity and differentiation, on innumerable particular responses or feedbacks without any central control, maintained by selection and transmission of numerous genetic instructions for the interacting species but no governing instructions for the system as a whole – is like no other organization that comes to mind. (Though there are points of partial resemblance with the increasing professional specialization of a civilization of high culture and technology like our own.) It is because of what is, to me, the distinctiveness of this synecological organization that I like to interpret the manner of its evolution through the metaphor of the evolutionary dance.

OBSERVED DIVERSITIES IN PLANT COMMUNITIES

On southern Long Island, on a relatively young land surface of limited topographic contrasts, forests of low alpha and beta diversity result from these evolutionary processes. Alpha measurements for the Brookhaven forests are 16 vascular plant species in a 0.1-hectare sample, and a Simpson[97] index of dominance concentration of 0.29. I think there is advantage, because of the partial independence of diversity (or richness in species) and dominance concentration in plant communities, in measuring these characteristics separately rather than relying on the Shannon-Wiener index alone.[106,cf.47,80,91.120] The forests combine a rather low species diversity for temperate forests with weak dominance concentration; the latter at least might be expected to increase with longer successional development. Broader observations on plant community diversities may be summarized in a few points.

1. The master latitudinal and altitudinal gradient of increasing diversity from cold climates toward the warm Tropics applies in a broad way to vascular plant communities on the alpha and gamma levels, at least. Alpha diversities of the tree stratum of tropical forests are several times higher than in the richest temperate forests.[3,89,14,4,28,71] I know of no effective comparisons of tropical and temperate beta diversities for plants, but arctic beta diversities seem, in some landscapes at least, to be low. Data of Janzen and Schoener[43] suggest that beta diversities of foliage insect communities, which should be closely related to vascular plant beta diversity, are higher in tropical than in temperate landscapes.

2. Different strata of plant communities show independent trends of diversity; the different strata are sometimes positively, sometimes negatively correlated along different ranges of environmental gradients. It is not uncommon in mountains to find tree species diversity increasing toward lower elevations, whereas herb species-diversity is maximum at middle elevations.[105,109,12] The middle-elevation forests richest in herbs can be richest also in total vascular plant alpha diversity, in departure from the overall latitudinal and altitudinal trend. One finds vegetation patterns in which beta diversities behave differently for the different strata; in some northern forests low beta diversity of the tree stratum is combined with higher beta diversity of the herb stratum along the same topographic gradients.

3. In the Temperate Zone both alpha and topographic beta diversities increase from maritime climates inland to continental climates (Figure 6). Such is the case in

mountains[105,109] and seems to be the case in boreal forests.[46] To offer only what seems the simplest explanation, maritime climates, though more stable with respect to temperature, may be unstable or unfavorable in consequence of drought stress during the summer season of maximum biological activity.

4. Temperate communities do not simply decrease in diversity along moisture gradients. Maximum diversity is in some cases in mesophytic communities[70,71] but in many cases it is in intermediate environments, neither the most mesophytic nor the driest.[104-5,109,9,62,63] When closed forests are compared with nearby woodlands of smaller trees in more open growth in a drier environment, it is usually the woodland which has the higher diversity. The observation relates to a point long known to plant geographers following Raunkiaer, that in temperate forest climates not the trees but the hemicryptophytes (herbs with their buds at the ground surface) predominate in numbers of species. A woodland may offer herbs an environment both more favorable in some respects – especially less limited in light – and more varied from place to place as regards effects of the trees on light and on soil conditions, than a forest. Some temperate grasslands, and even some less extreme deserts, have very high species diversities compared with temperate forests. Less extreme deserts may have, in addition to high diversity of perennial plants, a wealth of annual plant species or therophytes, a life-form absent or of modest representation in closed forests. The combined perennial and annual diversity of such deserts, notably the Sonoran desert of lower mountain slopes in Arizona[109] may exceed the diversity of almost any temperate forest.

5. Vascular plant species diversity in the Temperate Zone is thus not to be interpreted only by assumptions about favorableness, stability, and productivity.[cf.8] Environments in some sense favorable to the lower strata, relatively arid and unstable as the climate may be, are more conducive to high species diversity than the environments of forests. Highly productive communities – coastal redwood forests and saltmarsh, say – may be of low diversity; and less productive communities – woodlands, grasslands, and some deserts – of high diversity. Apart from the limitation of both production and diversity by extreme cold, drought, and salinity, there appears to be no significant correlation of diversity and productivity of temperate plant communities.

6. Diversity increases during plant succession, but may in many cases decrease from some stage before the climax into the climax itself.[103,58,59,69,24,25,99,74,112,120] Though moderate grazing may increase diversity, heavy grazing may produce retrogressive decrease in diversity,[41] as may exposure to irradiation.[118] Pichi-Sermolli[78,79] and Pielou[80] have observed that internal patterning or degree of patchiness of plant communities may decrease in later stages of succession. Dominance concentration may either increase or decrease during succession; and there is no necessary relation between low equitability or lognormal ("Type IV"[38]) distribution, and successional status or community stability.

7. Diversity is in some cases correlated with soil fertility.[22] When acid soils can be compared with calcareous soils the diversities are generally higher on the latter; the effect appears in Monk's[62,63] data on southeastern forests and Dahl's[10] on Norwegian alpine communities. In the Santa Catalina Mountains, however, the limestone communities were of comparable alpha diversity and lower beta diversity than the acid-soil communities.[110] Maarel[48,49] observed maximum diversity at inter-

mediate soil pH values of dune vegetation; Ramsay and De Leeuw[86] on soils of inter-mediate texture. In a semi-arid area Ionesco[40] found decreasing diversity with in-creasing depth of sand. It is paradoxical that serpentine soils, with their distorted representation of metallic elements and deficiency of calcium, should support vege-tation richer in species than adjacent more typical soils. Such is the case in the Siskiyou Mountains, where woodlands on serpentine were compared with forests of corresponding habitats on diorite[102,105] and in McNaughton's[61] California grasslands.

8. The low diversities of many evergreen coniferous forests contrast with higher diversities of many broadleaf deciduous forests.[106] Among five pairs of coniferous and deciduous forests of comparable habitats in the Great Smoky Mountains, mean species number per 0.1 ha were 15 for the coniferous, 41 for the deciduous stands. The coniferous forests have characteristically higher dominance concentration as well as lower diversity, but in samples from the Great Smoky Mountains diversity is lower in coniferous forests with dominance shared by two species, than in deciduous forests of strong concentration of dominance in one species.[106] Some other communi-ties of strong dominance by a single evergreen species are of low diversity, such as certain evergreen oak forests, eucalypt forests, and chaparral. Evergreen leaf litter decomposes more slowly, and soil conditions may be in a number of ways less favor-able than in deciduous forests. Evergreenness reduces the seasonal variability of light conditions on which a part of the herb diversity of deciduous forests (especially as re-gards vernal herbs) depends. More generally, however, there is reason to suspect that many of these cases of strong dominance and low diversity are cases of strong allelo-pathic effects by the dominant plants.[65,66,108] In these communities soil chemistry is much influenced by the secondary substance chemistry of the dominant plant spe-cies, and only those other species tolerant of this chemistry can occur there. The effect is illustrated in the fire cycle of the California chaparral, a community of low diversity in which the mature shrub canopy (dominated by *Adenostoma fasciculatum* and *Arctostaphylos* species) inhibits by allelopathy the germination and growth of an-nual herbs. When the canopy is removed by fire, the next rainy season produces a blooming of annual herbs whose seeds were already present in the soil.[67] There may be far more to dominance in plant communities than shading and competition for root space.

I would synthesize these observations thus: There are broad trends of decreasing diversities of vascular plant communities toward environments less stable, less favor-able, and more extreme, but these trends are in detail strongly modified by more complex relationships involving different plant strata and life-forms, and by the ef-fects of characteristics of dominant species on composition and diversity of the com-munities they dominate.

The crucial evidence for the saturation of alpha diversity of bird communities[54] is the relative convergence of diversity values in temperate and tropical forests (even if qualified in detail by treatment of forests as of three strata in the Temperate Zone, but four in the Tropics). The plant species diversity of these same forests, as repre-sented in the tree stratum, may climb from 3 species in a boreal forest, to 30 in a midlatitude cove forest, to 300 in a tropical rainforest.[82,83] No convergence or satura-tion is evident, whether along the latitudinal gradient or comparing temperate com-munities, or tropical communities, among themselves. This is not to deny the great interest of MacArthur's observation, but to deny transfer of application from bird to plant communities. We may then ask: How have plant communities evolved to produce the diversities we observe, including this difference from the phenomenon of relative saturation in bird communities?

EVOLUTION OF DIVERSITY

Organization I have characterized as the means by which functioning complexity is maintained through time. Diversity of a community is a form of complexity, of organized differentiation, and expresses the kind of organization – involving competitive interplay, role differentiation, and diverse interactions tending on the whole to modulate community fluctuation – we seem to observe in communities. Time is needed for the evolution of organization, whether that of protoplasm, the organism, a community, a culture, or the devices of high technology. There must be time during which the addition of functional details occurs gradually, the additions being each as it occurs integrated into the organization of the whole.

Time is involved also because additions to the complexity make possible further additions to the complexity. The existence of a community diversity, entailing n species in their niches, implies that there exist additional niche possibilities for the use in different combinations of these n species as resources by, say m species new to the community. At a later time $m+n$ species offer a still greater range of niche possibilities for ways of relating to different combinations of these. Diversification in natural communities can thus occur in a self-augmenting, or circularly self-inducing, or as Hutchinson puts it, an autocatakinetic manner.[37,8]

Plant populations are regulated by processes of which we know little, but which should involve internal differentiation of the community as it affects niche loci for seedlings. By niche loci I mean places on the ground where the particular combination of environmental characteristics determined by microtopography, soil, light, fungal biota, and effects of other plants, may permit germination and survival of seedlings of a given species. Density of a climax plant population may be governed by a circular steady-state operation involving density of seeds, frequency of these in suitable niche loci, survival of seedlings to reproductive age as affected by competition, predation, and environmental fluctuation in relation to relative favorableness of locus, and density of seeds that the population of surviving, mature plants produces.

The chemical differentiation of higher plants may be of great significance to diversity phenomena, as Ehrlich and Raven[16] have suggested. The higher plants are biochemical specialists: different compounds and groups of compounds of the secondary plant substances appear in different species and groups of species. This chemical differentiation, involving substances of no evident metabolic function, is an extraordinary fact which has long been known and has only recently begun to receive the evolutionary interpretation it requires. It seems evident now that secondary plant substances, though some are wastes, are in large part biochemical defenses of the plant against its enemies.[21,16,108] They are predominantly inhibitory, repellent, or toxic – to consuming animals, to bacteria and fungi, and, if sufficiently concentrated in the soil, to other plants as well.[65,66] As Ehrlich and Raven[16] observe, different plant families have made different "choices" of emphasis of secondary substances in their biochemistry. Animals feeding on the plants must accommodate to the choice, and consumption by the animals selects for intensification of the secondary substance defense of the plants. The effect has been a self-intensifying evolutionary process, from which result the biochemical differentiation of the higher plants, the different tolerances of plant chemistry by grazing animals, the balances between biochemical defenses of the plant and biochemical adaptation by the plant's enemies, and much of the niche differentiation among grazing animals (and, no doubt, dependent fungi and bacteria).

As regards diversities of plant communities, chemical differentiation may have these further implications. First, it implies differentiation of the community into partially separate complexes of chemically related organisms: plant species and the groups of animals and saprobes interacting with them.[42,108] One may argue that control by separate groups of predators implies that two species occupy different niches; in any case the partially separate control systems, by governing the populations at levels reducing competition between them, permit more effective division of niche space, hence increased numbers of niches and species which can coexist without competitive extinction.[31] The effect relates to Paine's[72,73] observations on the significance of predators for animal diversity. Second, it gives evolutionary meaning to the allelopathy which limits diversity in some plant communities of strong dominance. Third, it may contribute (along with instability effects and different inorganic nutrient requirements) to solution of the "paradox of the plankton" – the unexpectedly high diversities of communities of phytoplankton species all apparently competing as autotrophs for the same resources.[38,60] Some phytoplankton release toxic "secondary substances"; and most both release and absorb varied food molecules and vitamins.[94,11,15] Phytoplankton species may be chemically differentiated, with different patterns of partial or full dependence on absorption of different organic compounds, different positions in the network of biochemical exchanges in the community. Fourth, it may permit diversity to increase in terrestrial communities in a self-augmenting manner involving seedling survival. I shall assume that seedlings are sensitive, at least in a partial, relative way affecting differential survival, to allelopathic chemistry of the soil and to the microbiological community of the soil, which may be influenced in turn by biochemical adaptation of bacteria and fungi to the living plants and their dead remains. The larger the number of plant species present, the larger the diversity of niche loci as affected by combinations of microrelief, sun-flecking, and soil differences, with biochemistry of plant litter and microbial communities. The more diverse the plant community, the more complex the internal pattern of niche loci that may be utilized by seedlings of those species and that are potentially available as niche differences for other species that may enter the community.

I conclude that species diversity of plant communities is an evolutionary product, subject to self-augmentation through time. There is no ceiling or saturation level for the diversity which results; the chemical differentiation of the higher plants implies virtually unlimited potentialities for the addition of different species with different interactions with predators and seedling soil requirements. It is thus that plants and insects may differ from birds. The niche responses of birds are not to chemistry but to community structure and broad food categories, in which possibilities of niche division are limited. Bird communities can thus be saturated, determinate in their evolution of diversity; but plant and insect communities may show indefinite, indeterminate evolutionary increase in diversity.

Species diversity of plant communities is an evolutionary product the levels of which we observe may be influenced by time, environmental stability and favorableness, and characteristics of communities themselves. Time is difficult, since the species of communities enter from other communities, to treat as a measurable variable; but time of evolution since the last major environmental disturbance should contribute to diversity differences. Environmental stability and favorableness affect the number of species which have been able to adapt to and survive in an environment during the evolutionary time available. Stability has other implications in making possible greater emphasis of biological accommodation over adaptation to physical environment in the selective forces, and the actual genes and adaptations realized through them. It may determine the rate at which niche subdivision occurs and the degree of niche space division which has been possible in the time available. This is largely to restate the interpretation of tropical diversity by Dobzhansky and others[13,19,106,77] and Sanders'[92,93,33] time and stability hypothesis. To it one must add, for terrestrial plant communities, stratal relations and dominance effects as biological aspects of the community that affect the niche potentialities for other species, and thereby the rate at which species may be added to the community. The dance floor is always open, but open on different terms according to the kind of dance and the kinds of dancers that have already evolved.

CODA

I should like to add to this a coda, in a different key, since our concerns move today from three billion years of life to the brief evolution of Western civilization and its technology. It is possible to consider the matter most simply in terms of three combinations of determinate, versus indeterminate, increase of energy flow and complexity. The fourth combination seems of less interest. No natural community has indeterminate increase of energy flow, that is, productivity, but the complexity of species-nets through which the energy flows may be subject to either determinate or indeterminate evolutionary increase. The territorial birds represent the former case, the terrestrial plants and insects the latter. Diversity does not, of course increase to infinity, though biologists from the Temperate Zone visiting the Tropics may feel it is well on its way. An informal organization based on interplay of species and functional groupings evolves in conjunction with the diversity, and maintains it, and serves to prevent most disturbances from producing self-amplifying, major effects on the community.

Western and other modern societies accept as an ideal indeterminate increase in energy use, and thereby of wealth, to provide for and enrich an indeterminately increasing population. Thus provided for the population increases geometrically, and the product of geometric population increase and increase of energy use per individual is the accelerating, self-amplifying, and compulsive increase in energy use and other aspects of technology in the United States. To manage this system with its accelerating expansion requires increasing complexity of organization – increasing diversification and subdivision of corporations and public agencies, increasing speed and content of information communication and processing, increasingly narrow and diverse professional specialization of individuals who are increasingly parts of larger and more complex social organizations, and so on. An indeterminate increase in

192

complexity is thus occurring. We are still, with stunning technological skill and power, largely successful in an external, obstensible sense, at least, in keeping organized this system of expanding energy use and complexity. It is likely, though I cannot prove it, that the organizing devices are increasingly strained by that which they must organize, and may become increasingly vulnerable to self-amplifying disturbance.

There are, however, implications for human psychology. It is not by shortage of space and resources nor by massive biosphere toxicity that the limits of our civilization (if it will not limit itself) are most likely to be set, but by the effects of human numbers, technological and cultural change, and environmental degradation on the psychology of people. Crowding does not improve the quality of human beings and human relationships; and the provision of the individuals with technological power, notably the automobile, may increase the sense of subjugation to traffic, noise, pollution, and urban interminableness as products of crowding. The individual is to a degree diminished in his sense of a manageable world to which he can relate by the enormity of cities and their problems, by isolation from natural landscapes, by rapidity of cultural change, by participation in a perceptibly small role in perceptibly enormous social and corporate systems. One suspects that to bear well the conditions of urban life with its increasing scale, crowding, complexity, and bleakness there is increasing need of qualities of secure self-respect, patience and perspective, tolerance and maturity, that urban life does not now much encourage.

Nor, for that matter, may the life of the young in middle-class suburbs. As ideals comfort, pleasure, passive and commercial entertainment, insulation from adult problems, and relative freedom from real effort and discipline may have singularly unideal consequences. It seems they may impoverish the bases of growth of inner strength, sense of self, respect for work, and sense of community with others and the society and its purposes and cultural heritage. They can produce some youth who have never learned successful management of the angers endemic to the human condition, in whom this anger must direct itself toward destruction and self-destruction. The life of passive pleasure may deny the discipline of growth by which rationality and respect for others are superimposed on the easier ways of irrationality and antagonism. This life leads in an extreme few toward an ideological and emotional solipsism, combined with systematic hatred of restraint and authority, hatred of others' freedom, and antagonism toward understanding of complexity and toward the life of the mind. I describe the extreme, but do so because I believe the United States is, by self-augmenting processes now involving commerce and the communications industry, producing increasing numbers of youth with characteristics toward this extreme.

I will suggest the predicament of our society: A system of accelerating growth and increasing complexity is stretching ever tighter its means of organization, while producing social and environmental problems ever more difficult and beyond realistic prospects of solution, while increasing the tensions and frustrations of the human beings who must maintain the organization and try to deal with the problems, while producing increasing numbers who scorn the system and its complexities without a rational sense of the limitations on alternatives, while producing small but increasing numbers of human beings sufficiently damaged as such that they desire the ruin of the society which, for all they can understand, is responsible. I find this, if true, an unencouraging system of simultaneous, nonlinear, differential equations. One need blame no single development or factor which might reasonably have been dif-

193

ferent – whether overpopulation, or infatuation with technology, or wealth and parental indulgence, or decline of religion and moral restraint, or ruthless commercialization of young people's entertainment – for these processes interlink and intensify one another. Tragedy, or its potentiality, too, can be an evolutionary product.

I will observe in judgment only that our civilization is the first in history to have available scientific knowledge in the two areas of ecology and psychology that might have given warning, and that whatever counsel these fields might have offered a civilization based on intelligence was thought irrelevant. I am led to end on an unaccustomed note for a scientific paper. There may be reasons the problems we see mounting around us will not decrease. If, however, the implications for stability of our civilization are as Holling[34] and I have differently formulated them, then I wish us all – our great, open society and its professional workers in the fields of the rational mind – luck.

REFERENCES

1. ALECHIN, W.W., *Repert. Spec. Novar. Regni Veget., Beih.* **37**, 1 (1926).
2. BESCHEL, R.E. AND WEBBER, P. J., *Ber. Natwiss.-Med. Ver. Innsbruck* **53** (Festschr. Gams.) 9 (1963).
3. BLACK, G.A., DOBZHANSKY, T., AND PAVAN, C., *Botan. Gaz.* **111**, 413 (1950).
4. CAIN, S.A., DE OLIVEIRA CASTRO, G.M., PIRES, J.M., AND DA SILVA, N.T., *Am. J. Botany* **43**, 911 (1956).
5. CODY, M., *Am. Naturalist* **100**, 371 (1966).
6. CODY, M.L., *Ibid.* **102**, 107 (1968).
7. COHEN, J.E., *Ibid.* **102**, 165 (1968).
8. CONNELL, J.H. AND ORIAS, E., *Ibid.* **98**, 399 (1964).
9. CURTIS, J.T., in *The Vegetation of Wisconsin; an Ordination of Plant Communities*, 657 pp., Univ. Wisconsin Press, Madison, 1959.
10. DAHL, E., *Norske Vidensk.-Akad. Oslo, I. Mat.-Naturv. Kl., Skr.* 1956(3), 374 pp., 1957.
11. DANFORTH, W.F., in *Physiology and Biochemistry of Algae*, p. 99, R.A. Lewin, Editor, Academic Press, New York, 1962.
12. DAUBENMIRE, R. AND DAUBENMIRE, J.B., *Wash. Agr. Expt. Sta. Tech. Bull.* **60**, 1 (1968).
13. DOBZHANSKY. T., *Am. Scientist* **38**, 209 (1950).
14. DREES, E.M., *Vegetatio* 5/6, 517 (1954).
15. DROOP, M.R., in *Physiology and Biochemistry of Algae*, p. 141, R.A. Lewin, Editor, Academic Press, New York, 1962.
16. EHRLICH, P.R. AND RAVEN, P.H., *Evolution* **18**, 586 (1965).
17. ELTON, C., *Animal Ecology*, 3rd impression, 209 pp., Sidgwick & Jackson, London, 1947.
18. ENGELMANN, M.D., *Ecol. Monographs* **31**, 221 (1961); also in ref. 32, p. 332.
19. FISCHER, A.G., *Evolution* **14**, 64 (1960).
20. FISHER, R.A., CORBET, A.S., AND WILLIAMS, C.B., *J. Animal Ecol.* **12**, 42 (1943).
21. FRAENKEL, G.S., *Science* **129**, 1466 (1959).
22. FRYDMAN, I. AND WHITTAKER, R.H., *Ecology* **49**, 896 (1968).
23. GAUSE, G.F., *The Struggle for Existence*, 163 pp., Williams & Wilkins, Baltimore, 1934.
24. GOLLEY, F.B., *Ecol. Monographs* **35**, 113 (1965).
25. GOLLEY, F.B., PETRIDES, G.A., AND McCORMICK, J.F., *Bull. Torrey Botan. Club* **92**, 355 (1965).
26. GOULDEN, C.E., See paper in this Symposium.
27. GRANT, P.R., *Am. Naturalist* **100**, 451 (1966).
28. GREENLAND, D. J. AND KOWAL, J.M.L., *Plant Soil* **12**, 154 (1960).
29. HAIRSTON, N.G., *Ecology* **40**, 404 (1959); also in ref. 32, p. 319.
30. HAIRSTON, N.G., *J. Ecol.* **52** *(Suppl.)*, 227 (1964).
31. HARPER, J.L., CLATWORTHY, J.N., McNAUGHTON, I.H., AND SAGAR, G.R., *Evolution* **15**, 209 (1961).

32. *Readings in Population and Community Ecology*, 388 pp., W.E. Hazen, Editor, Saunders, Philadelphia, 1964.
33. HESSLER, R.R. AND SANDERS, H.L., *Deep-Sea Res.* **14**, 65 (1967).
34. HOLLING, C.S., See paper in this Symposium.
35. HUTCHINSON, G.E., *Proc. Acad. Nat. Sci. Phila.* **105**, 1 (1953); also in ref. 32, p. 2.
36. HUTCHINSON, G.E., *Cold Spring Harbor Symp. Quant. Biol.* **22**, 415 (1957).
37. HUTCHINSON, G.E., *Am. Naturalist* **93**, 145 (1959); also in ref. 32, p. 293
38. HUTCHINSON, G.E., *Am. Naturalist* **95**, 137 (1961).
39. HUTCHINSON, G.E., *The Ecological Theater and the Evolutionary Play*, 139 pp., Yale Univ. Press, New Haven, 1965.
40. IONESCO, T., *Bull. Serv. Carte Phytogeogr. Paris Ser. B* **3**, 7 (1958).
41. ITOW, S., *Japan. J. Botany* **18**, 133 (1963).
42. JANZEN, D.H., *Am. Naturalist* **102**, 592 (1968).
43. JANZEN, D.H. AND SCHOENER, T.W., *Ecology* **49**, 96 (1968).
44. KING, C.E., *Ecology* **45**, 716 (1964).
45. KOHN, A. J., *Ecol. Monographs* **29**, 47 (1959).
46. LA ROI, G.H., *Ibid.* **37**, 229 (1967).
47. LLOYD, M. AND GHELARDI, R. J., *J. Animal Ecol.* **33**, 217 (1964).
48. VAN DER MAAREL, E., *Rapp. Rijks Inst. Veldbiol. Onderz., Natuurbehoud, Zeist.*, 170 pp., 1966.
49. VAN DER MAAREL, E. AND LEERTOUWER, J., *Acta Botan. Neerl.* **16**, 211 (1967).
50. MACARTHUR, R.H., *Proc. Natl. Acad. Sci. U.S.* **43**, 293 (1957).
51. MACARTHUR, R.H., *Ecology* **39**, 599 (1958).
52. MACARTHUR, R., *Am. Naturalist* **94**, 25 (1960); also in ref. 32, p. 307.
53. MACARTHUR, R.H., *Am. Naturalist* **98**, 387 (1964).
54. MACARTHUR, R.H., *Biol. Rev.* **40**, 510 (1965).
55. MACARTHUR, R.H. AND MACARTHUR, J.W., *Ecology* **42**, 594 (1961).
56. MACARTHUR, R.H., MACARTHUR, J.W., AND PREER, J., *Am. Naturalist* **96**, 167 (1962).
57. MACARTHUR, R., RECHER, H., AND CODY, M., *Ibid.* **100**, 319 (1966).
58. MARGALEF, R., *Ibid.* **97**, 357 (1963).
59. MARGALEF, R., *Oceanogr. Mar. Biol. Ann. Rev.* **5**, 257 (1967).
60. MARGALEF, R., See paper in this Symposium.
61. MCNAUGHTON, S. J., *Ecology* **49**, 962 (1968).
62. MONK, C.D., *Ecol. Monographs* **35**, 335 (1965).
63. MONK, C.D., *Am. Naturalist* **101**, 173 (1967).
64. MOTOMURA, I., *Japan. J. Zool.* **44**, 379 (1932).
65. MULLER, C.H., *Bull. Torrey Botan. Club* **93**, 332 (1966).
66. MULLER, C.H., *Vegetatio* **16**, in press (1969).
67. MULLER, C.H., HANAWALT, R.B., AND MCPHERSON, J.K., *Bull. Torrey Botan. Club* **95**, 225 (1968).
68. NUMATA, M., NOBUHARA, H., AND SUZUKI, K., *Bull. Soc. Plant Ecol.* **3**, 89 (1953).
69. ODUM, E.P., *Ecology* **41**, 34 (1960).
70. OGAWA, H., YODA, K., AND KIRA, T., *Nature and Life in Southeast Asia* **1**, 21 (1961).
71. OGAWA, H., YODA, K., KIRA, T., OGINO, K., SHIDEI, T., RATANAWONGSE, D., AND APASUTAYA, C., *Ibid.* **4**, 13 (1965).
72. PAINE, R.T., *Am. Naturalist* **100**, 65 (1966).
73. PAINE, R.T., *Ibid.* **103**, 91 (1969).
74. PATRICK, R., *Proc. Natl. Acad. Sci. U.S.* **58**, 1335 (1967).
75. PATRICK, R., HOHN, M.W., AND WALLACE, J.H., *Notulae Naturae (Acad. Nat. Sci. Phila.)* **259**, 1 (1954).
76. PATRICK, R. AND STRAWBRIDGE, D., *Am. Naturalist* **97**, 51 (1963).
77. PIANKA, E.R., *Ibid.* **100**, 33 (1966).
78. PICHI-SERMOLLI, R.E., *Webbia* **6**, 378 pp. (1948).
79. PICHI-SERMOLLI, R.E., *J. Ecol.* **36**, 85 (1948).
80. PIELOU, E.C., *J. Theoret. Biol.* **10**, 370 (1966).
81. PIRES, J.M., DOBZHANSKY, T., AND BLACK, G.A., *Botan. Gaz.* **114**, 467 (1953).
82. POORE, M.E.D., *J. Ecol.* **52** *(Suppl.)*, 213 (1964).

83. Poore, M.E.D., *J. Ecol.* **56**, 143 (1968).
84. Preston, F.W., *Ecology* **29**, 254 (1948).
85. Preston, F.W., *Ecology* **43**, 185, 410 (1962).
86. Ramsey, D. McC. and De Leeuw, P.N., *J. Ecol.* **53**, 661 (1965).
87. Recher, H.F., *Am. Naturalist* **103**, 75 (1969).
88. Reiners, W.A., *Bull. Torrey Botan. Club* **92**, 448 (1967).
89. Richards, P.W., *The Tropical Rain Forest; an Ecological Study,* 450 pp., Cambridge Univ. Press, 1952.
90. Root, R.B., *Ecol. Monographs* **37**, 317 (1967).
91. Sager, P.E. and Hasler, A.D., *Am. Naturalist* **103**, 51 (1969).
92. Sanders. H.L., *Ibid.* **102**, 243 (1968).
93. Sanders, H.L., See paper in this Symposium.
94. Saunders, G.W., *Botan. Rev.* **23**, 389 (1957).
95. Schoener, T.W., *Evolution* **19**, 189 (1965).
96. Shinozaki, K., *Physiol. Ecol. Kyoto (Seiri Seitai)* **6**, 127 (1955).
97. Simpson, E.H., *Nature* **163**, 688 (1949).
98. Simpson, G.G., See paper in this Symposium.
99. Tagawa, H., *Mem. Fac. Sci. Kyushu Univ. Ser. E. (Biol.)* **3**, 165 (1964).
100. Udvardy, M.D.F., *Cold Spring Harbor Symp. Quant. Biol.* **22**, 301 (1957).
101. Vandermeer, J.H. and MacArthur, R.H., *Ecology* **47**, 139 (1966).
102. Whittaker, R.H., *Ecology* **35**, 275 (1954).
103. Whittaker, R.H., *Ecol. Monographs* **23**, 41 (1953).
104. Whittaker, R.H., *Ibid.* **26**, 1 (1956).
105. Whittaker, R.H., *Ibid.* **30**, 279 (1960).
106. Whittaker, R.H., *Science* **147**, 250 (1965).
107. Whittaker, R.H., *Biol. Rev.* **42**, 207 (1967).
108. Whittaker, R.H., in *Chemical Ecology,* E. Sondheimer and J.B. Simeone, Editors, Academic Press, New York, in press, 1969.
109. Whittaker, R.H. and Niering, W.A., *Ecology* **46**, 429 (1965).
110. Whittaker, R.H. and Niering, W.A., *J. Ecol.* **56**, 523 (1968).
111. Whittaker, R.H. and Woodwell, G.M., *J. Ecol.* **56**, 1 (1968).
112. Whittaker, R.H. and Woodwell, G.M., *J. Ecol.* **57**, 157 (1969).
113. Whitworth, W.A., *DCC Exercises in Choice and Chance,* p. 235, Deighton, Bell & Co., Cambridge, 1897 (reprinted by Hafner, New York, 1965).
114. Williams, C.B., *J. Ecol.* **34**, 253 (1947).
115. Williams, C.B., *J. Ecol.* **38**, 107 (1950).
116. Williams, C.B., *Patterns in the Balance of Nature, and Related Problems in Quantitative Ecology,* 324 pp., Academic Press, New York, 1964.
117. Woodwell, G.M., *Science* **156**, 461 (1967).
118. Woodwell, G.M. and Whittaker, R.H., *Quart. Rev. Biol.* **43**, 42 (1968).
119. Woodwell, G.M. and Whittaker, R.H., *Am. Zool.* **8**, 19 (1968).
120. Odum, E.P., *Science* **164**, 262 (1969).

[*Editor's Note:* The Discussion has been omitted.]

ERRATUM

Page 183, line 4 from the bottom of the page, "coenoclines" should be "coecoclines."

10

DIVERSITY AND PATTERN IN PLANTS AND INSECTS[1]

W. W. Murdoch, F. C. Evans, and C. H. Peterson

Abstract. The plants and Homoptera on three old fields in southeast Michigan were sampled. Within fields, correlations between plant and insect diversity were generally weak. But using all samples from three fields, evenness (J) and diversity (H) of the insects were highly correlated with plant evenness and plant diversity, respectively. For example, 72% of the variance in insect H could be accounted for by variation in plant H. Number of species (S) showed a positive but weaker correlation. When correlations were based on the pooled samples from each field, all three statistics for insects were highly correlated with those for the plants. Insect H was also highly correlated with plant structure (foliage height diversity, FHD) over all three fields. These two measures of plant diversity (H and FHD) were highly correlated and were equally good correlates of insect H. Together they accounted for 79% of the variance in insect H.

This extends to insects a correlation between plant and animal diversity, already well established for birds and possibly true for lizards and rodents. It leaves open the unresolved question as to whether plant structure or plant species diversity is more important.

The diversities of different components of a community seem to be correlated, in the few cases studied, and in particular, animal diversity has been correlated with aspects of the plant diversity. It is not surprising that a greater variety of plants should lead to a greater variety of plant-eaters. However, the reasons for observed correlations between plant structural diversity and animal species diversity are less obvious. In one group of animals the reason appears obvious; different bird species nest and forage at different heights and their diversity is related to the structural diversity of the vegetation (e.g., MacArthur and MacArthur 1961, Recher 1969).

In this paper we explore these relationships in insects. Plant-sucking bugs (Homoptera) of several old fields were studied because they form a dominant group of insect herbivores in these communities, and because their diversity might be expected to be closely tied to that of plants. Since these insects feed directly on the green plants, unlike the other groups that have been studied, and since at least some of them are host specific (DeLong 1948, Whitcomb 1957), we expected to find correlations with plant species diversity, but we also measured foliage height diversity (structure). To avoid geographic effects on diversity, and to keep in the same kind of habitat, we sampled several old fields, all within a 1-km radius and all abandoned for the same period of time.

METHODS

The study plots were in the University of Michigan's E. S. George Reserve in southeast Michigan.

[1] Received October 27, 1971; accepted May 9, 1972.

Each plot was a square, 30.4 m (100 ft) on a side, in the center of a large old field. The fields were abandoned about 1920. Each old field was surrounded by oak-maple-hickory woodland. Field 1 was dominated by dense grasses—particularly *Poa pratensis*—with some forbs scattered throughout. Field 2 was in the middle of "Evans Old Field" (Evans and Dahl, 1955) and was a typical "upland" dry field with incomplete vegetation cover, little grass, and dominated by forbs such as *Lespedeza* spp. and *Rumex acetosella*. *Poa compressa* and *Aristida purpurascens* were the dominant grasses. Field 3 was a damp meadow with a lush growth of both forbs (e.g., *Hieracium* spp., *Daucus carota*) and grasses (especially *Poa pratensis*). A fourth field, referred to at one point in the present paper, was a very sparsely vegetated, sandy "blow-out" area with the grass *Aristida purpurascens* dominant and *Lespedeza capitata* fairly abundant.

Sampling

The Homoptera were sampled by sweeping with a sweep net. Sampling was done in the morning, on warm sunny days when the vegetation was dry. Each sample was a set of sweeps taken while moving slowly forward in a line. The direction moved was always the same (north to south). In 1964 and 1967 each sample contained 10 sweeps taken along a transect 5 m long and 1 m wide. In 1966 each sample was three sweeps taken from 1 m². The insects were etherized, dried, and identified and counted later. All but a dozen or so species were identified, out of about 75 species collected. All 75 were used in the analyses.

Most sampling was done in July, and most of the analyses concern the July samples. July is the month in which nymphs are very rare, and nymphs were ignored in the analyses. Aphids also were excluded.

The plant-sampling technique was varied from year to year. In July 1964 a preliminary sampling was done in which nine regularly spaced samples of 1 m² each were taken in each field and the plants identified. Ten regularly spaced sweep samples of the insects were also taken in July 1964. In July 1966, 30 (22 in field 2) randomly placed 1 m² quadrats were sampled (see Table 2). Each plant species was identified and its cover estimated by eye as the number of square decimeters covered. For small areas this probably is as good a method of estimating cover as any other (D. Goodall, personal communication). There is some small error involved, and in samples with total cover, the estimated coverage of all species summed was sometimes over 100% (Appendix). In July 1966 these same quadrats were then sampled for Homoptera by sweep sampling. Thirty insect samples were taken in all three plots.

We were interested in the effects of scale on diversity, and in 1967 sample size was increased. In July, 10 randomly placed transects were sampled in each plot. Since we wanted to measure the structural diversity of the plants as well as species diversity, we used a "point-quadrat" method. Each sample consisted of a total of 100 points, points being placed at 5-cm intervals to give a 5-m-long transect. At each point a long metal rod was slowly passed down through the vegetation at an angle of 45° to the vertical and all contacts with plants were recorded. We recorded the species and in which of three heights the contact was made. The three levels were: 0–6 inches (0–2.4 cm), 6–18 inches (2.4–7.1 cm), and > 18 inches. These levels were used in calculating foliage height diversity. For each plant species, abundance in a sample was the total number of contacts, which could thus exceed 100. Each of these 10 transects was then sampled for Homoptera in July by taking 10 sweep samples along each transect. In addition, 10 similar random transects were swept in each of the three fields in June and August 1967.

To check on the plant-sampling methods, in July 1967 we estimated by eye the cover of each plant species in five 1-m² quadrats (giving a transect 5 m by 1 m), 0.5 m to the east of the point quadrat survey. Although the vegetation differs over this short space, these data gave the same conclusions as we have reached in this paper using the point data. In a few cases correlations with measures of the insects were even better than with the point data. So we are confident that the change of plant-sampling method between years (but not the change in sample size) is unimportant. We chose to use the point data because

they can be used to compute both plant species diversity and plant structural diversity.

Similarity index

As a measure of similarity between two samples, or between two sets of pooled samples, we used the index I (see Whittaker 1960 for references to earlier applications of this index). The proportionate similarity between two samples, A and B, is

$$I = 1 - 0.5 \left(\sum_i^S |a_i - b_i| \right),$$

where a_i is the proportion of the total individuals in sample A that belongs to species i, and b_i is the proportion in sample B belonging to species i, and there are S species. Complete similarity gives $I = 1$, and complete dissimilarity gives $I = 0$. This measure, like the diversity measures described below, is largely influenced by the few most abundant species.

Diversity indices

For each separate sample and for each pooled sampled, of both plants and insects, the Brillouin (1960) measure of diversity was used:

$$H = \frac{1}{N} \log_e \frac{N!}{N_1! \, N_2! \ldots N_s!},$$

where N is the total number of individuals, S is the total number of species, and N_i is the number of individuals in the ith species. Pielou (1966) has recommended the application of this Brillouin index to discrete collections such as the samples we have taken. In the case of the plants, the N_i's do not represent numbers of individuals, but rather the number of square decimeters of area covered in the sample.

As a measure of the evenness component of species diversity, we used the ratio of the observed diversity to the maximum possible diversity for that particular number of species and individuals (Pielou 1966):

$$J = H_{obs}/H_{max},$$

where

$$H_{max} = \frac{1}{N} \log_e \frac{N!}{\left\{ \left[\frac{N}{S} \right]! \right\}^{S-r} \left\{ \left(\left[\frac{N}{S} \right] + 1 \right)! \right\}^r},$$

where $[N/S]$ is the integer part of N/S, and $r = N - S[N/S]$.

Foliage height diversity was measured by the Shannon index (Shannon and Weaver 1963):

$$FHD = H' = \sum p_i \log_e p_i,$$

where p_i is the proportion of the total number of contacts with plants that occur in the ith stratum. This is the index that has been used in the past to measure FHD. S is the total number of strata. The use of *FHD* is an attempt to measure the structural diversity of the flora, and since forbs and grasses present two clearly distinct kinds of plant structure, the contacts in each stratum were further divided into two categories, depending on whether they were forbs or grasses; so *S* has a maximum value of 6.

RESULTS

Since we will compare samples from different times, we want to know if each field had a fairly similar community of plants and insects each July and over the summer season.

Table 1 shows comparisons (similarities) based on the pooled samples for each field for each month: (1) For both plants and insects the average similarity between Julys of different years, for a given field, is generally greater than the average similarity between that field and other fields in the same month. (2) Comparisons between months in the same year (1967), for a given field, are available only for insects. Again, average similarities between summer months are usually greater than average similarities between fields. Consequently, there is good reason for considering these three fields as rather distinct communities. They differ from each other, yet within each field there is good consistency of the assemblage over the years and, to a lesser extent, over the summer season.

Table 1 shows that the fields are all dissimilar, though the insect assemblages on different fields are more alike than are the plants. For both plants and insects, fields 1 and 3 are most alike. We will show below that field 1 is much simpler than the other two fields, as indicated by diversity indices, while fields 2 and 3 are about equally diverse although very dissimilar in species composition.

Comparison of plant and insect species diversity

In the following comparisons, each data point is a measure of some component of insect diversity for a sample, versus the same measure for plant diversity for the same sample. An example, shown in Fig. 1, is the evenness component for all samples in the three fields for July 1966. We consider both components of diversity—evenness (*J*) and number of species (*S*)—as well as the diversity index (*H*) itself.

Table 2 shows the basic diversity data and Table 3 shows the results of all correlations such as those described above. There are three types of analyses concerning *species* diversity.

In Table 3*a* note first the within-field comparisons of the three statistics—*S*, *J*, and *H*. In this first type of analysis we are looking at the smallest scale, since individual samples are the units. In 1966, seven out of nine correlations between plants and insects are significant, while in 1967 only one out of the corresponding nine correlations is significant. We believe

TABLE 1. Comparisons of the relative abundances of species between plots and times, using the similarity index I^a

(a) BETWEEN-FIELD COMPARISONS

Date	Plants			Insects		
Fields:	1x2	2x3	1x3	1x2	2x3	1x3
June 1967				.32	.46	.51
July 1967	.11	.05	.22	.09	.37	.43
Aug 1967				.14	.24	.45
July 1966	.07	.03	.44	.21	.28	.76
July 1964				.15	.41	.64

(b) BETWEEN-MONTH COMPARISONS

Field:	1	2	3	1	2	3
Julys of 3 years						
1966 x 1967	.71	.78	.67	.95	.41	.60
1966 x 1964				.83	.78	.69
1967 x 1964				.86	.37	.66
Same year (1967)						
June x July				.90	.38	.81
July x Aug				.78	.77	.50
June x Aug				.84	.36	.44

$^a I = 1 - 0.5 (\Sigma' a_i - b_i /)$ where a_i is the abundance of species i in sample a, and b_i is its abundance in sample b. I compares the species composition of two fields or of the same field in two months. The comparisons were based on the total numbers in the pooled samples for each field in each month.

that this difference between years is due to the different sample sizes in the 2 years. In any one field, the larger 1967 samples were all rather similar; that is, in 1967 we were close to taking a typical part of the community with each sample, since each sample extended over enough ground to include more different patches, so that the samples were representative. By comparison, the 1966 samples were small

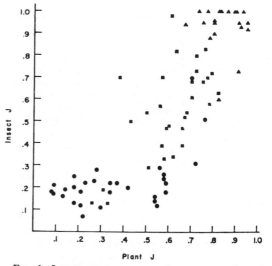

FIG. 1. Insect evenness versus plant evenness for all three fields in July 1966. Each point gives the data for one sample. Field 1 (●) had 30 samples, field 2 (▲) 22 samples, and field 3 (■) 30 samples. Samples with undefined evenness (only one species present) are omitted.

TABLE 2. Diversity statistics for three fields in July 1966 and 1967 for insects and plants[a]

Date	Variable		Field 1	Field 2	Field 3
July 1966	$\overline{H}$	Insects	0.33	1.02	0.93
		Plants	0.47	1.28	1.35
	H_p	Insects	0.43	2.48	1.17
		Plants	1.04	2.55	2.12
July 1967	$\overline{H}$	Insects	0.32	1.47	1.49
		Plants	0.95	2.05	2.02
	H_p	Insects	0.34	1.85	1.80
		Plants	1.61	2.48	2.57
July 1966	$\overline{J}$	Insects	0.22	0.87	0.55
		Plants	0.37	0.61	0.61
	J_p	Insects	0.16	0.85	0.34
		Plants	0.39	0.74	0.58
July 1967	$\overline{J}$	Insects	0.16	0.68	0.70
		Plants	0.57	0.81	0.79
	J_p	Insects	0.12	0.58	0.55
		Plants	0.55	0.75	0.71
July 1966	$\overline{S}$	Insects	4.6	7.5	7.6
		Plants	3.8	11.5	11.2
	S_p	Insects	14	22	32
		Plants	15	33	40
July 1967	$\overline{S}$	Insects	7.9	11.6	11.1
		Plants	6.3	14.6	15.4
	S_p	Insects	18	27	28
		Plants	19	29	38
July 1966	N_p	Insects	3634	274	1651
		Plants	2245	1181	2944
July 1967	N_p	Insects	3009	648	817
		Plants	1427	2135	2378
July 1967	$\overline{FHD}$	Plants	0.87	1.26	1.34
	FHD_p	Plants	1.06	1.33	1.40

[a] In July 1966, 30 samples were taken in fields 1 and 3, 22 were analyzed from field 2 for the plants, and 30 for the insects. In 1967, 10 samples were taken from each field. H is the diversity index, J is evenness, S is the number of species, and FHD is foliage height diversity. In each case the mean value of the samples in each field and the value for the pooled samples (subscript p) are given. N is the total abundance, measured for plants in square decimeters (1966) or number of contacts with a rod (see Methods).

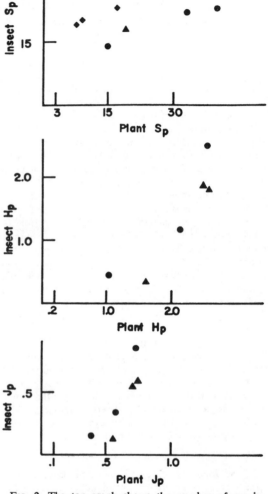

FIG. 2. The top graph shows the number of species (S_p) of insects and of plants for 3 years. There were four fields in 1964 (◆) and three in 1966 (●) and 1967 (▲). The middle graph shows insect species diversity versus plant species diversity (H_p) for three fields in 1966 and 1967. The bottom graph shows insect evenness versus plant evenness (J_p) for these three fields in 1966 and 1967. Each point shows data for the pooled samples from a field in July of 1 year.

enough to show large variations from patch to patch within a plot and each was therefore less "typical" of the field as a whole. This difference is illustrated in the Appendix, which shows the much greater range in values obtained from each field in 1966 than in 1967. However, even in 1966 when samples were small enough to produce variability, within fields the three aspects of insect diversity were not consistently correlated with plant diversity. We conclude that, within fields, when samples are small enough, the relation between insect and plant diversity is generally positive but weak.

The second analysis (Table 3b) uses samples from all three fields simultaneously, and again is looking at small-scale effects. This second type of correlation is largely influenced by differences in the average conditions between different fields. The poorest correlation is based on the numbers of species. This is true in both years. The correlation over all three fields is only barely significant in 1967. This may be be-

cause the true number of rare species on our sample spot is poorly estimated.

The correlation between insect evenness and plant evenness is striking. The 1966 relation over all three fields is shown in Fig. 1. The higher correlation in 1966 than in 1967 is caused in part by the fact that all three fields differed in 1966 in their average evenness, while in 1967 fields 2 and 3 were essentially indistinguishable (Table 2).

TABLE 3. Correlation coefficients (r) and F-ratios for comparisons between insect and plant diversity measures[a]

(a) WITHIN FIELD COMPARISONS—SPECIES DIVERSITY

Field:	July 1966			July 1967		
	1	2	3	1	2	3
S x S						
r	.56	.51	−.15	.24	−.29	.30
F	12.84**	7.11*	0.67	0.49	0.75	0.80
J x J						
r	.54	−.09	.64	−.06	.75	−.57
F	11.32**	0.14	19.24***	0.02	10.02**	3.84
H x H						
r	.66	.44	.59	.11	.20	−.04
F	21.29***	4.67*	15.01***	0.10	0.34	0.01

(b) COMPARISONS OVER ALL 3 FIELDS—SPECIES DIVERSITY

	S x S	J x J	H x H	S x S	J x J	H x H
r	.37	.82	.74	.44	.70	.85
F	12.42***	154.37***	96.29***	6.73*	27.04***	73.75***

(c) FOLIAGE HEIGHT DIVERSITY VS. INSECT DIVERSITY, 1967

Field:	Within fields			Over all 3 fields
	1	2	3	
r	.70	.25	.34	.84
F	7.65**	.55	1.04	67.83***

(d) FOLIAGE HEIGHT DIVERSITY VS. PLANT SPECIES DIVERSITY, 1967

Field:	Within fields			Over all 3 fields
	1	2	3	
r	.56	−.14	.16	.81
F	3.65	0.15	0.22	53.03***

*$P < 0.05$.   **$P < 0.01$.   ***$P < 0.001$.
[a]See text for explanation of the different kinds of comparisons. Symbols are as in Table 2.

The strong relation between H of plants and H of insects, over all fields, is due mainly to the evenness component. Indeed, the loose correlation in species number probably weakened the correlation for H in 1966. The rankings of fields 2 and 3 were different for mean number of species per sample than for mean evenness per sample (Table 2). When the two components of H vary differently in this way, it is more informative to consider them separately than combined as H. In 1967 in particular, the correlation between plant H and insect H is strong; 72% of the variance in insect H is accounted for by plant H (0.72 is the square of the r-value in Table 3b).

The third kind of analysis pools all samples from each field (Fig. 2) and is therefore concerned with larger-scale effects, and examines relations among fields. The pooled samples from each field constitute a datum. Each point here is more valuable as a measure of the diversity on the field since the "noise" among samples is eliminated and it is an even larger sample than each sample taken in 1967, which already was shown to have a much reduced variability. Unfortunately, we have no measure of sampling error, and there are few data points. The data for the

numbers of species (S_p) for all years (including those for an additional field for 1964) are presented in Fig. 2. Within each year there is a clear correlation between insects and plants, though it changes among years. Data for pooled evenness (J_p) and pooled diversity (H_p) are also shown in Fig. 2. Within each year the positive relation between insect J and plant J is obvious. The correlation using all six points is significant ($r = .86$, $F_{1,4} = 10.97$, $P < .05$).

The evenness index J is really a crude measure of the shape of the relative abundance curve. It would be better for some purposes to fit relative abundance data to some mathematical distribution having several parameters that describe the general slope of the line, the degree of concavity, Y-intercept, and so on. We have instead reproduced the curves themselves (Fig. 3). They stress a point that can be seen by comparing the evenness values for both years in Table 2: the abundances of the plant species are usually more even than those of the insects. Bearing this difference in mind, the variation in the form of the insect curve from field to field is clearly closely correlated with variations in the plant curves.

The relation for H_p, as might be expected, is also

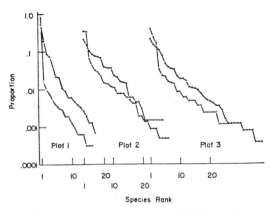

FIG. 3. Relative abundance curves for plants (•) and insects (o) from three fields in July 1967. The proportion that each species forms of the total abundance in the pooled samples from each field (plot in the figure) is located on a logarithmic scale against the species' rank.

close (Fig. 2, $r = .90$, $F_{1,4} = 16.91$, $P < .025$) and follows the pattern shown for J_p.

Plant structure and insect diversity

Foliage height diversity (*FHD*) was measured in 1967 only. Two sets of indices were calculated. The first used only the number of contacts made in each of the three horizontal layers. The second recognized six possible categories (see Methods): in each layer we distinguished additionally between grasses and forbs since these represent very different kinds of plant structure. No grasses occurred in the top layer, so there were actually five categories. Correlations were done using both indices, and similar results were obtained, with the five categories yielding stronger correlations with insect *H*. The two measures of *FHD* are highly correlated. The five-category *FHD* is the index discussed hereafter.

Table 3c shows that within fields, *FHD* is at best weakly correlated with insect species diversity, as was the case with measures of plant species diversity in 1967. These correlations are again based on the larger samples taken in 1967 and we do not expect strong correlations, if our previous explanation is correct. The plant samples from field 1 varied the most (Appendix) and this field shows a significant correlation between insect *H* and *FHD*.

Over all three fields there is a strong correlation between insect *H* and *FHD* (*FHD* explains 71% of the variance in insect *H*), about equal to that between insect *H* and plant *H* for the same samples.

To summarize—over all three fields insect *H* is highly correlated with both plant *H* and *FHD*. Although plant *H* and *FHD* are not significantly correlated within fields, based on the large samples, they

are strongly correlated over all three fields (Table 3d), though the relationship is slightly weaker than either statistic shows with insect *H*.

The combined effects of *FHD* and plant *H* upon insect *H* were analyzed by multiple regression, using them as the two independent variables and insect *H* as the dependent variable. The multiple correlation coefficient has the value 0.89. Thus the addition of either independent variable, to one considered alone, explains some 7%–8% additional variance in insect *H*, to give a total explained variance of 79%.

Seasonal changes in insect diversity

The number of insect species on each field increased from June to July and either declined in August (fields 2 and 3) or stayed high (field 1) (Table 4). Evenness changed consistently on all three fields. *J* was high in June, low in July, and high in August. *J* is affected by *S*, and the low values in July could simply reflect the addition of more rare species in that month. However, examination of the relative abundances in all 3 months showed that July had either a low or an intermediate proportion of rare species. Examination of the relative abundance curves in the 3 months showed that the drop in evenness in July is caused by one, two, or three species (in fields 1, 2, and 3 respectively) becoming numerically dominant in July while in the other 2 months the slope of the curve from abundant to moderately abundant species was gradual. The seasonal changes in evenness were consistent in all fields even though the absolute abundance of Homoptera declined throughout the season in fields 1 and 3 but was maximum in July in field 2.

Spatial homogeneity

We can measure how variable a field is from place to place by comparing pairs of samples. First, the mean similarity index ($\bar{I}$) for the field is computed by averaging the similarity indices obtained by making all possible comparisons between samples, taken two at a time. This number is hard to interpret because the number of species and their relative abundances vary from field to field, and therefore the opportunity for samples to differ also varies from field to field. We corrected for this by a method proposed by Peterson (1972). For each field, samples were drawn at random from the pooled sample data. The probability of an individual's belonging to a given species at each draw for each sample was simply the proportion that the species formed of the total. A new census was thus created in toto; *k* samples were drawn, where *k* was the number of samples that were actually taken in the real census of the field. Each of the *k* samples was always as large as the corresponding real sample, so that $n_1, n_2 \ldots n_k$ represents the string of sample sizes. A mean sim-

TABLE 4. Seasonal change in insect diversity in 1967. Based on pooled samples in each plot[a]

Field	S_p			J_p			H_p			N_p		
	June	July	Aug	June	July	Aug	June	July	Aug	June	July	Aug
1	10	18	18	.20	.12	.34	0.46	0.34	0.98	4004	3108	1386
2	17	27	21	.77	.58	.64	2.09	1.85	1.90	256	648	586
3	17	28	20	.61	.55	.65	1.69	1.80	1.85	1006	817	296

[a]Symbols are as in Table 2.

TABLE 5. Homogeneity among samples in a field[a]

	Plants in field no.			Insects in field no.		
Date	1	2	3	1	2	3
June 1967	—	—	—	1.00	.79	.71
July 1967	.56	.64	.60	.96	.85	.88
Aug. 1967	—	—	—	.90	.79	.76
July 1966	.66	.67	.58	:97	.85	.82
July 1964	—	—	—	.88	.91	.65

[a]Each number is the ratio $\bar{I}/\bar{I}'$, where $\bar{I}$ is the average similarity index obtained by comparing all possible pairs of samples on the field, and $\bar{I}'$ is a randomly generated average similarity index obtained for the field by assuming that the only differences between samples are due to statistical sampling error. See text for explanation.

ilarity index was then calculated between all possible pairs of the random samples, analogous to the mean similarity index calculated for the real census. Thus, this method attempted to calculate the average similarity for the field, under conditions where the only variance from sample to sample occurred as a result of random sampling error. Five such sets of samples were drawn for each field and the median ($\bar{I}'$) of the five mean similarity indices was chosen. We then divided the actual mean similarity index for the field by this generated index. Thus, each original index is corrected by dividing it by the index value that would be achieved if the species in the field were not clumped into groups. The randomly generated index essentially measures expected similarity in the absence of spatial heterogeneity. These quotients for all comparisons are given in Table 5. The generated index was always higher (more homogeneity) than the real index, except for the insects in one field in 1 month where real and random indices were equal.

The homogeneity of the plant samples falls within a very narrow range (.56–.67). Within this range the fields do not rank the same in the 2 years, but the differences between fields are so small that, lacking a statistical test, we may best regard the degree of spatial homogeneity as the same in all fields.

The range in insect homogeneity is greater (.65 — 1.00), though this partly reflects the larger number of sampling dates. Field 1 is usually the most homogeneous and field 3 the least homogeneous. The insects are more homogeneous in space than are the plants, as might be expected since insects

move around. The plant heterogeneity and insect heterogeneity do not vary together from field to field.

Spatial homogeneity is not correlated with the average diversity in a field ($\bar{H}$): when the fields are ranked with respect to plant $\bar{H}$, the rankings for spatial homogeneity are consistently different. When the fields are ranked with respect to spatial homogeneity of the insects, field 1 is always the most homogeneous; it is also the least diverse ($\bar{H}$), but on four of five dates, fields 2 and 3 rank differently with respect to $\bar{H}$ and spatial homogeneity. We conclude that these two aspects of diversity vary independently from field to field in both plants and insects.

Diversity and stability

A comparison of Tables 1 and 2 shows that, for the insects, field 1 is least diverse, on the average, and also has the highest similarity when Julys of different years are compared. Field 3 is intermediate in diversity and average similarity between years, and field 2 is most diverse and has the lowest similarity between years. In other words, these data suggest that, for the insects, the more diverse fields have more rearrangements of species and species abundances from one year to another. If we consider the amount of rearranging that occurs as a measure of community instability, it seems that diversity is negatively correlated with stability.

However, we have already noted that we expect differences in similarity between pairs of samples to be affected by the number of species in the community and by differences in their relative abundances. Our homogeneity index ($\bar{I}/\bar{I}'$) is designed specifically to correct for such differences that arise from random sampling. So we have corrected the similarities between Julys for each plot by dividing the average between-July similarity by the median of the series of randomly generated average similarities, as described above. In Table 6 these average values for each field are compared to the average diversity over the same 3 years in each plot. The negative correlation still holds. We have fewer data for the plants (Table 1), but this negative relationship across years does not seem to hold for the plants (we are grateful to D. Futuyma for first drawing our attention to these correlations).

203

TABLE 6. A comparison, for the insects, of the average diversity for each field with the homogeneity for the field, calculated between years, and between months in 1 year[a]

	Field	Average diversity	Homogeneity
Between years	1	0.57	0.90
	3	1.61	0.70
	2	2.11	0.58
Between months	1	0.60	0.86
	3	1.84	0.64
	2	2.02	0.57

[a]For the comparisons between years, we calculate the average pooled diversity for each field (H_D) of the July 1964, 1966, and 1967 values. For the comparisons between months we calculate the average of the June, July, and August, 1967 values. See text for an explanation of homogeneity. Each value is calculated by correcting the average similarity index between all three possible pairs of Julys, in one case, and all three possible pairs of summer months in the other case.

There is also a positive correlation between diversity and seasonal rearrangement in 1967, as indicated by the homogeneity index (Table 6).

DISCUSSION

MacArthur and MacArthur (1961) quantified a suspected relationship between bird species diversity and the amount of vertical complexity in different kinds of plant communities. As MacArthur has noted, this was a correlation that field ornithologists have always assumed to be true in some rough and ready fashion: different species of birds nest and forage generally at different heights, and tree-nesting species, for example, are unlikely to be found breeding in open fields. Subsequent work on birds has confirmed this correlation (for example, MacArthur, MacArthur, and Preer 1962, MacArthur, Recher, and Cody 1966, and Recher 1969). Cody (1968, 1970) and others have applied the idea successfully even to structurally very simple communities such as grasslands. As MacArthur (1964) pointed out, however, the correlation does not always extend to complicated habitats, or to all bird species (MacArthur et al. 1962).

In the early paper by MacArthur and MacArthur (1961:597), it was noted that plant species diversity alone was a good predictor of bird species diversity, but it was highly correlated with plant structural diversity and did not account for additional remaining variance in bird species diversity after bird diversity was correlated with plant structure. The subsequent work on birds has not dealt with plant species diversity, though the treatment of plant structure has become more sophisticated (Cody 1968). Since the birds studied have been mainly predators on insects, and seed-eaters, it may be the case that plant species diversity is not an important variable. However, the studies to date have largely ignored this hypothesis rather than tried to test it.

One would generally expect that where there is a greater diversity of food-stuffs, there will be a greater diversity of organisms that feed on them. In this respect, it is interesting that MacArthur and others have found a structural aspect of the habitat that is at least as important as the species component for birds. However, subsequent workers have tended to ignore the first, obvious, explanation for variations in H of the feeding species and have not looked at the diversity of the food. It is against this background that we would like to resurrect the habit of considering the obvious. Presumably, plant species diversity should become more important the more host specific the animal species are.

Pianka (1967) showed that the number of lizard species in various desert communities was highly correlated with the structural complexity of the desert shrubs and was not correlated with plant species diversity. There are several difficulties in interpreting the data. (1) Only the number of lizard species and not H is given. (2) The desert bushes varied in size (volume), but apparently each individual was given the same weight in calculating plant species diversity.

This particular example raises a point of general interest. The units in which plant species diversity is measured are important. If animals directly utilize the plants or a portion of the plants (whether for food, nesting, or some other reason), and if we believe that the number and relative abundance of different resources determine number and relative abundance of the animals (i.e., their diversity), then it is the abundance of these resources that we should measure. But what do we mean by plant "abundance"? MacArthur and MacArthur (1961) measured the amount of foliage belonging to each plant species and used this as the measure of abundance. Since individual plants, either of the same or different species, can vary enormously in size, some measure of cover, volume or biomass such as these authors used, is clearly preferable. It is inappropriate in such cases to compute plant species diversity simply by counting plant individuals if the individuals are of widely different sizes. (Similarly, one should then weight the insects by their consumption of the plants. Although desirable, this is generally impractical. In addition, in this study the insects are all about the same size.) Given that the plant species diversity in Pianka's study was not measured in the appropriate units, plant structural diversity and plant species diversity may indeed have been correlated. Then we would not be surprised if all three variables (plant species H, plant structure H, and lizard species) increased concomitantly along a latitudinal gradient. On the other hand, if the plant structure and plant species diversity were actually uncorrelated, then Pianka's data would provide good evidence that the number of lizard species was determined by plant structure and not by plant species diversity.

All but one of the lizard species studied by Pianka were carnivores, and generally they had diverse diets. As in birds, it may be quite reasonable that lizard diversity should not be greatly affected by plant species diversity. However, the case for plant structural diversity, as distinct from plant species diversity, as the determining factor of lizard diversity remains to be substantiated.

Rosenzweig and Winakur (1966) showed that different species of (mainly granivorous) desert rodents are most abundant in habitats that differ with respect to vegetation structure and sometimes possibly also soil structure. For example, some require open, sparsely vegetated areas while others require bushy areas. The authors measured rodent and plant diversity in 21 plots in Arizona. In both years of the study there was a weak positive relation between H for the rodents and their measure of vertical plant diversity. Rosenzweig and Winakur add two other aspects of habitat diversity to obtain a somewhat better correlation, but it is hard to interpret their compound structural diversity index. They show that the diversity of the rodents is not correlated with species diversity of the plants, though they seem again to have given equal weight to all plant individuals. To summarize: in this rodent study it seems clear that the rodent diversity was not related to plant species diversity, nor was it clearly related to plant structural diversity.

In the present study the species diversity *and* the structural diversity (foliage height diversity) of the plants on three old fields were good, and equally good, correlates of the species diversity of a large section of the herbivorous insects—the Homoptera (plant-sucking bugs). The correlation is weak when diversity of insects from one small patch (1 m²) to another nearby patch is examined, but it is especially strong in explaining the variation among different old fields. Thus, this study extends the relation between plant and animal diversity, well established for birds and tentative for lizards and rodents, to a group of herbivorous insects. However, it still leaves open the question of whether the important relation is between plant species diversity and animal diversity or between plant structural diversity and animal diversity. Since the Homoptera species probably are more host specific than carnivores (DeLong 1968, Whitcomb 1957), but are more general feeders than some leaf-eating insects, it may be that both plant species diversity and plant structural diversity are important in determining the insect species diversity in this case.

There is evidence, for vegetation at least, that the different components of diversity vary together along latitudinal gradients. J. H. Connell of the University of California at Santa Barbara has unpublished data showing that number of species (S), evenness (J), overall diversity (H), and spatial heterogeneity all increase in forest trees as they are sampled progressively closer to the equator. In the present study these variables did not vary together consistently. Spatial heterogeneity and H were not correlated. S and J sometimes did not vary together from field to field and varied differently from month to month in 1967 in two out of three fields. On the other hand, field 1 consistently had markedly fewer species of plants and insects and consistently lower J-values than fields 2 and 3. However, overall, the correspondence among the different aspects of diversity was poor.

To the extent that the above is generally true for areas at the same latitude, the best strategy is to seek separate explanations for the separate components of diversity. By contrast, an interesting question is: Is there a single explanation for the general increase in all components of diversity with decreasing latitude?

Finally, we showed (Table 6) that fields that were more diverse were also less stable in that, from year to year, there was more rearranging of species and species' abundances in the more diverse fields. We should stress that the measure of stability already is corrected for differences between communities with respect to number of species and relative abundances.

There is in the literature plenty of evidence that artificially simplified communities are less stable than more complex, coevolved communities. Our results do not contradict this generalization. However, insofar as our communities are coevolved, they go against the conventional wisdom that diverse coevolved communities are more stable than simpler coevolved communities.

ACKNOWLEDGMENTS

We are grateful to the following for their valuable comments and criticisms: M. L. Cody, J. H. Connell, D. Futuyma, and D. Janzen. S. J. Arnold kindly assisted in taking the 1967 plant and insect data. The work was supported by NSF grants G3223, G13521, and GB24026.

LITERATURE CITED

Brillouin, L. 1960. Science and information theory. 2nd ed. Academic Press, New York. 351 p.

Cody, M. L. 1968. On the methods of resource division in grassland bird communities. Am. Nat. **102**:107–147.

Cody, M. L. 1970. Chilean bird distribution. Ecology **51**:455–464.

DeLong, D. M. 1948. The leafhoppers or Ciccadellidae of Illinois (Eurymelinae-Balcluthinae). Ill. Nat. Hist. Surv. Bull. **24**:97–376.

Evans, F. C., and E. Dahl. 1955. The vegetational structure of an abandoned field in south-eastern Michigan and its relation to environmental factors. Ecology **36**:685–706.

MacArthur, R. 1964. Environmental factors affecting bird species diversity. Am. Nat. **98**:387–397.

MacArthur, R. H., and J. W. MacArthur. 1961. On bird species diversity. Ecology **42**:594–598.

MacArthur, R. H., J. W. MacArthur, and J. Preer. 1962. On bird species diversity. II. Prediction of bird census from habitat measurements. Am. Nat. **96**:167–174.

MacArthur, R., H. Recher, and M. Cody. 1966. On the relation between habitat selection and species diversity. Am. Nat. 100:319–332.

Peterson, C. H. 1972. Species diversity, disturbance and time in the bivalve communities of some coastal lagoons. Ph.D. Thesis. Univ. Calif. Santa Barbara. 229 p.

Pianka, E. R. 1967. Lizard species diversity. Ecology 48:333–351.

Pielou, E. C. 1966. The measurement of diversity in different types of biological collections. J. Theor. Biol. 13:131–144.

Recher, H. F. 1969. Bird species diversity in Australia and North America. Am. Nat. 103:75–79.

Rosenzweig, M. L., and J. Winakur. 1966. Population ecology of desert rodent communities: habitats and environmental complexity. Ecology 50:558–572.

Shannon, C. E., and W. Weaver. 1963. The mathematical theory of communication. Univ. Ill. Press, Urbana.

Whitcomb, R. F. 1957. Host relationships of some grass-inhabiting leafhoppers. M.S. Thesis. Univ. Ill., Urbana.

Whittaker, R. H. 1960. Vegetation of the Siskiyou Mountains, Oregon and California. Ecol. Monogr. 30:279–338.

APPENDIX

Diversity statistics and sample size for plants and insects from three fields in 1966 and 1967. Each number is the value for a single sample, and each line is a sample.

	Plants				Insects		
H	J	S	N	H	J	S	N
1966							
Field 1							
0.22	0.21	3	90	0.13	0.12	3	101
0.19	0.18	3	63	0.21	0.13	5	111
0.26	0.18	5	80	0.42	0.25	6	117
0.39	0.30	4	56	0.11	0.11	3	80
0.72	0.55	4	82	0.13	0.12	3	144
0.56	0.54	3	86	0.18	0.14	4	152
0.06	0.09	2	71	0.27	0.21	4	90
0.05	0.08	2	81	0.28	0.18	5	286
0.23	0.22	3	96	0.05	0.07	2	91
0.38	0.58	2	70	0.26	0.26	3	40
0.28	0.18	5	75	0.31	0.20	5	159
0.27	0.21	4	75	0.24	0.18	4	204
0.42	0.28	5	73	0.29	0.28	3	77
1.36	0.72	8	66	0.56	0.31	7	92
0.61	0.59	3	67	0.28	0.18	5	75
0.22	0.34	2	75	0.29	0.22	4	131
1.14	0.76	5	73	1.05	0.51	9	122
0.83	0.56	5	67	0.37	0.29	4	47
0.13	0.13	3	62	0.16	0.16	3	76
0.56	0.37	5	79	0.29	0.22	4	84
0.70	0.54	4	69	0.22	0.16	4	97
0.06	0.09	2	71	0.26	0.17	5	151
0.88	0.59	5	74	0.36	0.22	6	90
1.25	0.70	7	67	1.47	0.70	10	74
0.69	0.42	6	65	0.31	0.20	5	139
0.24	0.23	3	80	0.33	0.22	5	98
0.15	0.14	3	83	0.29	0.19	5	201
0.61	0.58	3	91	0.38	0.24	5	287
0.36	0.27	4	83	0.36	0.23	5	208
0.22	0.34	2	75	0.19	0.18	3	111
Field 2							
				1.11	1.00	5	9
				1.39	0.91	7	19
				1.37	1.00	8	10
				1.19	0.97	6	10
				1.27	1.00	7	9
				1.07	0.90	6	9
				1.42	0.90	8	17
				1.30	1.00	7	10
1.55	0.75	10	74	0.0	0.0	1	1
1.72	0.77	13	58	0.79	1.00	4	4
1.79	0.78	13	76	1.35	0.88	7	19
2.14	0.91	15	61	1.31	0.95	7	12
1.70	0.86	10	42	0.0	0.0	1	1
1.82	0.95	10	34	1.02	0.95	5	8
1.76	0.91	10	35	0.85	0.73	5	10
1.73	0.82	11	56	1.10	1.00	6	6
1.40	0.86	8	20	0.62	1.00	3	4
1.61	0.95	8	25	0.76	0.83	4	7
2.03	0.93	13	48	0.92	1.00	4	7
1.51	0.83	9	30	1.10	1.00	6	6
1.75	0.80	12	63	1.31	0.95	7	12
1.73	0.82	11	60	0.30	0.60	2	6
1.92	0.92	11	53	1.24	0.93	7	11
1.95	0.90	13	43	0.98	1.00	5	6
1.76	0.89	10	45	0.82	1.00	4	5
1.45	0.67	11	75	0.86	0.94	4	7
1.82	0.80	13	70	1.54	0.97	10	14
1.75	0.75	14	75	0.93	0.95	4	8
1.82	0.80	13	67	1.42	0.90	8	17
1.78	0.73	16	71	1.35	1.00	8	9
Field 3							
1.85	0.70	19	103	0.72	0.68	4	12

APPENDIX—Continued

	Plants				Insects		
H	J	S	N	H	J	S	N
1.14	0.57	9	75	0.96	0.70	6	15
1.39	0.63	11	94	1.01	0.82	5	14
1.88	0.75	16	102	1.05	0.68	6	32
1.70	0.72	14	77	1.01	0.80	5	15
1.75	0.67	18	108	1.04	0.52	10	45
1.78	0.77	13	92	1.39	0.70	10	45
1.68	0.72	13	97	1.17	0.73	7	26
1.52	0.79	8	99	1.48	0.72	10	63
1.31	0.60	11	98	1.04	0.48	11	72
1.68	0.76	11	99	1.60	0.83	11	30
1.67	0.82	9	98	1.42	0.63	14	50
1.23	0.58	10	94	0.71	0.33	11	79
1.30	0.57	12	98	0.56	0.39	5	34
0.87	0.43	9	100	0.92	0.50	8	55
1.18	0.56	10	100	1.16	0.57	10	57
1.64	0.68	14	96	1.00	0.54	8	52
1.32	0.62	10	97	0.65	0.34	8	97
0.86	0.39	11	102	1.06	0.70	6	28
1.51	0.66	12	104	0.85	0.47	8	47
1.46	0.66	11	105	0.74	0.39	8	79
1.30	0.59	11	99	0.96	0.47	9	97
1.04	0.51	9	95	0.43	0.29	5	69
1.61	0.70	12	106	0.85	0.61	5	29
1.30	0.61	10	100	1.49	0.98	9	13
1.01	0.50	9	105	1.07	0.54	9	77
0.66	0.33	9	98	0.14	0.13	3	60
0.56	0.31	7	105	0.37	0.19	8	165
0.50	0.26	8	102	0.17	0.13	4	159
1.72	0.78	11	96	0.89	0.58	6	35

	Plants					Insects		
H	J	S	N	FHD	H	J	S	N
1967								
Field 1								
1.26	0.64	8	162	1.03	0.23	0.12	7	248
0.49	0.29	6	116	0.59	0.20	0.09	9	576
0.52	0.78	2	106	0.35	0.08	0.06	4	438
1.46	0.67	10	140	0.92	0.39	0.20	8	259
1.14	0.61	7	178	1.16	0.38	0.18	9	203
0.54	0.35	5	129	0.80	0.24	0.13	7	270
1.63	0.75	10	150	0.97	0.32	0.16	8	273
0.77	0.45	6	137	0.76	0.39	0.17	11	404
0.71	0.53	4	194	1.16	0.81	0.37	10	140
0.97	0.62	5	172	0.81	0.19	0.11	6	297
Field 2								
2.03	0.79	15	217	1.37	1.43	0.67	11	71
2.10	0.84	14	185	1.11	1.61	0.75	12	55
1.98	0.81	13	193	1.31	1.73	0.73	15	64
2.16	0.82	16	260	1.39	1.73	0.75	14	65
2.10	0.82	15	166	1.18	1.39	0.71	10	41
2.38	0.92	16	180	1.14	1.39	0.75	8	52
1.97	0.80	13	264	1.29	1.47	0.66	13	57
1.55	0.65	12	267	1.21	1.30	0.59	11	105
2.07	0.80	15	215	1.32	1.25	0.60	10	72
2.14	0.80	17	200	1.20	1.42	0.64	12	66
Field 3								
2.15	0.76	19	293	1.37	1.47	0.68	11	72
2.14	0.76	20	241	1.26	1.47	0.74	9	71
1.93	0.72	17	257	1.13	1.54	0.79	9	55
2.17	0.79	18	270	1.47	1.62	0.69	13	96
1.78	0.81	10	220	1.36	1.20	0.59	9	101
1.71	0.74	11	261	1.36	1.63	0.65	15	120
2.10	0.85	13	313	1.41	1.59	0.65	15	87
1.94	0.79	13	238	1.37	1.82	0.81	13	59
2.09	0.86	13	195	1.22	1.13	0.58	8	108
2.21	0.78	20	232	1.40	1.48	0.77	9	48

11

Reprinted from *Ecology* **58**:711–724 (1977) by permission of the publisher, Duke University Press, Durham, North Carolina

SPECIES DIVERSITY AND COMMUNITY ORGANIZATION IN DESERT SEED-EATING ANTS[1]

DIANE W. DAVIDSON[2]

Department of Biology, University of Utah, Salt Lake City, Utah 84112 USA

Abstract. Patterns of species diversity and community organization in desert seed-eating ants were studied in 10 habitats on a longitudinal gradient of increasing rainfall extending from southeastern California, through southern Arizona, and into southwestern New Mexico. Local communities of harvester ants include 2–7 common species, and at least 15 species from five genera of Myrmecines compose the total species pool in these deserts. Ant species diversity is highly correlated with mean annual precipitation, an index of productivity in arid regions. Communities are structured on the basis of competition for food, and interspecific differences in worker body sizes and colony foraging strategies represent important mechanisms of resource allocation. Seed size preferences, measured for native seeds and in food choice experiments with seeds of different size but uniform nutritional quality, are highly correlated with worker body sizes. Species of similar body size can coexist within local habitats if they differ in foraging strategy. Interspecific aggression and territorial defense and microhabitat partitioning all appear to be relatively unimportant in these ant communities.

Patterns of species diversity and community organization in harvester ants are strikingly similar to those reported for communities of seed-eating rodents that occupy many of the same desert habitats. Separate regressions of within-habitat species diversity against the precipitation index of productivity for the two groups correspond closely in slope, intercept, and proportion of explained variation. Resource allocation on the basis of seed size characterizes local communities of both ants and rodents. Parallels between these two groups suggest that limits to specialization and overlap may be specified by parameters such as resource abundance and predictability that affect unrelated taxa similarly.

Key words: Ants; Arizona; California; communities; competition; desert granivores; diversity; insects; New Mexico; Novomessor; Pheidole; Pogonomyrmex; resource allocation; Veromessor.

INTRODUCTION

The southwestern deserts of the USA are inhabited by a remarkably diverse group of consumer species which have specialized to use the relatively nutritious and dependable seeds of desert plants. In an otherwise harsh and uncertain environment, a substantial proportion of the primary production exists as a persistent seed reserve in the soil. Numerous species of granivorous rodents, birds, ants, and other insects often coexist within local desert habitats, presenting a bewildering array of possibilities for competition and resource allocation. Here I analyze geographic patterns in the species diversity and coexistence of desert seed-eating ants in order to assess the interactions which determine the structural and functional organization of these communities. In this and a companion paper (Davidson 1977a), I present experimental and observational evidence bearing on mechanisms of resource allocation and consider the generality and significance of such mechanisms.

Desert ants have characteristics that facilitate the study of resource utilization and community organization. They are easily observed in their native habitats and manipulated in experiments, because of their relative insensitivity to the presence of observers. Unlike some other arthropods and many vertebrates, in which juveniles search for their own food and often utilize different resources than adults, ants have relatively fixed ecological roles based upon behavioral and morphological specializations of the adult workers. Workers are specialized food harvesters, and their behavior is not complicated by conflicting demands such as those related to courtship and mating. It is likely that natural selection on worker phenotypes operates at the level of the colony in these social insects, rather than individually on the sterile workers (Wilson 1971).

In the absence of recent perturbations, community diversity and organization should be determined by ecological interactions and reflect approximate equilibria between origination and extinction (Rosenzweig 1975). A total of at least 15 granivorous ant species inhabit the deserts studied here, and for any single habitat, the pool of potential colonists typically includes 5–7 species in addition to those coexisting in the habitat. The vagility of the winged reproductive forms should make ants excellent colonists, and if speciation rates have been broadly similar across these deserts, community diversity and structure should be determined primarily on the basis of extinction probabilities.

The distribution of diverse faunas of granivorous ants and rodents over the North American deserts presents a unique opportunity to compare geographic patterns of species diversity and community organization in unrelated taxa that exploit similar resources. Brown (1973, 1975) has analyzed rodent species diversity and community structure in relationship to both

[1] Manuscript received 10 June 1976; accepted 28 December 1976.

[2] Present address: Department of Biological Sciences, Purdue University, West Lafayette, Indiana 47907.

latitudinal and longitudinal gradients of precipitation, an estimate of seed production in these deserts. My study locations parallel one of Brown's (1975) transects from the extremely arid winter rain deserts of southern California to the more mesic regions of southeastern Arizona and southwestern New Mexico, which receive both summer and winter rains (Bryson 1957). I have compared ant communities over this geographic gradient in order to answer the following questions: (1) What determines species diversity in these ant communities? (2) How are communities organized and resources partitioned to permit coexistence of similar species? and (3) How do patterns of species diversity and community structure in desert harvester ants compare with those described for granivorous desert rodents?

METHODS

Census techniques and calculations

Ten desert study sites were selected on the basis of precipitation records from U.S. Weather Bureau stations. A broad spectrum of rainfall regimes was chosen, but habitat features such as slope and soil characteristics were deliberately held as constant as possible. Once weather stations had been selected, exact study locations were determined on the basis of their proximity in distance (within 20 km) and elevation to weather stations and the presence of uniformly sandy soils, not dissected by washes or disturbed by human

FIG. 1. Map of the study locations.

activity. Extremely rocky or sand dune substrates were avoided. Seven desert habitats were censused in spring (late April and early May) and summer (late July and early August) of 1974. Ants were substantially more active in the summer, and the remaining three habitats were censused only once, in August of 1975. Study sites are mapped in Fig. 1 and described in greater detail in Table 1.

Standard census procedures were used at all study sites and involved attraction of ants to seed baits. (Colony counts were attempted but found to be less reliable because of the difficulty of detecting cryptic nests.) At each locality 80 shallow glass ashtrays were placed in an 8 × 10 bait grid with approximately 5-m

TABLE 1. Descriptions of the study locations and characteristics of their ant faunas: mean annual precipitation ($\bar{x}$ ppt, mm), mean July temperature, species diversity of perennials (PSD), numbers of common and rare species of ants; ant species diversity (H) and mean distance from colony entrance to census bait ($\bar{D}$). See text for further explanation of these measurements

Location	$\bar{x}$ annual precipitation (mm)	$\bar{x}$ July temp. (°C)	PSD	Ant species			$\bar{D}$
				Common	Rare	Diversity (H)	
California							
1) San Bernardino County, 8.5 km SE Baker, 287 m	76	33.9	0.53	2	0	0.59	8.3
2) San Bernadino County, 11 km NNE Barstow, 659 m	88	29.6	1.57	2	1	0.78	6.7
3) Kern County, 1 km SE Mojave, 834 m	121	28.6	0.60	2	1	0.14	7.4
Arizona							
4) Yuma County, 15 km E Wellton (Tacna), 219 m	91	32.6	0.73	4	2	1.04	4.8
5) Maricopa County, 5 km SW Gila Bend, 225 m	142	34.4	0.00	3	0	1.08	2.9
6) Pima County, 8 km N Ajo, 537 m	216	32.9	0.33	6	1	1.43	5.0
7) Pinal County, 8 km SSW Casa Grande, 425 m	215	33.1	0.30	5	0	1.12	2.3
8) Cochise County, 10 km NNW Rodeo, New Mexico, 1,225 m (Rodeo A)	276	25.8	0.43	7	1	1.67	3.4
9) Cochise County, 9 km NW Rodeo, New Mexico, 1,225 m (Rodeo B)	276	25.8	0.59	7	1	1.77	2.4
10) New Mexico, Luna County, 9.4 km SE Deming, 1,320 m	224	27.0	1.29	5	2	1.54	4.1

209

spacing paced off among bait stations. The edges of the trays were flush with the surrounding soil surface. As in Brown et al. (1975), strips of masking tape were affixed to the trays to facilitate entrance and exit of ants, and one quarter-inch hardwarecloth was secured across the trays to prevent access by rodents; the screen did not interfere with the movements of ants. Baits consisted of barley seeds, ground in a grain mill to produce a variety of particle sizes ranging from entire seeds and large fragments to fine powder, and these were replenished continuously as foraging ants depleted them. Seed-eating ants were censused by direct observation of the numbers of individuals of each species removing seeds from each bait tray during a 60-s observation period, once every 2 h for a full cycle of diurnal and nocturnal surface temperatures. Bernstein (1974) and Whitford and Ettershank (1975) have demonstrated the importance of substrate temperatures in regulating activity patterns of the desert ants. Both air and soil temperatures were measured with a YSI telethermometer immediately before and after each census of the 80 baits.

Species diversity was calculated in two different ways. Shannon-Wiener diversities ($H = -\Sigma p_i \ln p_i$), incorporating a measure of species evenness as well as number, were computed on the basis of census data from seed baits. Because the values of p_i correspond to the proportion of the total seeds removed during the census that were taken by the ith species, this index measures diversity on the basis of resource exploitation rather than the relative abundances of colonies or workers of each species. In addition, the number of common species in each local habitat, a measure of species richness, was determined on the basis of criteria established prior to any of the censuses. Because seeds were supplied only in dense clumps, but interspecific differences in average colony size and foraging behavior were anticipated to influence the use of seeds in different patterns of dispersion (Davidson 1977a), common species were defined on the basis of both the numbers of baits which a species visited and its proportional contribution to seed removal. Common species are those which removed at least 10% of the total seeds taken during the combined observation periods and visited a minimum of 5% of the 80 baits, or those which took at least 5% of the seeds and visited at least 10% of the baits. For twice-censused habitats, species were designated as common if they met these requirements in either census, and diversity indices calculated for the two separate censuses were averaged.

For each species detected in a census, I recorded the distances from baits visited by workers to the entrances of their colonies. Mean distances were computed for each species, and these means, weighted by species abundances, were combined to calculate a community average foraging distance to baits. In this distance index,

$$D = \sum_i d_i S_i / S_T.$$

d_i corresponds to the mean foraging distance of the ith species, and S_i/S_T is the proportion of the censused seed removals that were attributable to species i. This estimate of mean foraging distance is analogous to that calculated by Bernstein (1975) for ants foraging on native seeds. For twice-censused sites, distance indices were averaged.

Several additional types of data were collected at each census location or compiled for each neighboring weather station. Seed harvesting ants of each species were collected and preserved in 70% ethanol and returned to the laboratory for measurement with a Wild M-5 binocular microscope and occular micrometer. The abundances of perennial plants were measured by the line intercept method, using ten 25-m transects at each census site. Species diversity indices ($H = -\Sigma p_i \ln p_i$) were calculated from these data on abundances. No attempt was made to quantify the abundance or diversity of annual plants, since these vary greatly over seasons and among years. Mean annual precipitation and mean July temperature were calculated from 20 yr of weather data (U.S. Climatological Data 1955 through 1974).

Ant species were categorized as either group or individual foragers based on their behaviors at baits and on numerous observations of ants collecting native seeds over a range of seasons and levels of resource abundance (Davidson 1977a). Individual foragers are those which were never noted to form distinct columns while gathering native seeds, although certain of these appeared to forage in groups while dismantling large insects or when seed baits chanced to be located very near their nest entrances. Species which form distinct columns to feed from seed baits and from reservoirs of native seeds in the soils are classified as group foragers, though certain of these may modify their behaviors somewhat seasonally (Whitford 1976; Davidson 1977a).

Overlaps between pairs of species on census baits were calculated and compared with values of overlap expected if species occurred on baits independently of one another's presence. The expected overlaps were generated as:

$$b_i \cdot b_j \cdot 80$$

where b_i and b_j are the proportions of baits used by species i and j respectively, and 80 baits were censused at each location. Overlaps were analyzed in this manner only for 51 pairs of species whose expected overlaps exceeded 5% of the census stations. Attractive seed baits may lure species from their normal foraging ranges and mask any existing interspecific microhabitat segregation. In order to test for these effects, a second census technique was employed in a 12-h census at Rodeo, New Mexico (site A). At 60 stations, separated from one another by distances of 5 m and not baited, ants were censused with a square-meter frame once every 2 h. Observed and expected overlaps were analyzed in the same manner as above. Finally, colonies on a 10,000-m² plot in this same habitat were accurately mapped in August of 1974 and compared with Poisson distributions for random spacing with respect to one another. Two distinct quadrat sizes (large = 100 m², n = 25; small = 25 m², n = 100) were employed to generate the Poisson distributions, because species differed considerably in their foraging ranges. For some ants, foraging was confined to within 3 m of the colony, whereas for others, workers frequently traveled up to 25 m, although the greatest densities of workers usually occurred near the colony entrances. Six of the seven species common to this site were sufficiently active to be mapped in August of 1974. Colonies of *Solenopsis xyloni* were not mapped, because they were only sporadically active during this period, and their use of multiple nest entrances made identification of individual colonies very difficult. *Novomessor cockerelli* also tended to forage from 2–3 nest entrances simultaneously, but colony boundaries were readily distinguished on the basis of the short (4–5 m) distances between related entrances of a single colony in contrast to the much more extensive spaces (approximately 20 m) between sets of nest mounds. In addition, I frequently observed workers commuting among associated nest entrances.

Relationship between worker body size and seed size

Both preliminary observations of body size patterns within communities and characteristics of the seed resources suggested that resources might be allocated in part on the basis of worker body size. Experimental and observational techniques were employed to test this hypothesis. Pearl barley seeds were ground in a grain mill and sieved in Tyler soil screens representing a geometric series of pore diameters. Four distinct size classes of seeds (size 1 median = 0.9 mm; size 2 = 1.5 mm; size 3 = 2.2 mm; and size 4 = 3.1 mm) were combined in equal quantities by weight (2.5 g of each size class), and presented in bait trays very near the nest entrances of seed-eating ants in habitats near Rodeo, New Mexico. For each of eight species, the sizes of the first 10 seeds removed by each of three colonies were noted and used to calculate a seed size index. This index was computed as

$$\sum_{i=1}^{4} p_i d_i,$$

where p_i is the proportion of seed choices of size i, and d_i equals the median diameter of seeds in the ith size class.

In order to test the relationship between seed size and body size for ants gathering native seeds, returning workers of six species of seed-eating ants at Rodeo, New Mexico (site A) were robbed of their prey items in September of 1974. The number of colonies sampled per species depended on the total number of colonies of that species present on the 10,000-m² study plot. The large and aggressive colonies of *Pogonomyrmex rugosus* have extensive foraging areas, and only three colonies actually occupied the study plot. To increase sample sizes, seeds were also collected from four additional colonies immediately adjacent to the study area. Thirty forage items were sampled from each colony except that, where colonies of *P. rugosus* had multiple foraging columns oriented in different directions, samples of 30 items were removed from each column and treated as measurements from independent colonies. The body size measurements of ants used in these comparisons are based on mean worker body lengths of ants sampled during the census of the Rodeo (A) study site.

RESULTS AND DISCUSSION
Species diversity and related patterns

The species diversity of granivorous ants in the southwestern deserts is strongly correlated with an estimate of seed production based on mean annual precipitation. Primary productivity is closely correlated with precipitation in arid regions (Rosenzweig 1968), and both seed production and germination in desert plants are known to depend on rainfall (Went and Westergaard 1949; Went 1955; Juhren et al. 1956; Tevis 1958a, b; Beatley 1967, 1974). Figure 2 illustrates the highly significant positive correlation between H, the index of ant species diversity calculated on the basis of exploitation of census baits, and the mean annual precipitation measured at nearby weather stations (r = .83; P < .01) for the 10 desert study localities. Any census procedure which measures the impact of species on baits risks overestimating those species which are most successful in exploiting high density resource patches. Measurements of the numbers of common species are less subject to this criticism, and this variable is even more strongly correlated with the estimate of productivity (r = .94; P < .01) than is the species diversity index. Table 2 gives the species compositions of the local ant communities.

At Mojave, California (mean annual precipitation equal to 121 mm), ant diversity falls considerably below the regression line. Strong winds prevailed during both censuses at this site, so that ants risked being carried away while foraging. If species differ in their degree of susceptibility to disruption of foraging by wind, the low measurement of exploitation diversity at this locality may simply be an artifact of censusing on windy days. Alternatively, if strong winds characterize Mojave, California, over much of the foraging season, seed production actually available to these ants may be significantly less than the precipitation value indicates.

The exploitation index of ant diversity (H) is poorly

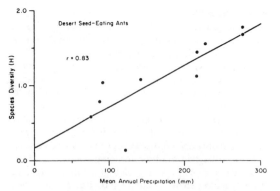

FIG. 2. Species diversity of seed-eating ants plotted against precipitation (index of seed productivity) for 10 census sites in the southwestern deserts. $H = -\Sigma p_i \ln p_i$ where p_i are the proportions of the total seeds removed by the ith ant species from seed baits.

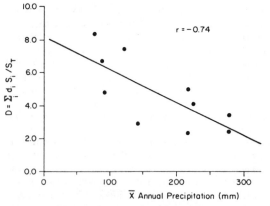

FIG. 3. The regression of mean foraging distance to baits (D) against mean annual precipitation. Here, d_i corresponds to the mean foraging distance (in meters) of the ith species, and S_i/S_T is the proportion of the censused seed removals attributable to species i.

correlated with both the diversity of perennial plants ($r = -.06$) and with mean July temperature ($r = -.40$), which varies with elevation along this latitudinal transect and may be an estimate of the length of the seed production period (Went 1948). Neither of these parameters contributes importantly as second variables to explaining the residual variance of the regression of H on the precipitation index of seed production.

The strong correlation between ant diversity and the estimate of productivity suggests that populations of desert granivorous ants are food-limited, and this interpretation is reinforced by the tendency for colonies to be more widely spaced in more arid habitats. The measure of mean foraging distance from colony entrances to census baits is significantly and inversely related to the mean annual precipitation in the habitat

(Fig. 3). Bernstein (1975) noted that foraging distances for ants gathering naturally available seeds in the Mojave Desert decreased similarly with increasing elevation, which she assumed to be correlated with enhanced seed production. The greater colony density in more mesic habitats is striking and is primarily attributable to an increase in abundances of species characterized by small body sizes (*see* Fig. 4 *and below*). Such species typically are absent from the most xeric habitats and occur with increasing frequency in the more productive deserts and arid grasslands. Bernstein's data on the abundances of *Pheidole xerophila* and *Pheidole gilvescens* in the Mojave Desert reveal a similar increase in the proportion of these

TABLE 2. Species composition of ant faunas at 10 study locations in the Mojave and Sonoran deserts. BL = mean worker body length; FB = foraging behavior (group or individual); c = common species; r = rare species. Locations are labelled as in Table 1

			Locations of census sites									
Species	BL	FB	1	2	3	4	5	6	7	8	9	10
Novomessor cockerelli	9.8	I			c			c		c	c	c
Veromessor pergandei	5.5	G	c	c		c	c	c	c			
Solenopsis xyloni	3.1	G			r	c		c	c	c	c	c
Pogonomyrmex rugosus	9.2	G						c		c		c
P. barbatus	9.1	G									c	
P. maricopa	7.6**	I					c*				r	
P. desertorum	7.0	I								c	c	c
P. californicus	6.8	I		c	c	c		r	c			
P. magnacanthus	5.8	I	c									
P. pima	4.5	I						c		r		
Pheidole militicida	3.8	G										r*
Ph. desertorum	3.4	I				r				c	c	c
Ph. xerophila	2.3	G						c*		c*	c	
Ph. gilvescens	2.2	G				r*			c*			
Ph. sitarches	1.8	G				c	c	c*		c*	c	r
Ph. sp.			r	r								

* Species determinations by Roy R. Snelling.
** *P. maricopa* averaged 6.1 mm in body length at location 5.

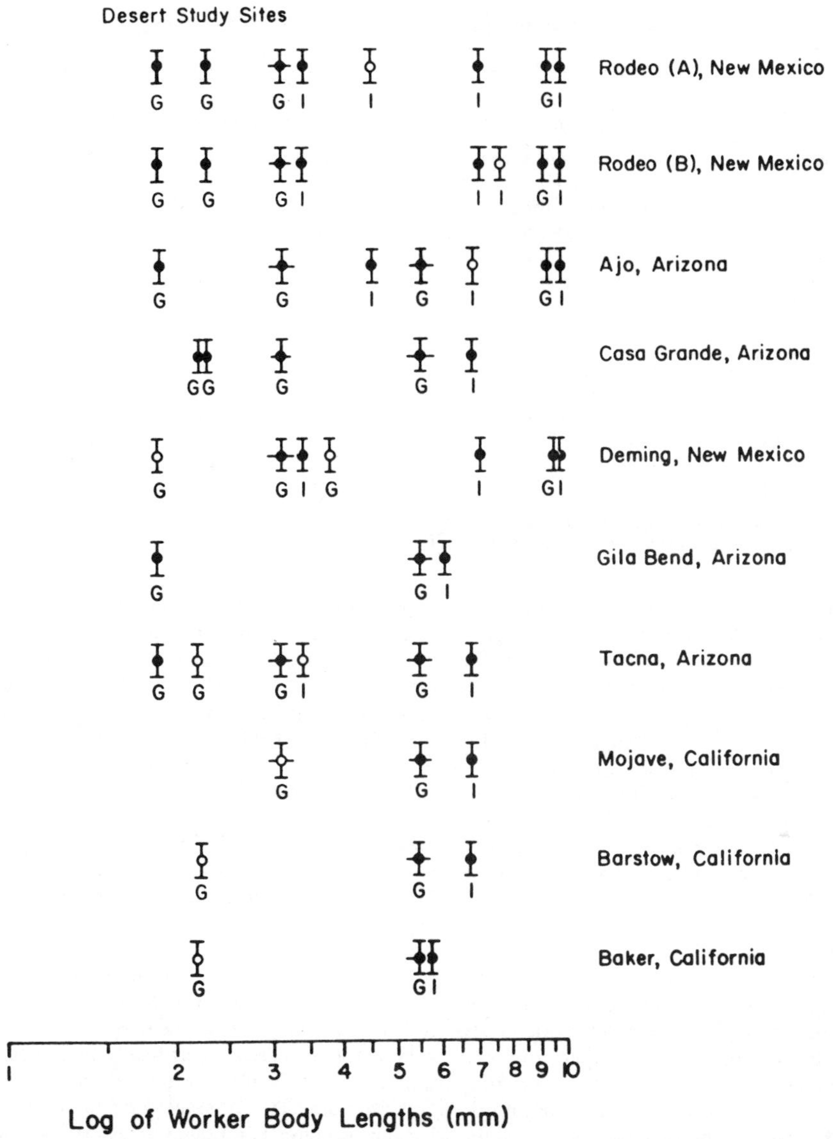

Desert Study Sites

FIG. 4. Mean worker body lengths (vertical bars) of ant granivores censused in 10 southwestern desert habitats. Closed circles depict common species and open circles represent those which are rare. (Criteria for commonness and rarity are specified in the text.) Horizontal bars denote worker size polymorphism, and methods of foraging (G = group and I = individual) are indicated beneath the vertical bars.

small species (worker body length ≈ 2.5 mm) with increasing elevation. Colonies of smaller species tend to have more limited foraging ranges and may, thus, have access to fewer resources. Among the ants encountered in this study, those with body lengths <3 mm typically forage no farther than 3–4 m from their nest entrances. This contrasts sharply with the distances of 40 m or more traversed by large (>9.0 mm) workers of *Pogonomyrmex rugosus* and *P. barbatus* (Hölldo-

bler 1974) and of the size-polymorphic *Veromessor pergandei* (3.0–8.5 mm).

Analyses of colony spacing and microhabitat overlap

If populations of desert seed-eating ants are food-limited, communities of these granivores should contain species that differ sufficiently in their resource

213

TABLE 3. Comparison of expected[a] and observed overlaps for species pairs in habitats censused by baiting (Rodeo B) and unbaited quadrat techniques (Rodeo A)

Species 1	Species 2	Bait census		Quadrat census	
		Expected	Observed	Expected	Observed
Novomessor cockerelli	Pogonomyrmex barbatus (or rugosus)	21.0	16	15.8	18
Novomessor cockerelli	Pogonomyrmex desertorum	19.4	22	32.1	31
Novomessor cockerelli	Solenopsis xyloni	24.7	23	10.5	10
Novomessor cockerelli	Pheidole xerophila	11.4	8	13.4	11
Pogonomyrmex rugosus (or barbatus)	Pogonomyrmex desertorum	18.5	6***	24.8	22
Pogonomyrmex rugosus (or barbatus)	Solenopsis xyloni	23.5	27	8.1	5
Pogonomyrmex rugosus (or barbatus)	Pheidole xerophila	11.0	10	11.0	8
Pogonomyrmex desertorum	Solenopsis xyloni	21.8	23	16.5	17
Pogonomyrmex desertorum	Pheidole xerophila	10.2	10	23.8	22
Solenopsis xyloni	Pheidole xerophila	12.9	10	6.9	6

[a] Overlaps were determined as $b_i \cdot b_j \cdot s$, where b_i and b_j are the proportions of baits or unbaited quadrats used by species i and j, and s equals the total number of stations.
*** This difference is significant at the .0001 level of probability.

requirements to permit coexistence. One potential mechanism of resource allocation is microhabitat partitioning, but competition between species of the granivorous ants studied here appears not to have been resolved primarily by interspecific territoriality or local differences in habitat preference. The results of the Poisson analysis of the spatial distribution of ant colonies at Rodeo, New Mexico (site A) are the first line of evidence supporting this conclusion. Colonies of only one pair of species, Pogonomyrmex rugosus and P. desertorum, were interspecifically spaced ($t = 3.8$; $P < .005$ with 25-m² quadrat size). Several more qualitative observations suggest interspecific territorial defense by P. rugosus against P. desertorum. In the fall of 1974, the former species ceased foraging somewhat earlier than did the latter. During late October, I observed P. desertorum workers attempting to establish colony entrances near P. rugosus mounds. Though not actively foraging, P. rugosus workers defended their territories against the intruders. However, by November P. rugosus was completely inactive and some P. desertorum colonies had successfully settled very near the nest mounds of their larger competitors (D. Hobbs, personal communication).

Demonstration of random interspecific colony spacing does not exclude the possibility that species feed in distinctly nonoverlapping foraging territories. Hölldobler (1974) reports that, within populations of Pogonomyrmex barbatus, foraging territories tend to be nonoverlapping, though regular spacing among nest mounds is not detected (but see Whitford et al. 1976). Analysis of interspecific overlaps of ants foraging on baits and in unbaited square-meter quadrats reveals little evidence for interspecific microhabitat partitioning of foraging grounds. Comparisons of observed and

expected overlaps for 51 pairs yield only two significant chi-square values. (Because of the large number of relationships tested and the consequent possibility of Type I error, the .001 probability level was required for statistical significance. Two other comparisons approached significance at probability levels of .05 and .01 respectively. A replicate test of spatial interactions between the latter pair in another habitat indicated random interspecific spacing.) Only two pairs of species apparently subdivide microhabitats. Pogonomyrmex rugosus and P. desertorum at Deming, New Mexico ($P < .001$) and P. barbatus and P. desertorum at Rodeo B ($P < .0001$) represent spatially segregating pairs.

The considerable interspecific spatial overlap detected during baiting censuses does not appear to be merely an artifact of luring species away from their normal foraging areas to artifically high accumulations of resource. In Table 3, expected and observed overlaps are compared for similar ant faunas censused by baiting (Rodeo B) and quadrat (Rodeo A) techniques. (Data from the bait census at Rodeo A could not be used because many of the expected overlaps fell short of the 5% requirement.) These ant communities are identical in species composition with one exception: Pogonomyrmex rugosus is abundant at the Rodeo A area, while P. barbatus occurs instead of P. rugosus at Rodeo B. These congeners do not usually coexist within local habitats and appear to act as ecological replacements for one another in many respects (Whitford et al. 1976). Both census techniques reveal extensive interspecific overlap in foraging ranges. The single instance in which baited and unbaited censuses gave significantly different overlaps involved the interactions between Pogonomyrmex desertorum, an individually foraging species of approximately 7.0 mm

worker body length and either *P. rugosus* or *P. barbatus*. In the unbaited quadrats, overlaps were significantly greater between *P. desertorum* and *P. rugosus*
than they were between *P. desertorum* and *P. barbatus* on census baits. This result is the reverse of the
expectation, were species being lured from their normal foraging areas by artificially large accumulations
of seeds. Together the data on foraging overlaps in
baited and unbaited quadrats, and those on colony
spacing imply that microhabitat partitioning is not a
major mechanism by which coexisting species of desert seed-eating ants avoid competition for food resources.

Body size and seed size selection

Resource partitioning on the basis of seed size plays
a much more pronounced role in the organization of
communities of desert seed-eating ants. This is
strongly implied by the combination of patterns of
worker body sizes among coexisting species and data
on the seed size preferences of these species. For each
of the 10 ant communities censused, mean worker
body lengths of constituent species are plotted on a
logarithmic scale in Fig. 4 (vertical bars). Several interesting patterns emerge from this graph. Coexisting
species with similar worker body sizes tend to differ in
their foraging behaviors. Within some communities,
such as Tacna, Deming, and Ajo, group and individual
foragers actually alternate along the body size gradient. Each of the three Mojave Desert habitats censused in southern California contains only two common species, and in each case the two species employ
different foraging strategies. Certain combinations
of species of similar size and foraging behavior never
occur. For example, the five relatively productive
communities which contain group foragers whose
workers exceed 9 mm in body length contain either
Pogonomyrmex rugosus or *P. barbatus* but never
both. Where Hölldobler (1974) studied *P. rugosus* and
P. barbatus in a zone of overlap in southeastern
Arizona, foraging territories were defended interspecifically, and aggressive encounters between species
were as intense as intraspecific aggressive encounters.
In the 5.8–7.0-mm body length range of individual
foragers, local habitats characteristically contain only
one of four possible species (*Pogonomyrmex califor-
nicus*, *P. maricopa*, *P. magnacanthus*, and *P. desertorum*) in sufficient abundances to be categorized as
common.

Finally, Fig. 4 illustrates that all species whose
workers average less than 3.4 mm in length are group
foragers. This foraging method may facilitate naviga-
tion and extend the feeding areas of these small workers.
Bernstein (1971) has shown that, for desert seed-eating
ants of a given body length, the maximum foraging
distance of trail-forming ants exceeds that of individually foraging species of similar body size. Like trends
are apparent in the present investigation.

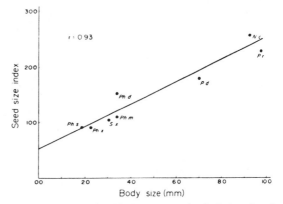

FIG. 5. The relationship between worker body length and
seed size index for experiments with 8 species of seed-eating
ants near Rodeo, N. Mex. Seed size indices were calculated
as $\Sigma p_i d_i$, where p_i is the proportion of the seeds chosen
from the ith size class of barley particles and d_i is the
median diameter of seeds in that size class. Species
designations are as follows: *Ph. x.* = *Pheidole xerophila*;
S. x. = *Solenopsis xyloni*; *Ph. m.* = *Pheidole militicida*; *Ph.
d.* = *Pheidole desertorum*; *P. d.* = *Pogonomyrmex desertorum*; *P. r.* = *Pogonomyrmex rugosus*; and *N. c.* = *Novomessor cockerelli*. All species except *Ph. m.* coexist at Rodeo,
N. Mex. (Site A).

The patterns of worker body sizes among coexisting
species are associated with resource subdivision on
the basis of seed size. In food choice experiments with
barley seeds and particles of several sizes, size indices
of seeds foraged are highly correlated with worker
body lengths (Fig. 5: $r = .93$, $P < .01$). For native
seeds collected by ants in September of 1974 at Rodeo
(A), New Mexico, an increase in average seed size
with worker body length gradually levels off at the
larger body sizes, presumably because large seeds are
more rare than small and medium-sized seeds. The
regression of the seed size index against the logarithm
of worker body size for these data is highly significant
(Fig. 6: $r = .76$, $P < .01$). On the average, larger
workers take larger seeds, and this relationship is consistent even within colonies of the size polymorphic
Veromessor pergandei (Davidson 1977b). The correlation between body size and prey size may arise in part
from purely mechanical limitations of "fit" between
seeds and ant mandibles, and in part from a reduced
capacity of small ants to locate large seeds. Species
with smaller workers forage over much shorter distances than larger species and may have fewer large
seeds available to them.

Foraging behavior and resource allocation

Although worker body sizes and seed sizes are significantly correlated, coexisting species of similar
body length overlap considerably in the ranges of seed
sizes that they utilize (Fig. 6). Resource allocation on
the basis of seed size alone appears insufficient to ac-

215

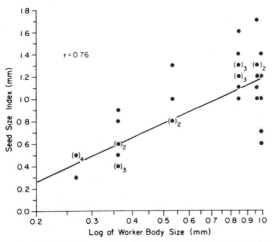

FIG. 6. Size index of native seeds foraged by 6 common species at Rodeo, N. Mex. (Site A) as a function of the logarithm of worker body length. Each point on the graph includes data from the seed component of 30 forage samples collected at one colony. Seeds were sized in a series of Tyler screen sieves. Indices were calculated as mean seed sizes, weighted by the abundances of seeds in each size class.

count for coexistence. The possibility that differences in foraging behavior may play a role in resource partitioning is suggested by the tendency for coexisting species of similar body size to forage in different ways (Fig. 4). In a companion paper (Davidson 1977a), I consider group and individual foraging techniques as specializations for exploiting seeds of different density or dispersion.

GENERAL DISCUSSION

Diversity and structure of ant communities

Differences in the species diversity of granivorous ants among 10 study sites in the southwestern deserts appear to be determined primarily by variability in the average productivity of the habitat. MacArthur (1972) hypothesized that, if the diversity of resources remains the same, consumer diversity should increase as productivity increases; previously scarce resources become sufficiently abundant to support new species, and greater specialization in the utilization of resources is also possible. Brown's (1973) studies of rodent communities in sand dune habitats of the Mojave and Great Basin deserts provided support for MacArthur's hypothesis, and Brown (1975) later confirmed this result independently for rodents in flatland sandy soil habitats of the Sonoran Desert. The demonstration of similar patterns for desert seed-eating ants attests to the generality of this relationship, at least for this relatively uncomplicated system in which groups of generalist consumer species utilize similar food resources in habitats that differ very little in structure, and where

recent historical perturbations appear not to have played a significant role.

Although investigations of very different ant communities have shown species diversity to be ecologically determined, none have explicitly related diversity to resource productivity. Levins et al. (1973) note the excellent colonizing abilities of ants and present evidence that distributions of ants on islands in the Puerto Rican bank are dynamically determined by migration and extinction events accompanied by high species turnover. Culver (1974) censused ants in a grassy field in West Virginia over two summers and found 6–7 species coexisting at any one time, although species composition was temporally variable, and as many as 10 species actually inhabited the field during some portion of his investigation. He postulated that this numerical constancy might reflect an equilibrium between migration and extinction. In comparing ant communities in a variety of habitat types in Puerto Rico, St. John (Virgin Islands), and West Virginia, Culver (1974) noted a positive correlation between ant species diversity and the structural complexity of the habitat but made no attempt to quantify the availability of food resources. Comparisons of species diversity in habitats of similar structure in Puerto Rico and West Virginia gave mixed results: second-growth forests in both regions supported approximately the same number of species, while grassy fields in West Virginia included more coexisting species than similar habitats in Puerto Rico (6–7 vs. 3). Because Culver's studies included predominantly omnivorous species, even qualitative estimation of differences in the resource regimes of the various habitats is difficult.

Bernstein (1971) found an increase in the species diversity of the total ant community (primarily insectivores and granivores with a few omnivores) with increasing elevation in the Mojave Desert, but both precipitation and temperature vary simultaneously over this gradient, and their effects are not easily distinguished. Increased precipitation at higher elevations should be correlated with greater production of both seeds and insects, and generally lower temperatures may both prolong the length of the seed production period (Went 1948) and curtail the foraging season for these ectothermic invertebrates. Although Bernstein interpreted the greater colony density at higher elevations as a response to the greater resource productivity, she attributed the accompanying enhancement of species diversity to the lengthening of the period over which food resources were produced. By comparison, in the present investigation, mean annual precipitation accounts for considerably more of the variation in species diversity than does mean July temperature.

The species diversity of consumers should depend to some degree upon the diversity as well as the productivity of resources. Although it would be difficult to quantify seed diversity with certainty for so many sites in which rainfall and seed production fluctuate dramatically over seasons and years, it is unlikely that the habitats studied here differ drastically in overall seed diversity. Ant species diversity is unrelated to the diversity of perennials, and various of the sites contain many of the same or closely related species of annuals and perennials. If the sizes of ants present at a given locality are indicative of the sizes of seeds produced there, the range of seed sizes may be similar across these deserts. The presence of the continuously size polymorphic *Veromessor pergandei* in habitats where species of relatively large and small body size are absent and the tendency for colonies of this species to be more polymorphic in habitats where interspecific competition is less intense (Davidson 1977*b*) results in the presence of a relatively similar range of worker body sizes within each of the study localities. Workers of this species ranged in body size from 3.5 to 8.4 mm, while the span of mean body sizes for all species encountered in this study is 1.8 to 9.8 mm.

Has competition played a major role in the organization of communities of seed-eating ants? Various criteria have been used as indirect evidence for competition. Schoener (1974) discussed two patterns that frequently characterize competitively structured communities: overdispersion of species in niche space and complementarity of niche dimensions. In communities of desert harvester ants, resource allocation is related to interspecific differences in both worker body sizes (associated with differential exploitation of different size classes of seeds) and in colony foraging behaviors (correlated with different efficiencies in gathering seeds from various density distributions [Davidson 1977*b*]). Figure 4 provides some suggestion that niche separation based on worker body sizes and colony foraging strategies of coexisting species is greater than would be predicted if combinations of coexisting species occurred at random with respect to these characteristics. However, without a more precise knowledge of seed abundance in each resource state, this interpretation remains relatively subjective. Stronger evidence for the role of competition in structuring these communities derives from other observations. The two dimensions of resource allocation considered here are consistently complementary: species of similar body size tend to coexist only if they differ in foraging behavior. A corollary is that species of similar body size and foraging behavior behave as ecological replacements for one another, never demonstrating local coexistence. Diamond (1975) has observed such replacement patterns among ecologically similar bird species on islands of the coast of Central America and interpreted these "checkerboard" distribution patterns as evidence for competition. Finally, character shifts, such as that occurring in *Veromessor pergandei* in response to changes in the competitive environment (Davidson 1977*b*), constitute some of the most powerful evidence for competition. The within-colony worker size polymorphism in this species is reduced as the intensity of interspecific competition increases, with an accompanying reduction in the variance of resource types utilized.

The presence of numerous seeds in desert soils after extensive periods of drought (Tevis 1958*c*) does not contradict the argument for food limitation of granivore populations. Food may be present without being available in sufficiently large packages or dense associations to be profitably harvested. Clearly the most desirable seeds will be depleted first, and, as this happens, species will be forced to expand their diets and patch utilization to encompass less rewarding prey. Tevis (1958*c*) has observed such changes in the diet of *Veromessor pergandei*. Among desert seed-eating ants, diet breadth is positively correlated with resource abundance both for experiments with "seeds" of identical quality but different sizes and for native seeds as their abundances change seasonally (D. W. Davidson, personal observation). Species that have high maintenance costs or that derive their resources predominantly from unproductive regions of the resource spectrum (e.g., very large seeds on the dimension of seed size) should be particularly sensitive to changes in resource density. Reichman (1974), working on an extensive plot in the Sonoran Desert, has demonstrated an approximately 40-fold change in average seed densities over only 2 yr. Fluctuations of so great a magnitude must dramatically alter granivore foraging efficiencies.

How do strategies of resource allocation in desert harvester ants compare with those for ants as a whole? Culver (1974) postulated that, in general, ants tend to be less constrained by their morphologies than are many other kinds of organisms (for example, birds), and that their interspecific interactions are frequently behaviorally based. Sociality may well increase the potential for such behavioral adaptations, and, in desert harvester ants, niche separation on the basis of foraging strategy is representative of such a behavioral solution to interspecific interactions. In contrast, resource partitioning on the basis of food particle size may be exceptional among ants. Culver (1972) anticipated this somewhat atypical form of niche separation in granivorous ants after calculating low microhabitat niche breadths and high microhabitat overlaps for granivorous Colorado species reported on by Gregg (1963).

What factors have caused the seed-eating ants to depart from the broad microhabitat niches and narrow microhabitat overlaps that characterize many other ant species (Culver 1972, 1974; Levins et al. 1973)? Mechanisms of resource subdivision should reflect the absolute heterogeneity of resources as well as the capacity of organisms to respond to this heterogeneity. Habitat structure is relatively simple in the 10 desert study sites, and while local variability in seed density may be associated with microtopographic features of the substrate (Reichman 1976), the pattern of this local

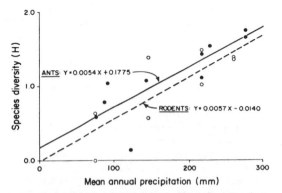

FIG. 7. Species diversities of seed-eating ants and rodents shown independently as functions of the precipitation index of productivity. Closed circles designate ant censuses, and open circles correspond to rodent censuses. Species diversities were calculated as $H = -\Sigma p_i \ln p_i$, where the p_i are the proportions of each species in the census or in the sample.

heterogeneity is probably not temporally stable. Culver (1974) noted a tendency for the frequency of aggression to be inversely related to habitat complexity and attributed this pattern to the greater ease of defending structurally simple habitats. Though living in relatively simple habitats, desert harvester ants may tend to use interspecific territorial defense and aggression very infrequently because of the high costs associated with defending so dispersed a resource as seeds. This generally low level of interspecific aggression may partially account for the high species diversity of granivorous ants in the southwestern deserts. MacArthur (1969) has pointed out that interference competition can reduce species packing of consumers that utilize noninteracting resources.

Parallels between communities of ants and rodents

Geographic patterns of species diversity and community organization in desert seed-eating ants may be compared in greater detail with the results of similar studies of desert rodent communities. Both the objectives and the methodologies of the present study closely resemble those of investigations by Brown (1973, 1975) and Brown and Lieberman (1973). Brown (1973, 1975) has described patterns of rodent species diversity along both latitudinal and longitudinal productivity gradients in the southwestern deserts. Over the longitudinal gradient, which parallels my own, Brown (1975) studied rodents in flatlands characterized by sandy soils and relatively uniform habitat structure. He demonstrated a significant positive correlation between rodent species diversity and an estimate of primary productivity by plotting the numbers of common species per habitat against the mean minus the standard deviation of the annual precipitation at nearby weather stations. Brown used this precipitation

statistic as an estimate of the amount of productivity that is predictable in 5 of every 6 yr. I have computed species diversity indices from Brown's rodent data and plotted them as functions of mean annual precipitation, with which they are equally well correlated, to make his data comparable to my own.

The relationships of ant and rodent species diversity to the precipitation index of productivity are illustrated in Fig. 7. Several attributes of the fitted regressions are strikingly similar: (1) the proportions of explained variation, (2) the slopes, and (3) the intercept values. Mean annual precipitation accounts for approximately 69% of the variation in species diversity among ant communities ($P < .01$) and 64% of this variation among communities of rodents ($P < .05$). The slopes of the two regressions are statistically indistinguishable, and their intercepts differ by less than 2%, but are slightly higher than expected considering that zero diversity values correspond to single-species communities (although the density of individuals may be very low). The precise translation of rainfall into seed productivity is not known. Seed transport by wind and water may import seeds into unproductive habitats from more productive areas nearby. Alternatively, the relationship between granivore diversity and precipitation may become nonlinear where precipitation levels are extremely low.

Mechanisms of resource partitioning that permit coexistence of potentially competing species should depend in part on the nature of the resources and might also be expected to show similar patterns in unrelated organisms. The extent of resource subdivision on various dimensions should reflect the absolute heterogeneity of resources as well as the capacities of the two groups to respond to this heterogeneity. For particulate food items composed of diverse size classes, subdivision by size is a common pattern. Coexisting species of both ants and rodents (Brown 1973, 1975; Brown and Lieberman 1973) characteristically differ in body sizes and utilize different sizes of seeds. In Fig. 8, seed size data are compared for ants (this study) and rodents (Brown 1975) feeding on native seeds in the vicinity of Rodeo, New Mexico. (Although *Dipodomys deserti* does not coexist in this habitat with *D. spectabilis*, the former species is included for comparison.) Within each category of granivores, species allocate seeds on the basis of size, but the two groups overlap broadly with one another in seed size utilization. The overall positive correlations between body size and seed size may depend on a number of different factors, including species' efficiencies at transporting or cracking seeds and the distances over which species of various sizes can travel in search of food. In this regard, it is interesting to note yet another similarity in patterns of community organization in ants and rodents (Brown 1973, 1975 for rodents). In neither taxon are species of small body size represented in the least productive of these sandy

DIANE W. DAVIDSON

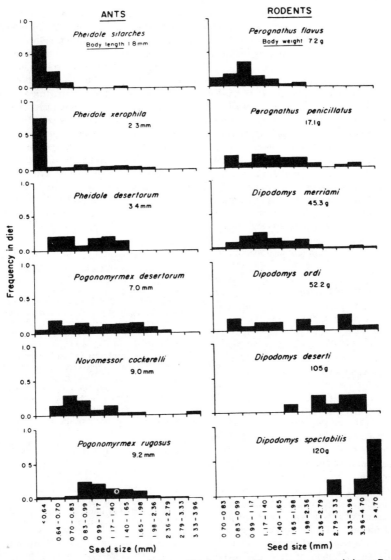

FIG. 8. Size frequency distributions for native seeds foraged by 6 species of harvester ants coexisting at Rodeo, N. Mex. and by 6 species of seed-eating rodents, 5 of which inhabit deserts near Rodeo, N. Mex. (*Dipodomys deserti* does not cooccur in the same habitats with *D. spectabilis*.) Data are not available for the cricetid rodents which lack cheek pouches. Seed size categories correspond to the geometric series of pore diameters in Tyler soil sieves which were used to classify seed sizes. Seeds from ants and rodents were sieved for different intervals of time, so that precise comparisons between taxa are not possible. Measurements correspond approximately to seed widths for ant seeds and to some value intermediate between width and length for rodent seeds. Rodent data are modified from Brown (1975).

soil habitats. The absence of small species from these communities may reflect their inability to harvest sufficient seeds because of limited mobility.

In both groups of granivores, resources also are subdivided on other dimensions as well, and two of these additional dimensions apparently reflect unique aspects of the biology of each taxon. Coexisting species of rodents that utilize similar size distributions

of seeds tend to forage in microhabitats that provide different amounts of cover (Rosenzweig and Winakur 1969; Rosenzweig 1973; Brown 1975), a factor likely to be correlated with exposure to avian predation. The failure of ants to distinguish microhabitats on the basis of cover is not surprising, as their risk of being eaten by such predators as *Phyrnosoma* species and spiders is less likely to vary with proximity to shrubs. In contrast, the behavioral plasticity conferred by sociality in ants permits resource allocation on the basis of seed density, or differential use of the same microhabitat because of differences in colony foraging behavior (Davidson 1977a).

The similarities observed between ant and rodent communities appear to require that precipitation affect the abundance and heterogeneity of seeds available to each taxon in approximately the same way. Although direct data bearing on this point are difficult to obtain, those that do exist suggest that this is likely (Pulliam and Brand 1975, Brown et al. 1975), because the two taxa overlap considerably in all parameters that have been studied except for temperature and season of foraging. These two factors do not appear to be important in climatic gradients such as the present one, where latitude and altitude are held relatively constant. However, along a latitudinal gradient in the Mojave and Great Basin deserts, where mean temperature declines significantly as precipitation increases, the diversity of seed-eating ants does not increase (Brown and Davidson 1977). Brown (1973, 1975) noted a more rapid increase in rodent species diversity with productivity over this same latitudinal gradient than in the transect across southern California and southern Arizona. In Brown's regressions, numbers of common species were plotted against the mean minus the standard deviation of annual precipitation. If species diversities ($H = -\Sigma p_i \ln p_i$) are regressed against mean annual precipitation as above, the slope of this relationship is .014, slightly over twice the slope calculated for the east–west transect. One explanation for the greater increase in rodent species diversity over the latitudinal gradient is that rodents are able to exploit a larger percentage of the enhanced production in areas where communities of ants are relatively impoverished. Brown and Davidson (1977) have evidence for reciprocal density compensations between Sonoran Desert faunas of ants and rodents in response to experimental exclusions of one or the other of the two taxa. These results imply that competition among distantly related taxa, although often overlooked in ecological studies, may be very important to the structuring of communities of consumers.

The strikingly similar patterns of species diversity in granivorous ants and rodents imply that resource limitations may determine the structure of communities even more precisely than would have been anticipated from existing theory. Several investigators (Thorson 1958; Cody 1968, 1975; Karr and James 1975; Brown 1975; Pianka 1975) have described parallels in structural and functional organization among geographically isolated communities of different but taxonomically related groups of species that are exposed to similar selective pressures. Comparable geographic gradients in species diversity in relation to latitude, altitude, and ocean depth (MacArthur 1972; Sanders 1968) characterize many unrelated groups of organisms and imply the operation of common mechanisms. The nearly exact correspondence of the slopes of the two regressions of granivore diversity is particularly noteworthy, because ants and rodents differ enormously in their morphological, physiological, ecological, and behavioral attributes. Although species in unrelated taxa partition resources in somewhat different ways, limits to specialization and overlap may be specified by general parameters such as resource distribution in space and time that affect unrelated taxa similarly. In this regard, one of the major problems confronting desert granivores is that seeds are produced in pulses of uncertain magnitude and frequency. It is probably not coincidental that both ants and rodents employ similar strategies of food storage, hibernation, and estivation to minimize the impact of environmental fluctuations.

ACKNOWLEDGMENTS

I am grateful for the assistance of a number of people during the course of these investigations. James H. Brown generously provided financial assistance, helpful counsel, and stimulating discussion throughout the study and read several earlier drafts of the manuscript. Karen Zemanek diligently assisted in the field in exchange for beans (literally). While providing companionship and moral support, Sam Davidson endured months of discomfort in the desert climate. Logistic assistance was given by H. Bond and E. Roos, and G. and A. Miller furnished shelter near the Rodeo, New Mexico study area. Conversations with T. C. Gibson and K. Zemanek were particularly helpful. I thank Roy R. Snelling for identifying some of the ants. This research was supported by the Desert Biome of the International Biological Program (Grant No. 556 to James H. Brown) and two grants to the author from the Society of the Sigma Xi, and was submitted in partial fulfillment of the doctoral degree at the University of Utah. The University of Utah provided a graduate research stipend for 1 year.

LITERATURE CITED

Beatley, J. C. 1967. Survival of winter annuals in the Mojave Desert. Ecology 48:745–750.

———. 1974. Phenological events and their environmental triggers in Mojave Desert ecosystems. Ecology 55:856–863.

Bernstein, R. A. 1971. The ecology of ants in the Mojave Desert: their interspecific relationships, resource utilization, and diversity. Ph.D. dissertation. UCLA.

———. 1974. Seasonal food abundance and foraging activity in some desert ants. Am. Nat. 108:490–498.

———. 1975. Foraging strategies of ants in response to variable food density. Ecology 56:213–219.

Brown, J. H. 1973. Species diversity of seed-eating desert rodents in sand dune habitats. Ecology 54:775–787.

———. 1975. Geographical ecology of desert rodents, p. 315–341. In M. L. Cody and J. M. Diamond [eds.] Ecology and evolution of communities. Belknap Press, Cambridge, Mass.

Brown, J. H., and D. W. Davidson. 1977. Competition between seed-eating rodents and ants in desert ecosystems. Science 196:880–882.

Brown, J. H., J. J. Grover, D. W. Davidson, and G. A. Lieberman. 1975. A preliminary study of seed predation in desert and montane habitats. Ecology 57:987–992.

Brown, J. H., and G. A. Lieberman. 1973. Resource utilization and coexistence of seed-eating desert rodents in sand dune habitats. Ecology 54:788–797.

Bryson, R. A. 1957. The annual march of precipitation in Arizona, New Mexico, and northwestern Mexico. Univ. Ariz. Inst. Atmos. Phys. Tech. Rep. No. 6.

Cody, M. L. 1968. On the methods of resource division in grassland bird communities. Am. Nat. 102:107–147.

———. 1975. Towards a theory of continental species diversities, p. 214–257. In M. L. Cody and J. M. Diamond [eds.] Ecology and evolution of communities. Belknap Press, Cambridge, Mass.

Culver, D. C. 1972. A niche analysis of Colorado ants. Ecology 53:126–131.

———. 1974. Species packing in Caribbean and north temperate ant communities. Ecology 55:974–988.

Davidson, D. W. 1977a. Foraging ecology and community organization in desert seed-eating ants. Ecology 58:711–724.

———. 1977b. Size variability in the worker caste of a social insect (Veromessor pergandei Mayr) as a function of the competitive environment. Am. Nat. In press.

Diamond, J. M. 1975. Assembly of species communities, p. 342–444. In M. L. Cody and J. M. Diamond [eds.] Ecology and evolution of communities. Belknap Press, Cambridge, Mass.

Gregg, R. E. 1963. The ants of Colorado. Univ. Colorado Press, Boulder. 792 p.

Hölldobler, B. 1974. Home range orientation and territoriality in harvesting ants. Proc. Natl. Acad. Sci. U.S.A. 71(8):3271–3277.

Juhren, M., F. W. Went, and E. Phillips. 1956. Ecology of desert plants. IV. Combined field and laboratory work on germination of annuals in the Joshua Tree National Monument, California. Ecology 37:318–330.

Karr, J. R., and F. C. James. 1975. Ecomorphological configurations and convergent evolution, p. 258–291. In M. L. Cody and J. M. Diamond [eds.] Ecology and evolution of communities. Belknap Press, Cambridge, Mass.

Levins, R., M. L. Pressick, and H. Heatwole. 1973. Coexistence patterns in insular ants. Am. Sci. 61:463–472.

MacArthur, R. H. 1969. Species packing, or what competition minimizes. Proc. Natl. Acad. Sci. USA 64:1369–1375.

———. 1972. Geographical ecology. Harper and Row, New York. 269 p.

Pianka E. 1975. Niche relations of desert lizards, p. 292–314. In M. L. Cody and J. M. Diamond [eds.] Ecology and evolution of communities. Belknap Press, Cambridge, Mass.

Pulliam, H. R., and M. R. Brand. 1975. The production and utilization of seeds in plains grassland of southwestern Arizona. Ecology 56:1158–1167.

Reichman, O. J. 1974. Some ecological factors of the diets of Sonoran Desert rodents. Ph.D. Dissertation. N. Ariz. Univ.

———. 1976. Seed distribution and the effect of rodents on germination of desert annuals. US/IBP Desert Biome Res. Mem. 76:20–26. Utah State Univ., Logan.

Rosenzweig, M. L. 1968. Net primary productivity of terrestrial communities: prediction from climatological data. Am. Nat. 102:67–74.

———. 1973. Habitat selection experiments with a pair of coexisting heteromyid rodent species. Ecology 54:111–117.

———. 1975. On continental steady states of species diversity, p. 121–140. In M. L. Cody and J. M. Diamond [eds.] Ecology and evolution of communities. Belknap Press, Cambridge, Mass.

Rosenzweig, M. L., and R. Winakur. 1969. Population ecology of desert rodent communities: habitats and environmental complexity. Ecology 50:558–572.

Sanders, J. L. 1968. Marine benthic diversity: a comparative study. Am. Nat. 102:243–282.

Schoener, T. 1974. Resource partitioning in ecological communities. Science 185:27–38.

Tevis, L., Jr. 1958a. Germination and growth of ephemerals induced by sprinkling a sand desert. Ecology 39:681–688.

———. 1958b. A population of desert ephemerals germinated by less than one inch of rain. Ecology 39:688–695.

———. 1958c. Interrelations between the harvester ant Veromessor pergandei (Mayr) and some desert ephemerals. Ecology 39:695–704.

Thorson, G. 1958. Parallel level bottom communities, their temperature adaptation, and the balance between predators and food animals, p. 67–86. In A. A. Buzzati-Traverso [ed.] Perspectives in marine ecology. Univ. Calif. Press, Berkeley, California.

U.S. Weather Bureau. 1955–1974. Climatological data. For Arizona, New Mexico, and California.

Went, F. W. 1948. Ecology of desert plants. I. Observations on germination in Joshua Tree National Monument, California. Ecology 29:242–253.

———. 1955. The ecology of desert plants. Sci. Am. 192:68–75.

Went, F. W., and M. Westergaard. 1949. Ecology of desert plants III. Development of plants in the Death Valley National Monument, California. Ecology 30:26–38.

Whitford, W. G. 1976. Foraging behavior in Chihuahuan Desert harvester ants. Am. Midl. Nat. 95(2):455–458.

Whitford, W. G., and G. Ettershank. 1975. Factors affecting foraging activity in Chihuahuan Desert harvester ants. Environ. Entomol. 4:689–696.

Whitford, W. G., P. Johnson, and J. Ramirez. 1976. Comparative ecology of the harvester ants Pogonomyrmex barbatus (F. Smith) and Pogonomyrmex rugosus. Ins. Soc. 23 (2):117–132.

Wilson, E. O. 1971. The insect societies. Belknap Press, Cambridge, Mass.

12

THE STRUCTURE OF
LIZARD COMMUNITIES

Eric R. Pianka
Department of Zoology, University of Texas, Austin, Texas

Strictly speaking, a community is composed of all the organisms that live together in a particular habitat. Community structure concerns all the various ways in which the members of such a community relate to and interact with one another, as well as community-level properties that emerge from these interactions, such as trophic structure, energy flow, species diversity, relative abundance, and community stability. In practice, ecologists are usually unable to study entire communities, but instead interest is often focused on some convenient and tractable subset (usually taxonomic) of a particular community or series of communities. Thus one reads about plant communities, fish communities, bird communities, and so on. My topic here is the structure of lizard communities in this somewhat loose sense of the word (perhaps assemblage would be a more accurate description); my emphasis is on the niche relationships among such sympatric sets of lizard species, especially as they affect the numbers of species that coexist within lizard communities (species density).

So defined, the simplest (and perhaps least interesting) lizard communities would be those that contain but a single species, as, for instance, northern populations of *Eumeces fasciatus*. At the other extreme, probably the most complex lizard communities are those of the Australian sandridge deserts where as many as 40 different species occur in sympatry (20). Usually species densities of sympatric lizards vary from about 4 or 5 species to perhaps as many as 20. Lizard communities in arid regions are generally richer in species than those in wetter areas; therefore, because almost all ecological studies of entire saurofaunas have been in deserts (18, 20, 25), this paper emphasizes the structure of desert lizard communities. As such, I review mostly my own work. Other studies on lizard communities in nondesert habitats are, however, cited where appropriate.

Historical factors such as degree of isolation and available biotic stocks (particularly the species pools of potential competitors and predators) have profoundly shaped lizard communities. Thus one reason the Australian deserts support such very rich lizard communities may be that competition with, and perhaps predation pressures from, snakes, birds, and mammals are reduced on that continent (20).

Climate is also a major determinant of lizard species densities. The effects of various other historical factors, such as the Pleistocene glaciations, on lizard communities are very difficult to assess but may be considerable.

One of the strongest tools available to ecologists is the comparison of ecological systems which are historically independent but otherwise similar. Observations on pairs of such systems allow one to determine the degree of similarity in evolutionary outcome. Moreover, under certain circumstances such natural experiments may even allow some measure of control over such historical variables as the Pleistocene glaciations. For example, faunas of independently evolved study areas with similar climates and vegetative structure should differ primarily in the effects of history upon them.

This paper consists of two major sections. In the first, "Patterns Within Communities," I briefly review fundamental aspects of community structure and lizard niches to establish a frame of reference and to lay the groundwork for the remainder of the paper. Next I discuss ways of quantifying these niche relationships. In the second section, "Comparisons Between Communities," I use these methods to examine and compare three independently evolved desert-lizard systems in some detail; this section is not a review of the literature but a quantitative summary of much of my own research over the last ten years.

PATTERNS WITHIN COMMUNITIES

The number of species coexisting within communities can differ in four distinct ways: (a) More diverse communities can contain a greater variety of available resources, and/or (b) their component species may, on the average, use a smaller range of these available resources (the former corresponds roughly to "more niches," "a larger total niche space," or "more niche dimensions," and the latter to "smaller niches"). (c) Two communities with identical ranges of resources and average utilization patterns per species can also differ in species density with changes in the average degree of overlap in the use of available resources; thus greater overlap implies that more species exploit each resource (this situation can be described as "smaller exclusive niches" or "greater niche overlap"). (d) Finally, some communities may not contain the full range of species they could conceivably support and species density might then vary with the extent to which available resources are actually exploited by as many different species as possible (that is, with the degree of saturation with species or with the number of so-called empty niches). MacArthur (11) summarized all but the fourth of the above factors with a simple equation for the number of species in a community N

$$N = \frac{R}{\overline{U}}\left(1 + C\frac{\overline{O}}{\overline{H}}\right) \qquad 1.$$

where R is the total range of available resources actually exploited by all species, $\overline{U}$ the average niche breadth or the range of resources used by an average species, C a measure of the potential number of neighbors in niche space, increasing more or less geometrically with the number of niche dimensions (below), and $\overline{O}/\overline{H}$ the relative amount of niche overlap between an average pair of species. MacArthur improved Equation 1 to handle situations in which resources are not distributed uniformly

$$D_s = \frac{D_r}{\overline{D}_u} \left(1 + \overline{C}\,\overline{a} \right) \qquad\qquad 2.$$

where D, is the diversity of species in the community, D_r is the overall diversity of the resources exploited by all species, $\overline{D}_u$ is the mean diversity of utilization or the niche breadth of an average species, $\overline{C}$ measures the average number of potential niche neighbors as before, and $\overline{a}$ is a measure of the average amount of niche overlap (MacArthur called this the mean competition coefficient). I return to Equation 2 below after considering various aspects of the niche relationships of lizards and how they can be quantified. Results presented here, however, depend in no way upon the validity of MacArthur's equation.

Niche Dimensions

Animals partition environmental resources in three basic ways: temporally, spatially, and trophically; that is, species differ in times of activity, the places they exploit, and/or the foods they eat. Such differences in activities separate niches, reduce competition, and presumably allow the coexistence of a variety of species (8, 11). Among lizards these three fundamental niche dimensions are often fairly distinct and more or less independent of each other, although they sometimes interact; for example, the mode of foraging can influence all three niche dimensions. For convenience I first treat each major niche dimension separately (below) and then briefly examine ways in which they interact. Rather than refer to "the trophic and temporal dimensions of the niche," etc, I use verbal shorthand and speak of the food niche, time niche, etc.

All else being equal (number of species, niche breadths, niche overlaps, etc), a greater number of effective niche dimensions results in fewer immediate actual neighbors in niche space; moreover, pairs of potential competitors with high overlap along one niche dimension may often overlap relatively little or not at all along another niche dimension, presumably reducing or eliminating competition between them.

TIME NICHE   To the extent that being active at different times leads to exploitation of different resources, such as prey species, temporal separation of activities may reduce competition between lizard species. Perhaps the most conspicuous temporal separation of activities is the dichotomy of diurnal and nocturnal lizards, which are entirely nonoverlapping in the time dimension. However, more subtle temporal differences in daily and seasonal patterns of activity are widespread among lizards, both within and between species. In the North American Sonoran desert, for example, *Uta stansburiana* emerge early in the day and comprise the vast majority of the lizards encountered during the cool morning hours (Table 1). Later, small *Cnemido-*

Table 1 Statistics on time of activity of four species of lizards in the Sonoran desert, expressed as time[a] since sunrise, during the period when temperatures are rising. All means are significantly different (*t* - tests, *P* < .01).

Species	$\bar{X}$[b]	S.E.[c]	s[d]	N[e]	95% Confidence Limits of Means
Uta stansburiana	3.67	0.06	1.39	470	3.55–3.79
Cnemidophorus tigris	4.11	0.05	1.33	669	4.01–4.21
Callisaurus draconoides	4.60	0.09	1.32	204	4.42–4.78
Dipsosaurus dorsalis	5.83	0.27	1.71	40	5.29–6.37

[a]in hundredths of an hour
[b]arithmetic mean
[c]standard error of the mean
[d]standard deviation
[e]sample size (number of lizards)

phorus tigris appear, while still later larger *C. tigris* emerge. As air and substrate temperatures rise with the daily march of temperature other species such as *Callisaurus draconoides* and *Dipsosaurus dorsalis* become active (Table 1). Similar patterns of gradual sequential replacement of species during the day occur in Australian skinks of the genus *Ctenotus* (21) and in lacertid lizards in the Kalahari desert of southern Africa (25). Daily patterns of activity also change seasonally with later emergence during cooler winter months than in warm summer ones (4, 13, 21, 23, 27, 30, 31, 46). Species with bimodal daily activity patterns during warm months (early and late in the day) often have a unimodal activity period during cooler months (13, 21, 30, 31, 46). Such seasonal changes in the time of activity presumably allow a lizard to encounter a similar thermal environment and microclimate over a period of time when the macroclimate is changing. Standardizing times of activities to "time since sunrise" (diurnal species) or "time since sunset" (nocturnal species) corrects for such seasonal shifts in time of activity and greatly facilitates comparison among species (Table 1) as well as comparisons between communities (below). Body temperatures of active individuals often reflect the time of activity reasonably well (21), although body temperature can be strongly affected by microhabitat(s) as well (4, 13, 14, 21, 26, 30, 32, 44). Thus species that emerge earlier in the day frequently have lower active body temperatures than those that emerge later; indeed, body temperature can sometimes be used as an indicator of time (21) or thermal (36, 41, 43) niche. The anatomy and size of a lizard's eyes are another useful indicator of its time niche; large eyes and elliptical pupils almost invariably indicate nocturnal activity (48).

PLACE NICHE   The use of space varies widely among lizard species. A few are entirely subterranean (fossorial), many others are completely terrestrial, while still others are almost exclusively arboreal. Various degrees of semifossorial and semiarboreal activity also occur. Microhabitat differences among species are often pronounced even within these groups. Thus some terrestrial species forage primarily in the open spaces between plants, whereas others forage mainly under or within plants, the plants sometimes having a particular life form. Similar subtle differences in the use of various parts of the vegetation also occur among arboreal lizard species, especially *Anolis* (35, 36, 39, 41, 43). Some lizard species are strongly restricted to a rock-dwelling (saxicolous) existence. In addition to such microhabitat specificity, various species have specialized in their habitat requirements. Thus different sets of species of Australian desert lizards are restricted to sandridge, sandplain, and shrubby habitats respectively (21, 28). As defined here the place niche is more inclusive than Rand's (35) structural niche, as it includes both habitat and microhabitat preferences. Exactly where in the environmental mosaic a lizard forages, as well as its mode of foraging in that space, is perhaps its most important ecological attribute.

Lizards that exploit space in different ways have evolved a variety of morphological adaptations for the use of space (21, 30, 33, 37); such anatomical traits are often accurate indicators of their place niche. Thus fossorial species typically have either very reduced appendages or none at all. Diurnal arboreal lizards are usually long-tailed and slender. Terrestrial species that forage in the open between shrubs and/or grass clumps generally have long hind legs relative to their size, while those that forage closer to cover or within dense clumps of grass usually have proportionately shorter hind legs (21, 30, 33). Lamellar structure often reveals arboreal or terrestrial activity as well as the texture of the substrate exploited (1). Moreover, terrestrial geckos have proportionately larger eyes than arboreal ones (33, 48).

FOOD NICHE   Most lizards are insectivorous and fairly opportunistic feeders, taking without any obvious preference whatever arthropods they encounter within a broad range of types and sizes. Smaller species or individuals, however, do tend to eat smaller prey than larger species or individuals (6, 21, 33, 38, 39, 43); also, differences in foraging techniques (below) and place and time niches often result in exposure to a different spectrum of prey species. Rather few lizard species have evolved severe dietary restrictions: among these are the ant specialists *Phrynosoma* and *Moloch* (17, 31, 32), termite specialists such as *Rhynchoedura* and *Typhlosaurus* (7, 33), various herbivorous lizards which include *Ctenosaurus, Dipsosaurus, Sauromalus,* and *Uromastix,* and secondary carnivores such as *Crotaphytus, Heloderma, Lialis,* and *Varanus* which prey primarily upon the eggs and young of vertebrates and the adults of smaller species (17, 19, 22, 24). All the above foods are at least temporarily very abundant making food specialization advantageous (12). Just as lamellar structure and hind leg proportions reflect the place niche of a lizard, head proportions, jaw length, and dentition frequently prove to be useful indicators of the food niche (6, 21, 38), especially of the sizes and kinds of prey eaten.

Another, somewhat more behavioral, aspect of a lizard's food niche concerns the way in which it hunts for prey. Two extreme types of foragers have been recognized (17, 40, 42): a lizard may either actively search out prey (widely foraging strategy) or wait passively until a moving prey item offers itself and then ambush the prey (sit-and-wait strategy). Normally the success of the sit-and-wait method requires a fairly high prey density, high prey mobility, and/or a low energy demand by the predator (40, 42). The effectiveness of the widely foraging tactic also depends on the density and mobility of prey and the predator's energy needs, but in this case the distribution of prey in space and the searching abilities of the predator may take on considerable importance (40, 42). Clearly, this dichotomy is artificial and these two tactics actually represent pure forms of a variety of possible foraging strategies. However, the dichotomy has substantial practical value because the actual foraging techniques used by lizards are often strongly polarized. Thus most teids and skinks and many varanid and lacertid lizards are very active and widely foraging, typically on the move continually; in contrast, almost all iguanids, agamids, and geckos are relatively sedentary sit-and-wait foragers. These differences in the mode of foraging presumably influence the types of prey encountered, thus affecting the composition of a lizard's diet.

INTERACTIONS BETWEEN THE TIME, PLACE, AND FOOD NICHES   Place niches and food niches of lizards change in time, both during the day and with the seasons. In the early morning, when ambient air and substrate temperatures are relatively low, lizards typically locate themselves in the warmer microhabitats of the environmental mosaic, such as depressions in the open sun or the sunny side of a rock, slope, sandridge, or tree trunk. Often an animal orients its body at right angles to the sun's beams, thereby maximizing heat gained from the sun. Later in the day as environmental temperatures rise the same lizards usually spend most of their time in the cooler patches in the environmental mosaic, such as shady spots underneath shrubs or trees (4, 26, 27, 41). Finally, as the surface gets still hotter many lizards retreat into cool burrows; certain species, such as *Amphibolurus inermis,* climb up off the ground into cooler air and face into the sun, minimizing their heat load due to solar irradiation (4, 26). Thus time of activity strongly affects a lizard's place niche and its habitat and microhabitat requirements may dictate periods when the animal can be active.

Similarly, the composition of the diet of many lizards changes as the relative abundances of different types of prey fluctuate with the seasons (and probably within a day). Nocturnal lizards clearly encounter a different spectrum of potential prey items than diurnal lizards, and those that forage in different places usually encounter different prey. The mode of foraging or the way in which a lizard uses space can influence both its place and food niches; thus widely foraging species typically have broader place niches than sit-and-wait species, while the latter type of foragers often tend to have broader food niches than the former. Recall that pairs of lizard species with high overlap along one niche dimension, say microhabitat, may have low overlap along another niche dimension such as foods eaten, effectively reducing interspecific competition between them.

Niche Breadth and Niche Overlap

In addition to the differences in times of activity and use of space and foods noted above, lizard species differ in the spans of time over which they are active as well as the ranges of spatial and trophic resources they exploit. As outlined above, such differences in niche breadth may have a considerable impact upon the structure and diversity of lizard communities. Following MacArthur (11), niche breadth along any single dimension is here quantified using Simpson's index of diversity

$$B = 1 / \sum_i^n p_i^2 \qquad \qquad 3.$$

where p_i represents the proportion of the ith time period (or microhabitat or food type) actually used; B varies from unity to n depending upon the p_i values. Niche breadths based on a different number of p_i categories can be compared after standardizing them by dividing by n. Overall niche breadth along several niche dimensions can be estimated either as the product or the geometric mean of the breadths along each component dimension (recall that the lower bound on B is one) or by the arithmetic mean of the latter breadths.

Niche overlap also varies among lizard species and between communities. Overlap along any single niche dimension can be quantified in a wide variety of ways (2, 5, 10, 21, 34, 39, 47). Here I use still another measure of overlap, based upon Levins' (10) formula for α

$$a_{jk} = \frac{\sum_i^n p_{ij} \, p_{ik}}{\sum_i^n p_{ij}^2} \qquad \qquad a_{kj} = \frac{\sum_i^n p_{ij} \, p_{jk}}{\sum_i^n p_{ik}^2} \qquad \qquad 4.$$

where p_{ij} and p_{ik} are the proportions of the ith resource used by the jth and the kth species respectively. The above equations have been used to estimate the so-called competition coefficients (10, 11, 47), and give different α values for each partner in a niche overlap pair provided that niche breadths (the inverse of the denominators in Equation 4) differ. Here I use the following multiplicative measure of overlap

$$O_{jk} = O_{kj} = \frac{\sum_i^n p_{ij} \, p_{ik}}{\sqrt{\sum_i^n p_{ij}^2 \sum_i^n p_{ik}^2}} \qquad \qquad 5.$$

where the p_{ij} and p_{ik} are defined as before (I am indebted to Selden Stewart for suggesting this equation). Equation 5 is symmetric and gives a single overlap value for each niche overlap pair; it can never generate values less than zero or greater than one [Equation 4, however, does give one α value (of a pair) that is greater than unity provided niche breadth and overlap are high]. Overall niche overlap along several niche dimensions can be estimated by the product of the overlaps along each component dimension (10, 21), although this procedure may either overestimate or underestimate overall overlap (H. S. Horn, personal communication; R. M. May, unpublished). Thus if niches are completely separated along any single niche dimension both niche overlap along that dimension and overall niche overlap are zero.

COMPARISONS BETWEEN COMMUNITIES

During the last decade I have studied in some detail three independently derived and evolved, but otherwise basically comparable, sets of desert lizard communities at similar latitudes in western North America, southern Africa, and Western Australia. Here I use data from these studies to quantify and compare various parameters of lizard niches. Although lizards were studied on 32 different study areas (below) I lump data from various study areas within each continental desert-lizard system here for brevity and clarity (a more detailed area by area analysis will be undertaken elsewhere). A few allopatric species pairs are thus treated as though they are sympatric, but the vast majority of the species considered are sympatric on one or more study areas.

The number of sympatric lizard species on 14 North American desert study areas varies from 4 to 11, with either 4 or 5 sympatric species in the northernmost Great Basin desert, 6–8 species in the more southern Mojave and Colorado deserts, and 9–11 species in the still more southerly Sonoran desert (16–18). (The analysis to follow includes only 10 southern North American desert study areas.) Ten study areas in the Kalahari desert of southern Africa support 12–18 sympatric species of lizards (25). In the Western Australian desert 18–40 species of lizards occur together in sympatry on eight different study sites (20, 21, 33). In addition to such censuses of lizard species densities, I gathered supporting data on the physiography, climate, vegetation, and faunas of each of the 32 desert study areas (15–18, 20, 21, 25, 28–31).

The actual diversity of lizards observed on all sites within each desert-lizard system, estimated using the relative abundances of the various species in my collections (below) as p_i's in Equation 3, are: North America = 3.0 (28% of the maximum possible diversity of 11), Kalahari = 12.5 (60% of the maximum possible diversity of 21), and Australia = 19.0 (32% of the maximum possible diversity of 59). (These are crude approximations of the actual lizard diversities, both because real relative abundances doubtless differ somewhat from the relative abundances in my samples and because not all species actually occur in sympatry.)

Time of activity and microhabitat were recorded for most active lizards encountered. Table 2 lists the average numbers of species in five basic time and/or place niches in each desert system (see also below). Wherever possible, lizards were collected; these specimens[1] allowed analysis of stomach contents. Twenty basic prey categories, corresponding roughly to various orders of arthropods, were distinguished. Both the numbers and volume of prey items in each category were recorded for every stomach.

I used these data on time of activity, microhabitat usage, and stomach contents for the following analyses of the time, place, and food niches of desert lizards. The numbers of lizards active at different times were grouped by species into 22 hourly categories expressed in time since sunrise for diurnal species (14 categories) and time since sunset for nocturnal ones (limitations on human endurance allowed only 8

[1]Some 5000 North American lizards, over 6000 Kalahari lizards, and nearly 4000 Australian ones, all of which are now lodged in the Los Angeles County Museum.

Table 2 Average numbers of species of lizards in five basic niche categories on study areas in the three desert systems. The percentage of the average total number of species in each system is also given.

Niche Category	North America[a]		Kalahari[b]		Australia[c]	
	$\overline{X}$	%	$\overline{X}$	%	$\overline{X}$	%
diurnal terrestrial	5.7	69	6.3	43	14.4	51
diurnal arboreal	1.2	14	1.9	13	2.6	9
nocturnal terrestrial	1.4	17	3.5	24	7.6	27
nocturnal arboreal	0.0	0	1.6	11	2.6	9
fossorial	0.0	0	1.4	10	1.1	4
totals	8.3	100	14.8	101	28.3	100

[a]10 different southern study areas

[b]10 study areas

[c]8 study areas

nocturnal hourly categories); these 22 time categories were used as p_i's in the above equations. Fifteen basic microhabitat categories were recognized and used as p_i's. Time and place niche breadths and overlaps were calculated for desert lizards in these three independently evolved systems of lizard communities using Equations 3 and 5 and the above data on the numbers of lizards active at different times and in different microhabitats. The overall span or diversity of time of activity of all the lizards in each continental desert system (D_r in Equation 2), as well as the microhabitats used by them, were estimated using Equation 3 and the proportions of each time period or microhabitat type as computed from grand totals summed over all lizard species. Stomach content data (prey items by volume[2]) allowed similar calculations of food niche breadths and overlaps, as well as the average and overall diversity of foods eaten by all lizards, $\overline{D}_u$ and D_r, in each of the above deserts. Mean niche breadths of all the species in a given community ($\overline{D}_u$ in Equation 2) were also calculated for the time and place niches. Average niche overlap along each niche dimension in any particular community was calculated as the arithmetic mean of all interspecific overlaps (calculated from Equation 5); products of these values were also computed to estimate overall niche overlap.

Diversity of Resources Used by Lizards

The overall diversity of times of activity of all lizards (D_r for the time niche) in each desert-lizard system was computed using Equation 3 and the proportional representation of the 22 hourly time categories among all species (recall that these categories are expressed in hours since sunrise or sunset and that they therefore correct

[2]Prey items in the same 20 categories by numbers of items, rather than their volumetric importance, and prey in 34 size categories (irrespective of type) were also examined, but are not considered further here because there is very little niche separation in either of these two aspects of the food niche.

somewhat for seasonal shifts in activity patterns). Overall diversity of time of activity thus computed is quite low in North America (5.9 or only 27% of the maximum possible value of 22) and nearly twice as large in the Kalahari and Australia (11.6 and 11.7 respectively, or about 53% of the possible maximum). A major factor contributing to the greater diversity of time of activity in the Kalahari and Australia is the increased numbers of nocturnal lizards in the southern hemisphere (Table 2), although the diversity of time of activity of diurnal lizards is also somewhat higher in these two deserts than in North America. Lizards are active year around in the Kalahari and Australia and they were sampled over the entire year, while the seasonal period of activity is shorter in North America and lizards were sampled only over a six-month period. Whatever the reason(s) for this difference between the desert systems, the more diverse communities of the Kalahari and Australia certainly exhibit much greater temporal variation in their times of activity on both a daily and a seasonal basis than the less diverse North American lizard community.

Overall microhabitat diversity, computed using Equation 3 and the 15 basic microhabitat categories as exploited by all the lizards in each system, represents D_r for the place niche; again, it is very low in North America (3.3 or only 22% of the maximal value of 15), where the vast majority of lizards were first sighted in the open sun, and considerably higher in the Kalahari (8.8 or 59% of maximum) and Australia (8.2 or 55% of maximum). These differences in the diversity of microhabitats actually used by lizards are due partly to an increased incidence of arboreal and subterranean lizards in the two deserts of the southern hemisphere (Table 2), although more animals are also first sighted in the shade of various types of plants (Table 3). Nocturnality is much more prevalent in the Kalahari and Australia (Table 2) and contributes to the increased use of shade in these lizard communities (nocturnal lizards were arbitrarily assigned to shade categories in Table 3, although this somewhat confounds place and time niches).

Somewhat surprisingly, the overall diversity of foods eaten by all the lizards[3] in a community, or D_r for the food niche, is lowest in the Kalahari (4.4 or 22% of the maximal value of 20), intermediate in Australia (7.4 or 37% maximum), and highest in the least diverse lizard communities of North America (8.7 or 44% of maximum). The low diversity of foods eaten by Kalahari lizards stems from the preponderance of termites in the diets of these lizards (Table 4). Examination of Table 4 shows that the proportions of various prey categories actually eaten by lizards differ markedly among the desert systems. For example, although termites are a major food item in all three deserts, their fraction of the total prey eaten by all lizards is considerably higher in the Kalahari (41.3%) than in either of the other deserts (16.5 and 15.9%). Prominent prey in the Australian desert are vertebrates (24.8%), especially lizards, and ants (16.4%). By volume, beetles constitute 18.5% of the food eaten by North American desert lizards, 16.3% of that eaten by Kalahari lizards, but only 7.3% of the Australian desert lizard diet.

[3]Computed using Equation 3 and the proportion of the total volume of food in each of 20 prey categories in the stomachs of all the lizards collected in a series of communities from each desert-lizard system.

Table 3 Microhabitats actually used by all lizards in three different desert systems. Nocturnal lizards assigned to shade categories. Numbers (*N*) and percentages (%).

Microhabitat Category	North America		Kalahari		Australia	
	N	%	*N*	%	*N*	%
Subterranean	0	0.0	579	12.1	17	0.5
Terrestrial						
open sun	1335	45.3	890	18.6	596	19.0
grass sun	92	3.1	155	3.2	314	10.0
bush sun	883	30.0	547	11.4	192	6.2
tree sun	103	3.5	126	2.6	31	1.0
other sun	95	3.2	6	0.1	14	0.4
open shade	49	1.7	546	11.4	547	17.4
grass shade	2	0.1	274	5.7	525	16.6
bush shade	165	5.6	765	15.9	221	6.9
tree shade	30	1.0	179	3.7	81	2.6
other shade	72	2.4	18	0.4	43	1.3
Arboreal						
low sun	12	0.4	125	2.6	56	1.5
low shade	6	0.2	109	2.3	224	7.0
high sun	50	1.8	200	4.2	91	2.0
high shade	51	1.8	276	5.8	250	7.7
TOTALS	2945	100.1	4795	100.0	3202	100.1

Table 4 Major prey items in the stomachs of all lizards in three different desert systems by volume in cubic centimeters.

Prey Category	North America		Kalahari		Australia	
	volume	percentage	volume	percentage	volume	percentage
spiders	50	1.6	36	3.1	54	3.4
scorpions	23	0.7	33	2.9	22	1.4
ants	307	9.7	155	13.6	261	16.4
locustidae	364	11.5	70	6.1	138	8.7
blattidae	100	3.2	4	0.4	37	2.3
beetles	587	18.5	187	16.3	117	7.3
termites	525	16.5	473	41.3	253	15.9
homoptera-hemiptera	31	1.0	15	1.3	30	1.9
lepidoptera	68	2.1	16	1.4	9	0.5
all larvae	384	12.1	41	3.6	80	5.0
Miscellaneous arthropods	225	7.0	76	6.6	107	6.7
vertebrates	246	7.8	26	2.3	395	24.8
plants	262	8.3	13	1.2	89	5.6
TOTALS	3172	100.1	1145	100.0	1592	99.9

To give each niche dimension equal weight the above estimates of D_r were standardized by dividing by the number of p_i categories and multiplying by 100, thus expressing the diversity of use of resources as a percentage of the maximal possible resource diversity along a given niche dimension. The overall diversity of resources used by all lizards in all three niche dimensions was then computed as the product of the above three standardized D_r values divided by 1000. So estimated, overall diversity of resources used is lowest in North America (25.9), intermediate in the Kalahari (68.9), and highest in Australia (107.5); moreover, these estimates of the size of the lizard niche space are directly proportional to observed lizard diversities in the various deserts (above).

Differences in Niche Breadth

Niche breadths for the food, place, and time niches, as well as their products (overall niche breadth) were calculated for 91 species of desert lizards in 10 families on the three continents. Frequency distributions and averages of all the species in each desert-lizard system are shown for each niche dimension in Figure 1; these mean niche breadths represent the average diversity of utilization of each niche dimension, or $\bar{D}_u$ in Equation 2, by the lizards in a given system. In all three deserts average time niche breadths are very similar, though their frequency distributions differ

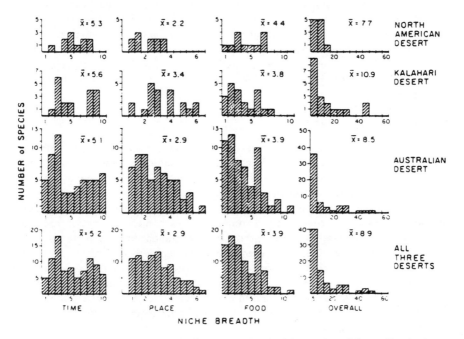

Figure 1 Frequency distributions of niche breadths of 91 species of desert lizards along three major niche dimensions in three deserts. Overall niche breadths, computed as the products of the standardized breadths along each component dimension, weight each niche dimension equally. See text for discussion.

(Figure 1). The frequency distribution of time niche breadth of North American lizards is fairly continuous, but these distributions are distinctly bimodal in the Kalahari and Australia where most nocturnal species have relatively narrow time niches while diurnal ones generally have comparatively broader time niches. (The narrow time niches of nocturnal lizards are probably an artifact due to the shorter nighttime sampling period; however, this bias is similar in all three deserts and should not generate differences between the desert systems.) Place niche breadths are more evenly distributed than time niche breadths, although the distributions are skewed with more narrow place niches than broad ones (Figure 1); place niches are smallest in North America ($\bar{x} = 2.2$, or 15% of maximal value), intermediate in Australia ($\bar{x} = 2.9$, or 19% of maximum), and broadest in the Kalahari ($\bar{x} = 3.4$, or 23% of maximum). In all three deserts food niche breadths appear to be distinctly bimodal, suggesting a natural dichotomy of food specialists versus food generalists (Figure 1). Average food niche breadth is fairly similar in all three deserts and is largest in North America.

Because species with broad niches along one dimension often, though by no means always,[4] have narrow niches along another dimension, overall niche breadths are strongly skewed with the majority of species having rather narrow overall niches (Figure 1). Nevertheless, a few species in the Kalahari and Australia with broader than average niches along all three niche dimensions have extremely broad overall niches (Figure 1). Average overall niche breadth is smallest in North America (7.7), intermediate in Australia (8.5), and largest in the Kalahari (10.9). However, overall niche breadths, as well as average overall niche breadths, do not differ strikingly between the desert systems; indeed, if anything, overall niches tend to be slightly larger in the more diverse communities, rather than smaller as might have been anticipated.

Niche Dimensionality

Any given niche dimension's potential to separate niches, and thus its potential effectiveness in reducing interspecific competition, should be roughly proportional to the ratio of the overall diversity of use of that niche dimension divided by the diversity of utilization by an average species, or $D_r/\bar{D}_u$. Table 5 summarizes much of the above discussion and lists the ratios of $D_r/\bar{D}_u$ for each major niche dimension in the three desert-lizard systems. Estimates for each niche dimension are also multiplied to give overall estimates (products of the standardized estimates for each component dimension). Thus measured, the dimension with the greatest apparent potential to separate niches in North America is food, which, by the same criteria, is a comparatively negligible niche dimension in the Kalahari; conversely, by these standards place and time niches seem to have a much greater potential to separate niches of Kalahari lizards than North American ones (Table 5). All three niche dimensions, especially place and time, appear to have the potential to separate niches of Australian lizards. The products of the $D_r/\bar{D}_u$ ratios for all three dimen-

---

[4]Product moment correlation coefficients among niche breadths along various dimensions range from −0.38 to 0.40 and are generally weak and seldom statistically significant.

Table 5 Estimates of various niche parameters (see text and Table 6).

Desert and Niche Dimension	D_r	$\overline{D}_u$	$D_r / \overline{D}_u$	$\overline{C}$	Mean Overlap (all pairs)	Mean Overlap (nonzero pairs)
North America						
time	25.4	24.2	1.05	3.0	0.58	0.86
place	22.0	14.6	1.51	3.0	0.34	0.55
food	43.7	22.0	1.98	1.2	0.46	0.49
overall	25.9	7.7	3.34	-9.5	0.09	0.23
Kalahari						
time	52.7	25.4	2.07	11.7	0.43	0.78
place	58.9	22.8	2.58	13.3	0.29	0.38
food	22.2	18.8	1.18	14.9	0.64	0.64
overall	68.9	10.9	6.34	12.2	0.08	0.27
Australia						
time	53.3	23.1	2.31	22.9	0.32	0.54
place	54.8	19.1	2.87	19.2	0.29	0.35
food	36.8	19.3	1.90	28.4	0.32	0.36
overall	107.5	8.5	12.62	17.3	0.03	0.13

sions (Table 5), which should be proportional to the overall potential for niche separation, increase from North America (3.3) to the Kalahari (6.3) to Australia (12.6), as might be expected. Hence, as measured by $D_r/\overline{D}_u$, the potential for niche partitioning seems to be greater in more diverse lizard communities; moreover, this potential is directly proportional to actual lizard diversities observed.

Differences in Niche Overlap

Figure 2 shows the frequency distributions of niche overlap values for all inter-specific pairs along each niche dimension in the three desert systems (calculated using Equation 5). Estimates of overall overlap, computed as the products of the overlap along the three niche dimensions, are shown at the right of the figure. Although there are some striking differences and trends in overlap patterns,[5] among both niche dimensions and deserts, overall overlaps are uniformly low in all three deserts (Figure 2 and Tables 5 and 6). The vast majority of interspecific pairs overlap very little or not at all when all three dimensions are considered. This is demonstrated by low overall overlap values and by the size of the "zero" classes of overall overlap in the various deserts (Tables 5 and 6). Table 5 gives averages both for all overlap pairs and for only those pairs which overlap somewhat (that is, all pairs other than those with zero overlap) for each niche dimension and for overall overlap estimates. Provided average niche breadth $(\overline{D}_u)$ remains relatively constant, the number of possible nonoverlapping pairs increases markedly as overall niche space (D_r) increases. Hence the average niche overlap of pairs with some overlap is of interest as it should reflect the limiting similarity and/or maximal tolerable overlap

[5]For instance, distributions of time niche overlap are distinctly bimodal in all three deserts (particularly North America and the Kalahari), reflecting the nonoverlapping times of activity of nocturnal and diurnal species.

Table 6 Summary of overall niche overlap patterns (see text and Table 5).

Desert System	Total Number of Overlap Pairs	Number of Zero Overlap Pairs	Zero Overlap Pairs as % of Total	Number of Nocturnal-Diurnal (ND) Pairs	ND Pairs as % of Zero Overlap Pairs	Number of Non-ND Pairs with Zero Overlap	Non-ND Pairs with Zero Overlap as % of Zero Overlap Pairs	Non-ND Pairs with Zero Overlap as % of all Ovelap Pairs
North America	55	37	67%	18	49%	19	51%	35%
Kalahari	171	101	59%	78	77%	23	23%	13%
Australia	1596	1255	78%	680	54%	575	46%	36%

in each desert system. Although a substantial number of nonoverlapping pairs are nocturnal-diurnal species pairs, many non-nocturnal–diurnal pairs also do not overlap (Table 6). The proportion of such zero overlap pairs is distinctly lower in the Kalahari desert, where only 23% of the non-nocturnal–diurnal pairs do not overlap, than in North America and Australia (51 and 46% respectively). Furthermore, the average overlap among all nonzero overlap pairs tends to be somewhat greater in the Kalahari and North America than in Australia, suggesting that maximal tolerable niche overlap is lower in the latter desert (Table 5).

Although niche overlap values are far from normally distributed (Figure 2), arithmetic means [especially of the nonzero overlap values (Table 5)] do reflect differences between the various niche dimensions and deserts. Average overlap in microhabitat is low and generally similar in all three deserts, while average overlaps in the time and food niches are considerably more variable (Figure 2 and Table 5). Average time niche overlap is high in North America, while both average food and time niche overlaps are high in the Kalahari. In Australia, average niche overlap values are low along all three niche dimensions (Table 5). As a result, overall overlap is distinctly lower in Australia than in the other two desert systems. Thus overall niche overlap seems to vary inversely with lizard species diversity.

Numbers of Neighbors in Niche Space

By far the most difficult parameter to estimate in Equation 2 is the number of neighbors in niche space C (indeed, MacArthur did not indicate how one might

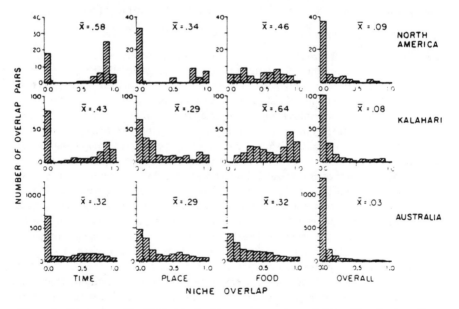

Figure 2 Frequency distributions of niche overlap values of desert lizards along three major niche dimensions in three deserts. See text for discussion.

attempt to estimate C). This quantity cannot be estimated satisfactorily from my data in an independent way; however, $\overline{C}$ can be calculated by simply rearranging Equation 2 to solve for $\overline{C}$

$$\overline{C} = \frac{\left\{ (\overline{D}_u / D_r) D_s - 1 \right\}}{\overline{O}}$$

6.

Values of $\overline{C}$ estimated by substituting various estimates of other parameters (above) into Equation 6 are listed in Table 5. These values appear to be reasonable for any single niche dimension. However, the estimate of the number of neighbors in overall niche space (all three niche dimensions) is actually negative for North America. Estimates of the number of neighbors in overall niche space are much higher and more reasonable in the Kalahari and Australia (Table 5).

As indicated earlier, communities can differ in species diversity with differences in the extent to which they contain as many different species as they can support. The negative estimate of the number of neighbors in overall niche space in North America suggests that lizard diversity in these deserts may actually be lower than it could potentially be, or that these deserts may not be truly saturated with species. Further, the complete absence of any fossorial lizards or any which are both nocturnal and arboreal in North America (Table 2) suggests that these niches either (a) do not exist, (b) are unoccupied, or (c) are occupied but by another kind of animal (see next section). (Indeed, I would be quite surprised if a successful climbing gecko such as the Australian *Gehyra variegata* were unable to invade the North American desert without a simultaneous extinction of another nocturnal animal.)

Reciprocal Relations With Other Taxa

The ecological roles of lizards and various other taxa, especially birds and mammals, are strongly interdependent (9). Thus lizards may capitalize on variability of primary production, and this might be a factor contributing to their relative success over birds in desert regions (18, 20, 25). There are proportionately more species of ground-dwelling insectivorous birds in the Kalahari than there are in Australia (29), suggesting that competition between birds and lizards may be keener in southern Africa than it is in Australia. Figure 3 plots the number of bird species against the number of lizard species on 27 study areas representative of each desert system. As the total number of species increases, the numbers of bird species increase faster than lizard species in North America and the Kalahari, whereas in Australia lizards increase faster than birds. This figure suggests a sharp upper bound on the number of sympatric lizard species in North America and the Kalahari, but no such limit in Australia. Exactly the reverse seems to be true of birds in the three continental desert systems; that is, a distinct upper limit on bird species diversity appears to exist in Australia, but not in either North America or the Kalahari. The reasons for such differences between the three desert systems are elusive and must remain conjectural (9). There are very few migratory bird species in Australia, whereas a number of migratory birds periodically exploit the North American and Kalahari deserts; competitive pressures from these migrants must have their effects upon the lizard

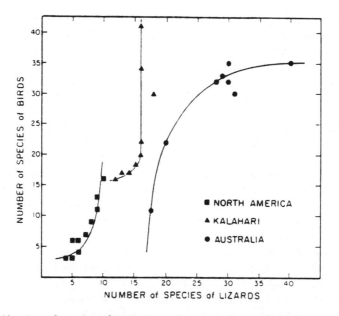

Figure 3 Number of species of birds plotted against the number of lizard species on various study areas within three desert systems. See text.

communities in the latter two desert systems. The higher incidence of arboreal, fossorial, and nocturnal lizard species in the Kalahari and Australia, as compared with North America (Table 2), are probably related to fundamental differences in the niches occupied by other members of these communities such as arthropods, snakes, birds, and mammals (20, 25). These differences in the composition and structure of the various communities presumably have a historical basis. Thus southern Africa has an exceptionally rich termite fauna, which in turn may have allowed the evolution of termite-specialized subterranean *Typhlosaurus* species (7). The prevalence of nocturnality among Kalahari and Australian lizards may arise from variations among systems in either or both of the following: (*a*) differences in the diversity of available nocturnal resources, such as nocturnal insects, or (*b*) differences in the numbers and/or densities of insectivorous and carnivorous nocturnal birds and mammals. The mammalian fauna of the Australian desert is conspicuously impoverished, and the snake fauna less so; in this desert system varanid and pygopodid lizards are ecological equivalents of carnivorous mammals and snakes, respectively, in North America and the Kalahari (20, 25). Such usurpation of the ecological roles of other taxa in the other deserts has expanded the diversity of resources exploited by Australian desert lizards (20).

Within-Habitat and Between-Habitat Diversity

Overall species diversities in an area (as opposed to point diversities) can differ in a way that is included neither in Equation 2 nor in the above analysis of niche

239

relationships. Thus only the so-called "within-habitat" component of diversity (11, 25) was considered above (indeed, for brevity and clarity, data from various different study areas within each desert-lizard system were lumped for the above analyses). The other way in which communities can differ in species diversity is through differences in species composition from area to area or habitat to habitat within a study area (no study area is perfectly homogeneous); such horizontal turnover in species composition represents the so-called "between-habitat" component of diversity (11). To estimate the amount of between-habitat diversity in each of the above desert-lizard systems I calculated coefficients of community similarity[6] for every pair of lizard communities within each continental desert system (25). Community similarity values are high and rather uniform in the North American desert ($\bar{x}$ = 0.67, S. E. = 0.019, s = 0.153, N = 66) and the Kalahari desert ($\bar{x}$ = 0.67, S. E. = 0.015, s = 0.127, N = 66), indicating little difference between study areas in species composition (i.e. a low between-habitat component of diversity). However, community similarity values are significantly lower (t-tests, $P < 0.01$) in the Australian desert ($\bar{x}$ = 0.49, S. E. = 0.027, s = 0.144, N = 28), demonstrating that this component of diversity is greater in that desert system. Habitat specificity is much more pronounced in Australian desert lizards than it is in North American or Kalahari desert lizards (20, 25, 28). For example, although both the Kalahari and the Australian deserts are characterized by long stabilized sandridges, only a single species [*Typhlosaurus gariepensis* (7)] is specialized to Kalahari sandridges whereas ten lizard species are sandridge specialists in Australia (20, 28).

TAXONOMIC COMPONENTS OF LIZARD SPECIES DENSITY

Because closely related species are often ecologically similar and therefore in strong competition when they occur together, Elton (3) suggested that competitive exclusion should occur more frequently between pairs of congeneric species than between more distantly related pairs of species. Moreover, he reasoned that if this argument is valid fewer pairs of congeneric species should occur within natural communities than in a random sample of species and genera from a broader geographic area which includes several to many different communities. Frequent cases of abutting allopatry (parapatry) of congeners seem to support this argument. Elton examined the numbers of congeneric species in portions of many different natural communities and found evidence for such a paucity of congeners, even in spite of the bias towards an increased number of congeneric pairs due to the possibility of inclusion of two or more communities (and thus abutting allopatric congeneric pairs) in his samples. Although his numerical analysis has since been shown to be incorrect (49), his argument is still reasonable and worthy of consideration. Using a corrected statistical approach, Williams (49) failed to find fewer congeners than expected in a variety of natural communities (indeed, he found more than expected in many). Terborgh

[6]Community similarity (CS) is simply X/N, where X is the number of species common to two communities and N is the total number of different species occurring in either; thus CS equals one when two communities are identical, and zero when they share no species.

& Weske (45) also used this corrected method to calculate the expected numbers of congeneric species pairs in Peruvian bird communities, and found that these communities were not impoverished with congeneric pairs, thus refuting any increased incidence of competitive exclusion among congeners in this particular avifauna. Similar analyses of the saurofaunas of the Kalahari and the Australian deserts are summarized in Figures 4 and 5. Again, the observed numbers of congeneric pairs are not conspicuously or consistently lower than expected.

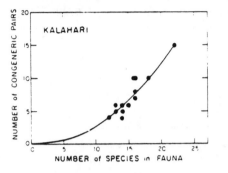

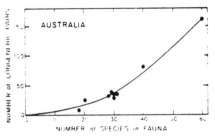

Figure 4 Dots represent the actual numbers of pairs of congeneric species of lizards observed on ten study areas in the Kalahari desert. Curve is the expected number of such pairs in a random subsample of the entire fauna.

Figure 5 Dots are the actual numbers of pairs of congeneric species of lizards observed on eight Australian desert study areas. Curve represents the number of such pairs expected in a random subsample of the entire fauna.

CONCLUDING REMARKS

Interpretation of the structure of desert lizard communities has become steadily more difficult as the amount of information increases. Early in these studies, I expected to find much more pronounced similarities between these independently evolved, but otherwise basically similar ecological systems. Although a few crude ecological equivalents can be found among the different desert-lizard systems (26, 27, 30, 32), the ecologies of most species are quite disparate and unique. As seen above, the diversity of resources actually used by lizards along various niche dimensions, as well as the amount of niche overlap along them, differs markedly among the desert systems; moreover, the relative importance of various niche dimensions in separating niches varies. Thus food is a major dimension separating the niches of North American lizards, whereas in the Kalahari food niche separation is slight and differences in the place and time niches are considerable. All three niche dimensions are important in separating the niches of Australian desert lizards. Overall niche overlap is least in the most diverse lizard communities of Australia. Differences in diversity between the three continental systems stem from differences in the overall diversities of resources used by lizards or the size of the lizard niche space, as well as from differences in overall niche overlap, but are not due to conspicuous differences in overall niche breadths. Factors underlying these observed differences in diversity of utilized resources and niche overlap are poorly understood at present, but probably involve some of the following: (*a*) the degree to which any given system is truly saturated with species, (*b*) differences in the available range of resources among deserts that stem from historical factors, such as diversification of termites, reciprocal relations with other taxa, and the usurpation of their ecological roles, (*c*) differences between desert systems in the extent of spatial heterogeneity and habitat complexity which alter the degree of habitat specificity and the between-habitat component of diversity, and (*d*) other factors, such as possible differences in climatic stability and predictability, which might affect tolerable niche overlap.

Acknowledgments

My research has benefited from contacts with numerous persons and parties, too many to enumerate here. I am grateful to Henry Horn, Chris Smith, Robert Colwell, Raymond Huey, William Parker, Larry Gilbert, and my wife Helen for reading this manuscript, and to Glennis Kaufman for much help in data analysis. Finally, the project would have been impossible without the financial assistance provided by the National Institutes of Health and the National Science Foundation.

Literature Cited

1. Collette, B. B. 1961. Correlations between ecology and morphology in anoline lizards from Havana, Cuba and southern Florida. *Bull. Mus. Comp. Zool. Harvard Univ.* 125:137–62

2. Colwell, R. K., Futuyma, D. J. 1971. On the measurement of niche breadth and overlap. *Ecology* 52:567–76

3. Elton, C. S. 1946. Competition and the structure of ecological communities. *J. Anim. Ecol.* 15:54–68

4. Heatwole, H. 1970. Thermal ecology of the desert dragon, *Amphibolurus inermis. Ecol. Monogr.* 40:425–57

5. Horn, H. S. 1966. Measurement of overlap in comparative ecological studies. *Am. Natur.* 100:419–24

6. Hotton, N. 1955. A survey of adaptive relationships of dentition to diet in the North American Iguanidae. *Am. Midl. Natur.* 53:88–114

7. Huey, R. B., Pianka, E. R., Egan, M. E., Coons, L. W. 1974. Ecological shifts in sympatry: Kalahari fossorial lizards (*Typhlosaurus*). *Ecology* 55:In press

8. Hutchinson, G. E. 1957. Concluding remarks. *Cold Spring Harbor Symp. Quant. Biol.* 22:415–27

9. Lein, M. R. 1972. A trophic comparison of avifaunas. *Syst. Zool.* 21:135–50

10. Levins, R. 1968. *Evolution in Changing Environments.* Princeton: Princeton Univ. Press. 120 pp.

11. MacArthur, R. H. 1972. *Geographical Ecology: Patterns in the Distribution of Species.* New York: Harper and Row. 269 pp.

12. MacArthur, R. H., Pianka, E. R. 1966. On optimal use of a patchy environment. *Am. Natur.* 100:603–09

13. Mayhew, W. 1968. Biology of desert amphibians and reptiles. *Desert Biology,* ed. G. W. Brown, 195–356. New York: Academic

14. Parker, W. S., Pianka, E. R. 1973. Notes on the ecology of the iguanid lizard, *Sceloporus magister. Herpetologica* 29:143–52

15. Parker, W. S., Pianka, E. R. Comparative ecology of populations of the lizard *Uta stansburiana.* Unpublished

16. Pianka, E. R. 1965. *Species diversity and ecology of flatland desert lizards in western North America.* PhD thesis. Univ. Wash., Seattle. 212 pp.

17. Pianka, E. R. 1966. Convexity, desert lizards, and spatial heterogeneity. *Ecology* 47:1055–59

18. Pianka, E. R. 1967. On lizard species diversity: North American flatland deserts. *Ecology* 48:333–51

19. Pianka, E. R. 1968. Notes on the biology of *Varanus eremius. West. Aust. Natur.* 11:39–44

20. Pianka, E. R. 1969. Habitat specificity, speciation, and species density in Aus-

tralian desert lizards. *Ecology* 50:498–502

21. Pianka, E. R. 1969. Sympatry of desert lizards *(Ctenotus)* in western Australia. *Ecology* 50:1012–30

22. Pianka, E. R. 1969. Notes on the biology of *Varanus caudolineatus* and *Varanus gilleni. West. Aust. Natur.* 11:76–82

23. Pianka, E. R. 1970. Comparative autecology of the lizard *Cnemidophorus tigris* in different parts of its geographic range. *Ecology* 51:703–20

24. Pianka, E. R. 1970. Notes on the biology of *Varanus gouldi flavirufus. West. Aust. Natur.* 11:141–44

25. Pianka, E. R. 1971. Lizard species density in the Kalahari desert. *Ecology* 52:1024–29

26. Pianka, E. R. 1971. Comparative ecology of two lizards. *Copeia* 1971:129–38

27. Pianka, E. R. 1971. Ecology of the agamid lizard *Amphibolurus isolepis* in Western Australia. *Copeia* 1971:527–36

28. Pianka, E. R. 1972. Zoogeography and speciation of Australian desert lizards: an ecological perspective. *Copeia* 1972:127–45

29. Pianka, E. R., Huey, R. B. 1971. Bird species density in the Kalahari and the Australian deserts. *Koedoe* 14:123–30

30. Pianka, E. R., Parker, W. S. 1972. Ecology of the iguanid lizard *Callisaurus draconoides. Copeia* 1972:493–508

31. Pianka, E. R., Parker, W. S. Ecology of the Desert Horned Lizard, *Phrynosoma platyrhinos.* Unpublished

32. Pianka, E. R., Pianka, H. 1970. The ecology of *Moloch horridus* (Lacertilia: Agamidae) in Western Australia. *Copeia* 1970:90–103

33. Pianka, E. R., Pianka, H. Comparative ecology of twelve species of nocturnal lizards (Gekkonidae) in the Western Australian desert. Unpublished

34. Pielou, E. C. 1972. Niche width and niche overlap: a method for measuring them. *Ecology* 53:687–92

35. Rand, A. S. 1964. Ecological distribution in anoline lizards of Puerto Rico. *Ecology* 45:745–52

36. Rand, A. S., Humphrey, S. S. 1968. Interspecific competititon in the tropical rain forest: ecological distribution among lizards at Belem, Para. *Proc. US Nat. Mus.* 125:1–17

37. Sage, R. D. 1973. Convergence of the lizard faunas of the chaparral habitats in central Chile and California. *The Convergence in Structure of Ecosystems in Mediterranean Climates,* ed. H. Mooney. New York: Springer-Verlag

38. Schoener, T. W. 1967. The ecological significance of sexual dimorphism in size in the lizard *Anolis conspersus. Science* 155:474–77

39. Schoener, T. W. 1968. The *Anolis* lizards of Bimini: resource partitioning in a complex fauna. *Ecology* 49:704–26

40. Schoener, T. W. 1969. Models of optimal size for solitary predators. *Am. Natur.* 103:277–313

41. Schoener, T. W. 1970. Nonsynchronous spatial overlap of lizards in patchy habitats. *Ecology* 51:408–18

42. Schoener, T. W. 1971. The theory of foraging strategies. *Ann. Rev. Ecol. Syst.* 2:369–404

43. Schoener, T. W., Gorman, G. C. 1968. Some niche differences in three lesser antillean lizards of the genus *Anolis. Ecology* 49:819–30

44. Soulé, M. 1968. Body temperatures of quiescent *Sator grandaevus* in nature. *Copeia* 1968:622–23

45. Terborgh, J., Weske, J. S. 1969. Colonization of secondary habitats by Peruvian birds. *Ecology* 50:765–82

46. Tinkle, D. W. 1967. The life and demography of the side-blotched lizard, *Uta stansburiana. Misc. Publ. Mus. Zool., Univ. Mich.* No. 132:1–182

47. Vandermeer, J. H. 1972. Niche theory. *Ann. Rev. Ecol. Syst.* 3:107–32

48. Werner, Y. L. 1969. Eye size in geckos of various ecological types (Reptilia: Gekkonidae and Sphaerodactylidae). *Isr. J. Zool.* 18:291–316

49. Williams, C. B. 1964. *Patterns in the Balance of Nature and Related Problems in Quantitative Ecology.* New York: Academic. 324 pp.

13

Reprinted from *Ecology* **58**:1007–1019 (1977) by permission of the publisher, Duke University Press, Durham, North Carolina

BIRD SPECIES DIVERSITY ON AN ANDEAN ELEVATIONAL GRADIENT[1]

JOHN TERBORGH

Department of Biology, Princeton University, Princeton, New Jersey 08540 USA

Abstract. This paper analyzes patterns of bird species diversity on an elevational transect of the Cordillera Vilcabamba, Peru. Major changes in climate and vegetation are encompassed by the transect which extended from the Apurimac Valley floor at 500 m to the summit ridge of the range at > 3,500 m. Four vegetation zones are easily discerned—lowland rain forest, montane rain forest, cloud forest, and elfin forest. In progressing upwards there is a monotonic trend toward decreasing canopy stature and reduced number of plant strata.

The vegetation gradient provided the opportunity to examine the relation between bird species diversity and habitat complexity in an entirely natural setting. The decrease in forest stature with elevation was closely paralleled by decreasing avian syntopy (the total number of bird species cohabiting the forest at a given elevation). Bird species diversity was shown to be highly correlated with foliage height diversity, using either four or five layers in the foliage height diversity calculation ($r = .97$), and less well correlated using three layers, as defined previously by MacArthur ($r = .84$). At this superficial level the trend in bird species diversity seemed to be adequately explained as a response to the vegetation gradient.

This preliminary conclusion was found to be illusory when the elevational trend in syntopy was reexamined separately for three major trophic subdivisions of the fauna. The number of insectivores decreased 5.2-fold from the bottom to the top of the gradient, frugivores decreased by a factor of 2.3, and nectarivores showed no change. It was now clear that the diversity in each of these trophic categories was responsive to environmental influences other than, or in addition to, the gradient in habitat structure. Additional factors implicated by the available evidence were competitive interactions with other taxa at the same trophic level, changing composition of the resource base as a function of elevation, and declining productivity at high elevations.

Analysis of netted bird samples revealed an unexpected diversity maximum in the lower cloud forest zone. The immediate cause of this was a relaxation of the vertical stratification of foraging zones, such that an anomalously large fraction of the species present entered the nets. The excess diversity was found to consist almost entirely of insectivores. Several factors appear to contribute to the ultimate causes of the diversity maximum: greater patchiness of the montane forest due to the rugged topography, a higher density of foliage near the ground, and possibly increased resource productivity. A correlation between diversity and density in the netting results suggested a causal connection mediated via resource levels.

The conclusion is that diversity is a complex community property that is responsive to many types of influences beyond simply the structure of the habitat.

Key words: Andes; birds; ecotone; environmental gradient; foliage height diversity; Peru; species diversity.

INTRODUCTION

The species diversity problem has been under intensive investigation by ecologists and biogeographers for the past 20 yr. Yet we are still far from a global understanding of the many diversity patterns in nature. In studies involving birds, success has been limited to exposing empirical relationships that serve to predict diversity patterns within but not between biogeographic provinces. A knowledge of the foliage height profile of a habitat in eastern North America, for example, allows a fairly accurate prediction of its bird species diversity (MacArthur 1964), but the same empirical relationship fails if we use it to try to predict the bird species diversity of a tropical habitat (Terborgh and Weske 1969). Similarly, one may be able to predict the number of bird species inhabiting an oceanic island from the species-area and species-distance re-

[1] Manuscript received 5 August 1976; accepted 21 January 1977.

gressions of other nearby islands (Diamond 1973), but the prediction fails if we apply it to islands in another ocean. Clearly we have a better feeling for the environmental regulation of diversity within biogeographic provinces than we do for what causes the differences between them.

The most thoroughly documented noninsular pattern is the positive relation between bird species diversity and measures of habitat complexity, a relation that has been shown to hold, with scaling adjustments, within several biogeographic regions (MacArthur et al. 1966; Recher 1969; Karr 1971). The notion that complex habitats provide greater opportunities for resource subdivision has intuitive appeal. Nevertheless, the causal chain that links habitat structure with consumer diversity is by no means simple, and may involve a number of branching or independent connections. Furthermore, the causal links may differ from one set of species to another. Much of the work on diversity to date has concentrated on establishing the empirical relations between habitat structure and diversity: the causal connections have been relatively little explored. It is this last aspect of the diversity problem that motivates the present article.

The measurements of habitat structure and species diversity come from an elevational transect of the eastern Peruvian Andes. The stature and vertical layering of the vegetation on the transect change in a systematic fashion with elevation, providing a monotonic natural gradient of habitat complexity. I will first examine the relation between avian diversity and two measures of habitat complexity: forest stature and foliage height diversity. I will then show that a strong correlation between species diversity and these measures of habitat complexity is partly fortuitous and masks probable responses to a changing trophic organization of the community, the presence and absence of other taxa in the same trophic levels, and variations in resource productivity along the gradient.

THE VILCABAMBA ELEVATIONAL GRADIENT

With the help of several colleagues and Peruvian assistants, I conducted an intensive survey of the elevational distribution of birdlife along a transect of the West-facing slope of the Cordillera Vilcabamba in Central Peru (approximate location, 12°35′S, 73°40′W). This was accomplished in a series of six expeditions during the period of 1965 to 1972. Descriptions of the climate and vegetation along the transect have been published in a number of previous articles, and so I will reiterate only essential points here (Terborgh 1971, 1973a; Terborgh and Dudley 1973).

Temperature is the climatic variable most closely associated with elevation ($T = -0.56°C/100$ m gain in elevation), though the amount of cloudiness and the frequency of rainfall increase noticeably toward the upper end of the transect. Superimposed on the smooth temperature gradient are four structurally and (to a lesser degree) floristically distinct vegetation formations: lowland rain forest, montane rain forest, cloud forest, and elfin forest. They are characterized, in this order, by declining stature and reduced vertical stratification.

I concur with Richards (1952) that lowland rain forest on good sites contains five vertically distinct strata. ranging from a 50- to 60-m A-story of giant emergents down to a 0.5-m herbaceous D-story. (One can satisfy himself of this by counting the vertically superimposed crowns beneath emergents. The most frequent number is five.) On our transect, lowland rain forest occupied only the flat Apurimac Valley floor. (Virtually none remains at present.) Montane rain forest (650–1,385 m) is floristically similar but lacks the emergent stratum. Gradual trends toward reduced canopy height and increased foliage density in the understory are discernible as one climbs upward. This continues through the cloud forest belt (1,385 to roughly 2,500 m), which in the Vilcabamba is structur-

ally chaotic due to the extremely steep and irregular terrain. Frequent treefalls and landslides result in an irregular canopy that admits a great deal of light into the understory, with a consequent proliferation of climbing bamboo (*Chusquea* spp.). Elfin forest, the uppermost vegetation zone, first appears on ridge tops at ≈2,500 m in the form of stunted, nearly impenetrable thickets. On good sites, elfin forest may include three strata, corresponding roughly to trees, shrubs, and herbs, but on steep slopes or in poorly drained areas where sphagnum peat accumulates, there are only scattered low trees and a grassy understory. Elfin forest and cloud forest interdigitate on ridges and slopes between 2,400 and 2,700 m. In many places the transition between them is marked by an abrupt drop in canopy height from 20 m or so down to 8 m or less. Although the elfin forest zone is visibly patchy due to the effects of irregularities in exposure and soil conditions, the overall trend toward structural simplification and reduced stature of the vegetation continues to timberline (3,500 m).

AVIAN SYNTOPY AND HABITAT COMPLEXITY

We recorded syntopy as the total number of bird species regularly using the forest within ±30 m elevation of our 9 bush camps. The lists were compiled from sight, sound, and netting records, as described previously (Terborgh 1971). Only forest-dwelling species were counted. Thus the totals for each site omit a few species that were obvious vagrants from other habitats or that use the forest discontinuously along stream margins or in treefall openings. Although the syntopic species at any elevation differ greatly in abundance, from large eagles with territories of several square kilometres to common small passerines, all coexist in the matrix of the forest. In compiling the results my colleagues and I invested a total of 13 party-months on the transect, including 3 wk or more at each of the 9 bush camps. The thoroughness of the effort can be judged by the fact that the final expedition in 1972 added only two new species to a cumulative total of 600.

Figure 1 reveals that forest stature and avian syntopy vary with elevation in a strikingly parallel fashion. The parallelism falters only at the lowland forest-montane forest ecotone where the giant 60-m emergent stratum drops out of the vegetation. However, 60-m trees are rare, even in the best lowland stands, and nowhere does a closed canopy form at heights >30–40 m. A biologically more realistic index of the stature of the lowland forest would be a weighted mean of the heights of unshaded canopy trees. This would fall in the range of 30–45 m and would preserve the parallelism with the syntopy curve over the entire gradient.

As discussed above, the number of vertically superimposed plant strata in the Vilcabamba vegetation formations is closely correlated with the overall

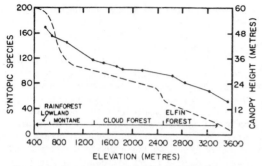

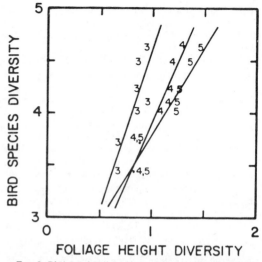

Fig. 1. Avian syntopy (solid line) and forest stature (dashed line) vs. elevation in the Cordillera Vilcabamba. Vegetation zones indicated above the abscissa. Syntopy is the number of species regularly using the forest within ± 30 m of each station. Canopy height represents the mean height of the tallest tree stratum.

Fig. 2. Bird species diversity vs. foliage height diversity in the Cordillera Vilcabamba. Numbered points refer to the number of foliage layers used in computing the foliage height diversity. Lines represent least squares regressions with correlation coefficients. $r_3 = .84$; $r_4 = .97$; $r_5 = .97$. Further methodological details in text.

stature of the forest. Thus, the data associate the richer low-elevation communities with a taller, more complex and highly stratified habitat. Expressed in this form, the results suggest nothing new.

Bird species diversity vs. foliage height diversity

To see whether the conclusions reached using a crude and easily obtained measure of habitat complexity—stature—would hold up under more conventional methods of analysis, we determined the foliage height profile of the forest at seven elevations. Foliage densities were measured with MacArthur's checkerboard technique (MacArthur and MacArthur 1961) up to 6 or sometimes 9.14 m [20 or 30 feet], which is as high as we could elevate the checkerboard. Educated guesswork served to fill in the rest of the profile. (For some examples, see Terborgh and Weske 1969.) Foliage height diversities were then computed on the basis of 3 layers (0–2', 2–25', >25'), 4 layers (0–2', 2–10', 10–50', >50'), and 5 layers (0–2', 2–10', 10–30', 30–90', >90'). The intervals for the 3- and 4-layer calculations were those used by MacArthur et al. (1966), while those for the 5-layer calculations were based on Richard's (1952) description of the 5 strata of lowland rain forest.

Bird species diversity is not directly measurable in tropical forest because of the presence of many cryptic species that are rarely observed, and because there is little diurnal or seasonal overlap in the vocal activity periods of many species. Thus it was necessary to estimate bird species diversity from syntopy values by assuming that community equitability is constant at all elevations. Some confidence in this procedure can be obtained from the fact that the species equitabilities of netted samples cluster tightly in the range of 0.7 to 0.8 with no apparent elevational trend (Fig. 6, and see Table 1 in Terborgh and Weske 1975). Because large birds, which are poorly sampled by nets, tend to be

less common than small ones, a lower value, arbitrarily set at 0.6, was used to convert syntopy values to estimates of the number of equally common species. The natural logarithms of these estimates represent bird species diversity. (Letting $H = \Sigma\, p_i \ln p_i$, $\exp^H$ gives the number of equally common species and $\exp^H/S$ = equitability, where S = the total number of species in the sample.) We have just obtained estimates of H by proceeding through this backwards, i.e., $H = \ln (0.6S)$. Even if equitability does vary somewhat from station to station, most of the variation will be wiped out by taking the logarithm.

Bird species diversity correlates well with foliage height diversity on the Vilcabamba gradient, regardless of how many layers are used in computing the latter (Fig. 2). A better fit is obtained with 4 or 5 layers because foliage height diversity measurements based on 3 layers do not discriminate between forests in which most of the foliage is above 7.6 m.

Variation in habitat complexity explains a large fraction of the variance in avian syntopy/diversity on the Vilcabamba elevational gradient, whether one takes into account the details of the vertical organization of the vegetation, as with the foliage height diversity index, or uses only the simple metric of overall forest stature. Investigations in other parts of the world have frequently, though not invariably, produced similar results, except that values of bird species diversity corresponding to a given foliage height diversity have in general been much lower (MacArthur 1964; MacArthur et al. 1966; Recher 1969; Terborgh and Weske

1969; Karr and Roth 1971). It is likely that regional differences are better explained by evolutionary histories of their samples than by postulating as yet undiscovered distinctions in the quality of habitats (Vuilleumier 1972; Cody 1975; Karr 1975; Pearson 1976, and later discussion).

At this point one could easily close the paper with the statement that the Vilcabamba results have merely confirmed a common empirical relationship. When one examines the underlying details, however, a much more complicated picture emerges.

Trophically independent subdivisions of the avifauna

Instead of looking at the very broad pattern described by the entire forest-dwelling avifauna, it is of interest to examine its component subdivisions to see whether they all respond similarly to the gradient. Discounting raptors and vultures, tropical avifaunas can be fairly discretely partitioned into three trophically distinct subdivisions: insectivores, frugivores (including granivores), and nectarivores. A minority of species that feed on nearly equal mixtures of insects and fruit (mainly tanagers and some honeycreepers), or of fruit and nectar (certain honeycreepers) were split evenly between the respective categories. Species that feed on markedly uneven mixtures were assigned to the category representing the major component of their diets. For example, flycatchers that seasonally take some fruit were nevertheless included as insectivores, and some manakins and certain cotingids that occasionally catch insects were classified as frugivores. To a first approximation at least, the subdivisions are trophically, and therefore competitively, independent of one another. Now, if structural complexity of the habitat were the controlling factor in the diversity of birds in general, the three major subdivisions of the avifauna should all respond in a like manner to the vegetational gradient. But they do not (Fig. 3). The trends are strikingly different. Syntopy among insectivores undergoes the greatest change with elevation (>5-fold). Frugivores decline less sharply (2.3-fold) while syntopy in nectarivores is entirely independent of the gradient.

The simple parallelism between vegetational complexity and avian syntopy has vanished, leaving in its place a trio of contrasting cases to be examined individually in the light of the special circumstances pertaining to each. My interpretations, while largely anecdotal, are nevertheless based on what I regard as sound biological evidence.

Insectivores.—Why should insectivores decline most precipitously with elevation? The answer to this seems to be compounded of several trends. First, it could be expected that insectivores would be most severely affected by structural simplification of the habitat. While fruit and flowers can be carried on a plant in only a limited number of ways, insects can

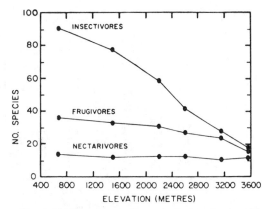

FIG. 3. Number of syntopic species in three tropic guilds vs. elevation in the Cordillera Vilcabamba. From the bottom of the gradient to the top, syntopy in the guilds decreases by the following factors: insectivores 5.1-fold; frugivores (including granivores) 2.3-fold; nectarivores 1.2-fold. Further details in text.

conceal themselves or escape by a great variety of means. Diamond (1973) has shown, for example, that fruit-eating birds in the southwest Pacific sort mainly by size, while, in contrast, it is routine to find several like-sized insectivores sharing the same habitat and segregating by subtle behavioral differences (MacArthur 1958). The simple fact that most avifaunas contain much larger numbers of insectivorous species and families than of taxa specialized on other types of resources testifies to the vast array of behavioral and morphological specializations that can be effectively employed in pursuit of insect prey (Lein 1972; Keast 1972).

Beyond the obvious generalization that structurally complex vegetation offers more opportunities for specialized techniques of harvesting insects, one can make a few more particular statements about how such opportunities decrease on the Vilcabamba gradient. One clear case is that of bark-feeding birds (woodpeckers, Picidae, and woodcreepers, Dendrocolaptidae). As many as 30 species in these two families can be found in a single locality in the lowlands where the variety of bark substrates is impressive: giant trunks, tiny twigs, swinging vines, bamboo canes, termite nests, bromeliads, rotting knot holes, etc. Near timberline the situation is drastically different. The trees are low, spindly, and nearly uniform in size, and moreover, because of the festoons of mosses and lichens that cling to most trunks and branches, there is very little exposed bark. Here the bark-feeding "guild" is comprised of a single small woodpecker. Terrestrial insectivores that forage in the leaf litter are another case. A dozen or more species may coexist in the lowlands, while above 2,400 m there is only one, a species of antpitta (*Grallaria*). The floor of the upper montane forest is a cold damp carpet of sphagnum;

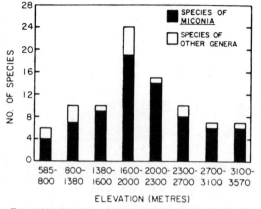

FIG. 4. Number of species in the ornithochoric plant family Melastomataceae occurring in eight arbitrary elevation zones in the Cordillera Vilcabamba. *Miconia* is the predominate genus at all elevations. Redrawn from Mardres (1970).

there scarcely is any leaf litter. One can search through the sphagnum at length without finding any arthropods at all (Terborgh and Weske 1972).

This brings up another probable cause for the decline in insectivores: a scarcity of insects at high elevations. Several major groups including ants and termites appear to drop out altogether above 2,500 m, while others (Lepidoptera, Orthoptera, Diptera, Coleoptera, Hymenoptera) are conspicuously less abundant. Netting yields indicate that the biomass of insectivorous birds drops some 20- to 30-fold between midelevations ($\approx$1,500 m) and timberline, as will be shown later. Reduced syntopy among insectivores at high elevations can thus be explained as a response both to simplified vegetation and to a reduced food resource base.

Frugivores.—Syntopy among fruit- and seed-eating birds shows a more moderate 2.3-fold decrease with elevation. It could be that this merely reflects reduced opportunities for vertical partitioning of resources, though I believe that other factors are involved as well. For one, the availability of fruit crops is considerably more seasonal in the lowlands (Smythe 1970) than it appears to be at high elevations where many plants bear fruits and flowers continuously for long periods. For another, the bulk of the lowland fruit crop is harvested by mammals. At one site in the Peruvian lowlands we found that the biomass of frugivorous primates was approximately 400 kg/km², a value that is well above any reasonable estimate of the biomass of frugivorous birds in the same forest (Janson 1975). But this is by no means the whole picture, as other less easily censused mammalian groups may consume more fruit than do the primates, e.g., bats, rodents, peccaries, marsupials, procyonids. Most of these animals drop out of the Andean fauna below 2,000 m, and many of them below 1,000 m. In spite of their powers

of flight, frugivorous bats respond more like other mammals than like birds. Below roughly 1,300 m our mist nets (on the few occasions we left them open) captured more frugivorous bats at night than birds of all species by day. Above this level the reverse was true and we routinely left nets open at night. These observations suggest that birds, by default, harvest a much greater proportion of the fruit crop at middle and upper elevations than they do in the lowlands. Moreover, it is possible that more suitable fruit is available, as the incidence of several ornithochoric plant families increases sharply in the upper montane forest (e.g., Ericaceae, Rubiaceae, Melastomataceae—Fig. 4).

Why then, if all these things are true, does the number of frugivorous bird species not increase with elevation? Although the proportion of fruit-eating birds in the community does increase markedly, the number of species does not. Part of the answer may lie with the reduced potential for spatial partitioning of the resource in structurally simplified vegetation. (Terrestrial frugivores, for example, drop out entirely above 2,800 m—because of the sphagnum?—though lowland forests may accommodate as many as 6 or 8.) There may be other influences as well, such as declining plant productivity at high elevations. Though the elevational variation in syntopy among frugivorous birds plots as a deceptively simple monotonic trend, it is clear that the underlying causality is complex and still largely unresolved.

Nectarivores.—The situation with regard to nectar-feeding birds is also complex. In the lowlands there is a pronounced vertical stratification of species. Nets capture little else than hermits (subfamily Phaethorninae) while several genera of trochiline hummingbirds and a number of honeycreepers occupy the canopy. Conditions in the lowland forest militate against high densities of nectarivores. Flowering tends to be markedly seasonal, and only a few species are in bloom at any time (Janzen 1967; Frankie et al. 1974). A large majority of the plants are entomophilous (Faegri and Pijl 1971). All these circumstances are reversed at high elevation. The climate is much less seasonal, a large proportion of the plants have flowering periods that last for months (Nevling 1971), and the flora is rich in ornithophilous genera and families (e.g., Ericaceae, Loranthaceae, Loganiaceae, Onagraceae, Bromeliaceae, Verbenaceae, etc.). Whereas hummingbirds and honeycreepers usually comprise <10% of the individuals in lowland net samples, these 2 families frequently constitute half of the catch near timberline. The roughly constant number of nectarivores over the whole gradient may result from a fortuitous balance of two countervailing influences: greater opportunities for vertical partitioning of nectar resources in the lowland forest vs. more abundant and more constant supplies of these resources at high elevations.

In summary, we first found that the overall trend in avian syntopy along the elevational gradient closely paralleled a gradual telescoping of the forest. However, a very rudimentary trophic breakdown of the fauna clearly demonstrated the folly of taking a good correlation too seriously in the absence of compelling *a priori* logic. Each trophic subdivision of the fauna responds to the gradient in a strikingly individualistic fashion. In every case the causal mechanisms seem to be compounded of several more or less independent influences. The structure of the habitat seems to play a role in every instance, but superimposed on it are effects that derive from major decreases (insects) or increases (nectar) in the availability of food resources, or from the dropout of a major class of competitors (frugivorous mammals). The structure of the habitat *per se* is by no means the only determinant of avian diversity.

DIVERSITY IN NETTED POPULATION SAMPLES

Now that we have examined the pattern of avian syntopy on the elevational gradient, we can ask whether a similar pattern holds for the component of the fauna that frequents the forest understory. First, a few methodological comments.

Netting method

Mist nets capture birds using the airspace between 0.1 and 2 m above the ground. The size we use (12 m long, 36 mm mesh) is maximally efficient for birds in the 8- to 60-g range; smaller species occasionally slip through the mesh and larger ones frequently bounce out without becoming entangled. Our standard procedure is to construct long end-to-end lines of 20 or more nets (up to 50) and to operate them for periods of 4 to 8 consecutive days. At the end of such a run, >80% of the resident population will have been captured, as can easily be verified by checking for bands on birds observed in the vicinity. Population density estimates can be obtained by making use of the fact that daily catch rates (of previously unmarked individuals) decline in a log-linear fashion with time. Regression analysis then yields a statistic that we call the projected population per net (see caption of Fig. 8 and Terborgh and Faaborg 1973 for further details).

As a sample accumulates the number of species captured increases, rapidly at first, and then gradually to a plateau because of the many inefficiently trapped terrestrial and canopy-dwelling species present at most sites. Diversity estimates that are statistically independent of the density measurements are obtained by basing them on the first 100 individuals captured. (More than 100 individuals are always contained in the density estimates.)

Species diversity in net samples

Now let us attempt to anticipate how diversity, as sampled with mist nets, might vary with elevation. Re-

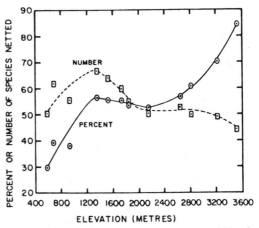

FIG. 5. Number of species contained in samples of 300 mist netted birds (dashed line), and the percent this represents of total syntopy (solid line) vs. elevation in the Cordillera Vilcabamba.

call first that the total number of species present (syntopy) declines from 170 in the lowland forest to 52 at timberline, a 3.3-fold decrease. However, the stature of the habitat also decreases by an equal or greater factor, suggesting that the concentration of species in a standard vertical slice of the forest may not change very much. Moreover, we can expect the nets to capture a much higher proportion of the whole community toward the upper end of the gradient where the vegetation is low.

This does indeed happen, but with some unanticipated kinks. The fraction of the entire syntopic community that is captured in a standard net sample of 300 individuals increases from ≈30% in the lowlands to >85% at timberline. But instead of rising as the mirror image of the forest stature curve, as might be expected, the curve displays a prominent shoulder between 1,200 and 2,000 m (Fig. 5). In the absence of any major changes in forest stature in this region, the result is puzzling. An anomalously large fraction of the community is captured in the lower cloud forest zone with the consequence that netted species diversities rise to a pronounced peak (Fig. 6). The high diversities are not due to the presence of additional species, as the syntopy curve (Fig. 1) clearly shows, but rather to an unexpected local relaxation in the vertical stratification of the community. Neither is the high diversity explained simply by an increased incidence of canopy species in the samples, because the inclusion of inefficiently captured species results in reduced equitabilities and little or no change in diversity. To the contrary, equitabilities are high in the region of the diversity peak. Elsewhere on the transect the results are much as anticipated; comparing the bottom and the top of the gradient, the net samples contain roughly equal diversities, in spite of the large difference in syn-

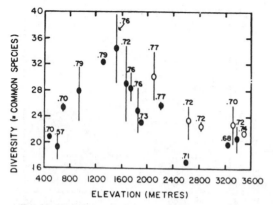

FIG. 6. Number of equally common bird species (=exp − $\Sigma p_i \ln p_i$) contained in netted samples of 100 individuals vs. elevation in the Cordillera Vilcabamba. Vertical bars represent the range of values obtained at stations sampled on 2 or more yr. Open points indicate lines where nets ran along exposed ridgetops (cejas) as distinguished from broad slopes or level ground (closed points). Numbers give the mean equitability (=exp − $\Sigma p_i \ln p_i/S$) of the sample(s) at each station.

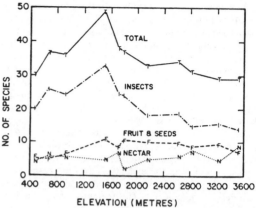

FIG. 7. Trophic composition of net samples vs. elevation in the Cordillera Vilcabamba. Criteria are the same as in Fig. 3.

topy. The problem that remains is how to account for the anomalous midelevation peak.

The midelevation diversity peak

An increased concentration of foliage at the level of the nets might contribute to greater coexistence, as has been demonstrated in some other studies (Karr and Roth 1971; Willson 1974). There may be some evidence for this, as shall be discussed later. However, the elevational patterns do not accord. Netted bird diversity is at a maximum between 1,400 and 1,600 m while the thickest understory occurs between 2,000 and 3,000 m.

Though the structure of the habitat may be of some ultimate significance in regulating the diversity pattern, it is possible that better insights may be derived by examining the more proximal clues offered by the organization of the avian community itself. The community undergoes pronounced compositional shifts as its trophic structure changes, and as a result of a nearly constant rate of species turnover with elevation (Terborgh 1971). Whole guilds wax and wane, or drop out altogether. Obligate army ant followers are a good example. Members of this guild are among the commonest species in lowland net samples. Their relative abundance and the number of species decline steadily with elevation until the last of them drops out near 2,000 m. Earlier I mentioned the cases of the bark-gleaning guild and terrestrial insectivores. There are many more examples, including ones that entail increases with elevation, as well as decreases.

Given that such marked compositional shifts involve guilds that depend on a wide range of distinct food and habitat resources, it is at least conceivable that there

might be some point on the gradient at which the spectrum of available resources allowed a maximum number of guilds to coexist.

Equitability of faunal components

Using the same rough breakdown of the fauna into insectivores, frugivores, and nectarivores, we note the surprising result that the midelevation diversity peak is almost entirely attributable to an increased concentration of insectivores (Fig. 7). At this crude level at least, enhanced equitability of trophic guilds is not the answer.

To pursue the matter further we focus on insectivores, in particular, the six most important families: Tyrannidae (flycatchers)—59 spp.; Formicariidae (antbirds)—43 spp.; Furnariidae (ovenbirds)—26 spp.; Dendrocolaptidae (woodcreepers)—16 spp.; Troglodytidae (wrens)—8 spp.; and Parulidae (warblers)—8 spp. (Numbers are all species inhabitating the gradient; many are not readily captured in nets.) Significantly, each of the four largest of these families possess a distinctive and characteristic type of foraging behavior. There are a few exceptions in each family, but to a first approximation woodcreepers glean the bark of trunks and major limbs, antbirds are foliage gleaners, ovenbirds have a creeping habit and are partial to searching through epiphytes and dead leaves, and flycatchers hawk or hover-snatch. The deployment of such conspicuously different foraging techniques probably implies differential resource utilization, as Hespenheide (1975) has shown for a number of tropical and temperate insectivores. It is not immediately obvious, however, why the proportions of these families should vary so strongly with elevation (Table 1). Antbirds predominate in the lowlands, contributing more species to the net samples than the other five families combined. But by 2,500 m they have virtually dropped out of the fauna. Flycatchers and ovenbirds, on the other hand, prevail near the top

TABLE 1.   Representation of major insectivorous families in Vilcabamba net samples (first 300 individuals). Number of species

Family	Elevation (m)									
	585	685	930	1,350	1,520	1,730	1,835	2,215	2,640	3,510
Dendrocolaptidae	4	4	4	3	2	1	3	3	0	0
Furnariidae	4	4	5	10	9	8	7	4	5	3
Formicariidae	15	12	9	8	9	8	6	4	1	0
Tyrannidae	4	9	11	10	11	8	9	6	10	6
Troglodytidae	1	1	1	3	2	3	2	3	1	0
Parulidae	0	2	2	1	3	3	2	3	3	1
Total	28	32	32	35	36	31	29	23	20	10

of the gradient. The unexpectedly high net diversity at midelevations is clearly a consequence of the good and strikingly equitable representation of all the families of insectivores. Since each of the distinctive family foraging modes requires appropriate substrates, the key to the problem lies in understanding how the midelevation forest provides such a wealth of foraging opportunities.

Structural heterogeneity of the midmontane forest

Mid- and high-elevation forests are far more heterogeneous in the horizontal plane than are lowland forests. In part this is an inevitable consequence of mountain topography, with its ridges, slopes, and ravines. The continuity of the forest is frequently interrupted by landslide tracks and deep stream gorges; tree heights vary greatly between sheltered ravines and exposed ridgetops. All these irregularities create a variety of "edge" situations that are exploited by flycatchers. Climbing bamboos (*Chusquea* spp.) invade old landslides and treefalls, forming impenetrable thickets that are the home of certain ovenbirds, antbirds, and wrens. Where rising air currents are deflected by slopes and ridges, the forest becomes choked with a profusion of mosses and other epiphytes that carpet all exposed surfaces, both vertical and horizontal. The thick mats of mosses, lichens, and ferns offer a novel substrate that is exploited by a great array of creeping birds, notably ovenbirds and certain wrens. In addition, the forest harbors a full spectrum of the more conventional types of bark and foliage gleaners that are so prevalent in the lowlands. In sum, the steep terrain, irregular canopy, and an extraordinary variety of arboreal substrates all contribute to the microspatial heterogeneity of the midmontane forest of the Cordillera Vilcabamba. Lowland forests, in contrast, are more regularly stratified in the vertical plane, and less heterogeneous in the horizontal plane. (Such structural differences between forests, while potentially important to bird diversity, are not registered in foliage height diversity measurements because horizontal variation in foliage density is averaged in the computation.)

Is the elevational pattern of bird species diversity adequately explained by the structural heterogeneity of the midmontane forest? No. What does seem to be accounted for is the more equitable representation of the principal families of insectivores. But there are other features of the pattern that are not explained, such as why there are decidedly fewer insectivores at 2,200 m than at 1,500 m (Table 1), even though at the higher elevation the forest still contains large trees and remains extremely heterogeneous, as is affirmed by the high family equitability. For the answer to this question we must inquire into the causes of the declining role of insectivory toward the upper end of the gradient. This will be taken up in the next section.

AVIAN DENSITY, SPECIES DIVERSITY, AND RESOURCES

Population density estimates derived from 4-day net runs yield an elevational pattern that closely parallels the trend in species diversity (Fig. 8). A high degree of correlation between the two sets of results is confirmed by plotting them together in a scatter diagram (Fig. 9). Maximum levels of both density and diversity coincide in a narrow elevational belt corresponding to lower cloud forest. Two quite independent questions are posed by the findings: (1) is a causal interrelationship implied by the correlation of density and diversity, and (2) what features of the changing conditions on the gradient could account for the pattern of avian abundance in the understory? These questions will be taken up in order.

Density and diversity

Our earlier experience with the parallelism between syntopy and forest stature provided a trenchant reminder to proceed with caution in the interpretation of correlations. In the present case it is not even clear which is the independent variable. There are plausible arguments on both sides. At one extreme, it could be held that the number of individuals in a community is a function of the number of species it contains. This is what one could expect if avian niches were rigidly unresponsive to variations in the intensity of interspecific competition. Though something like this may occur in the impoverished Pacific islands studied by Diamond (1971b), a large body of evidence indicates that in gen-

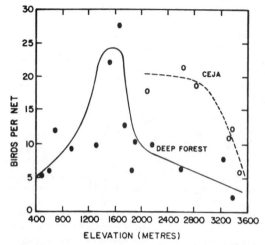

FIG. 8. Avian population density vs. elevation in the Cordillera Vilcabamba. The ordinate gives the projected population per net, computed from the first sample taken at each station as $[N/n]/[1 - (Cf/Co)]$, where N = the number of different individuals in the sample, n = the number of nets used, Cf = the final and Co = the initial capture rates (in birds per net-day). Cf and Co are taken from the log-linear regression of capture rate vs. accumulated net-days of sampling. Further details on this method are given in Terborgh and Faaborg (1973). Open points represent ceja lines, as described in the caption of Fig. 6. Single values only are given for each station because banded birds are adept at avoiding nets, hence lower values are obtained when repeat measurements are made on subsequent years.

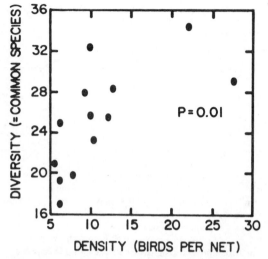

FIG. 9. Scatter plot of bird species diversity vs. population density in Vilcabamba net samples. Values were taken from the closed (deep forest) points in Figs. 6 and 8. The correlation is highly significant, $P = .01$.

eral birds are capable of almost unlimited competitive release (Crowell 1962; Diamond 1971a; MacArthur et al. 1972, 1973). Indeed, our own efforts to determine the limits of competitive release among Caribbean birds have been notably unsuccessful. Tiny islands with only 10 or 12 species support avian biomasses that equal or exceed those of much larger species rich islands (Terborgh and Faaborg 1973; Terborgh et al. 1977). Among Caribbean bird communities at least, densities are more or less constant over a wide range of diversities. Thus it seems doubtful that diversity could be the independent variable in Fig. 9.

An intermediate proposition would be the following. Since we have already noted that the equitability of insectivorous families increases sharply from the lowlands to middle elevations, apparently in response to a greater structural heterogeneity of the habitat, it is possible that a broader spectrum of foraging opportunities would contribute to an increase in both density and diversity. Though this cannot be rejected as a partial explanation, there are additional facts that remain unresolved. One is the decline in number of insectivores between 1,500 and 2,200 m, a zone in which family equitability is high and within which there is little perceptible change in the character of the forest.

Another is that the relative increase in density in Fig. 9 exceeds the relative increase in diversity. If the high diversity of the lower cloud forest region were simply a response to the presence of additional foraging opportunities, one would indeed find more species, but the average abundance per species would remain the same or decrease; certainly it would not increase. But in our results it does. To account for this it seems necessary to postulate that at least some of the extra diversity is a consequence of increased density.

An increased density of birds in the understory could result from a greater concentration of foraging substrate (foliage), reduced mortality (predation, disease, etc.), or a higher productivity of food resources. Greater amounts of foliage in the understory might result in increased foraging activity at net level, though, as mentioned above, maximum foliage densities near the ground are attained above 2,000 m where both density and diversity are declining. We have no information on mortality factors, nor is it clear what effect local variations in predators, parasites, etc. would have on diversity. Specialized bird hawks (*Micrastur, Accipiter*) occur in low numbers over the entire gradient. To me the most plausible cause of the midelevation "hot spot" is a greater abundance of food resources.

A more productive environment could support both a more numerous and more diverse community because it could provide essential food resources in greater amounts and in greater variety. But why should avian resources, especially insects, reach a peak in the lower cloud forest zone? Most of what follows is admittedly *a posteriori* rationalization,

252

though some of the corroborative evidence is independently derived.

Productivity and the environmental gradient

The problem has two parts: why do bird densities increase from the lowlands upward into the cloud forest, and why do they then decrease at still higher elevations. In discussing this I shall assume only that there is some proportionality between the density of bird populations and the productivity of the resources that sustain them.

Taking the second part of the problem first, what is the evidence that the availability of insects may decline toward the upper part of the gradient? (1) Subjective impressions. Ants, termites and flying insects—hymenoptera, flies, cicadas, lepidoptera—are omnipresent in low-elevation forests. At night there is a constant din of insect noise, and heaps of scorched bodies accumulate under the pressure lantern. In contrast, the nighttime silence at high elevation is broken only by an occasional owl or frog. The lantern attracts nothing more than a few moths. During the day one sees no ants or termites and only an occasional butterfly during the brief periods of sunshine. (2) The number of syntopic insectivores decreases by a factor of 5 between the bottom and top of the gradient. (3) The rate at which nets capture insectivorous birds decreases by more than 10-fold between 1,500 and 3,500 m (Fig. 8). If we take into account the fact that nets sample a much larger fraction of the high-elevation community (Fig. 5), the disparity becomes something like 20- to 30-fold. (4) Janzen (1973b) has shown that arthropod density and diversity decrease sharply at high elevations in Costa Rica. While these are all statements of circumstantial evidence, taken together they argue plausibly that the decrease in diversity of insectivorous birds above 1,500 m is due at least in part to a declining resource base.

The evidence that insect densities increase upwards from the lowlands to a midelevation maximum is far less compelling. (1) There is no corresponding increase in the number of avian insectivores, only in the overlap of their foraging zones. (2) The increased capture rates could be at least partly an artifact of this and other circumstances, such as the irregular terrain. (3) Greater concentrations of foliage near the ground could result in increased foraging activity at net level in the absence of any real increase in bird numbers. (4) Still another possibility is the preemption of resources that are harvested by other taxa at lower elevations. Highly insectivorous squirrel monkeys (*Saimiri sciureus*) abound in the undisturbed lowland forest, but these animals drop out at the base of the mountains around 600 m. In the one locality for which we have reliable census data, the squirrel monkey population ($\approx$80/km^2) has a biomass roughly equal to that of all avian insectivores in the same forest (Janson 1975). Thus, though the densities of insectivorous birds do

appear to increase at midelevations, there is little in our experience to suggest that insects do also. (5) The only direct evidence comes from Janzen's (1973a,b) sweep net survey of an elevational transect in Costa Rica in which he found maximum numbers of arthropod species and individuals in lower montane forest between 1,000 and 1,500 m. Indeed, the elevational pattern of arthropod density and diversity in his sweep net samples parallels that of our mist net results so closely as to merit further discussion.

I agree with Janzen in the feeling that the elevational pattern in arthropod abundance reflects an underlying variation in plant primary productivity. Neither of us has any direct evidence; yet there are sufficiently compelling circumstantial arguments to excuse a few lines of speculation. Janzen (1973b) maintains that net productivity can be expected to increase at midelevation because of reduced nighttime temperatures and the attendant savings in energy lost to respiration. At higher elevations, diurnal as well as nocturnal temperatures are severely depressed, resulting in lower productivity. Whether one accepts this argument or not, there are a number of additional grounds for rationalizing a midelevation productivity maximum.

First, several aspects of the lowland environment are ameliorated at midelevations in ways that would be expected to enhance photosynthetic activity.

1) Prolonged dry spells (>1 wk) are rare to nonexistent at midelevations, and accordingly there is no deciduousness, even in the tallest trees. On the other hand, droughts of a month or more are a routine annual event in the lowlands where most of the taller trees drop their leaves for some portion of the dry season, and many subcanopy species reduce their leaf volumes.

2) Middle elevations characteristically experience a propitious daily weather regimen in which the mornings and late afternoons tend to be sunny, and middays and early afternoons cloudy or rainy. Light intensity and temperature in the canopy are thus moderated through the day, mitigating the conditions that produce a midday photosynthetic depression. The Apurimac Valley lowlands, at least during the dry season, tend to be shrouded in mists for the early morning hours, and hot, dry, and sunny through the middle of the day.

3) Retarded rates of organic decomposition at middle and higher elevations may mean a more gradual release of mineral nutrients. This should imply a more continuous rate of growth than is possible in the lowlands where nutrients are released from the leaf litter in a burst at the beginning of the rainy season (Richards 1952).

4) The duff mat (25 cm thick at 1,800 m) that develops on the forest floor above 1,000 m soaks up water like a sponge, retarding runoff and buffering the superficial layers of the soil against desiccation.

Even if one were satisfied that these circumstances

pointed to rising productivity above the lowlands, it would remain to be explained why there seems to be a decline at still higher elevations. Again, there appear to be a number of contributory circumstances.

1) Reduced light intensity. The amount of cloudiness becomes extreme toward the top of the range. Even in mid dry season the summit ridge receives <20 h of sun a week, and most of this comes before 0800 when there may still be frost on the ground.

2) Low temperatures, especially low root temperatures, may adversely affect growth via retarded nutrient uptake.

3) Permanently high humidities imply low transpiration rates, another factor that could hamper nutrient uptake (Baynton 1969; Gates 1969; Weaver et al. 1973).

4) A ubiquitous carpet of sphagnum above 2,000 m lowers the pH to 4–5 in the root zone (our measurements), perhaps further exacerbating a scarcity of nutrients.

5) Drastically retarded organic decomposition at high elevations leads to the accumulation of thick peat deposits that could sequester nutrients in forms that are not readily taken up by roots.

Finally, I wish to present a piece of anecdotal evidence that to me very convincingly illustrates the existence of an optimal productivity zone in the lower montane region. It is the pattern of recent agricultural settlement in the eastern Andean sectors of Colombia, Ecuador, Peru, and Bolivia. The first colonists to arrive in an unexploited area invariably choose sites within the elevational span of 500 to 1,200 m. In most places the terrain above this level becomes too steep to cultivate easily. But even so, as the pressure of immigration continues, late-arriving settlers prefer to hack out pitiful little plots on 30 to 40 degree slopes, rather than to invade the expansive Amazonian flat lands. These people know what they are doing. It is common knowledge, gained by trial-and-error experience, that it is futile to attempt to cultivate food and cash crops much below 500 m. In Colombia, where the pressure of overpopulation is most severe, agricultural settlement has spread practically to timberline, while vast expanses of lowland forest remain virtually uninhabited. The behavior of peasant farmers is probably as good a bioassay for inherent productivity (with a correction for steepness of the terrain) as any we have.

I would not have dwelt so long on these speculative matters had I not been impressed by the striking parallels between Janzen's results and ours. In both cases maximum species diversities occurred in areas supporting maximum densities of individuals. It is tempting to imagine that an increased abundance of resources in a complex environment like a tropical forest could allow a greater array of foraging specializations to achieve profitability, thus accounting for enhanced diversity. While this interpretation has a commonsen-

sical appeal it is only suggested, not proven, by the data on hand.

BIRD SPECIES DIVERSITY IN PERSPECTIVE

The primary lesson of our examination of bird species diversity patterns in the Cordillera Vilcabamba is that diversity is a complex community phenomenon, not just a number which can be "explained" with one or another competing hypothesis. Contributory factors overlie in an intricate mosaic, so that one cannot even be assured of unidirectional trends in moving along major environmental gradients. Carefully conducted comparisons are at present our most effective means for dissecting out the important variables.

Influences of history and geography may dominate when the geographical scale of a comparison is large. Thus the factors which appear to control the trends within the 30-km Vilcabamba transect—habitat structure, abundance and qualitative balance of food resources, presence or absence of other taxa at the same trophic levels—may play minor or negligible roles in interregional comparisons. The richest mature forests in North America, for example, harbor barely half as many bird species as the stunted 4-m elfin shrubland astride the cold, mist-shrouded Vilcabamba summit ridge. Clearly the differences in habitat complexity and productivity run the wrong way to explain this seemingly paradoxical contrast. I side with MacArthur (1969) in thinking that evolutionary processes (speciation, immigration, and extinction) are largely at the root of such interregional differences, as I have already expounded elsewhere (Terborgh 1973*b*).

The possible causal connections between productivity and diversity remain problematical. An increased abundance of resources could provide sustenance for additional species through either or both of two mechanisms: the availability of broadened resource spectra or the availability of the same basic spectrum of resources in a greater variety of sites in a complex environment. It goes without saying that a necessary condition for a positive response to a high productivity site is the presence of appropriately adapted species in adjacent habitats. Merely fertilizing a pond or a piece of grassland may not provide an adequate test, because it creates a novel situation to which many of the species present are not adapted. Resource spectra may consequently narrow instead of broaden.

Diversity has proven difficult to study because it lies at the end of a number of interacting causal chains, often several steps removed from the things we are able to measure. Even if we could measure the whole gamut of proximate variables—the precise habitat needs of individual species, resource spectra, the pressure of interspecific competition, etc.—we would still have to contend with historical/evolutionary influences in accounting for interregional contrasts. The contribution of this work has not been in resolving issues so much as in demonstrating the composite nature of the diversity problem and the essentiality of reducing it to its component parts as a precondition for further understanding.

ACKNOWLEDGMENTS

This article is based on results that were gathered over an 8-yr period on 6 expeditions to the Apurimac Valley and Cordillera Vilcabamba. So many people contributed to the undertaking that it is impossible to name them all individually, but a few deserve special mention. I am especially grateful to John Weske, my closest collaborator on 5 of the expeditions, for major contributions to the fieldwork and for assisting in the preparation of the manuscript. Many of the data presented here were included, in preliminary form, in his doctoral thesis (Weske 1972). Perhaps our greatest assets in the exploratory work were the strength, resourcefulness, and reliability of our two perennial assistants, Klaus Wehr and Manuel Sanchez. Our base camp for all the expeditions was at Hacienda Luisiana, where we received extraordinary assistance and hospitality from its owner, Jose Parodi V. Invaluable storage facilities at the Museo de Historia National were provided by the director, Ramon Ferreyra.

Financial support of one or more of the expeditions was received from the American Philosophical Society, the Chapman Fund of the American Museum of Natural History, the National Geographic Society, and the National Science Foundation (GB-20170). We are grateful for their sustained interest in the research.

LITERATURE CITED

Baynton, H. W. 1969. The ecology of an elfin forest in Puerto Rico. 3. Hilltop and forest influence on the microclimate of Pico del Oeste. J. Arnold Arbor. Harv. Univ. 50:80–92.

Cody, M. L. 1975. Trends toward a theory of continental species diversities: bird distributions over Mediterranean habitat gradients, p. 214–257. In M. L. Cody and J. M. Diamond [eds.] Ecology and evolution of communities. Belknap, Cambridge, Mass.

Crowell, K. 1962. Reduced interspecific competition among the birds of Bermuda. Ecology 43:75–88.

Diamond, J. M. 1971a. Ecological consequences of island colonization by southwest Pacific birds, I. Types of niche shifts. Natl. Acad. Sci. Proc. 67:529–536.

——. 1971b. Ecological consequences of island colonization by southwest Pacific birds, II. The effect of species diversity on total population density. Natl. Acad. Sci. Proc. 67:1715–1721.

——. 1973. Distributional ecology of New Guinea birds. Science 179:759–769.

Faegri, K., and L. van der Pijl. 1971. The principles of pollination ecology (2nd ed.). Oxford.

Frankie, G. W., H. G. Baker, and P. A. Opler. 1974. Comparative phenological studies of trees in tropical wet and dry forests in the lowlands of Costa Rica. J. Ecol. 62:881–919.

Gates, D. M. 1969. The ecology of an elfin forest in Puerto Rico, 4. Transpiration rates and temperatures of leaves in cool humid environment. J. Arnold Arbor. Harv. Univ. 50:197–209.

Hespenheide, H. A. 1975. Prey characteristics and predator niche width, p. 158–180. In M. L. Cody and J. M. Diamond [eds.] Ecology and evolution of communities. Belknap, Cambridge, Mass.

Janson, C. H. 1975. Ecology and population densities of primates in a Peruvian rainforest. Undergraduate thesis, Princeton University. 96 p.

Janzen, D. H. 1967. Synchronization of sexual reproduction of trees within the dry season in Central America. Evolution 21:620–637.

——. 1973a. Sweep samples of tropical foliage insects: description of study sites, with data on species abundances and size distributions. Ecology 54:659–686.

——. 1973b. Sweep samples of tropical foliage insects: effects of seasons, vegetation types, elevation, time of day, and insularity. Ecology 54:687–708.

Karr, J. R. 1971. Structure of avian communities in selected Panama and Illinois habitats. Ecol. Monogr. 41:207–233.

——. 1975. Production, energy pathways and community diversity in forest birds, p. 161–176. In F. B. Golley and E. Medina, [eds.] Tropical ecological systems. Springer-Verlag, N.Y.

Karr, J. R., and R. R. Roth. 1971. Vegetation structure and avian diversity in several new world areas. Am. Nat. 105:423–435.

Keast, A. 1972. Ecological opportunities and dominant families, as illustrated by the Neotropical Tyrannidae (Aves). Evol. Biol. 5:229–277.

Lein, M. R. 1972. A trophic comparison of avifaunas. Syst. Zool. 21:135–150.

MacArthur, R. H. 1958. Population ecology of some warblers of northeastern coniferous forests. Ecology 39:599–619.

——. 1964. Environmental factors affecting bird species diversity. Am. Nat. 98:387–397.

——. 1969. Patterns of communities in the tropics. Biol. J. Linn. Soc. 1:19–30.

MacArthur, R. H., and J. W. MacArthur. 1961. On bird species diversity. Ecology 42:594–598.

MacArthur, R. H., J. M. Diamond, and J. R. Karr. 1972. Density compensation in island faunas. Ecology 53:330–342.

MacArthur, R. H., H. Recher, and M. Cody. 1966. On the relation between habitat selection and species diversity. Am. Nat. 100:319–332.

MacArthur, R. H., J. MacArthur, D. MacArthur, and A. MacArthur. 1973. The effect of island area on population densities. Ecology 54:657–658.

Mardres, J. H. W. 1970. Distribution of Melastomataceae along environmental gradients in Peru and the Dominican Republic. Master's thesis, Univ. of Maryland. 73 p.

Nevling, L. I., Jr. 1971. The ecology of an elfin forest in Puerto Rico, 16. The flowering cycle and an interpretation of its seasonality. J. Arnold Arbor. Harv. Univ. 52:586–613.

Pearson, D. L. 1976. The relation of foliage complexity to ecological diversity of three Amazonian bird communities. Condor 77:453–466.

Recher, H. F. 1969. Bird species diversity and habitat diversity in Australia and North America. Am. Nat. 103:75–80.

Richards, P. W. 1952. The tropical rainforest. Cambridge Univ. Press, Cambridge. 450 p.

Smythe, N. 1970. Relationships between fruiting seasons and seed dispersal methods in a Neotropical forest. Am. Nat. 104:25–35.

Terborgh, J. 1971. Distribution on environmental gradients: theory and a preliminary interpretation of distributional patterns in the avifauna of the Cordillera Vilcabamba, Peru. Ecology 52:23–40.

——. 1973a. Vilcabamba Birdlife. Explor. J. 51:48–56.

——. 1973b. On the notion of favorableness in plant ecology. Am. Nat. 107:481–501.

Terborgh, J., and T. R. Dudley. 1973. Biological exploration of the Northern Cordillera Vilcabamba, Peru. Natl. Geogr. Soc. Res. Reports, 1966 Projects, p. 255–264.

Terborgh, J., and J. Faaborg. 1973. Turnover and ecological release in the avifauna of Mona Island, Puerto Rico. Auk 90:759–779.

Terborgh, J., and J. S. Weske. 1969. Colonization of secondary habitats by Peruvian birds. Ecology 50:765–782.

255

————. 1972. Rediscovery of the Imperial Snipe in Peru. Auk **89**:497–505.

————. 1975. The role of competition in the distribution of Andean birds. Ecology **56**:562–576.

Terborgh, J., J. Faaborg, and H. J. Brockmann. 1977. Island colonization by Lesser Antillean birds. Auk. *In press*.

Vuilleumier, F. 1972. Bird species diversity in Patagonia (temperate South America). Am. Nat. **106**:266–271.

Weske, J. S. 1972. The distribution of the avifauna in the Apurimac Valley of Peru with respect to environmental gradients, habitat, and related species. Ph.D. thesis, Univ. of Oklahoma. 137 p.

Weaver, P. L., M. D. Byer, and D. L. Bruck. 1973. Transpiration rates in the Luquillo Mountains of Puerto Rico. Biotropica **5**:123–133.

Willson, M. F. 1974. Avian community organization and habitat structure. Ecology **55**:1017–1029.

Part IV

DIVERSITY IN EXPERIMENTAL COMMUNITIES

Editor's Comments
on Papers 14 Through 17

14 **WILSON and SIMBERLOFF**
    *Experimental Zoogeography of Islands: Defaunation and
    Monitoring Techniques*

15 **CAIRNS et al.**
    *The Relationship of Fresh-Water Protozoan Communities to
    the MacArthur-Wilson Equilibrium Model*

16 **HAIRSTON et al.**
    *The Relationship Between Species Diversity and Stability:
    An Experimental Approach with Protozoa and Bacteria*

17 **PATRICK**
    *The Effect of Invasion Rate, Species Pool, and Size of Area
    on the Structure of the Diatom Community*

Much of the data on which conclusions about diversity are
based are derived from field observations. This is natural since ecology
is fundamentally a field science. In addition, for many groups of
organisms it would be very difficult to raise many species at one
time in the laboratory and to replicate their environmental relationships. Nevertheless, the laboratory does provide opportunity to control the nature of the environment and the types of biological
interactions. Diversity has been investigated with a variety of experimental laboratory and field approaches that have ranged from manipulation of discrete natural areas—that is, the denuding of an island
by Wilson and Simberloff (Paper 14)—to placing artificial habitats
in a natural body of water by Cairns and coworkers (Paper 15; Patrick
et al., 1954). Some experiments were carried out using a few species
(Paper 16; Vandermeer, 1969). Patrick (Paper 17) conducted experiments on effects of invasion rates on diversity in microcosms in a
greenhouse.

The papers in this section also illustrate different methods of
approach toward studying diversity. It is gratifying that many of the
field and laboratory experiments confirm each·other.

REFERENCES

Patrick, R., M. H. Hohn, and J. H. Wallace, 1954, A New Method for Determining the Pattern of the Diatom Flora, *Acad. Nat. Sci. Phila. Not. Nat.* **259:**1–12.

Vandermeer, J. H., 1969, The Competitive Structure of Communities: An Experimental Approach with Protozoa, *Ecology* **50:**362–371.

14

EXPERIMENTAL ZOOGEOGRAPHY OF ISLANDS: DEFAUNATION AND MONITORING TECHNIQUES

EDWARD O. WILSON AND DANIEL S. SIMBERLOFF[1]

The Biological Laboratories, Harvard University, Cambridge, Massachusetts 02138

(Accepted for publication December 16, 1968)

Abstract. In order to facilitate experiments on colonization, a technique was developed that permits the removal of the faunas of very small islands. The islands are covered by a tent and fumigated with methyl bromide at concentrations that are lethal to arthropods but not to the plants.

Seven islands in Florida Bay, of varying distance and direction from immigrant sources, were censused exhaustively. The small size (diameter 11–18 m) and ecological simplicity of these islands, which consist solely of red mangrove trees (*Rhizophora mangle*) with no supratidal ground, allowed the location and identification of all resident species. The terrestrial fauna of these islands is composed almost exclusively of arboreal arthropods, with 20–50 species usually present at any given moment. Surveys of these taxa throughout the Florida Keys, with emphasis on the inhabitants of mangrove forests, were made during 1967 in order to estimate the size and composition of the "species pool."

The seven experimental islands were defaunated in late 1966 and early 1967, and the colonists were monitored for 17–20 man-hours every 18 days. Precautions were taken to avoid artificial introductions during the monitoring periods.

INTRODUCTION

The first attempt to formulate a quantitative theory which would unite an ever-increasing mass of insular biogeographic data was made by Mac-

[1] Present address: Department of Biological Science, Florida State University, Tallahassee, Florida 32306

Arthur and Wilson (1963, 1967), who postulated an equilibrium number of species on an island determined by the intersection of immigration and extinction curves drawn as a function of the number of species already present. In addition to speculation on the forms of these curves and effects of varying both island area and distance from the source, they derived the equation

$$t_{0.90} = 1.15 \frac{\text{mean } \check{S}}{X} \qquad (1)$$

where $\check{S}_i =$ the hypothetical equilibrium number of species on an island

mean $\check{S} =$ the mean of $\check{S}_i$ for a number of very similar islands i

$X =$ the extinction rate (or turnover rate, since an equilibrium requires that extinction = immigration) in number of species per time at equilibrium

$t_{0.90} =$ time required for number of species present to increase from 0 to 90% of $\check{S}$.

Later the same authors elaborated upon the general shapes of immigration and extinction curves, discussed colonization rates and curves, and predicted the distribution of survival times of species that succeed in colonizing an island (MacArthur and Wilson 1967).

Data to test these hypotheses are scarce. Most biogeographic information, when combined with information furnished by the scant fossil record, can at best lead to descriptions of broad patterns of distribution and zoogeographic regions and to hypotheses about the essentially long-term processes responsible for them (e.g., Darlington 1957). Such theories, though occasionally suggestive for specific groups, fail to explain the existing taxonomic distributions and the underlying colonization mechanisms for most islands. They neglect the short-term, even daily events that largely determine the parameters of colonization on many islands. Such events become decisively more important than evolution as distance of island from source area decreases. In particular, the classical zoogeographic methods do not provide a test for the existence of an equilibrium number of species or for the accuracy of equation (1).

The literature on biotic dispersal, both anecdotal and systematic, provides information on the relative importance of active and passive transport and a wealth of data on the specific agents of passive transport (e.g., Wolfenbarger 1946). Certain cases of overseas dispersal have even been traced to specific meteorological events (French 1964). For no island, however, is the time course of colonization by even a large fraction of the inhabitants known. The large number of records of long distance dispersal implies that immigration rates to islands are probably high, especially for organisms capable of passive transport by wind. But the shape and determinants of the immigration curve cannot be deduced. At best a lower limit might be given.

In the recent studies of Surtsey, a new volcanic island near Iceland, there has been a deliberate attempt to document the colonization process from

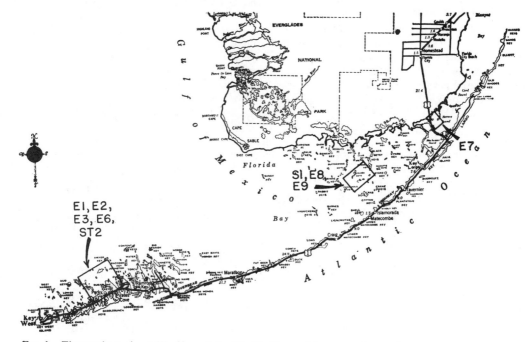

FIG. 1. The southern tip of Florida and the Florida Keys. The rectangles enclose the experimental areas shown in detail in Figures 3–5.

Fig. 2. *Upper:* Island E1, the second smallest island in the experimental series. *Lower:* Island E9, the largest island in the experimental series; note also the presence of supratidal mud.

the birth of the island (Fridriksson 1967, Hermannsson 1967, Lindroth et al. 1967). But the infrequent natural occurrence of such events tends to render quantitative biogeographic theory the product of abstract speculation and of enlargement and sophistication of untestable hypotheses. What is needed, clearly, is a method either of producing new islands similar to natural ones or of sterilizing preexisting islands.

THE EXPERIMENTAL ISLANDS

Along the Overseas Highway of southern Florida (U. S. Route 1) there exist a vast array of small, approximately circular islands situated in shallow bay water (Fig. 1). Together, they constitute a potential natural "laboratory" for experimental biogeography. The islands with diameter 10–20 m (Fig. 2) are 5–10 m tall and usually

consist entirely of one to several red mangrove trees (*Rhizophora mangle*), with rarely a small black mangrove bush (*Avicennia nitida*). Occasionally, in areas of weak tide, the trees are surrounded by small areas of supratidal mud and sand. Such land is only intermittently supratidal, since it is completely flooded during prolonged winds of 10 knots or more. Individual islands may differ in several minor respects such as the number of large trunks and amount of dead bark they contain, but on the whole they are remarkably similar to each other in physical appearance.

A fruitful defaunation experiment with these islands requires the following conditions:

i) that there be enough of the islands for replication and variation in distance to the nearest source area;

ii) that the animal diversity be sufficient to allow statistical treatment, and the organisms be physically large enough to insure recording inconspicuous forms as well as conspicuous ones;

iii) that the extremely small size of the islands compensates for their relative nearness to source areas and therefore produces a distance effect.

During an exploratory trip to the Florida Keys in 1966, we found that conveniently accessible small islands were numerous at most distances up to 200 m from the nearest large islands. Beyond 200 m they were rare, although somewhat larger islands (diameter 50 m and greater) were plentiful. A very few small islands suitable for our purposes were located 0.2–1.4 km from the nearest source area.

A set of such islands were chosen for experimentation and their faunas carefully surveyed. Since the breeding fauna of islands this size consists almost entirely of species of insects and spiders, we made a reference collection of land ar-

TABLE 1. Parameters of experimental islands

	Island name	Diameter (m)	Distance from nearest source (m)	Initial no. arthropod species	Location
Series 1	E1	11	533	25	off Squirrel Key
	E3	12	172	31	in Rattlesnake Lumps
	ST2	11	154	29	off Saddlebunch Key
	E2	12	2	43	off Snipe Keys
	E6[a]	15[a]	73[a]	30[a]	in Johnston Key Mangroves
	E7	25	15	29	near Manatee Creek
Series 2	E8	18	1,188[b]	29	between Calusa Keys and Bob Allen Keys
	E9	18	379	41	between Calusa Keys and Bob Allen Keys
	E10[a]	11[a]	37[a]	20[a]	near Bottle Key

[a] Control island
[b] "Stepping Stone" (SI) 572 m away.

thropods of the Florida Keys with the assistance of Robert Silberglied. This collection enabled us to identify the arthropods on the experimental islands, and it provided a measure of the size and composition of the species pool of the presumed source area, the entire Florida Keys.

Animal diversity on the small mangrove islands proved to be adequate for statistical purposes. About 75 insect species (of an estimated 500 that inhabit mangrove swamps and an estimated total of 4,000 that inhabit all the Keys) commonly live on these small islands. There are also 15 species of spiders (of a total Keys complement of perhaps 125 species) and a few scorpions, pseudoscorpions, centipedes, millipedes, and arboreal isopods. At any given moment 20–40 species of insects and 2–10 species of spiders exist on each of the islands.

Many birds visit the islands either to roost or to forage, but very few nest there. The only birds that bred on our islands during the course of the experiment (1966–68) were one or two pairs each of the Green Heron, White-crowned Pigeon, and Gray Kingbird. Snakes (mostly water snakes of the genus *Natrix*) occasionally swim to small mangrove islands, and raccoons visit islands located on shallow mud banks. These vertebrates were not included in the censuses.

On the basis of their distance and direction from the nearest source area, their size, and their accessibility, the islands listed in Table 1 were chosen for defaunation. Two islands (E6 and E10), one each in the vicinity of the two groups of experimental islands, were selected to serve as control islands; and they were censused at the same time as the experimental islands. In order to ascertain whether any long-range change had occurred in the general fauna of small mangrove islands in Florida Bay during the course of the

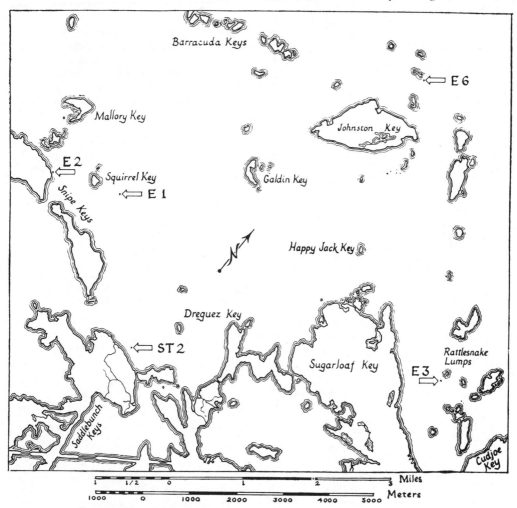

Fig. 3. Map showing the location of the experimental islands of series 1.

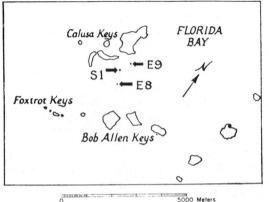

Fig. 4. Map showing the location of the experimental islands of series 2.

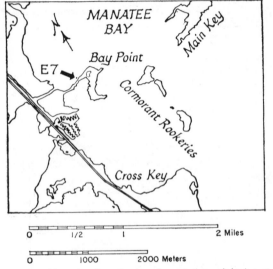

Fig. 5. Map showing the location of the original test island E7.

experiment, a second census of the control islands was made at the close of the experiment.

The two series of experimental islands form two widely separated groups. Series 1 lies within the Great White Heron National Wildlife Refuge, in a region north of Sugarloaf Key (Fig. 3). Series 2 is in the Everglades National Park near the Calusa Keys north of Islamorada (Fig. 4). E7, which was chosen primarily to test a method of defaunation, is located in Barnes Sound south of the Glades Canal (Fig. 5).

The islands were also selected to show as much variation as possible in the degree of isolation from the immigrant sources. The coefficient of correlation for island series 1 between "distance from nearest source" and "number of arthropod species" is −0.83, indicating the operation in the

original faunas of the distance effect as defined by MacArthur and Wilson (1967). Qualitative differences in faunas of increasingly distant islands were even more striking than quantitative ones. For many groups—especially ants and spiders— the distance effect is so regular that one can guess not only the species number but also the approximate species composition. For example, centipedes and pseudoscorpions were found almost exclusively on islands of less than 200 m distance from the nearest large island.

Islands of this minute size can be remarkably long lived. All *Rhizophora mangle* trees, and particularly those on low, small islands, have numerous aerial roots: both brace roots growing out from the main stem and prop roots growing down from branches. The latter can emerge from as high as 4 m above the water (LaRue and Muzik 1954) and in time can become as thick as the main stem. *Rhizophora* trees reproduce viviparously, their seeds growing on the tree until they become large, long, pointed seedlings. When they drop, frequently from the upper canopy, some evidently plant themselves where they fall (Davis 1940). Mangrove swamps are often quite thick, and on small islands there may be several trees. Stems, thickened prop roots, and brace roots in-

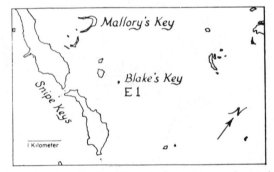

Fig. 6. Map showing E1 and surrounding region as it appeared in 1851–57.

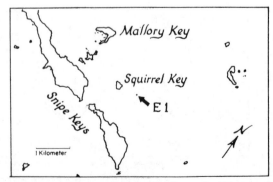

Fig. 7. The same region shown in Figure 6 as it appeared in 1964.

termingle and reproduce themselves so that it is impossible to distinguish how many trees are present, which are which, which is the initial one (even if it is still present), and which are the main stems.

Maps of the Florida Keys, especially of the Sugarloaf Key-Key West area, have been remarkably detailed and accurate with respect to small mangrove islands for at least a century. These tell us that E7 and all the islands in series 1 have existed at least 25 years. The other experimental islands were not mapped in sufficient detail to determine longevity. The size and shape of individual islands may change, even drastically, but in general the islands survive the climatic catastrophes to which they are repeatedly subjected. For example, Figure 6 shows E1 (Blake's Key) and surrounding areas as it appeared in 1851–57, while Figure 7 shows the same island region on the Coast and Geodetic Survey map of 1964. It therefore seems reasonable to assume that the faunas found initially on the experimental islands were the product of many years' colonization. Our experiments have subsequently shown that the minimum ages of the islands greatly exceed the time required to reach species equilibrium from a start of zero species (Simberloff and Wilson 1969).

DEFAUNATION

Our first attempt to defaunate islands made use of an insecticide spray of short-lived residual effect. On July 9–10, 1966, two experimental islands (E1 and E2) were sprayed until dripping with 60 g parathion, 240 g diazinon and 180 g Pylac sticking agent per 100 l fresh water. When the islands were examined 1 day later, all the surface and bark fauna and most borers were dead. However, the following live animals were found in thin hollow twigs. E1: one adult wasp, *Scleroderma macrogaster* (Bethylidae); workers and larvae of *Crematogaster ashmeadi* (Formicidae). E2: two larvae of ?*Styloleptus biustus* (Cerambycidae); two colonies of *Paracryptocerus varians* (Formicidae); and one colony of *Tapinoma littorale* (Formicidae). A more thorough examination 9 days after spraying produced the following live and apparently healthy insects. E1: two larvae of ?*Styloleptus biustus* (Cerambycidae). E2: one larvae of ?*Styloleptus biustus* (Cerambycidae); one lepidopteran larva; and two colonies of *Xenomyrmex floridanus* (Formicidae). The large populations of the ant colonies and advanced development of the beetle larvae precluded these insects having immigrated during the 9 days following spraying.

The results on E1 and E2 showed that a spray cannot be expected to penetrate all the hollow twigs of a *Rhizophora* island. Some twig dwellers will probably survive, particularly if they inhabit narrow twigs (*Tapinoma* and *Xenomyrmex*) or else pack their tunnels tightly with powdery excreta (*Styloleptus*). Had we attempted to kill these survivors by using a spray with a long-lived or more powerful residue, we would have postponed the beginning of the colonization curve, since immigrants may be killed by the residue. Because we had to be certain that all colonists, or at least all but those belonging to a small number of known taxa, would be destroyed immediately, we abandoned the spray technique.

We turned next to the more difficult alternative technique of fumigation. The method has the obvious advantages that the residual effect of a gas is negligible, and even the inhabitants of hollow twigs are contacted. This is the standard method used by professional exterminators on termites and other wood-boring insect pests.

The fumigation of living plants is a relatively uncharted area. The biochemical effects of fumigants on plants are poorly known, damage usually being described in terms of the physical appearance of the plant shortly after fumigation (Page and Lubatti 1963). To our knowledge, the only prior fumigation of red mangrove trees was by B. P. Stewart (pers. comm.), who used methyl bromide on small plants. With 32 kg/1000 m³ for 2 hr at 32°C and 40 kg/1000 m³ for 2 hr at 27°C he found that all insects apparently were killed, although some did not die until 48 hr after fumigation. Mortality of cockroach eggs was unrecorded in Stewart's study. Damage to the plant was limited primarily to young leaves and twigs, a result that is in accord with the general finding of Page and Lubatti (1963) that the growing stages of both plants and insects are the most susceptible to fumigation.

In our preliminary trials we were limited by the fact that fumigants highly soluble in water, such as hydrogen cyanide, could not be used. These substances would be expected either to leave the fumigation tent through the water or to concentrate to an unknown extent in the water under the island. Either effect would greatly complicate the maintenance and measurement of gas concentration. Consequently, the following four relatively insoluble fumigants were tried: Pestmaster Soil Fumigant-1 (methyl bromide 98% wt, chloropicrin 2% wt); Acritet 34-66 (acrylonitrile 34% vol, carbon tetrachloride 66% vol); Dow Ethylene Oxide (ethylene oxide 100%); and Vikane (sulfuryl fluoride 95% wt, inert ingredients 5% wt). Field tests were performed on small *Rhizophora* trees in Matheson

Hammock Park, Coral Gables, by National Exterminators of Miami, Florida, under the direction of Steve Tendrich in consultation with the authors. Additional tests on the mortality of mangrove colonists were performed at National Exterminators' laboratory in Miami.

With the temperature outside the fumigation tent remaining at 24°–29°C, we determined which duration-concentration intensities caused no more than bearable damage to the tree yet were lethal to all the mangrove arthropods. Acritet, Vikane, and ethylene oxide caused extensive, apparently irreversible damage to red mangrove and were rejected. However, methyl bromide at 22 kg/1000 m³ for 2 hr caused browning of only about 10% of the leaves 1 week after treatment, with no visible later effects. This duration and concentration killed all the mangrove inhabitants (including such resistant forms as roach eggs and lepidopteran pupae) with one possible exception. When data from all tests at this duration-concentration are lumped together, only 95% of cerambycids and 80% of weevils were dead when dug out of their burrows 1–10 days after fumigation. The remainder all died within 2 days of removal. It is tempting to ascribe the death of these few larvae to a delayed effect of the fumigant. However, a control experiment consisting of the removal of five unfumigated cerambycid larvae (*Styloleptus biustus*) from their respective burrows caused death within 5 days without any additional treatment. Monro and Delisle (1943) report that many insects are active soon after exposure to methyl bromide but succumb in time (some lepidopterous larvae surviving for 2 months), and that this delayed effect is especially common in fumigations at the low concentrations used in our own field tests. This peculiarity is confirmed by Chisholm (1952) for Japanese beetle larvae and other insects. Of at least equal importance, the results from our experimental islands (Simberloff and Wilson 1969) strongly suggest that the cerambycids were eliminated by the fumigation treatment. We therefore conclude that even the deep-boring beetle larvae were very probably killed by the treatment.

The overt physical damage to the trees consisted of a browning of leaves and shoots and occasional heavy oozing of sap. At 61 kg/1000 m³ the leaf browning was manifest upon tent removal, but at 35 kg/1000 m³ and less there was little or no immediate browning. The effect became evident at about 24 hr and worsened in terms of fraction of leaves browned through 72 hr, by which time up to 90% of the leaves were brown and after which no further browning occurred. Browning did not occur randomly throughout the canopy. Instead, whole sections of the tree became completely brown while others remained almost unscathed. The upper canopy was always more severely browned, but there seemed to be no other consistent damage pattern.

Abscission of the dead leaves began within a week and, often aided by wind, was usually complete within 8 weeks. At this time damaged trees looked superficially dead except for remaining green sections. Closer examination showed that the damaged trees could be conveniently divided into three classes, as follows:

For those trees where browning of leaves and shoots appeared complete by 72 hr the damage was irreversible. Within a few months bark began to peel and twigs with green wood could not be found. A year later there was no evidence of life. Island E8 is in this class. In the second category, which includes only E7, a large majority of the leaves and shoots were browned and died, while a small section remained green. In the third category, the trees were less severely damaged (60% or fewer browned leaves). Green sections remained unchanged, while leaf abscission in most of the damaged sections was accompanied by the sprouting of new shoots and leaves from almost all branches. Within 2 months the trees appeared normal except for occasional leafless patches in tne extreme upper canopy. Island E1, which 2 days after fumigation appeared to have about 50% brown leaves, recovered after 1 month to the extent that it did not appear to have incurred more damage than similar trees do during heavy storms, and 2 months after defaunation seemed quite normal except for a small bare patch in the upper canopy. At no time did its bark crack or peel.

In most damaged trees of the first two classes—where either the whole tree or all but a small section was killed—there was copious oozing of sap for up to a week after fumigation. Such oozing was never observed in untreated trees, nor in those fumigated trees which showed good leaf replacement.

The action of methyl bromide on living plants has been studied rather extensively by several authors. Mainwaring (1961) examined the physiological responses of several plants to methyl bromide, and his numerous observations imply that its major effect is auxin inhibition. The damage we observed seemed too drastic and certainly too immediate to be attributed solely to auxin inhibition, although the presence of such a biochemical action cannot be ruled out.

Whatever the effect of methyl bromide on *Rhizophora mangle*, it apparently had the expected Q_{10} of 2–3. Our initial tests in Matheson Hammock were conducted on small plants well shaded by large trees, and the temperature inside the fumigation tent remained within 3° of the air temperature (24°–29°C). The experimental islands could not be shaded, and even on overcast days the temperature inside the dark tent was at least 32°C. On sunny days the inner temperature was so high that the presence of the covering alone caused browning of leaves. It seems probable that the generally more severe damage to the upper canopy throughout this experiment was caused by higher temperature, especially since methyl bromide gas has a density (3.974 g/liter at 20°C) considerably greater than that of air.

Air humidity is much less important than temperature in its effect on fumigation damage to plants (Page and Lubatti 1963) ; our experiments were almost all conducted within a narrow humidity range (65%–90%) and the effects of varying this parameter were not studied.

The first two experimental islands (E7 and E1) were fumigated during daytime and damaged. In order to reduce heat damage we therefore fumigated all other islands at night. The concentration was increased from 22 kg/1000 m³ to 28–30 kg/ 1000 m³ with no observable additional damage.

The chemical means for defaunation now having been chosen and tested, we were left with an overwhelming physical problem of how to fumigate an entire island. Assuming that we could somehow get the fumigation tent (a 25-m by 25-m piece of plastic-impregnated canvas with addi-

TABLE 2. Arthropod species found on E7 just prior to and following fumigation

INSECTS	
Embioptera:	gen. sp.[a]
Orthoptera:	Latiblattella n. sp.
	Cycloptilum sp.
	Tafalisca lurida[a]
Coleoptera:	Pseudoacalles sp.[b]
	Leptostylus sp.
	Styloleptus biustus[b]
	Tricorynus sp.
Psocoptera:	Archipsocus panama
Hemiptera:	Pseudococcus sp.
Lepidoptera:	Nemapogon sp.[a]
Hymenoptera:	Casinaria texana
	Camponotus floridanus[a]
	Camponotus planatus[a]
	Paracryptocerus varians[a]
	Pseudomyrmex elongatus[a]
	Xenomyrmex floridarus[a]
ARACHNIDS	
Araneae:	Ariadna arthuri
	Eustala sp.
	Leucauge venusta[a]
Acarina:	Galumna sp.
OTHER	
Chilopoda:	Orphnaeus brasilianus[a]
Isopoda:	Rhyscotus sp.[a]

[a] discovered dead after fumigation
[b] discovered mostly dead after fumigation (see text)

tional side pieces for larger islands) onto an island, the tree could not support the weight without major limb breakage, since even our special half-weight tent weighed 275 kg. It was evident that some sort of superstructure was needed, one which would surround and rise over the island to remove most or all of the tent weight from the tree. Constructing a permanent framework and lowering it over the islands by helicopter was too expensive.

An alternative method employing a temporary full framework constructed on the site was used on October 10, 1966 to defaunate E7. The procedure was as follows (see also Fig. 8) : First, planks to prevent the framework from sinking into the mud were placed underwater in a square around the islands and held manually until the first tier of scaffolding (which was sufficiently heavy to keep them from floating) was placed on top. Then the remaining tiers were added, one side at a time, until the island was surrounded by a cube of scaffolding. Fig. 8 (upper) shows the framework complete except for the upper tier of scaffolding in the right rear corner and the square walkway of planks around the very top of the structure.

The tent was then raised by block and tackle to the top of the framework and lowered over the island. Methyl bromide was introduced from 45 kg cylinders through the tent wall at opposite sides of the island, 2.5 m above the water, and into a metal tub mounted with an electric fan to

FIG. 8. *Upper:* The scaffolding constructed around E7, complete except for the top walkway. *Lower:* the fumigation tent over E7.

disperse the gas. Chloropicrin (tear gas), added to the odorless methyl bromide as a safety measure, was allowed to drip into burlap sheets at the bottom of the tubs. Concentration of the methyl bromide was measured frequently by a Gow-Mac thermal conductivity meter with a self-contained pump attached to a testing station 2.5 m above the water and well away from the two shooting stations.

The methyl bromide concentration was gradually built up from 0 to 22 kg/1000 m³ over 30 min, then kept at 22–25 kg/1000 m³ for 2 hr. Outside air temperature during this time remained steady at about 27°C. At the end of the fumigation the gas was released through seam openings for 45 min, and the tent was then removed.

Immediately following the dissipation of the gas, E7 was explored for a total of 6 man-hours. The results generally confirmed the evidence from the Matheson Hammock tests. Dead individuals representing 11 of the 23 species observed before fumigation were found, frequently in large numbers and in all life stages. The details are given in Table 2. The only live animals encountered were larvae of two deep-boring beetles. These formed only a small fraction of the original population, as shown by the following recovery data: of 59 larvae of *Styloleptus biustus* (Cerambycidae), 2 were alive and 57 dead; of 10 larvae of *Pseudoacalles* sp. (Curculionidae), 2 were alive and 8 dead. One of the two live cerambycids was extremely sluggish and died an hour after removal. The other cerambycid and the two weevils seemed healthy, though one of the weevils became torpid and died 9 hr after removal.

The tree was severely damaged as described previously. Within 2 months 85% of the foliage was dead, although the single living section has remained green until at least May 1968. Thus the procedure worked essentially as expected, and this aspect of the fumigation was left unchanged in subsequent defaunations, except that the work was henceforth done at night in all but one instance (E1). The cubical scaffolding framework had disadvantages, chief among them being the large crew necessary to erect it (at least five men) and the number of pieces and weight of the equipment. This necessitated numerous boat trips between E7 and our staging area near Manatee Creek, since the shallow water along the route precluded the use of large craft. This problem was greatly aggravated on the other islands, which are located considerably further from a convenient staging area. We therefore set out to devise a tower to be located in the center of the island and rising high enough so that most of the tent weight rested on it. This device proved to be more vul-

FIG. 9. *Upper:* the fumigation tent being opened over E2 from a tower support. *Middle:* E2 partially covered. *Lower:* E9 covered by the fumigation tent on a tower support.

nerable to high winds, but it has the great advantage of requiring less equipment and a smaller crew to assemble it. National Exterminators contracted a steeplejack to perfect the tower method. After a month of experimenting on islands near Key Largo, the following procedure was adopted and used for the remaining islands.

A collapsible 10-m triangular steel tower was erected in the center of the island, either propped on a large root or forced deep into the ground. The steeplejack then climbed to the top and threw out three guy wires which were fastened down

TABLE 3. Extent of damage to foliage and subsequent recovery on experimental islands

Island	Fumigation date	Time	Extent of damage (burnt leaves)	Damage location	Recovery of foliage
E1.............	3/13/67	day	50%	general	complete in 2 months, except for extreme upper canopy
E2.............	3/7/67	night	10%	upper canopy	complete in 2 months
E3.............	3/17/67	night	5%	upper canopy	complete in 2 months
E7.............	10/10/66	day	85%	all except one section	none
E8.............	4/17/67	night	100%	complete	none
E9.............	4/7/67	night	20%	upper canopy and one other section	little in upper canopy complete elsewhere in 2 months
ST2.............	4/20/67	night	5%	scattered	complete in 2 months

with mudscrews or stakes, depending on the substrate. Using a block and tackle, one man on the tower and one on the island lifted the rolled tent until it formed a triangle over the island. The tent was then carefully unrolled down the guide wires until the island was covered (Fig. 9). It was impossible to unfurl the tent from the tower in winds greater than 8 knots. Sandbags and stakes were used to keep the tent edges submerged.

Equipment used in the tower method was identical to that described already for the scaffolding method, and was similarly situated. The procedure was identical except that gas was administered from 0.5-kg cans and concentration and duration were increased to 28–33 kg/1000 m³ for 2.5 hours on all islands but E1. After fumigation the methyl bromide was released through tent seams as with scaffolding, but in the tower system tent removal was more difficult. At low wind speeds the tent could be laboriously refurled and gradually lowered to the boat. We fortuitously discovered that high winds, instead of causing extensive damage as we expected, could actually aid tent removal. The windward side of the tent was lifted by poles to about 3 m above water, whereupon sufficient wind entered the tent to lift it up over the tree and to drop it, partly extended, into the water on the leeward side of the island.

E1 was fumigated in the daytime. For all remaining islands the tent was unrolled no earlier than 60 min before sundown, the fumigation was conducted between 8:00 and 12:00 pm, the gas was removed by 1:00 am, and the tent was either off the island or refurled within 90 min of dawn. Effects on the mangrove trees of the fumigated islands are noted in Table 3.

All islands were examined immediately after tent removal. No living animals at all were found on E1, E2, E8, and E9. A single live curculionid larva was found on E3. On ST2, 1 living and at least 100 dead polyxenid millipedes (*Lopho-*

proctinus ?bartschi) were found. Although we have no further direct evidence and know of no work on methyl bromide fumigation of millipedes, we suspect that all millipedes not destroyed immediately were killed by the delayed effect of methyl bromide discussed earlier.

In sum, we feel that this fumigation technique killed all mangrove arthropods with the single remotely possible exception of deep-boring beetle larvae.

MONITORING TECHNIQUE

Precautions were taken at all monitorings[2] to prevent contamination of the experimental islands. Boats were never brought in contact with the islands and only seldom tied to them; usually they were anchored at least 12 m away. Before wading to an island, experimenters examined clothing, equipment, and persons for animals, which were destroyed. To lessen the chance of phoresy, investigators simply immersed themselves briefly or sprayed themselves with "Off" insect repellent. All equipment used on the island was sprayed weekly with a short-lived insecticide.

The absence of supratidal land, except around E9, greatly simplifies the environment of the experimental islands. The arboreal substrate may be arbitrarily divided into hollow twigs, living branches and twigs, dead bark and tree holes, leaves and flowers, green shoots, and fruits. Because of the extremely small size of the experimental islands it was possible to study most of these microhabitats all but exhaustively for the pre-defaunation censuses. A typical example is the survey of E3, where we broke and examined 3,500 of an estimated 5,000 small hollow twigs, collected

[2] The term "monitoring" is used to designate one or more of the censuses of species, conducted for a period of 17–20 hr during the course of the experiment. In the present series of articles, monitoring, census, and censusing are used interchangeably to refer to the listing of species.

90% of the fruits, examined virtually all of the leaves, and looked under at least 80% of the pieces of dead bark. *Rhizophora* has numerous limbs and our islands were relatively low, so that we were able, by climbing carefully, to examine the canopy right to the top.

The fumigation procedure itself gave added evidence of the completeness of our original surveys. The fumigant included 2% chloropicrin, which drives out even deep-boring and hollow-twig-dwelling arthropods (cf. Lauck 1965). The animals either fell into the water or remained hanging from branches by spines or silk, tenuously and conspicuously. Those which fell generally remained afloat and, when a tent seam was opened after fumigation, flowed out in masses where they were collected on a screen.

The chloropicrin technique was especially effective on E8, E9, and ST2, and the results indicated greater densities of both insects and spiders than we had suspected. Only on E8, however, were animals found that had not been noted in the original surveys. These consisted of four beetles:

Staphylinidae: gen. sp.
Carabidae: *Bembidion* sp. nr. *contractum*
Carabidae: *Tachys occulator*
Oedemeridae: *Oxacis* sp.

Of these the first three were apparently living in a patch of temporarily supratidal mud, which was submerged again 2 weeks later during a strong wind. The *Oxacis* were encountered only on the fumigation tent, and even their status as "transients" is doubtful.

Defaunated islands were censused for approxi-

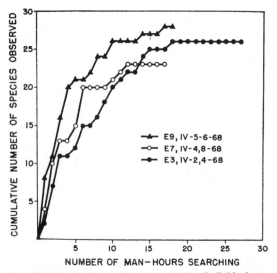

Fig. 10. Cumulative species counts during individual censuses on three of the experimental islands.

mately 2 days every 18 days. At each census period approximately 10% of all hollow twigs were broken, and with this treatment the number of hollow twigs seemed to remain constant throughout the experiment. Only when gentle lifting failed to allow vision was dead bark removed, and then only a part was peeled off. Small amounts of carbon tetrachloride and ammonia, and tapping with a steel probe were used to drive animals out from underneath bark and inside hollow twigs without damage to tree or colonist. Although the pre-defaunation searching methods had been drastic, care was taken following defaunation to prevent the censusing from being destructive and from altering the proportions of the respective microhabitats on an island.

Most colonists could be identified in the field because of the small subset of the Keys fauna which invades small mangrove islands and the completeness of our reference collection. Those that could not be recognized with certainty were treated in one of two ways. If the animal in question was obviously part of a sizable population, as with many psocopterans and thrips, a small collection was made that did not exceed 2% of the estimated minimum size of the population. When an unknown arthropod was not part of a large population, we photographed it.

It was quickly discovered that the rate of discovery of new species during each censusing period, although at first very high, declined approximately linearly to near 0 at about 14 man-hours of searching. A very few species are expected to appear from immigration within the monitoring period. Figure 10 depicts the cumulative number of species observed vs. time for three representative monitorings. Different sizes of islands, weather conditions, and especially varying population sizes caused minor differences in the curves from island to island and from time to time; but despite great changes in the numbers of species through the course of the experiment, the curves remained remarkably similar in shape. The censusing period was therefore chosen to last between 17 and 20 man-hours.

Because of the simplicity and very small size of the experimental islands, and because several nocturnal censuses located no new species, we felt that all colonists having crepuscular or nocturnal activity were being recorded during the diurnal monitoring. The anyphaenid spider *Aysha* sp. and ant *Paracryptocerus varians* are both strictly nocturnal, yet both were recorded during the day in their retreats. Similarly the crepuscular braconid wasp *Macrocentrus* sp. and largely nocturnal ants *Camponotus floridanus* and *Tapinoma littorale* were repeatedly observed.

ACKNOWLEDGMENTS

We thank the United States Department of the Interior for permission to fumigate the islands of this experiment. In particular, Jack Watson of the Great White Heron National Wildlife Refuge, Roger W. Allin and William B. Robertson of the Everglades National Park, and Robert M. Linn of the National Park Service examined the plans of the experiment and concluded, with the authors, that the total mangrove faunas would be unaffected. Steve Tendrich, Vice-President of National Exterminators, Inc. (Miami) coped imaginatively and effectively with the unique problems presented by the project. W. H. Bossert and P. B. Kannowski assisted in the original censuses. Maps and tables for all parts of this series were done by Mrs. Helen Lyman. The work was supported by NSF grant GB-5867.

LITERATURE CITED

Chisholm, R. D. 1952. Nature and uses of fumigants, p. 358. *In* Insects, The Yearbook of Agriculture 1952.

Darlington, P. J. 1957. Zoogeography: the geographical distribution of animals. Wiley, New York. 675 p.

Davis, J. H. 1940. The ecology and geologic role of mangroves in Florida. Papers from Tortugas Lab., Carnegie Inst. Wash. 32: 307–412.

French, R. A. 1964 (1965). Long range dispersal of insects in relation to synoptic meteorology. Proc. Int. Congr. Entomol. 12th (London). 6: 418–419.

Fridriksson, S. 1967. Life and its development on the volcanic island, Surtsey. Proc. Surtsey Research Conf., 1967: 7–19.

Hermannsson, S. 1967. Introduction. Surtsey Research Progress Report, III: 1.

LaRue, D. C., and T. T. Muzik. 1954. Growth, regeneration and precocious rooting in *Rhizophora mangle*. Pap. Mich. Acad. Sci. Arts Lett. (I) 39: 9–29.

Lauck, D. R. 1965. Chloropicrin for fast action with a Berlese funnel. Turtox News 43: 115.

Lindroth, C. H., H. Andersen, H. Bodvarsson, and S. H. Richter. 1967. Report on the Surtsey investigation in 1966. Terrestrial invertebrates. Surtsey Research Progress Report, III: 59–67.

MacArthur, R. H., and E. O. Wilson. 1963. An equilibrium theory of insular zoogeography. Evolution 17: 373–387.

MacArthur, R. H., and E. O. Wilson. 1967. The theory of island biogeography. Princeton University Press. 203 p.

Mainwaring, A. P. 1961. Some effects of methyl bromide on aphids and whitefly and their host plants. Doctoral thesis, London Univ., London, England.

Monro, H. A. U., and R. Delisle. 1943. Further applications of methyl bromide as a fumigant. Sci. Agr. 23: 546–556.

Page, A. B. P., and O. F. Lubatti. 1963. Fumigation of insects. Ann. Rev. Entomol. 8: 239–264.

Simberloff, D. S., and E. O. Wilson. 1969. Experimental zoogeography of islands. The colonization of empty islands. Ecology 50: 278–296.

Wolfenbarger, D. O. 1946. Dispersion of small organisms. Amer. Midland Naturalist 35: 1–152.

15

Reprinted from *Am. Nat.* **103**:439–454 (1969) by permission of The University
of Chicago Press

THE RELATIONSHIP OF FRESH-WATER PROTOZOAN COMMUNITIES TO THE MACARTHUR-WILSON EQUILIBRIUM MODEL

John Cairns, Jr.,* M. L. Dahlberg,† Kenneth, L. Dickson,*
Nancy Smith,‡ and William T. Waller*

INTRODUCTION

These studies involved the colonization of originally sterile, submerged plastic substrates by fresh-water protozoans. Two series of substrates exposed to presumably comparable ecological conditions, but 70 yards apart, were used to study (1) the rate of colonization and (2) similarities and differences between the two groups. The equilibrium model for island faunas proposed by MacArthur and Wilson (1963) may also be tested with the data obtained.

A number of investigators have used artificial substrates for collecting aquatic organisms (e.g., Butcher 1946; Patrick et al. 1954; Grzenda and Brehmer 1960; ZoBell and Allen 1933). An excellent review of the earlier literature in this field has been prepared by Cooke (1956). A recent paper by Hargraves and Wood (1967) discusses periphyton algae on selected host plants. Immigrant or invading species which colonize an uninhabited artificial substrate may come from the atmosphere (e.g., see Berger 1927; Darwin 1863; Gislen 1948; Huber-Pestalozzi 1937; Hudson 1889; Maddox 1870; Messikommer 1943; National Research Council 1941; Pouchet 1860; Schlichting 1961, 1964; and, of course, Pasteur 1860), from dispersal by other organisms (e.g., see Maguire 1963; Maguire and Belk 1967), or being carried to the substrate by water currents.

Protozoans may be borne to a new site by water currents, a possibility too obvious to need documentation. In all cases species may arrive in an inactive or encysted state, or as fully functioning active organisms. Protozoans probably differ in their dispersal capacity, since many species that tolerate a broad range of aquatic environmental conditions are often

* Department of Biology, Virginia Polytechnic Institute, Blacksburg, Virginia 24061.

† Department of Forestry and Wildlife, Virginia Polytechnic Institute, Blacksburg. Virginia 24061.

‡ Department of Botany, University of Michigan, Ann Arbor, Michigan 48104.

not found in habitats which are within the measurable tolerance limits of that organism (Cairns 1964, 1965).

There is some evidence that species composition and number of Protozoa may vary on homogeneous substrates kept under laboratory conditions (Cairns and Yongue 1968). However, the total number of species seems to be remarkably constant through time in a mature community in a complex natural habitat which is under no stress (Patrick 1961). This relative stability in the total number of species suggests that interrelationships comparable to those described by MacArthur and Wilson (1963) may be in operation for fresh-water protozoan communities.

MATERIALS AND METHODS

The Artificial Substrates

Twenty floats made from Isothane (a polyurethane foam manufactured by Bernel Foam Products, Inc., Buffalo, N.Y.) were used as substrates. All 20 of the synthetic sponges were 3cm × 3cm × 15cm. Each float was tied by nylon line to an anchor and placed along the west shore of South Fishtail Bay, Douglas Lake, Michigan. The floats were anchored in 4–5 ft of water (by the end of the experiment, the water level in Douglas Lake had dropped about 1 ft), and the lines were long enough so the substrates floated at the water surface. The floats were set out in two groups of 10 each, and the floats in each group were placed about 4–5 ft apart. The two groups of floats were separated by about 70 yards of open water. Five floats were carried away the first week, leaving eight floats in the first group (that closest to the camp) and seven floats in the second group; subsequently one float was accidentally eliminated on the twenty-second day and two floats disappeared in a storm the thirty-sixth day. By the end of the experiment, seven floats remained in the first group and five in the second group.

Sampling Procedure

Samples of the protozoan populations on the floats were obtained by squeezing the synthetic sponges over a bottle so that about 10 ml of material were collected. The floats were sampled biweekly initially and weekly during the latter part of the experiment. In the cases where data are given for two consecutive days, half of the samples were obtained and studied on each day, and the samples were collected from alternate substrates (i.e., floats 1, 3, 5, 7, and 9 were sampled one day, and floats 2, 4, 6, 8, and 10 were sampled the next day). Most of the samples were collected between 8:00 A.M. and 10:00 A.M. and were examined immediately upon bringing them into the laboratory, so that the protozoans could be identified while alive. Throughout the study equal time was spent examining each sample. Representative subsamples from the top and bottom of each collection jar, both on the side toward and the side away from the light, were studied. In every case, three standard-sized glass slides with at least four drops of water on each slide were examined.

Recording Procedure

Two systems of recording the composition of the populations were used in the course of this study. On first examination (day 8), plus (+) and minus (—) signs were used to indicate presence or absence, respectively, of a given species in a sample, since in the initial examination there were generally fewer than three individuals of any one species in any one sample. As densities increased, a more discriminating density record was kept.

The data for the rest of the sampling periods (days 11, 15, 19–20, 22–23, 29–30, 37, 42–43) were recorded by the rating system shown below, which is similar to that proposed by Sramek-Husek (1958).

Rating	Number of Individuals per Slide
+	(1–3 individuals per 3 slides)*
1	1–3
2	4–6
3	7–10
4	11–15
5	16 and up

* First series of samples only.

Whenever possible, identifications were made to species. In instances where this was not possible, the genus was noted and usually accompanied by a sketch or note on morphology so that this particular taxon could be separated from other taxa. Less effort was expended trying to identify the small flagellates and helizoans as a group; the ciliates were studied more carefully, partially because they could be identified and counted more rapidly. No attempt was made to identify testaceans unless pseudopodia were observed; many empty tests were seen, however.

RESULTS

The number of protozoan species for each sampling location and day of sampling is given in table 1. Using the number of species present, we analyzed the data first considering each station (a substrate in a particular location) as a point invaded (table 1) and second considering each of the two areas as a point invaded (table 2). In the latter case all stations in an area were grouped together and the total number of different species for all substrates in each area determined. In each case a two-way mixed-model analysis of variance was used to test for differences between areas and between days. In the first case the analysis of variance was conducted using the average number of species per day per station in an area, and in the second case the number of species found per day per area was used.

TABLE 1

Number of Protozoan Species for Each Artificial Substrate on Specific Sampling Days

DAY OF SAMPLING	Location															Average
	Area 1 Substrates								Area 2 Substrates							
	1	2	3	4	5	6	7	8	9	10	11	12	13	14	15	
8	5	4	3	6	1	3	8	7	5	6	10	9	5	5	9	5.7
11	14	14	11	11	1	12	12	2	14	1	17	13	1	16	15	10.3
15	15	23	21	20	22	18	19	22	13	19	1	21	20	17	23	18.3
19–20	23	21	22	18	21	17	18	23	22	21	22	20	20	21	27	21.1
22–23	24	24	20	22	22	17	23	21	24	19	16	19	20	19	24	20.9
29–30	x*	29	x	30	34	x	20	26	31	x	33	28	x	27	28	28.6
37	33	x	x	x	33	x	x	38	x	29	x	33	x	x	x	33.2
42–43	29	x	33	36	x	28	38	36	35	39	40	31	x	x	x	34.4

* x = no reading taken.

TABLE 2
THE NUMBER OF PROTOZOAN SPECIES FOR EACH
AREA ON SPECIFIC SAMPLING DAYS

DAY OF SAMPLING	AREA 1		AREA 2	
	No. of Species	No. of Stations	No. of Species	No. of Stations
8	12	8	16	7
11	25	8	28	7
15	32	8	31	7
19–20	34	8	33	7
22–23	33	8	30	7
29–30	39	5	44	5
37	44	3	40	2
42–43	53	6	49	4

Analysis of variance tests indicated no significant difference between the two areas, but there was a highly significant difference ($P = .005$) between sampling days. The total number of species found at each sampling date for each substrate, the number of new species, the number of recurring species, and the number of species eliminated are shown in table 3. New species were those that had never been recorded on a previous sampling date for a particular substrate. It is quite likely that species recorded in sequential sampling of a particular location may be missing for one or more sampling periods but later be recorded again. This may be the result of either (1) the number of individuals per species declining to a point where the species were easily missed or (2) actual elimination and subsequent reinvasion of the species. Recurring species were considered in this study to be those that were eliminated and subsequently recurred. The number of species eliminated was found by adding the number of new species plus those recurring to the total number present at the preceding sampling date and then subtracting the present total number from this figure, for example, days 15 and 19 for substrate 1 (table 3). At day 19 there were 23 species present, of which 10 were new and five were recurring. If all of the species present at day 15 had remained on the substrate until day 19, then there should have been 30 species at day 19. However, only 23 were present. Therefore, seven species became extinct or eliminated between the two dates.

In order to evaluate the equilibrium model for island faunas proposed by MacArthur and Wilson (1963), colonization *rates* and extinction *rates* were calculated from the data presented in table 1. The number of new species plus the number of recurring species divided by the time in days between sampling periods equals the colonization rate expressed in species per day. Extinction rate was determined by dividing the number of species eliminated by the days between sampling periods. This also was expressed in species per day. Both colonization rates and extinction rates were calculated for each sampling period for each substrate.

Figure 1 presents the relationships between time in days and (1) mean number of species and (2) colonization rate, both calculated from the data

TABLE 3
TOTAL NUMBER OF PROTOZOAN SPECIES PRESENT, NUMBER OF NEW SPECIES,
NUMBER OF RECURRING SPECIES, AND NUMBER OF SPECIES ELIMINATED
FROM EACH SUBSTRATE ON EACH SAMPLING DAY

Substrate No.	Date	Total No. of Species	New Species	Recurring Species	Species Eliminated
1	8	5	5	0	0
	11	14	11	0	2
	15	15	6	0	5
	19–20	23	10	5	7
	22–23	24	5	1	5
	29–30	...	...	...	...
	37	33	9	2	2
	42–43	29	6	1	11
2	8	4	4	0	0
	11	14	10	0	0
	15	23	10	0	1
	19–20	21	3	0	5
	22–23	24	6	0	3
	29–30	29	10	1	6
	37	...	...	...	...
	42–43	...	...	...	...
3	8	3	3	0	0
	11	11	8	0	0
	15	21	12	0	2
	19–20	22	5	0	4
	22–23	20	2	1	5
	29–30	...	...	...	...
	37	...	...	...	...
	42–43	33	15	4	6
4	8	6	6	0	0
	11	11	8	0	3
	15	20	10	3	4
	19–20	18	5	1	8
	22–23	22	3	6	5
	29–30	30	11	3	6
	37	...	...	...	...
	42–43	36	14	2	10
5	8	1	1	0	0
	11	1	1	0	1
	15	22	21	1	1
	19–20	21	6	0	7
	22–23	22	5	3	7
	29–30	34	14	2	4
	37	33	11	0	8
	42–43	...	...	...	...
6	8	3	3	0	0
	11	12	10	0	1
	15	18	10	2	6
	19–20	17	6	0	7
	22–23	17	3	3	6
	29–30	...	...	...	...
	37	...	...	...	...
	42–43	28	13	5	7
7	8	8	8	0	0
	11	12	6	0	2
	15	19	11	0	4
	19–20	18	3	1	5
	22–23	23	6	3	4
	29–30	20	6	2	11
	37	...	...	...	...
	42–43	38	20	3	5

TABLE 3—(Continued)

Substrate No.	Date	Total No. of Species	New Species	Recurring Species	Species Eliminated
8	8	7	7	0	0
	11	2	2	0	7
	15	22	16	6	2
	19–20	23	5	0	4
	23–23	21	3	1	6
	29–30	26	9	2	6
	37	38	18	2	8
	42–43	36	8	4	14
9	8	5	5	0	0
	11	14	9	0	0
	15	13	6	0	7
	19–20	22	9	2	2
	22–23	24	4	1	3
	29–30	31	9	3	5
	37	...	...	...	...
	42–43	35	14	0	10
10	8	6	6	0	0
	11	1	0	0	5
	15	19	15	3	0
	19–20	21	8	1	7
	22–23	19	3	3	8
	29–30	...	...	...	...
	37	29	15	3	8
	42–43	39	11	4	5
11	8	10	10	0	0
	11	17	10	0	3
	15	1	0	0	16
	19–20	22	7	14	0
	22–23	16	0	1	7
	29–30	33	16	5	4
	37	...	...	...	...
	42–43	40	14	1	8
12	8	9	9	0	0
	11	13	9	0	5
	15	21	15	3	10
	19–20	20	4	1	6
	22–23	19	4	1	6
	29–30	28	10	2	3
	37	33	14	2	11
	42–43	31	11	3	16
13	8	5	5	0	0
	11	1	0	0	4
	15	20	16	3	0
	19–20	20	5	1	6
	22–23	20	5	1	6
	29–30	...	...	...	...
	37	...	...	...	...
	42–43	...	...	...	...
14	8	5	5	0	0
	11	16	12	0	1
	15	17	10	1	10
	19–20	21	4	3	3
	22–23	19	5	0	7
	29–30	27	8	4	4
	37	...	...	...	...
	42–43	...	...	...	...
15	8	9	9	0	0
	11	15	9	0	3
	15	23	9	1	2
	19–20	27	6	1	3
	22–23	24	3	0	6
	29–30	28	6	3	5
	37	...	...	...	...
	42–43	...	...	...	...

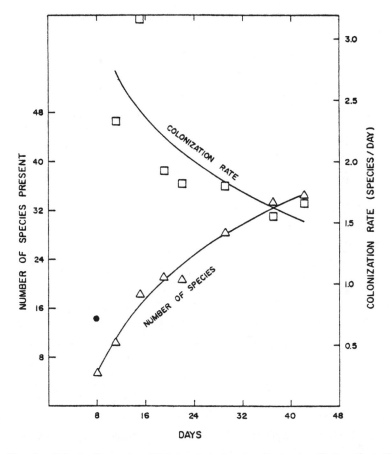

FIG. 1.—Colonization of artificial substrates by Protozoa. Notice that the
form of the curve is consistent with the general form predicted by theory
(MacArthur and Wilson 1967, fig. 21*A*).

in table 1. A simple exponential curve was fitted by regressing the mean
number of species on the logarithm of time in days; the hypothesis of
zero slope was strongly rejected at the .005 significance level ($F = 422.0$
with 1, 6 df). Thus, the colonization rate can be represented as a simple
exponential function of time. The data from day 8 were omitted in the
regression analyses of colonization rate because it appeared that a lag time
occurred during which bacteria and other microorganisms invade the
substrate making it "acceptable" for colonization by Protozoa. The same
model and fitting technique were also applied to the data for colonization
rate; again, the hypothesis of zero slope was rejected, but at the .10
significance level ($F = 5.72$ with 1, 5 df).

Figure 2 shows the relationship between extinction rate and time in
days. The simple exponential model was again tested with these data; the
hypothesis of zero slope could not be rejected ($P = .25$; $F = 1.68$ with 1,

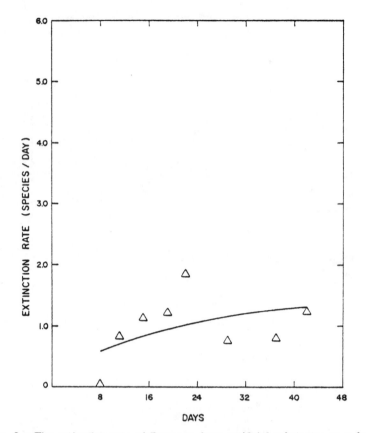

Fig. 2.—The extinction rate of Protozoa from artificial substrates as a function of time in days.

6 df). The data do not appear to deviate from the simple curvilinear relationships as depicted by MacArthur and Wilson (1967, fig. 21A).

Figures 3, 4, and 5 show the number of species at sampling locations 5, 8, and 12 in each of five ratings based on the approximate number of individuals per species as well as the percentage of species in each of the five categories. Sampling locations 5, 8, and 12 were used because these were the only ones with continuous observations throughout the study.

Another point of interest is the similarity and differences between species composition found at various locations. The following information indicates the nature of these differences:

Absent at location 5 but present at location 8 = 4 species
Absent at location 5 but present at location 12 = 1 species
Absent at location 8 but present at location 5 = 5 species
Absent at location 8 but present at location 12 = 5 species
Absent at location 12 but present at location 5 = 4 species
Absent at location 12 but present at location 8 = 6 species

Locations 5, 8, and 12 were the only ones with uninterrupted sampling.

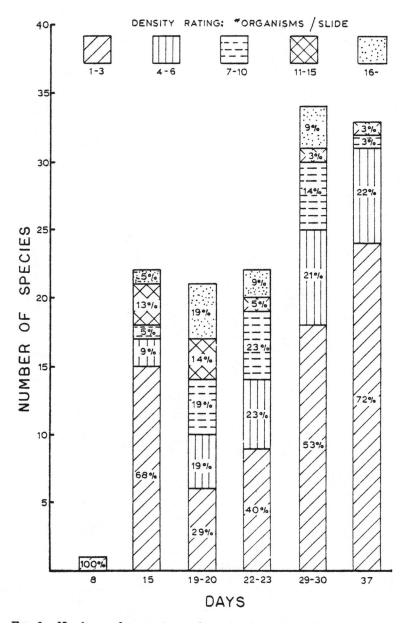

FIG. 3.—Number and percentage of species in each of five density-rating groups at sampling location 5.

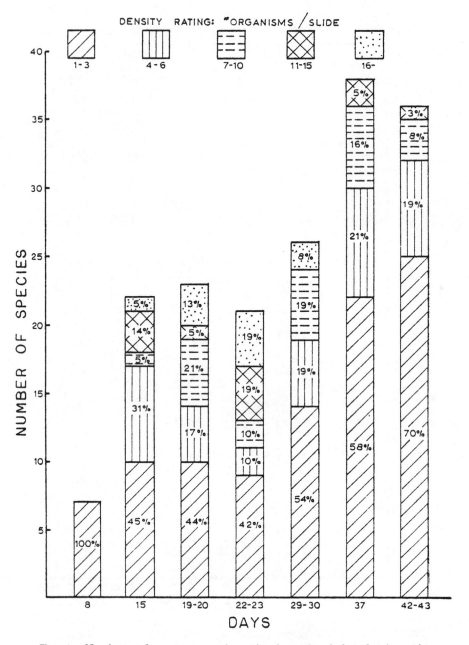

FIG. 4.—Number and percentage of species in each of five density-rating groups at sampling location 8.

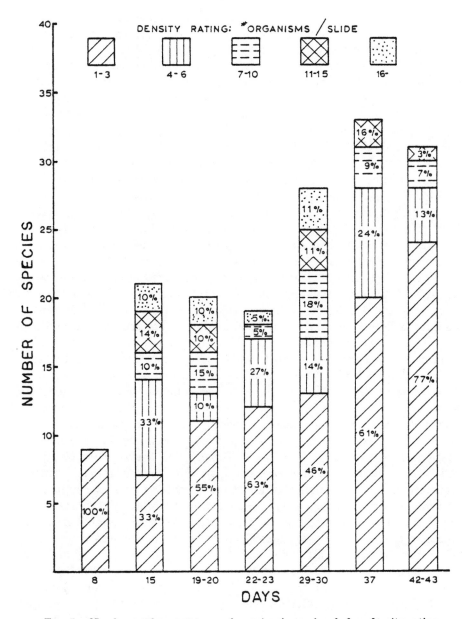

FIG. 5.—Number and percentage of species in each of five density-rating groups at sampling location 12.

DISCUSSION OF THE RESULTS

Probably the most striking feature of this study is expressed in figure 1. This colonization curve shows a good fit to a simple exponential function of time. As suggested by MacArthur and Wilson (1963), a new habitat has a high invasion rate of new species and a low extinction rate of those that become established. Invasion and extinction of species will continue until an equilibrium point is reached (in a relatively stable environment). In the early stages of invasion of a new substrate, the colonization rate greatly exceeded the extinction rate, but as the number of species increased, extinction and colonization rates approached equilibrium. The curves in figures 1 and 2 support this hypothesis for fresh-water protozoan communities; however, this may be a series of temporary invasion-extinction equilibrium points through time which become increasingly more stable. If one assumes a fairly constant invasion rate, then the increasing establishment of new species will be accompanied by a gradually increasing extinction rate for established species and a gradually decreasing colonization rate until a relatively steady-state equilibrium is reached. In this study, equilibrium was not achieved but only approached. Colonization rate at the end of 42 days was greater than extinction rate. However, the two rates were converging, as indicated by figures 1 and 2. Equilibrium and stability in terms of numbers of species is only obtained when colonization rate is equal to extinction rate. Apparently a stable community of organisms is able to maintain this balance unless exposed to unusual stress.

The number of new species and the number of species eliminated are given in table 3. It is interesting that the number of species eliminated is low at first and then increases, as one would expect from the model of MacArthur and Wilson (1963). However, this is followed by a decline in the percentage of species eliminated, with a secondary low for sampling locations 5, 8, and 12 on days 29–30. During this period there was a general increase in the number of species present, so that if the extinction rate were solely a function of the number of species present coupled with invasion pressure (which is here assumed to be relatively constant) one would expect a further rise in the percentage of species eliminated on days 29–30. In this case it appears that a relatively stable community was established temporarily at all three locations, followed by an "elimination" (i.e., species not observed) percentage that exceeded previous levels. At location 5 this peak occurred on day 37 and at locations 8 and 12 on days 42–43. There was not sufficient evidence to determine the cause of these changes, but they do appear to be a significant variation from the Mac-Arthur-Wilson model in that the extinction curve may not be a simple exponential as shown in figure 2.

The lack of a significant difference between sampling areas is not a startling result, but it does show that biological-environmental interactions appear to be comparable when two areas are similar. This suggests that the formation and composition of protozoan communities are the result of the same complex interactions that determine aggregations of higher organisms.

The results shown in figure 3 (sampling location 5) indicate that initial invasions are characterized by low numbers of individuals per species (e.g., at 8 days). This phase was followed by an increase in numbers of individuals per species of some species at 15 days, accompanied by a further increase in the number of species with few individuals. The two sampling periods following this (i.e., days 19–20 and 22–23) were characterized by an increase in the number of individuals per species without much change in the number of species present. This was followed by a second surge in the number of species present on days 29–30, caused primarily by an increase in species represented by a few individuals.

The results from sampling location 8 (figure 4) indicated nearly a comparable pattern. The pattern at sampling location 12 (figure 5) was comparable with location 5 in that the two major increases in the number of species occurred between days 8 and 15 and between days 22–23 and 29–30. Sampling location 12 differed from location 5 by having (except for day 15) an increase in the number of species with few individuals throughout the sampling period. Sampling location 8 seems to be somewhat intermediate between these two. These results indicated that density relationships may be quite different in developing communities with comparable numbers of species.

The fate of species that occur intermittently at a single sampling location was not clear. It is possible that some of these were present continuously but sometimes had densities below detectability with the methods used. Another possibility is that they were actually eliminated (or they encysted) and subsequently reinvaded the area; both possibilities are probably operative.

Each sampling location was to a certain degree unique in species composition. These differences could be the result of (1) sampling error, (2) microenvironmental differences between areas, or (3) qualitative or quantitative differences in immigration or extinction of species. Since sampling problems for very low density, fragile species are enormous, these apparent differences will not be easily resolved.

The results in table 2 suggest there is an upper asymptotic limit in diversity or number of species. That is, the numbers of species present in each area are remarkably similar at any one point in time despite different numbers of "surviving" substrates (i.e., different unit area for colonization).

CONCLUSIONS

The invasion of artificial substrates by fresh-water protozoans is a very complex process which in a general way appears to fit the model proposed by MacArthur and Wilson (1963, 1967). Favorable interactions among fresh-water protozoan species are strongly indicated in the experiments reported in this paper. Following an initial period of colonization, there was an increase in the number of individuals per species but no marked increase in the number of species. This was followed by a secondary increase in the number of species, possibly based on microenvironmental changes caused by the protozoans already inhabiting the area. If this hypothesis is correct, it would mean that presence of some species may enhance the suitability of the environment to some of the immigrant species. Since microorganisms may produce extracellular products (Fogg 1965) of use to other species, this is not at all unlikely. Obviously future studies should involve the influence of other organisms such as bacteria and algae upon the suitability of the substrate.

SUMMARY

Two series of 10 artificial substrates each were placed in Douglas Lake, Michigan, and their colonization by species of fresh-water protozoans studied. Identifications were made to species whenever possible at intervals of approximately one week, and rough estimates of density were made as well. Although the aggregations of species colonizing each of the substrates were not identical, the colonization process.itself was remarkably similar for the entire series. When the number of species was plotted against time in days, a simple exponential curve adequately described the relationship. Colonization rates and extinction rates were compared with the equilibrium model for island faunas proposed by MacArthur and Wilson. These results suggest that the formation and composition of protozoan communities on artificial substrates are the result of interactions comparable to those proposed by MacArthur and Wilson.

ACKNOWLEDGMENTS

This work was carried out while the senior author was a visiting professor at the University of Michigan Biological Station, Pellston, Michigan. We are indebted to the late Dr. Alfred H. Stockard, then director of the biological station, for his help and many courtesies during the course of the investigation. Dr. Robert H. MacArthur and Dr. Daniel Janzen offered many valuable suggestions during the preparation of this manuscript.

LITERATURE CITED

Berger, H. 1927. Beiträge zur Ökologie and Soziologie der luftlebigen (atmophytischen) Kieselalgen. Ber. Deut. Bot. Ges. 45:385–407.

Butcher, R. W. 1946. The biological detection of pollution. J. Inst. Sewage Purification 2:92–97.

Cairns, J., Jr. 1964. The chemical environment of common fresh-water Protozoa. Acad. Natur. Sci. Philadelphia, Notulae Naturae, no. 365. 6 p.

———. 1965. The environmental requirements of Protozoa, p. 48–52. In Biological Problems in Water Pollution, 3d seminar 1962, USPHS Pub. no. 999–WP-25 (abstr. p. 385–386).

Cairns, J., Jr., and W. H. Yongue, Jr. 1968. The distribution of fresh-water Protozoa on a relatively homogeneous substrate. Hydrobiologia 31(1):65–72.

Cooke, W. B. 1956. Colonization of artificial bare areas by microorganisms. Bot. Rev. 22:613–638.

Darwin, C. 1863. Yellow rain. Gardeners Chron. 1863 (July 18):675.

Fogg, G. E. 1965. The importance of extracellular products of algae in the aquatic environment, p. 34–37. In Biological Problems in Water Pollution, 3d seminar 1962, USPHS Pub. no. 999–WP-25.

Gislen, T. 1948. Aerial plankton and its conditions of life. Biol. Rev. 23:109–126.

Grzenda, A. R., and M. L. Brehmer. 1960. A quantitative method for the collection and measurement of stream periphyton. Limnol. Oceanogr. 5(2):190–194.

Hargraves, P. E., and R. D. Wood. 1967. Periphyton algae in selected aquatic habitats. Int. J. Oceanogr. Limnol. 1(1):55–66.

Huber-Pestalozzi, G. 1937. Das Phytoplankton des Süsswassers. Thienemann Binnengewässer 16:62–72.

Hudson, C. T. 1889. Presidential address, Royal Microscopical Society. Roy. Microscop. Soc., J., p. 169–179.

MacArthur, R., and E. O. Wilson. 1963 An equilibrium theory of insular zoogeography. Evolution 17:373–387.

———. 1967. The theory of island biogeography. Princeton Univ. Press, Princeton, N.J.

Maddox, R. L. 1870. On the apparatus for collecting atmospheric particles. Monthly Microscop. J. 3:286–290.

Maguire, B., Jr. 1963. The passive dispersal of small aquatic organisms and their colonization of isolated bodies of water. Ecol. Monogr. 33:161–185.

Maguire, B., Jr., and D. Belk. 1967. Paramecium transport by land snails. J. Protozool. 14(3):445–447.

Messikommer, E. L. 1943. Unterschungen über die passive Verbreitung der Algae. Schweiz. Z. Hydrol. 9:310–316.

National Research Council, Committee on Apparatus in Aerobiology. 1941. Techniques for appraising air-borne populations of microorganisms, pollen, and insects. Phytopathology 31:201–225.

Pasteur, L. 1860. Expériences relatives aux générations dites spontanées. Acad. Sci. (Paris), Compt. rend. 50:303–307.

Patrick, Ruth. 1961. A study of the number and kinds of species found in rivers in eastern United States. Acad. Natur. Sci. Philadelphia, Proc. 113(10):215–258.

Patrick, Ruth, M. H. Hohn, and J. H. Wallace. 1954. A new method for determining the pattern of the diatom flora. Acad. Natur. Sci. Philadelphia, Notulae Naturae, no. 259. 12 p.

Pouchet, F. 1860. Atomosphérique-Moyen de rassembler dans un espace infinitement petit tous les corpuscules normalement envisibles contenus dans un volume d'air déterminé. Acad. Sci. (Paris), Compt. rend. 50:748–750.

Schlichting, H. E., Jr. 1961. Viable species of algae and Protozoa in the atmosphere. Lloydia 24(2):81–88.

———. 1964. Meteorological conditions affecting the dispersal of air-borne algae and Protozoa. Lloydia 27(1):64–78.

Sramek-Husek, R. 1958. Die Rolle der Ciliatenanalyse bei der biologischen Kentrolle von Flussverunreinigungen. Verhandlungen Int. Verein Limnol. 13:636–645.

ZoBell, C. E., and E. C. Allen. 1933. Attachment of marine bacteria to submerged slides. Soc. Exp. Biol. Med., Proc. 30:1409–1411.

16

Reprinted from *Ecology* **49**:1091–1101 (1968)

THE RELATIONSHIP BETWEEN SPECIES DIVERSITY AND STABILITY: AN EXPERIMENTAL APPROACH WITH PROTOZOA AND BACTERIA[1]

N. G. Hairston,[2] J. D. Allan, R. K. Colwell, D. J. Futuyma, J. Howell, M. D. Lubin, J. Mathias, and J. H. Vandermeer

Department of Zoology, The University of Michigan, Ann Arbor, Mich. 48104

(Accepted for publication August 26, 1968)

Abstract. Small experimental communities of bacteria and Protozoa were designed to test the widely held hypothesis that higher species diversity brings about greater stability. Three species of bacteria, three species of *Paramecium* and two species of protozoan predators, *Didinium* and *Woodruffia*, were used. The communities were maintained by regular additions of the appropriate combinations of species of bacteria. Stability was measured as persistence of all species and as a tendency to maintain evenness of the species abundance distribution. The measures were in essential agreement. Stability at the *Paramecium* trophic level was increased by increasing diversity at the bacterium level, but three specis of *Paramecium* were less stable than two. An important finding was that one pair of *Paramecium* species consistently showed greater stability without the third species than with it. This finding indicates that there were significant second-order effects, with two species having an interaction that was detrimental to the third species. We conclude that much more experimental and observational work is necessary before the nature of any functional relationship between diversity and stability can be claimed with confidence.

INTRODUCTION

A widely held opinion is that increased species diversity leads to increased stability within ecological systems. This opinion has almost reached the status of an axiom since the demonstration of its theoretical validity by MacArthur (1955). We do not challenge MacArthur's conclusions

for the restricted conditions that he postulated and agree that they follow logically and mathematically from his premises. It is not clear, however, that the same formal properties would obtain in a more realistic community in which there is overlap in food species among different species of the same trophic level. Such an extension of MacArthur's conclusions would have to be reconciled with the equally logical and experimentally verifiable conclusions of Volterra (1926), Lotka (1932), and Gause (1934) concerning the outcome of competition.

[1] Research supported by National Science Foundation Grant GB 2364. We thank G. W. Salt for cultures of *Woodruffia*.
[2] Present address: Museum of Zoology, The University of Michigan.

We have not attempted a formal solution of the problem raised, but have made a direct experimental approach using communities built up from known numbers of species of Protozoa and bacteria.

Synthetic communities of this kind have the obvious disadvantage of impermanence. Against this disadvantage are set the advantages that they can be of manageable size and complexity, and that they permit replication and other benefits of experimental design which are difficult or impossible to achieve with natural systems.

Protozoan populations were followed in all of the experiments. We recognize the equal desirability of having information about the bacterial populations, but we chose to follow the Protozoa in detail instead of attempting to sample the bacteria, which would have reduced greatly the total amount of data because of the difficulties of sampling. This choice also permitted us to maintain the experimental communities by the regular addition of bacteria (see below).

Diversity and stability have intuitive meanings that are clear and distinct, but when sophisticated measures are attempted, certain difficulties of definition and of distinction arise.

Considering diversity first, most of the measures proposed have been for large collections from nature (Fisher, Corbet, and Williams 1943; Preston 1948; Simpson 1949; MacArthur 1957; Margalef 1957). For reasons which have been explained elsewhere (Hairston 1959, 1964), we reject indices of diversity which depend upon the assumption that the distribution of individuals among species follows a particular mathematical form. Of those remaining, the index of diversity devised by Simpson (1949) is the probability that any two individuals taken at random belong to the same species. Its only disadvantage is that combinations larger than two are ignored. More general are the statistics H, H', and H" derived from information theory (Margalef 1957; Lloyd and Ghelardi 1964; MacArthur 1964; Pianka 1966; Pielou 1966a, b, c):

$$H = (1/N)(\log N! - \sum_1^s \log N_i!)$$

$$H' = -\sum_1^s P_i \log P_i$$

$$H'' = -\sum_1^s (N_i/N) \log (N_i/N)$$

The notation used is that of Pielou, who points out the problems associated with equating H" with H'. In these equations, N is the total number of individuals, s is the total number of species, N_i is the number of individuals in the ith species, and

P_i is the proportion of the total number of individuals that belong to the ith species.

The indices are all related to the probability that any group of individuals contains different species. The formulae are conventionally stated to express the uncertainty regarding the species of any individual selected at random. They have come to be widely accepted as measures of diversity. A serious problem with the use of information statistics as a measure of diversity in our work is the close resemblance between them and the definition of stability (S) by MacArthur (1955). The formulations are identical in the case of H', the only difference being that in the diversity index P_i is the proportion of all individuals that belong to the ith species, while in the formula for stability P_i is the proportion of total energy following one of a series of converging trophic paths through a community. Moreover, since both measures are strongly influenced by the total number of species present, they cannot be considered independent variables. Thus it appears to us to be circular to use H' or H" to establish diversity and MacArthur's S to measure stability, even assuming that S can be estimated in any practical way. In addition, if it were possible to evaluate the relative magnitude of energy flow through the respective converging paths in a food web, a good measure of the complexity of the system could be obtained, but unless all of the assumptions listed by MacArthur are met, the measure would not necessarily be one of stability. We have therefore used the word stability in the ordinary sense of ability to withstand change.

The exact abundance of each species in each community at its theoretical stable state is of course not known. In the absence of this knowledge, each community was begun with equal numbers of each species. The observations indicated that a rapid departure from equality of numbers resulted in early elimination of species from the community. We consider a slow change in the relative abundance of species to represent greater stability than a rapid change. In our experiments, then, more nearly equal abundances of species at any given time represent greater stability than do abundances that are widely different.

Accepting change in relative abundance as an inverse measure of stability means that our only independent measure of diversity is that which was established initially by experimental design. Essentially, this is the number of species introduced into the experimental communities.

One further point requires clarification. In some sense, "abundance" is qualified by the sizes of the organisms. Events in the experimental communities being at least partly determined by the relative degrees to which the different species exploit their environment, relatively few members of a large species may be considered to have utilized as large a part of the resources as large numbers of a smaller species. In order to compensate for this factor, we have cast the data in terms of equal volumes of the different species, in addition to making calculations based on the actual counts.

The choice among information formulae seems to depend more upon the nature of the data than upon the kind of interpretation that we wish to make (Margalef 1957; Pielou 1966 a, b, c). The ecological literature is not entirely clear on this point, but for data consisting of numbers of whole organisms, the parameter H is preferred over its estimators, whereas for data that are more nearly continuous, such as units of biomass, H', as estimated by H", seems more appropriate from a theoretical standpoint. The numerical solutions are not greatly different, but H' is always greater than H for the same set of data. Therefore, we have used H" to estimate diversity in the analysis in which changes in the relative volumes of the *Paramecium* species provide the data. The parameter H has been used as the measure of diversity in the analysis of changes in relative numerical abundance.

METHODS

Techniques

The experimental communities consisted of one to three species of bacteria and one to three species of *Paramecium;* in some cases, one or two species of predaceous Protozoa were added. The bacteria were *Aerobacter aerogenes,* hereafter designated as A, and two unidentified bacilliform species isolated from a natural habitat. Both of the unidentified species, referred to below as B2 and B3, are larger than A, and B3 grows in chains, forming a loose net-like mat in the bottom of the depressions of the spot-plates used.

Paramecium species were selected for ease of visual identification. They were Varieties 3 and 4 of *Paramecium aurelia* and an unidentified variety of *Paramecium caudatum.* Henceforth, the three species of *Paramecium* will be referred to as Pa3 and Pa4, and Pc respectively. They are easily separated on the basis of size, being approximately 150 μ, 110 μ, and 250 μ long, respectively, and there are other morphological characters that distinguish them from each other. The two species of predators were *Woodruffia metabolica* and *Didinium nasutum.*

Bacteria were maintained on nutrient agar slants; bacterial cultures for maintaining para-

mecia and for the experimental communities were grown in an extract of Cerophyl (1.5 g/liter H₂O). Mixtures of bacterial species were prepared by adding together equal parts of Cerophyl cultures saturated with the different species.

Stocks of *Paramecium* were maintained in test tube cultures (Sonneborn 1950), individuals being isolated from the stocks for the initiation of experimental communities. Each community was started with four animals of each of the species of *Paramecium* called for in that community by the experimental design. For each species of *Paramecium* to be introduced, an initial volume of 0.1 ml of bacterial culture was first measured into a depression. Additional bacteria were added to the communities at the rate of 0.05 ml of the appropriate culture on alternate days throughout the experiment. There was thus a tendency for the composition of the lowest trophic level to be restored periodically.

In each community involving predators, the rest of the species called for were started 2 days before the addition of the predators; two *Didinium* and/or four *Woodruffia* were then added, depending upon the experiment.

All experimental communities were established in the deep depression slides (spot-plates) that are standard in much protozoological work (Sonneborn 1950). The capacity of a depression is approximately 1.0 ml. The experiments were carried out at a constant temperature of 25°C.

Experimental design

The species of the two lower trophic levels (three species of bacteria and three of *Paramecium*) were put together in all 49 combinations having at least one species of each trophic level. Communities containing only one species of *Paramecium* were observed in a different way from those containing two or three species, since short-term stability of a single *Paramecium* population is not particularly relevant to the question posed. In the case of these communities, observations were made of daily division rates and of the carrying capacity of a unit volume of bacterial culture. Daily isolation lines were maintained over a 4-day period. Each of the 21 possible combinations of one to three bacterial species with one species of *Paramecium* was replicated three times by each of five observers. There were thus 60 observations of the daily division rate for each set of conditions. Estimates of the carrying capacity of the bacterial cultures were obtained by establishing 12 cultures for each condition; 12 paramecia were placed in 0.1 ml of each bacterial culture. Three of these cultures were sacrificed on each of

4 successive days, all paramecia being counted. The highest daily average was considered to be the maximum carrying capacity for 0.1 ml of that bacterial culture.

The 28 communities established with 2 or 3 species of *Paramecium* (plus one to three species of bacteria) were set up in sufficient replicates to be observed four times each after 2, 6, 10, 14, and 20 days. Two replicates were observed on each occasion by each of two observers. For four of the combinations, the results were soon obvious, and only two replicates were observed; the effort saved was used to add extra replicates for combinations in which the outcome was especially doubtful. Combinations with only two replicates were Pa4 and Pc with A, AB3, B2B3, and AB2B3. Combinations with six replicates were Pa3 and Pa4 with AB2, AB3, and AB2B3, and Pa3 and Pc with B2B3.

For each observation, a volumetric aliquot of 0.05 ml was removed after thorough mixing of the contents of the depression. All paramecia in the aliquot were then counted and sorted. Early in the experiment, total counts were made of some communities. The process of counting destroyed the replicate. Therefore, sufficient replicates were started to allow all observations to be made on independent communities. Laboratory accidents prevented observations in 23 replicates out of 560 (4.1%). As might be expected, it was not always the case that data from the different replicates showed homogeneity of results. Of the 138 different sets of observations, 20 (14.5%) showed significantly different proportions of the *Paramecium* species between replicates, as compared with an expected number of seven cases (5%). Differences between observers accounted for 14 of the 20 cases; differences between the replicates of single observers accounted for six. There is no internal evidence that the 20 cases upset the overall results in any way.

Communities including predaceous Protozoa were established on a limited basis. Only 21 out of the 147 possible combinations were set up. Each predator species and the combination were established with Pa4 and Pc on A, AB3, AB2B3; with Pa3 and Pa4 on AB3; and with Pa3, Pa4, and Pc on A, AB3, and AB2B3. These communities were very impermanent, as will be described below. Counts were made on them after 1, 2, 5, 9, and 13 days from the time the predators were introduced.

RESULTS

Growth rates and carrying capacities

Each species of *Paramecium* showed a characteristic set of daily division rates on the different

TABLE 1. Growth rates (as average number of divisions of *Paramecium* per day) of the three species of *Paramecium* on seven different bacterial foods. All experiments conducted at 25°C

Paramecium	Bacteria						
	A	B2	B3	AB2	AB3	B2B3	AB2B3
Var 4......	3.28	3.22	3.02	3.33	3.44	3.30	3.21
Var 3......	2.25	2.75	2.67	2.59	2.53	2.75	2.58
P.c.........	2.03	2.19	1.98	2.25	2.24	2.34	2.23

bacterial cultures (Table 1). There is a marked negative relationship between daily division rate and size. Analysis of variance of the detailed data reveals that the differences between the species are highly significant ($F_{2, 1050} = 341$), as are the differences between different bacterial foods ($F_{6, 1050} = 5.42$) and the interaction between *Paramecium* species and bacterial foods ($F_{12, 1050} = 2.44$). All probabilities are less than 0.01. The interaction comes largely from the inverse effect of A on the division rates of Pa3 and Pa4. Pa4 grew more rapidly in the cultures containing A than in those from which A was absent; exactly the reverse was true for Pa3. It was also observed that Pa3 and Pc were favored by the presence of B2, whereas Pa4 was not.

There did not appear to be a strong relationship between growth rate and the number of species of bacteria used. All three species of *Paramecium* grew more rapidly on two species of bacteria than on one or three.

A large amount of variation was observed in the carrying capacity (K) for each species of *Paramecium* on the different bacteria and their combinations (Table 2). Variance analysis of the

TABLE 2. Saturation levels for three species of *Paramecium* in 0.1 ml of each of seven different bacterial foods at 25°C

Paramecium	Bacteria						
	A	B2	B3	AB2	AB3	B2B3	AB2B3
Pa4........	162	329	174	126	280	271	319
Pa3........	118	166	84	147	106	133	137
Pc.........	33	52	47	40	61	62	64

data shows differences between saturation levels of the different species to be highly significant ($F_{2, 30} = 532$), as are differences within each of the three species on different bacterial foods ($F_{4, 30} = 213$). The interaction between species of *Paramecium* and food is also highly significant ($F_{8, 30} = 33$). Although there is a tendency for the carrying capacity for all three species to increase with the number of species of bacteria supplied as food, the correlation is significant only for Pc ($r = 0.72$, $N = 15$). The information obtained from the single species cultures is highly relevant to our general problem of the relationship between stability and diversity, since it shows that the three species of *Paramecium* were quite different in their utilization of the different bacteria.

It is worth noting that no significant relationship can be demonstrated between mean division rate and saturation value for any of the three species of *Paramecium*.

COMMUNITIES WITH TWO TROPHIC LEVELS

Trends

The *Paramecium* populations in the various experimental communities changed in general agreement with expectations based on the assumption that the species were in competition for a common resource. In every case, one species increased in abundance, while the one or two other species eventually decreased relative to the first (Table 3). In 12 of the 28 kinds of communities, the less successful species had disappeared from all replicates by the 20th day. Irregularities in the recorded data result from the fact that complete counts were made of some communities early in the experiment. Thus attention should be focused more on the relative abundances than on the absolute numbers, especially for days 2 and 6. Adjustments can be made, of course, because total populations can be calculated from the aliquots. Statements of statistical probability, however, require the use of the actual counts, as given in Table 3.

The rate of change in proportions of the different species shown in Table 3 is exaggerated by the fact that it was the smallest species that always showed an increase. If consideration is given to the proportion of the resource that has been acquired, the larger and less abundant species should be given more weight than is apparent from the numerical abundances alone. This point will be pursued further below.

The effect of a lower trophic level on persistence

Although there might be some debate over any criterion chosen to measure stability, it will probably be agreed that the complete loss of a species from a community indicates lack of stability. The number of individual communities from which species were lost therefore provides an acceptable basis for comparison of stability under the different experimental conditions.

The data for the 20th day in Table 3 represent totals for 110 different replicates (two replicates of P3 and Pc on B2 were lost). In addition to the 12 sets in which one or more species disappeared from all replicates, there were eight sets in which species were lost from some of the repli-

TABLE 3. The changes in abundance of *Paramecium* species in the experiments with two trophic levels. The numbers represent the sums of counts from all replicates. (Species of Paramecium are shown in italics)

Days after start	A		B2		B3		AB2		AB3		B2B3		AB2B3	
	Pa4	*Pa3*	*Pa4*	*Pa3*	*Pa4*	*Pa3*	*Pa4*	*Pa3*	*Pa4*	*Pa3*	*Pa4*	*Pa3*	*Pa4*	*Pa3*
2	112	58	62	36	68	15	235	84	354	95	208	41	274	116
6	103	79	228	117	166	25	339	144	371	177	340	46	418	119
10	70	29	131	39	163	28	298	121	263	109	391	54	388	122
14	73	30	191	41	49	0	213	59	142	61	181	30	196	68
20	122	13	200	1	364	12	232	49	166	74	212	21	174	74

Days after start	A		B2		B3		AB2		AB3		B2B3		AB2B3	
	Pa4	*Pc*	*Pa4*	*Pc*	*Pa4*	*Pc*	*Pa4*	*Pc*	*Pa4*	*Pc*	*Pa4*	*Pc*	*Pa4*	*Pc*
2	43	8	154	33	58	23	53	26	47	7	38	13	132	9
6	110	1	252	14	237	19	311	6	301	12	157	18	320	10
10	115	2	369	6	340	4	340	6	185	6	198	10	271	1
14	160	0	274	0	217	3	400	1	210	5	128	2	No data	
20	157	0	417	1	277	0	404	0	176	0	176	0	263	0

Days after start	A		B2		B3		AB2		AB3		B2B3		AB2B3	
	Pa3	*Pc*	*Pa3*	*Pc*	*Pa3*	*Pc*	*Pa3*	*Pc*	*Pa3*	*Pc*	*Pa3*	*Pc*	*Pa3*	*Pc*
2	57	25	43	32	60	18	42	60	31	28	60	68	102	40
6	44	28	128	46	68	15	52	23	54	44	115	133	171	32
10	60	29	92	24	86	18	53	5	95	46	112	76	119	33
14	74	7	90	38	79	6	52	3	72	7	148	110	167	34
20	49	1	44	10	217	23	68	4	112	0	190	49	202	13

Days after start	A			B2			B3			AB2			AB3			B2B3			AB2B3		
	Pa4	*Pa3*	*Pc*	*Pa4*	*Pa3*	*Pc*	*Pa4*	*Pa3*	*Pc*	*Pa4*	*Pa3*	*Pc*	*Pa4*	*Pa3*	*Pc*	*Pa4*	*Pa3*	*Pc*	*Pa4*	*Pa3*	*Pc*
2	105	11	23	282	48	49	95	15	33	106	46	19	117	42	25	98	15	32	158	31	13
6	141	14	10	251	8	16	130	11	16	156	45	22	188	62	28	234	31	11	241	84	22
10	94	0	3	246	19	27	217	6	8	155	39	18	190	45	1	199	20	9	142	58	12
14	278	0	1	202	21	2	240	1	0	210	19	12	111	33	1	197	17	8	No data		
20	194	0	0	379	17	0	261	1	0	224	23	7	164	26	0	200	28	6	91	58	0

292

cates, while remaining in low numbers in the others. At least one species remained numerous in every replicate examined.

In counting the number of replicates from which a species of *Paramecium* had disappeared, each one starting with three species was counted twice, there being the possibility that two species could have disappeared. The total number of possible disappearances was therefore 138; 68 of these were realized by the 20th day of the experiment. The distribution of these losses among the communities with one, two, and three species of bacteria is of special interest here. The data show a clear relationship between the persistence of *Paramecium* species and the number of species of bacteria present in the community. Among the 56 possible cases containing one species of bacteria, 18 (32.1%) showed persistence of the *Paramecium* species of less abundance; among the 62 possible cases with two species of bacteria, 38 (61.3%) showed this persistence; and among the 20 possible cases with three species of bacteria, the *Paramecium* species of less abundance persisted in 14 (70.0%). The differences are highly significant; $\chi^2 = 13.478$ with 2 DF ($P < 0.01$). An increase in diversity at the lower trophic level therefore did lead to an increase in stability in the upper trophic level, at least in the sense that all species persisted longer.

The effect of other species at the same trophic level on persistence

The same data can be analyzed for the relationship between persistence of the species of *Paramecium* and the number of species of *Paramecium* with which the community was started. The question being asked here is whether or not an increase in diversity in one trophic level brings about an increase in stability at the same trophic level. On the 20th day, 47 of 82 replicates (57.3%) showed persistence of both species when only two were used. In the community started with three species of *Paramecium*, 23 cases of persistence were recorded out of 56 possible (41.1%). The difference is not significant statistically ($\chi^2 = 2.895$, $0.05 < P < 0.10$). Hence we reject the hypothesis that increasing diversity increases stability. Indeed, on the 14th day, persistence was significantly more frequent in the 2-species communities than in the 3-species communities ($\chi^2 = 11.85$, DF = 1).

The persistence of a higher trophic level (*Paramecium*) was directly related to the species diversity of a lower trophic level (bacteria), but inversely or not at all related to its own diversity. Consequently, a complete absence of any relationship between stability and diversity results when the initial diversity of both trophic levels is considered simultaneously.

Changes in diversity

We have already stated our reasons for maintaining that a slow change in diversity represents a more stable community than does a rapid change in diversity. The numerical abundances of the species of *Paramecium* in each community can be used to calculate H as defined in the Introduction

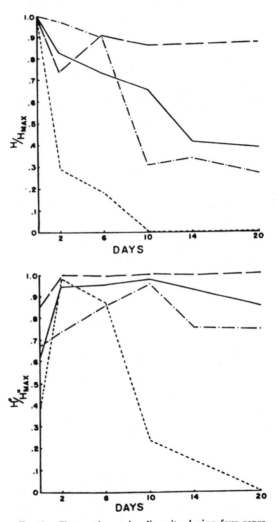

FIG. 1. Changes in species diversity during four representative experiments. Upper graph based on numerical abundance; lower graph shows results after correcting for size differences between the species. Combinations of *Paramecium* and bacterial species shown as follows: solid lines, Pa3 Pa4 Pc on AB_2; broken lines, Pa3 Pa4 on AB_3; dotted lines, Pa4 Pc on AB_2B_3; dot-dash lines, Pa_3 Pc on AB_2.

(above). This parameter, however, varies with both N, the total number of individuals in the system, and S, the total number of species. Inasmuch as the different communities were set up with different numbers of species, direct comparisons of changes in H would be difficult to evaluate. In order to make the figures easily comparable, H for each observation was made a fraction of the maximum value that it could have had, given the total numbers of individuals and of species which were actually present. Changes in this value, H/H_{max}, can then be compared among the different experiments, because all experiments were started with H equal to H_{max}. Graphs made with lines connecting the successive values of H/H_{max} describe the course of events in the communities (upper graph of Fig. 1). The slopes of these lines would provide a suitable basis for comparing the stability under different initial conditions, except that the calculations would require an assumption of linearity, which probably does not hold here. We have avoided this problem by including in the computer program for calculating H/H_{max} a calculation of the area under the line on each graph. The slower the descent of the line from its initial value of 1.0, the greater the area under the line and the more stable the community. The areas for all experiments are given in Table 4. The figures reflect the whole history of the experiments more accurately than does simple per-

TABLE 4. The areas under the lines connecting calculated values of H/H_{max} for each experiment

Species of bacteria used	Combination of *Paramecium* species			
	Pa4-Pa3	Pa4-Pc	Pa3-Pc	Pa4-Pa3-Pc
A.	16.72	3.21	13.13	5.26
B₂.	13.84	4.72	16.64	7.91
B₃.	8.04	5.84	11.23	6.33
AB₂.	16.07	4.89	11.21	12.24
AB₃.	17.27	4.95	13.78	11.57
B₂B₃.	11.24	6.75	18.93	9.43
AB₂B₃.	16.44	2.63	12.86	13.78
Avg.	14.23	4.71	13.97	9.50

sistence, and results are not identical. Two species of bacteria gave greater stability than one in all experiments, but adding a third species of bacteria resulted in a further increase in stability only for the experiments combining Pa4 with Pa3 and Pa4 with Pa3 and Pc. Examination of the details in Table 4 leads to the suggestion that the presence of Bacterium A decreased the performance of Pc relative to both Pa4 and Pa3. Thus, when Bacterium A was present, even with the other two

species, the stability of the system tended to be low.

The figures in Table 4 confirm the conclusion that adding diversity at the *Paramecium* trophic level did not increase the stability of the experimental community. Indeed, the average stability for all experiments with three species of *Paramecium* was lower than the average for two species.

Changes in diversity after correcting for biomass

The three species of *Paramecium* were of different sizes, as pointed out under Methods. We have used an inexact but satisfactory method of expressing the abundance of the larger species in terms of an equal volume of the smaller. The method used was to take the ratios of the cubes of the average lengths, as given under Methods above. The implied assumption that the different species are exactly the same shape is not completely valid, but the differences, although recognizable (Hairston and Kellermann 1965) are not great enough to upset the calculations in any serious way. Pc is found to be 4.63 times as large as Pa3 and 11.74 times as large as Pa4; Pa3 is 2.54 times as large as Pa4. These factors were applied to the calculated total populations in each community, giving figures that express the respective volumes in terms of the smallest species in the community. The estimate of diversity H″ was then calculated for each set of observations.

Comparisons of the values of H″ are difficult for experimental communities that were started with different numbers of species. Therefore we have calculated the maximum value that H″ could have (H''_{max}) for the number of species with which the community started. The ratio H''/H''_{max} was then used in all comparisons. The approximate range of results obtained is shown in the lower graph of Figure 1. Typically, diversity rose to a peak by the 10th day and declined thereafter with varying rapidity. In a few cases, the communities would be rated completely stable up to the 20th day, inasmuch as no appreciable decline was observed.

The typical pattern is readily explained in terms of competition and of the manner in which the data were handled. Starting with equal numbers of all species of *Paramecium* means that the value of H″ was at less than its maximum possible value, because the larger species were given more weight in the calculations than were the smaller species, and H''_{max} is found when all species are equally abundant. As already noted, the smaller species always won, or showed a tendency to do so, and as it increased in relative abundance, the value of H″ increased also. It is obvious from this consideration that for the two-species communities the value of H''/H''_{max} must have been 1.0 at some instant, whether observed or not, and that, although not necessarily true for the three-species communities, the ratio must have approached 1.0 closely for them as well. After the point was reached where the two or three species present filled approximately equal volumes, the competitive process continued, with the larger species decreasing relative to the smaller. Thus, the value of H'' tended to decline after the peak was reached.

The increases in H''/H''_{max} shown in Figure 1 thus create an impression of a trend toward greater stability that is spurious because the upward trend reflects the same ecological process as the downward trend. The difference in initial values of H''/H''_{max} is caused by differences in relative size in the various combinations of *Paramecium* species. This factor partially invalidates comparison of the upward trends, and in this analysis we have concentrated on the downward trends after the peak had been reached. Quantitatively we express these trends as the slope of the linear regression of H''/H''_{max} against time, ignoring points at times before the occurrence of the maximal value of H''/H''_{max}. These slopes for the different experimental communities provide an acceptable basis of a comparison of stability following the initially established diversity. Because it is negative the slope is directly proportional to stability, as can be visualized from our preceding arguments.

Table 5 gives the slopes for all 28 experimental communities. If the results from all combinations

TABLE 5. Slope of linear regression of H''/H''_{max} on time excluding those values occurring before the time of maximal value of H''/H''_{max}

Bacteria used	Combination of Paramecium species				Avg. ($\bar{x}$)	Grand avg ($\bar{x}$)
	Pa4-Pa3	Pa4-Pc	Pa3-Pc	Pa4-Pa3-Pc		
A	−.0422	−.0518	−.0612	−.0662	−.0553	
B₂	−.0452	−.0538	−.0004	−.0283	−.0317	−.0430
B₃	−.0356	−.0347	−.0077	−.0875	−.0414	
AB₂	−.0038	−.0596	−.0194	−.0124	−.0238	
AB₃	.0001	−.0507	−.0946	−.0101	−.0388	−.0267
B₂B₃	−.0130	−.0741	+.0158	+.0008	−.0176	
AB₂B₃	+.0002	−.0572	−.0142	−.0299	−.0252	−.0252

of *Paramecium* are pooled, the mean slopes for experiments on different numbers of bacterial species are interesting and highly relevant to our central problem. For communities with one species of bacteria the average slope is —0.043; with two species of bacteria, the average slope is —0.027; and with three species of bacteria, it is —0.025.

It thus appears that stability as defined here

increases with diversity, when diversity is considered only as the number of species of bacteria. Again the difference is detectable only between one and more than one bacterium, the difference between two and three species of bacteria being negligible.

The experiments with two species of *Paramecium* showed wide variations in average slope of H''/H''_{max} with time. The seven experiments with Pa4 and Pa3 gave an average slope of —0.0199; those with Pa3 and Pc gave an average slope of —0.0258, and those with Pa4 and Pc gave an average slope of —0.0545. The experiments with all three species of *Paramecium* gave an average slope of —0.0319, which clearly does not demonstrate increased stability over the two-species communities. Therefore we also conclude from this analysis that increasing diversity does not increase stability at the same trophic level.

The effect of a third species of Paramecium on competition

The experimental communities can be thought of as being three sets of experiments, each set consisting of a comparison of competition between two species of *Paramecium* in the presence and absence of a third species. These comparisons provide the opportunity to make direct statistical tests.

We set up the null hypothesis that for any day in any bacterial food the proportions of, for example, Pa4 and Pa3 are the same in the two-way experiment as in the presence of Pc. If increasing diversity increases stability, it should frequently be the case that the two species in the three-way competition were in a significantly more equal proportion than they were in the two-way competition. If diversity has no effect on stability, at the same trophic level, the null hypothesis should be accepted in about 100 of the 105 possible tests. If increasing diversity decreases stability, as suggested by the analysis of persistence, there should be many experiments in which the pair of species in two-way competition were in a significantly more equal proportion than they were with the third species present.

The results of the 105 chi-square tests are summarized in Table 6. At least one expected number was too low in 32 comparisons for the test to be valid. In only 36 of the remaining 73 cases was the null hypothesis accepted. Therefore, we can state with confidence that the competitive relationships were significantly different with two species alone as compared with competition in the presence of the third species.

Detailed examination of the results of this analy-

TABLE 6. Summary of chi-square tests of proportions of each pair of *Paramecium* species alone and in combination with the third species. Rejections are at the 0.05 level at least

Day of observation	Number of tests			
	Rejecting null hypothesis		Accepting null hypothesis	Expected numbers too low for valid test
	More equal abundance in 2-way competition	More equal abundance in 3-way competition		
2.........	5	3	13	
6.........	5	6	10	
10.........	5	4	6	6
14.........	3	1	4	13
20.........	4	1	3	13
Totals......	22	15	36	32

sis shows that the combination of Pa4 and Pa3 was responsible for 18 of the 22 comparisons showing a significantly more uniform abundance in the two-way than in the three-way competition experiments. By contrast the combination of Pa3 and Pc was responsible for 9 of the 15 comparisons in which the pair showed a significantly more uniform abundance in the three-way competition experiments. Of the remaining six such comparisons, five were for the combination Pa4 and Pc. Thus different combinations of species gave very different results.

Adding Pc to the combination of Pa4 and Pa3 has the effect of decreasing the performance of Pa3 relative to Pa4, in spite of the fact that Pa3 increased at the expense of Pc in all experiments. Adding Pa4 to the combination of Pa3 and Pc increases the performance of Pc relative to Pa3.

The analysis reveals important second order effects. There is evidently an interaction between Pa4 and Pc that is detrimental to Pa3. The mechanism by which this result is obtained is not obvious.

The most important conclusions from this analysis are that the effect of adding diversity at the same trophic level is specific to the particular combination of competing species, and that the relationship between stability and diversity at the same trophic level cannot yet be predicted on *a priori* grounds.

COMMUNITIES WITH THREE TROPHIC LEVELS

When predators were added to the experimental communities the commonest result was rapid and complete destruction of the system. Among the 50 replicates observed, only three contained any Protozoa 13 days after the introduction of

TABLE 7. Average duration in days of *Paramecium* and predators in experimental communities of different initial composition

No. of species of predators	Duration of	No. of species of *Paramecium*	
		2	3
1...............	*Didinium*	4.44	3.30
	Paramecium	1.44	1.57
	Woodruffia	5.60	5.20
	Paramecium	8.60	9.00
2...............	*Didinium*	4.33	3.71
	Woodruffia	3.18	2.86
	Paramecium	2.40	3.57

the predators. All of these were experiments in which the predators had died, and the remaining Protozoa were Pa4.

The data are best represented by tallying the day on which each species was last observed and calculating the average number of days over which each species persisted for a given set of replicates. The results are summarized in Table 7. The diversity among bacteria has been omitted, as the number of species had no consistent effect on the persistence of the Protozoa. Differential persistence of the species of *Paramecium* was similar to that found without predators, in that Pa4 was the last to disappear. "*Paramecium*" in the table thus refers to Pa4, the most persistent species.

Only two of the differences shown by inspection of Table 7 can be shown to be statistically significant. *Paramecium* persisted significantly longer in the presence of *Woodruffia* than with *Didinium*, and *Woodruffia* persisted significantly longer alone than it did in competition with *Didinium*.

The apparent trends that would be relevant to our central problem are not significant statistically. That is, the paramecia cannot be shown to persist longer when three species were used than when two were used, and having only one predator does not lend any more stability to the paramecia than having two predator species.

DISCUSSION

A theory of the determination of the stability of natural communities is one of the primary objectives of community ecology as a science, and is rapidly becoming a focus of attention in a number of historically distinct subdisciplines within ecology and biogeography (Watt 1968). The hypothesis of a positive relation between the stability of a community and its trophic complexity is an old one, originating in nonrigorous observational and anecdotal evidence, but which has repeatedly been invoked, and affirmed as valid, in the ecological and evolutionary literature. Precise definitions of the terms "stability" and "complexity" are seldom used, however, and the frequent vagueness of these discussions suggests a reliance by their authors more on the intuitive feeling that this positive relationship must hold, than on concrete empirical demonstrations of its validity. Appeal to MacArthur's (1955) formalization of this concept seems frequently to serve as a reassurance that the vague verbal argument does not entirely lack a theoretical basis.

It must be recognized, however, that Mac-Arthur's model, like any model, is no stronger in its predictive ability than is allowed by its assumptions. MacArthur points out that the model is based on several unrealistic simplifications, which are probably unrealistic enough to matter. It should be noted, moreover, that the model fails to consider a variety of topologically different food webs, including the highly realistic ones in which more than one species exists at the highest trophic level. MacArthur's model may well be an important component of future, more comprehensive models of community dynamics, but in its present form is arguable on several bases, some of which we have noted.

It is not our intent to deny the possibly real relationship between stability and trophic complexity in natural communities. It would obviously be unreasonable to argue strongly from highly artificial, simplified, multispecific systems in the laboratory to even the simplest natural communities. We maintain, however, that acceptance of this important hypothesis must have some basis in empirical evidence, and that little rigorous evidence exists at the present time. Our evidence suggests that stability on a higher trophic level is increased by diversity on a lower trophic level, but that stability is not predictable from consideration only of the number of energy pathways in the system. Hence these model systems cannot be said to support any glib statements to the effect that increased species diversity results in increased community stability. Watt (1964) has examined the relationship between fluctuations in populations of forest insects and the number of locally sympatric species of insects, and has failed to find a positive relationship between stability and diversity. Patten and Witkamp (1967), in analyzing the dynamics of the flow of cesium in laboratory systems, found no evidence that increasing the complexity of their systems increased the stability of cesium concentrations in the systems' compartments. It would, then, seem premature to accept this hypothesis with any degree of confidence.

The formulation of predictive models in ecology has been hampered by the conflicting requirements of generality and realism (Holling 1966). It is clear that the construction of models of the internal dynamics of communities will have to take into account, to at least some degree, the precise nature of the relationships among species. In our experimental communities, interspecific competition seems to have been the most important determinant of the change in state of these systems. The increase in stability with an increased number of bacterial species seems most reasonably explained by a reduction in competition among the *Paramecium* species, by virtue of trophic specialization. It seems likely that this kind of interspecific interaction must be described and incorporated into a model for community stability if such a model is to be adequately predictive of events in either experimental or natural communities.

Assumptions such as linearity, and the insignificant contribution of higher order interaction terms to the variance in dependent variables are often found convenient in constructing mathematical models. Such assumptions, however, often prove invalid (Slobodkin 1965). The nonadditive effect of *Paramecium caudatum* and *P. aurelia* var. 4 in suppressing *P. aurelia* var. 3 is a case in point, which, by virtue of the paucity of studies on competitive interactions among more than two species, has few parallels in the existing literature. A comparable case in warblers (*Dendroica*) has been described by MacArthur (1968), and the relative fitness values of karyotypes in *Drosophila pseudoobscura* have been shown to depend on the presence or absence in the population of a third chromosome sequence (Levene, Pavlovsky, and Dobzhansky 1954). The extent to which such seemingly exceptional phenomena must be accounted for in general theory is of course difficult to predict (Lewontin 1967), but the frequent violation of mathematical theory by biological peculiarity suggests that models must make allowance for such effects (Slobodkin 1965). It would be surprising for interaction effects to be found in such seemingly simple organisms as ciliates, yet not be significant in complex natural communities.

Recent ecological research has shown a growing tendency to test hypotheses not only by observational methods that rely on correlation to derive statements about causation, but upon the more powerful methods of experimental manipulation as well. Such manipulation, both in model laboratory systems and in the field, will probably prove necessary if the complexities of communities are to be resolved into a set of more elementary generalities.

297

LITERATURE CITED

Fisher, R. A., A. S. Corbet, and C. B. Williams. 1943. The relation between the number of species and the number of individuals in a random sample of an animal population. J. Animal Ecol. 12: 42–58.

Gause, G. F. 1934. The struggle for existence. Williams & Wilkins, Baltimore. 163 p.

Hairston, N. G. 1959. Species abundance and community organization. Ecology 40: 404–416.

———. 1964. Studies on the organization of animal communities. J. Animal Ecol. 33 (Suppl.): 227–239.

Hairston, N. G., and S. L. Kellermann. 1965. Competition between Varieties 2 and 3 of *Paramecium aurelia*: The influence of temperature in a food-limited system. Ecology 46: 134–139.

Holling, C. S. 1966. The strategy of building models of complex ecological systems, p. 195–214. *In* K. E. F. Watt (ed.) Systems analysis in ecology. Academic Press, New York. 276 p.

Levene, H., O. Pavlovsky, and T. Dobzhansky. 1954. Interaction of the adaptive values in polymorphic experimental populations of *Drosophila pseudoobscura*. Evolution 8: 335–349.

Lewontin, R. C. 1967. Population genetics. Ann. Rev. Genetics 1: 37–70.

Lloyd, M., and R. J. Ghelardi. 1964. A table for calculating the "equitability" component of species diversity. J. Animal Ecol. 33: 217–225.

Lotka, A. J. 1932. The growth of mixed populations: two species competing for a common food supply. J. Washington Acad. Sci. 22: 461–469.

MacArthur, R. 1955. Fluctuations of animal populations, and a measure of community stability. Ecology 36: 533–536.

———. 1957. On the relative abundance of bird species. Proc. National Acad. Sci. U. S. 43: 293–295.

———. 1964. Environmental factors affecting bird species diversity. Amer. Naturalist 98: 387–397.

———. 1968. The theory of the niche. *In* R. C. Lewontin (ed.) Symposium on population biology. Syracuse Univ. Press, Syracuse. In press.

Margalef, R. 1957. La teoría de la información en ecología. Mem. Real Acad. Ciencias y Artes de Barcelona 32: 373–449. (Reprinted 1968 as: Information theory in ecology. General Systems 3: 36–71).

Patten, B. C., and M. Witkamp. 1967. Systems analysis of ^{134}Cesium kinetics in terrestrial microcosms. Ecology 48: 813–824.

Pianka, E. R. 1966. Latitudinal gradients in species diversity. Amer. Naturalist 100: 33–46.

Pielou, E. C. 1966a. Species-diversity and pattern-diversity in the study of ecological succession. J. Theor. Biol. 10: 370–383.

———. 1966b. Shannon's formula as a measure of specific diversity: its use and misuse. Amer. Naturalist 100: 463–465.

———. 1966c. The measurement of diversity in different types of biological collections. J. Theor. Biol. 13: 131–144.

Preston, F. W. 1948. The commonness, and rarity, of species. Ecology 29: 254–283.

Simpson, E. H. 1949. Measurement of diversity. Nature 163: 688.

Slobodkin, L. B. 1965. On the present incompleteness of mathematical ecology. Amer. Scientist 53: 347–357.

Sonneborn, T. M. 1950. Methods in the general biology and genetics of *Paramecium aurelia*. J. Exp. Zool. 113: 87–147.

Volterra, V. 1926. Variazioni e fluttuazioni del numero di individui in specie animali conviventi. Mem. R. Accad. Naz. dei Lincei. Ser. VI, vol. 2: 31–113.

Watt, K. E. F. 1964. Comments on fluctuations of animal populations and measures of community stability. Can. Entomol. 96: 1434–1442.

———. 1968. Ecology and resource management. McGraw-Hill, New York. 450 p.

17

Reprinted from *Natl. Acad. Sci. (USA) Proc.* **58:**1335-1342 (1967)

THE EFFECT OF INVASION RATE, SPECIES POOL, AND SIZE OF AREA ON THE STRUCTURE OF THE DIATOM COMMUNITY

By Ruth Patrick

Many papers have been written concerning the diversity of communities of naturally occurring species. Various diversity indices (Fisher, 1943; Shannon and Weaver, 1948) have been formulated, and various types of models (Preston, 1948; MacArthur, 1957) have been used to describe the structure of the community in terms of its diversity. More recently, MacArthur and Wilson (1963) have emphasized the importance of size of area and invasion rate in the maintenance of a diversified community.

In the present studies three series of experiments were performed. One was to show what effect the size of area and the number of species in the species pool which were capable of invading an isolated area had on the numbers of species which composed the diatom community. The second series of experiments was to show what effect altering the invasion rate had on the structure of the community. The third series of experiments was a comparison of the structure of diatom communities in structurally and chemically similar streams on the island of Dominica and in the United States. The temperature of the water in the two rivers was similar when the studies were made.

Methods and Procedures.—In order to carry out the first series of experiments, duplicate small glass squares 9 mm², 36 mm², and 625 mm² were each erected on a small plastic pedicel about 10 mm in length which was fastened onto a glass slide. These glass slides were placed in plastic boxes which were opened so that the current would pass across the glass squares, giving similar conditions for the attachment of the diatoms across the surface. Having the current strike the slides would have created strong, uneven current patterns.

The slides were placed in clean boxes each day and the pedicels and slides carefully cleaned. This was to minimize the chance that diatoms from sources other than the passing water might invade the glass squares on the ends of the pedicel. The experimental design was to simulate the invasion of species from distant sources onto islands of varying size ranges. These slides were placed in the flowing water from a spring (Roxborough Spring) and in a eutrophic stream (Ridley Creek). Previous studies had shown that the total number of species in Roxborough Spring in the area studied was about 60 at any one time, and in Ridley Creek about 250. It should be noted that the chemical characteristics of the water and the temperature were more variable in Ridley Creek than in Roxborough Spring. Since the spring has a deep source, the chemical characteristics and temperature of the water in it remain almost constant over long periods of time.

The second series of experiments was designed to simulate what might happen if the invasion rate were lowered as happens on an island fairly distant from a highly diversified species pool, such as pools that exist on continents. An island of any significant size would have almost infinite size as far as the diatom community was concerned.   The main factor limiting diversity would be the invasion rate.

In this series of experiments, as contrasted with the first series, the invading species in each case were from the same water and the same species pool was available, but the rate of invasion was different.

Glass slides (3 in. × 1 in.) were placed in boxes and water flowed over their surfaces at 550–600 liters/hour.   This rate of flow had previously been found to produce on such slides diatom communities which were typical of a natural stream. In another box the same rate of flow was maintained by recycling filtered stream water which furnished most of the flow.   During the first three days diatoms were allowed to invade at a rate of 550–600 liters/hour.   The reason for allowing the full flow during the first few days was to develop some semblance of a diatom flora on the slides.   It is usually about two days before the diatom flora starts to develop. After the first three days, the invasion rate of new diatoms was at 1.15 liters/hour. Two sets of these experiments were carried out in a eutrophic stream, one in late September to early October, and one in late October to early November.

The third series of experiments was carried out under natural conditions.   Diatom communities were studied in similar oligotrophic streams on the island of Dominica and in the state of Maryland.

Discussion of Results.—The results of the first series of experiments are shown in Table 1*A*.   These experiments show that size of area influences the number of species established in an area, that the number of species increases greatly at first and is not very different between four days and one week, and that subsequently the pattern of increase is irregular.   On one of the 36-mm^2 slides exposed in the fall in Roxborough Spring there seems to be a slight decrease of species at eight weeks.   Other experiments carried out in a similar manner in Roxborough Spring in the summer, when growth is many times as rapid, showed this decrease more distinctly at the end of two weeks (Table 1*B*).   In each case the species which disappeared were those represented by very small populations.

TABLE 1

(A) EXPERIMENTS IN SEPTEMBER–OCTOBER 1964

Size of slide:	625 mm²		Roxborough Spring 36 mm²		9 mm²		Ridley Creek 36 mm² No. of species,
	Box 1	Box 2	Box 3	Box 4	Box 5	Box 6	box 7
4 days	46	37	23	23	1	3	—
1 week	40	32	28	24	7	—	—
2 weeks	54	35	—	22	10	10	—
8 weeks	—	—	29	14	19	14	160

(B) EXPERIMENTS IN ROXBOROUGH SPRING DURING SUMMER 1964

	1 Week, 144-mm² slide		2 Weeks, 144-mm² slide		1 Week, 625-mm² slide		2 Weeks, 625-mm² slide	
	Box 1	Box 2	Box 3	Box 4	Box 5	Box 6	Box 7	Box 8
No. of species	32	28	23	22	47	44	29	28

When we compare slides of the same size (36 mm²) from Roxborough Spring and from an area in Ridley Creek which had a much larger species pool and a more rapid invasion rate because of swifter current, we see that at the end of eight weeks 160 species were established on the Ridley Creek slides whereas 14–29 species were established on the Roxborough Spring slides (Table 1*A*).

The second series of experiments in which the invasion rate was reduced but the area was large produced the results shown in Figures 1–4. The number of species composing the community was reduced and the sizes of the populations were more variable than in the communities with a high invasion rate. This is shown by the reduction in the height of the mode, the reduction in the observed and calculated species in the community, the increase in σ^2, and the number of intervals covered by the curve. The diversity index (Shannon and Weaver, 1948) is also less. This decrease in the diversity index, particularly in the October–November study, is due to the fact that two species, *Nitzschia palea* and *Navicula luzonensis*, are excessively common in the reduced-flow-rate community. In the September–October reduced-flow study, the commonest species is represented by about one seventh the number of specimens as in the October–November reduced-flow experiments, and more species are represented by fairly large populations. Therefore, the diversity index is not as greatly reduced. This difference in sizes of populations of species is probably due to the fact that conditions for growth were better for more species in the September–October period than in the October–November period.

When one compares the total biomass of the September–October community that had a lower invasion rate and lower number of species with the biomass of the community that had a higher invasion rate and larger number of species, it is only 17 per cent greater in the community with fewer species (21.2 mg as compared to 18 mg). This is probably within the range of natural variation, and the biomasses of the two communities are not really different. It would appear that the numbers of species composing the communities does not significantly affect the total biomass of the community.

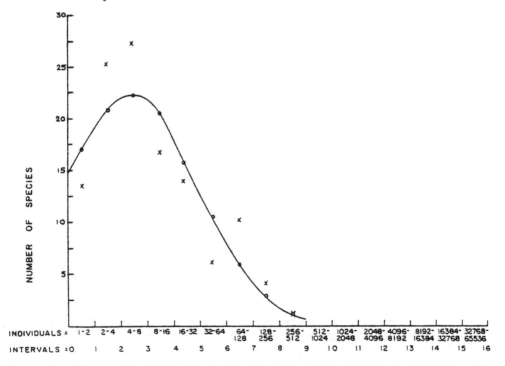

Fig. 1.—Invasion rate 550–600 liters/hr, October–November, 1964. Height of mode, 22.4 species; observed species, 123; σ^2, 6.2; intervals covered by the curve, 9; diversity index, 3.805.

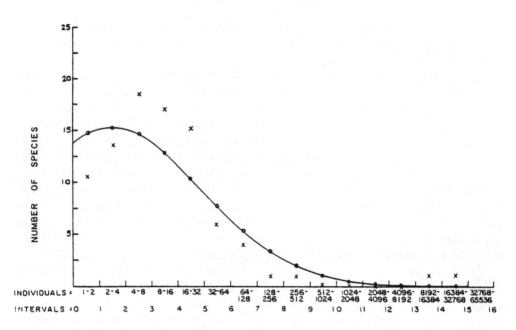

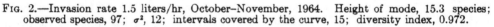

FIG. 2.—Invasion rate 1.5 liters/hr, October–November, 1964. Height of mode, 15.3 species; observed species, 97; σ^2, 12; intervals covered by the curve, 15; diversity index, 0.972.

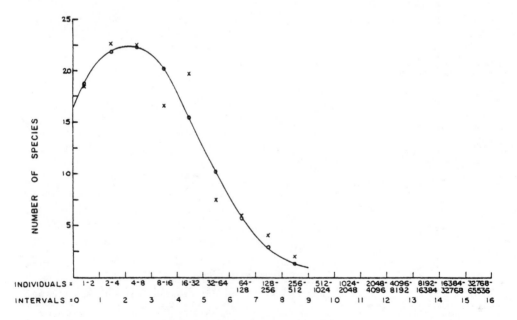

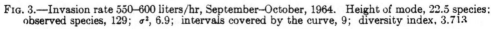

FIG. 3.—Invasion rate 550–600 liters/hr, September–October, 1964. Height of mode, 22.5 species; observed species, 129; σ^2, 6.9; intervals covered by the curve, 9; diversity index, 3.713

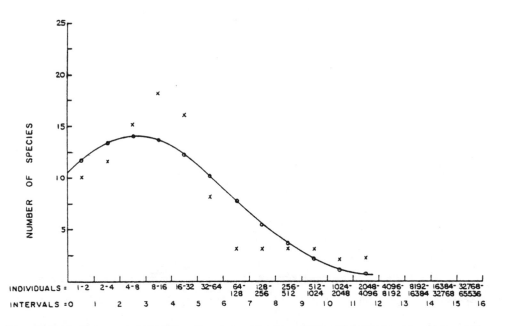

FIG. 4.—Invasion rate 1.5 liters/hr, September–October, 1964. Height of mode, 13.9 species; observed species, 100; σ^2, 12.6; intervals covered by the curve, 12; diversity index, 2.522.

The results of the third series of studies are set forth in Figures 5–7. It will be seen that the number of species in the two Dominica diatom communities is considerably smaller than in the diatom community in the United States. The sizes of the populations of the most common species in all three communities fell in the same interval of the truncated normal curve and were represented by 14,400 specimens in the Layou River, 9,525 specimens in Check Hall River, and 15,975 specimens in Hunting Creek. The main difference in the population sizes of the various species in the streams was that although there were fewer species in the Dominica streams, more of them had fairly large populations than those in Hunting Creek. Most of the species in Hunting Creek had moderate to very small populations. Therefore σ^2 is larger in the Dominica streams than in Hunting Creek. Likewise the Shannon-Weaver diversity index is larger in the Dominica streams. However, Fisher's α (1943), which effectively indicates species numbers even if the populations are small, is much larger in Hunting Creek (25.99) than in the Dominica streams (Layou River, 5.594; Check Hall River, 5.667). This latter index seems more clearly to indicate the differences in species numbers in these two types of communities, while the Shannon-Weaver index more clearly indicates the unevenness of the distribution of individuals in the various species populations.

Conclusions.—From these studies, which were controlled, semilaboratory experiments as well as actual field studies, it is evident that size of area, number of species in the species pool which are capable of invading the area, and the rate of invasion by the species greatly influence the numbers of species and the diversity of the community.

A reduced invasion rate (size of area and number of species in the species pool remaining the same) reduced the total number of species in the community, par-

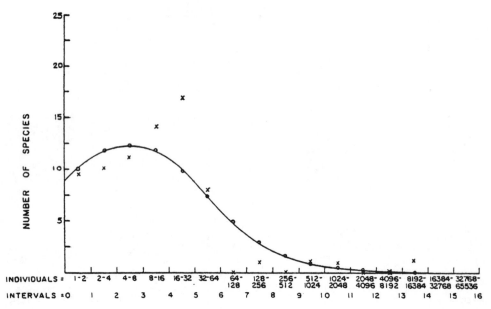

Fɪɢ. 5.—Hunting Creek, Maryland; truncated curve for a diatom community. Height of mode, 12 species; observed species, 79; σ^2, 9.1; intervals covered by the curve, 14; diversity index, 0.789.

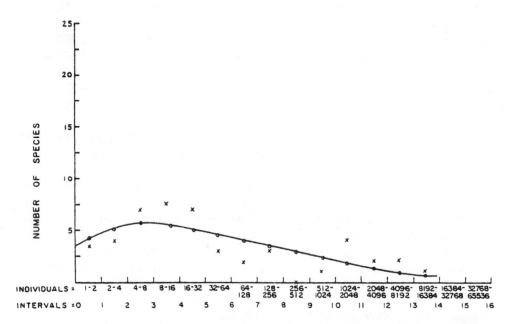

Fɪɢ. 6.—Check Hall River, island of Dominica; truncated curve for a diatom community. Height of mode, 5 species; observed species, 46; σ^2, 21.6; intervals covered by the curve, 14; diversity index, 1.919.

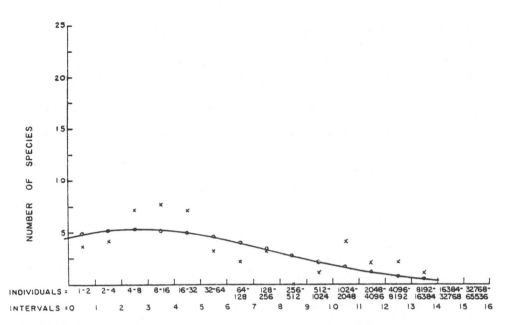

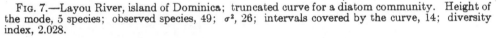

Fig. 7.—Layou River, island of Dominica; truncated curve for a diatom community. Height of the mode, 5 species; observed species, 49; σ^2, 26; intervals covered by the curve, 14; diversity index, 2.028.

ticularly those species with small populations which are typically part of a natural continental community. There is an increase in species with fairly large populations although the total number of species decreases.

Similar results were found when diatom communities on the island of Dominica were compared with a diatom community in a similar type of stream in the state of Maryland. The island communities had much smaller numbers of species with larger populations than the continental community. They also had fewer species with small populations.

One of the main results of a high invasion rate is to maintain in a community a number of species with relatively small populations. The value of these species to the community may be similar to the presence of relatively rare genes in a gene pool. They may function to preserve the diversity of the community under variable environmental conditions. Such rare species at any one point of time under changed environmental conditions might be better adapted than the presently common species and might be able to increase rapidly.

The size of the species pool which may potentially invade an area also has a great effect on the number of species composing the community, as seen in the comparison of the species numbers in the communities in the Roxborough Spring studies and the Ridley Creek studies. The size of the area to be invaded also affects the diversity of the community. Once the area becomes filled, the first species to be eliminated were those represented by very small populations.

The author wishes to express her gratitude to Dr. G. E. Hutchinson and Dr. Robert MacArthur for their helpful advice and criticism, and to Miss Noma Ann Roberts, Mr. Raymond Cummins, Mr. Roger Daum, and Miss Lee Townsend, who have been most helpful in the carrying out of these studies.   The author also wishes to thank the U.S. Public Health Service (WP-00475) for their financial support for part of this work.

Fisher, R. A., A. S. Corbet, and C. B. Williams, "The relation between the number of species and the number of individuals in a random sample of an animal population," *J. Animal Ecol.*, **12,** 42–59 (1943).

MacArthur, R., "On the relative abundance of bird species," these PROCEEDINGS, **43,** 293–295 (1957).

MacArthur, R., and E. O. Wilson, "An equilibrium theory of insular zoogeography," *Evolution*, **17,** 373–387 (1963).

Preston, F. W., "The commonness and rarity of species," *Ecology*, **29,** 254–283 (1948).

Shannon, C. E., and W. Weaver, *The Mathematical Theory of Communication* (Urbana: University of Illinois Press, 1948), pp. 3–91.

Part V

MEASUREMENTS OF DIVERSITY

Editor's Comments
on Papers 18, 19, and 20

18 **PRESTON**
   The Commonness, and Rarity, of Species

19 **MAY**
   Patterns of Species Abundance and Diversity

20 **SUGIHARA**
   Minimal Community Structure: An Explanation of Species Abundance Patterns

Various approaches have been used to express diversity in various groups of organisms. The earliest methods were by histograms and graphics of various types to show the relative abundance of species. Indices have been used, of which the earliest were the similarity indices in which the amount of species overlap (the number of species that are the same in two communities) was determined— that is, the Jacard index. These have been elaborated by Patten (1962), MacIntire (1978) and MacArthur (1965) to yield more information.

Measurements of diversity using numbers of species and numbers of individuals are of many types. Among the earliest was the Simpson index (1949), primarily a measure of dominance. Margalef (1958) was the first to apply the equations from the information theory to diversity. In these applications the evenness of distribution of specimens within a group of species is of main concern. A somewhat different index was developed by Fisher and coworkers (1943) in which the number of species was given greater importance than the relative population sizes of species. A simple index of diversity that is often used is a count of the species present.

Various attempts have been made to express diversity as it relates to numbers of taxa and the relative sizes of their populations by the use of models. The best known are the MacArthur model sometimes referred to as the "broken stick" model (Paper 1) and the truncated log-normal curve (Paper 19; Patrick et al., 1954).

A measurement of diversity useful in relating the relative abundance of species and the number of species is the importance curve developed by Whittaker (1965). For a student interested in knowing the various mathematical approaches toward diversity that have been used, *Ecological Diversity* by Pielou (1975) is recommended. An important evaluation of various indices is given by Whittaker (1972). Also of interest is the paper by Kempton and Taylor (1974) on log-series and log-normal parameters.

Paper 18 is included because it has challenged scientists to determine how widely applicable the truncated normal curve is to expressing the relationships of species in a community; Paper 19 discusses many different mathematical approaches to determining the diversity and similarity of communities.

Preston (1962) stated that all log-normal curves were canonical and stated the theoretical reasons why this was true. This statement has been challenged by May (Paper 19) and others. In Paper 20, Sugihara sets forth a different theoretical approach as to why log-normal curves are canonical and presents data from the real world to substantiate his conclusions.

REFERENCES

Fisher, R. A., A. S. Corbet, and C. B. Williams, 1943, The Relation Between the Number of Species and the Number of Individuals in a Random Sample of an Animal Population, *J. Anim. Ecol.* **12**:42–58.

Kempton, R. A., and L. R. Taylor, 1974, Log-Series and Log-Normal Parameters as Diversity Discriminants for the Lepidoptera, *J. Anim. Ecol.* **43**:381–399.

MacArthur, R. H., 1965, Patterns of Species Diversity, *Cambridge Philos. Soc. Biol. Rev.* **40**:510–533.

MacIntire, C. D., 1978, The Distribution of Estuarine Diatoms Along Environmental Gradients in a Canonical Correlation, *Estuarine Coastal Mar. Sci.* **6**:447–457.

Margalef, D. R., 1958, Information Theory in Ecology, *Gen. Syst. Yearb.* **3**:36–71. (Translated from *R. Acad. Cienc. Artes Barcelona Mem.* **32**:373–449, 1957.)

Patrick R., M. H. Hohn, and J. H. Wallace, 1954, A New Method for Determining the Pattern of the Diatom Flora, *Acad. Nat. Sci. Phila. Not. Nat.* **259**:1–12.

Patten, B. C., 1962, Species Diversity in Net Phytoplankton of Raritan Bay, *J. Mar. Res.* **20**:57–75.

Pielou, E. C., 1975, *Ecological Diversity,* Wiley, Toronto, 165p.

Preston, F. W., 1962, The Canonical Distribution of Commonness and Rarity, *Ecology* **43**:185–215, 410–432.

Simpson, E. H., 1949, Measurement of Diversity, *Nature* **163**:688.

Whittaker, R. H., 1965, Dominance and Diversity in Land Plant Communities, *Science* **147**:250–260.

Whittaker, R. H., 1972, Evolution and the Measurement of Species Diversity, *Taxon* **21**:213–251.

18

Reprinted from *Ecology* **29**:254-283 (1948)

THE COMMONNESS, AND RARITY, OF SPECIES

F. W. PRESTON

SECTION I. GENERAL

The purpose of this paper is to deduce, from a number of examples and from theoretical considerations, some plausible general law as to how abundance or commonness is distributed among species. Experimentally, this could be done by making a complete *census* of every species, but, with rare exceptions, this procedure is quite impractical. We therefore attempt to deduce the *"universe"* from a *sample*.

Commonness, as understood by ecologists, has several rather different meanings: we are here concerned with (1) the total number of living individuals of a given species, which might be called its global abundance, (2) the total number of individuals living at any instant on a given area, such as on an acre or a square mile, which might be called its local abundance, (3) the ratio which the number of individuals or one species bears to that of another species, i.e., its relative abundance, and (4) the number of individuals *observed*, for example, the number of a moth species counted in a sample from a light trap, its "observed," "apparent," or "sample" abundance. There will not usually be any difficulty in deciding which phase of the subject is under discussion at any time. The Raunkiaer "Index of Frequency" is a measure of ubiquity rather than of commonness as above defined: its relation to our other concepts is discussed later.

As a rule, we are interested in a sample only in so far as it throws light upon the "universe" that is being sampled. The sample will be a sufficiently accurate replica of the universe provided (1) it is a perfectly "random" sample, and (2) no species is represented in the sample by less than 20 or 30 individuals.

In most ecological work condition (2) will never obtain, and much of the present paper will center on this difficulty. Condition (1) will not usually obtain in the broadest sense, and needs a moment's consideration.

A geologist sampling an ore-body, whose boundaries have been delimited accurately by previous exploration, has a known "universe" and merely needs information on composition. His universe is permanent. But in ecological work, the "universe" changes rapidly. The moths flying tonight are not those that flew a month ago, or will fly a month hence. Those flying this year are a vastly different association from those that flew last year in the same area. The same thing is true of rodent populations, of birds, and of plants. We are dealing with a fleeting and fluctuating assemblage, a "universe" continually expanding, contracting, and changing in composition. Thus it is important to recognize at the outset that, for the purposes of our present investigation, the "universe" from which the sample is drawn is that universe declared to us by the sample itself, and not our preconceived notion of what the universe ought to be.

Further, it is important to recognize that the randomness we seek is merely randomness *with respect to commonness or rarity*. A light trap is satisfactory in this respect and samples its own universe appropriately. It is definitely *selective* in respect of phototropism, but it is *random* in respect of commonness, i.e., it does not care which of two moths, equally phototropic, it catches, though one may be a great rarity and the other of a very common species. On the other hand, an entomologist, or even an intelligent boy, with a net, is not a satisfactory collector, for he will go after the rarity. For this reason we have to reject Corbet's ('41) collection of Malayan butterflies, as not being a *random* sample.

Previous discussions. In the early years of the present century an extensive literature accumulated in connection with Raunkiaer's Law of Frequency. This was apparently intended as an index of commonness or rarity, more particularly in plant associations but not necessarily confined thereto. and, subject to some assumptions, it *is* a measure of commonness or rarity. The method of expressing the results, though it appears simple, is in reality very inconvenient, and gives no direct clue to the nature of the universe being sampled.

An analysis of Raunkiaer's findings must be postponed, in the present paper, until after we have established the relationship between sample and universe.

More recently Williams ('43) and his collaborators have taken up the subject from a different point of view.

Williams was struck by the fact that in a "random" collection of butterflies or moths there were usually great numbers of species represented by a single specimen (hereinafter called "singletons"), a much less number of species represented by two specimens ("doubletons") and still fewer species represented by three, four, and so on. He noticed further than if N be the number of species represented by singletons, the number represented by doubletons was $N/2$, the number represented by three was $N/3$ and so on.

Hence in the collection, if the observation were valid indefinitely, we should have a series

$$N \left(1 + \frac{1}{2} + \frac{1}{3} + \frac{1}{4} + \frac{1}{5} + \cdots \right)$$

as the number of species represented by

$$1 \quad 2 \quad 3 \quad 4 \quad 5 \quad \cdots$$

specimens.

This is a harmonic series, and it has the grave theoretical disadvantage that the sum of its terms is infinite, calling for an infinite number of *species* in the collection. As if that were not bad enough, the number of *specimens* is of the order of infinity to the second power.

Williams, aware of these difficulties, apparently sought the help of Fisher ('43) who introduced an arbitrary modifying term into the series, so that it was no longer strictly harmonic, and both species and specimens were reduced to finite numbers. The degree of agreement thereby produced may be judged from the group of papers by Corbet, Williams and Fisher ('43). Even on the collections on which it was tested it does not work too well, and Corbet, in an earlier paper ('41), was aware that as collecting continued, it would represent the facts increasingly poorly.

Method of graphing. It has often been a matter of comment by ecologists that one or two species are extraordinarily abundant at a particular time and place: all others seem rare in comparison. Thus in Seamans' moths of Alberta, discussed later, out of a total of 303,251 specimens. 91,502 belonged to a single species; while in Dr. King's moths of Saskatchewan, 38 species accounted for 77,260 specimens, and 237 more accounted for only 9,850 additional individuals.

Under these circumstances it would be logical, merely as a matter of convenience, to try plotting commonness (number of individuals per species) on a logarithmic base. But there is a more cogent basis for so doing. Commonness is a *relative* matter. One species, we say, is twice as common as another, and a natural series of groups representing commonness would run:

Species group	A	B	C	D	E	F	G	H	I	J	etc.
Approximate specimens observed of that species	1	2	4	8	16	32	64	128	256	512	etc.

These groups are a sequence of *octaves* of frequency. It is the most natural grouping possible, but our ultimate justification for adopting it is not its naturalness, but the fact that it works. It is a geometric, or logarithmic series, and this paper produces evidence that commonness of species appears to be a simple Gaussian curve on a geometric base (i.e. "lognormal" curve). The intervals of an octave turn out to be a convenient size of interval, as well as the most readily grasped ones.*

Boundaries of the octaves. An octave is simply an interval of two-to-one. On the piano it may be from C to C' or G to G': similarly in our ecological studies, it is the interval 4-to-8 or 6-to-12. We might choose the numbers 1, 2, 4, 8, etc., as the mid-points of our octaves, but it is more convenient to choose them as the end-points, or boundaries between octaves. Thus denoting the octave 1 to 2 as A, and so on, we have octave D comprising the interval whose boundaries are 8 and 16.

Thus if a given species is represented by 9, 10, 11, 12, 13, 14 or 15 specimens, it clearly falls in octave D. All species falling in octave D may be thought of as having roughly the same degree of commonness, in comparison with those falling for instance in octave J, which are represented by from 513 to 1023 specimens.

If a species is represented by 8 specimens, octave D is credited with half a species, and octave C is credited with the other half. Similarly a species with 16 specimens is credited half to D and half to E. Octave B is comprised of all species having 3 specimens in the sample, plus half the species having 2 and half the species having 4. Octave A is composed of half the species represented by singletons, and half those represented by dou-

bletons. Half the singletons have to be assigned to octaves below A, which we designate by Greek letters and discuss later.

Procedure. We obtain a sample (say a catch of moths in a light trap), and make a complete tally of the relevant material (e.g., macrolepidoptera, noctuids, all lepidoptera, or all insects), determining the "representation" (number of individuals) of each species. This assigns the species to its appropriate octave of frequency. We count the number of species in each octave. This is the ordinate of our curve, and the abscissa is a scale of octaves of increasing commonness.

The experimental points thus obtained might, as the reader will doubtless observe, be "graduated" quite well as a Poisson distribution. Indeed, it is possible that that might be the most formal approach, and in the long run the most accurate. However, experience shows that an ordinary Gaussian curve (the "normal curve of errors" $y = e^{-(ax)^2}$) is very satisfactory also, provided the curve is "decapitated" at the veil line as discussed later. The use of this curve has special advantages. Singleton ('44) has worked out a method of fitting a decapitated Gaussian curve in accordance with the principle of Maximum Likelihood: this method was used by Dr. J. L. Glathart (of the Preston Laboratories) to graduate many of the curves we have used. Others are graduated by a graphical method of our own. Both give comparable results, and the methods, for lack of space, will not be discussed in the present paper.

Relation between Williams' harmonic series and the octave plot

If Williams' unmodified law were true, viz. that if there are N species represented by one specimen each, there will be $N/2$ represented by two specimens, and so on, we could calculate how many species would fall in each of our octaves. It comes out as follows:

---

* The above procedure is equivalent to taking the ordinary logarithms of the number of specimens per species and grouping the species into the groups whose logarithms extend over an interval of 0.3010 i.e., such groups as 0–0.3010, 0.3010–0.6021, 0.6021–0.9031, etc.

Octave	Number of species in that octave
A (1–2)	0.7500 N
B (2–4)	0.7083 N
C (4–8)	0.6971 N
D (8–16)	0.6941 N
E (16–32)	0.6934 N
F (32–64)	0.6932 N
G (64–128)	0.6932 N

All subsequent octaves have approximately 0.6931 N, where 0.6931 = $\log_e 2$. Except for the first two octaves, the number of species in each octave would be constant to within one per cent. The data are graphed in figure 1.

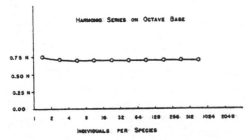

FIG. 1. The harmonic series on an octave base. If the commonness of species followed the (unmodified) Williams' law, then the number of species in each octave would be essentially constant.

Experimental evidence shows that this state of affairs is far from the truth. Theoretical considerations also require its categorical rejection. There are an infinite number of octaves to the right, and if they were all equally filled with species, there would be an infinite number of them, and excessively common species would be as numerous as very moderately common ones, which we know is not the case. Therefore the curve must descend to the right. Williams himself recognizes this, and this is the basis of Fisher's modification of the Harmonic Series Law, which modification produces a descent to the right.

But as will appear later, the real curve also descends to the left. It is a humped curve. This requires a totally different approach.

We are now ready to consider the experimental evidence.

SECTION II. EXAMPLES

We have one or two examples of bird-counts that meet our essential requirements, and four or five of moths in light traps. Most counts do not meet our requirements, but some throw light on the central problem and seem to extend our findings to other biological universes.

I. Saunders ('36). The birds of Quaker Run Valley, western New York State

Saunders attempted a complete census of the breeding birds on a tract of 16,967 acres. His main reliance was upon a count of singing males, aided by a few auxiliary methods.

Table I-A below is the number of breeding pairs of each species, from commonest to rarest, in the order in which Saunders reported them, but with the names of the species omitted.

Table I-B condenses this information ready for graphing, and the plot is given in Fig. 2.*

The total number of species here reported is 80, but Saunders states:

"At the present date 141 species of birds are known to occur in Allegany Park and its vicinity during the months of July and August. Of these birds, 27 occur as fall migrants or late summer wanderers. Thirteen breed outside the park itself, but evidently in the near vicinity. Eleven more breed in the park but not, so far as is known, in the Quaker Run Valley. The remaining 90 occur in their breeding season and undoubtedly breed or have bred in the Quaker Run Valley in the past 12 years. Of these 90 species, 11 are irregular in their occurrence and have not been found every year. The remaining 79 are regular breeders and, judging by past experience, can be expected to breed in the area every year."

In the graph (fig. 2) the curve drawn is a Gaussian curve, with its mode (crest) at $i = 30$ specimens (breeding pairs) per species, that is, near the end of the 16–32 octave: the general equation of such a curve is

$$n = n_0 e^{-(aR)^2},$$

* The Veil Line shown in this and other graphs will be discussed later.

Birds of Quaker Run Valley

TABLE I-A. *Values of i (specimens per species)*

1670	1656	1196	868	723	723	675	506	477	389	367
324	311	310	288	282	280	270	220	188	181	179
161	160	158	152	138	111	109	91	90	88	79
60	57	56	50	46	46	43	43	35	34	33
32	32	30	28	28	26	24	23	22	17	15
14	12	10	10	10	10	10	8	8	7	6
6	5	5	4	4	4	4	3	3	3	3
2	1	1	?	?	?	?	?	?	?	?
?	?	?								

TABLE I-B. *Values of n (species per octave)*

Octave	<1	1 to 2	2 to 4	4 to 8	8 to 16	16 to 32	32 to 64	64 to 128	128 to 256	256 to 512	512 to 1024	1024 to 2048
Species per octave	>1	1½	6½	8	9	9	12	6	9	11	4	3

where n_0 is the number of species in the modal octave, n is the number in an octave distant R octaves from the mode, and a is a constant to be calculated from the experimental evidence.

In this particular case

$$n = 10e^{-(0.194R)^2}.$$

The curve theoretically extends to infinity, asymptotic to the x-axis, both to left and right. The area "under" the curve (between curve and x-axis) from minus infinity to plus infinity, is theoretically the *universe being sampled,* i.e. the total number of species theoretically available for observation.

$$N = \int_{-\infty}^{+\infty} n\,dR = n_0\sqrt{\pi}/a,$$

where $\pi = 3.1416$ and $\sqrt{\pi} = 1.77$. Putting $n_0 = 10$ and $a = 0.194$ we get $N = 91$, which agrees very closely with Saunders' own estimate (quoted above) of what it is.

II. *Preston and Norris ('47). Breeding birds of the Frith (Preston Laboratory Grounds)*

From table II of the paper by these authors, we may obtain data of the total pairs of breeding birds, and the total number of "expected" nests (remembering that some species make more nests than one, even if successful with the first, and most species make a second nest if the first is unsuccessful). The count for the combined years 1944 and 1945 gives:

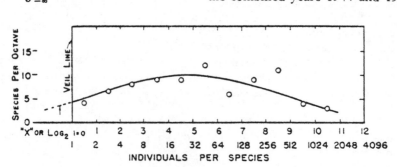

FIG. 2. Saunders' breeding birds. The octaves are definitely not equally filled, and the curve looks as if it is humped up in the middle.

Octave i	— <1	A 1–2	B 2–4	C 4–8	D 8–16	E 16–32	F 32–64	G 64–128	H 128–256	I 256–512
Species (n) Nests	>3½	5	6½	10	7	5	4	4	2	0
Species (n) Bird-Pairs	>5	6	7	11½	8½	4	7	1	0	0

TABLE II. *Values of n (species per octave)*

Octave	<1	1 to 2	2 to 4	4 to 8	8 to 16	16 to 32	32 to 64	64 to 128	128 to 256	256 to 512	512 to 1024	1024 to 2048	2048 to 4096	4096 to 8192	8192 to 16384
Species per octave	>19	37	42	49½	45½	42	28½	26½	30	14	9	2	0	2	2

These results are not graphed, as only a few hundred bird-pairs and nests were involved, and statistically this is hardly sufficient. So far as the data go, however, they support a humped distribution with its crest in octave C, both for birds and for nests.

III. *Dirks ('37). Moths in a light trap at Orono, Maine*

Dirks operated a light trap over a period of four years (1931–34). Exclusive of microlepidoptera he caught 56,131 specimens. These he classified as to species and sex, and gives three tabulations: total moths, female moths, gravid female moths. Since most females were gravid, the third tabulation closely resembles the second and will not concern us here.

Condensing his tabulation for total moths produces table II.

This is graphed in figure 3. The first number (for the octave less than 1) is indicated by an upward pointing arrow, meaning that the true number for this range is above the tip of the arrow, a point to which we shall return later. The other points are plotted as circles. The smoothed curve drawn through the points appears to have its crest at about 7 or 8 specimens per species. The left-hand end of the curve is, of course, quite incomplete. It represents the rarer species which were absent from the catch. The general shape of the curve, however, appears clear enough, and we may assume that it is approximately a "normal" probability curve with its mode at, or very near, 8 specimens per species.

The equation of the curve is $n = 48e^{-(0.207R)^2}$, and on this basis the total number of species in the universe being

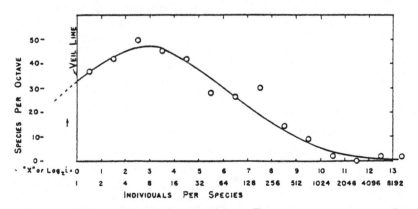

FIG. 3. Dirks' moths of Orono, Maine. The octaves are not equally filled; the curve is humped; we see the right-hand tail, but the left-hand one is hidden behind a veil.

315

260   F. W. PRESTON

sampled is

48 $\sqrt{\pi}/(0.207) = 410$ species
approximately.

Dirks actually obtained 349 species, so his sample represents about 85% of the species that were theoretically available at Orono, during the four years in question, for trapping by means of such a light trap as he used.

IV. *Dirks ('37). Female moths in a light trap at Orono, Maine*

The condensed statement of results follows:

The corresponding graph is given in figure 5. Its mode is at about 8 specimens per species and its equation is

$$n = 35e^{-(0.227R)^2}.$$

The total population being sampled is $N = 35 \sqrt{\pi}/(0.227) = 273$ species. Williams actually trapped, in the four years covered by his report, 240 species, involving 15,609 specimens (macrolepidoptera).

Williams comments that the total Hertfordshire fauna (of this type of lepidoptera) amounts to 461 species, and tries to figure how long it would take to catch

TABLE III. *Values of n (species per octave)*

Octave	<1	1 to 2	2 to 4	4 to 8	8 to 16	16 to 32	32 to 64	64 to 128	128 to 256	256 to 512	512 to 1024	1024 to 2048	2048 to 4096	4096 to 8192
Species per octave	>27½	42	39½	27½	28	23	19½	9	4	4	0	0	1	1

This is graphed in figure 4, whose mode is at or near 1½ specimens per species, and equation is

$$n = 42e^{-(0.205R)^2}.$$

On this basis, the available universe comprised 363 species, of which 226 were represented in the sample, in the form of 12,799 individuals. Females in Dirks' sample were only 22.8% of the total, which probably means, not that they were rarer in the outer world, but that they were travelling less than the males. Dirks obtained a sample of 85% of the "total" species (410) in the total universe, but only 62% of the females (226/363 = 62%) which constituted a smaller universe (363 species as against 410).

V. *Williams ('43). Moths in a light trap at Rothamsted, England*

The condensed tabulation is:

examples of all of them. But if our present theory is valid, the "universe" he was sampling was not the whole Hertfordshire fauna, but a local population comprising about 60% of the Hertfordshire total. This may be compared with the experience of Dirks, who was sampling not the fauna of Maine, but a local population near Orono.

VI. *King (unpublished). Moths in a light trap at Saskatoon, Saskatchewan*

This is a collection of macrolepidoptera, numbering 87,110 individuals, 277 species, caught in Dr. King's "Trap A," over a period of 22 years, from 1923 to 1944 inclusive.

The data are graphed in figure 6, and the curve, calculated by Singleton's method, has its mode at 4 specimens per species and the equation

$$n = 33.0e^{-(0.152R)^2}$$

TABLE IV. *Values of n (species per octave)*

Octave	<1	1 to 2	2 to 4	4 to 8	8 to 16	16 to 32	32 to 64	64 to 128	128 to 256	256 to 512	512 to 1024	1024 to 2048	2048 to 4096
Species per octave	>17½	23	27½	36	27½	33	31	13½	19	5	6	0	1

316

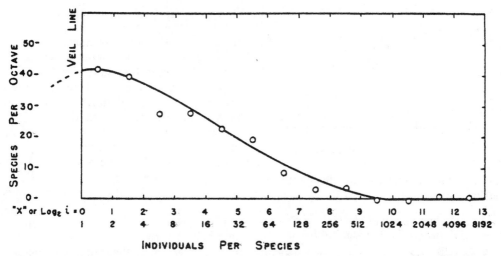

FIG. 4. Dirks' female moths. The hump seems to be very close to the veil. In Sweadner's moths of Wyoming (not illustrated), the veil has moved to the right of the hump, and all we see is the descending side.

The "universe" accordingly comprises 384 species, and since Dr. King caught 277, his trapping is seemingly 72% complete.

It may seem strange that after 22 years the collection should be so incomplete, when 4 years of trapping at other places yield values over 80%. But it must be remembered that what is being sampled during the 22 years is a 22-year universe, not a 4-year universe, and that the moths available for trapping, in Saskatoon at any rate, vary enormously from year to year. This will be dealt with, briefly, in the later discussion.

Since we have Dr. King's data year by year, it is quite possible to make up subsamples and plot the results year by year or quadrennium by quadrennium. The curves retain their general shape, though the height at the mode tends to fall to about 24 species per octave instead of the 33 species per octave in the 22-year aggregate. This alone, of course, shows a likelihood that we are dealing with a vastly more extensive universe if we trap

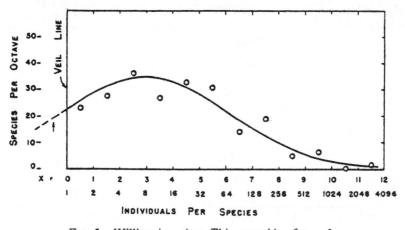

FIG. 5. Williams' moths. This resembles figure 3.

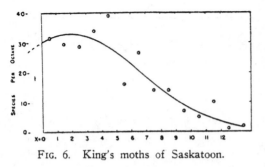

FIG. 6. King's moths of Saskatoon.

for a longer period— at any rate at Saskatoon.

VII. *Seamans (unpublished). Moths in a light trap at Lethbridge, Alberta*

Seamans' records cover the period 1921–43 inclusive, with the exception of the year 1938, when the collections were damaged by mice before they could be catalogued. The number of specimens is 303,251; of species, 291. This is the most extensive collection at present available to me in a completely counted condition. The graph is given in figure 7: its mode is at about 32 specimens per species, and its equation

$$n = 30.0e^{-(0.160R)^2}.$$

The theoretical universe is accordingly 332 species and the collection is about 88% complete.

With the Lethbridge data, as with those of Saskatoon, it is possible to make up samples for individual years, or for quadrennia. For individual years the modal octave tends to hold about 20 species per octave: for quadrennia it averages about

25: for the 22 years, about 30, but the general shape is very similar. The graphs are accordingly not reproduced here.

These few examples show what the experimental plottings look like, and the degree of fit between the Gaussian curve and the data. In all these examples we have a sample with more than half the universe represented, i.e., the crest of the curve is in sight. Fragmentary collections are available where the crest has not been reached, and probably Raunkiaer's collections (discussed later) were of this type.

With the aid of these actual examples,* thus very briefly presented, we now pass to a discussion of the general principles, the veil line, the relation between sample and universe, and some deductions about the nature of biological universes in general.

SECTION III. DISCUSSION

The veil line

If the universe contains species so rare that the sample, if perfectly proportioned to that universe, should theoretically contain only a fraction of a specimen, it is most likely that the species will be unrepresented in the collection. It cannot be represented by less than one whole specimen, and by definition this would be

* Hinton ('40) gives data on a collection of adult Mexican water-beetles, and some others. This graduates very well as a decapitated Gaussian curve. The internal evidence suggests it was a random collection, but since Hinton gives no comment on that point, discussion is omitted in the present paper.

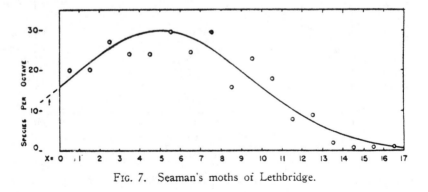

FIG. 7. Seaman's moths of Lethbridge.

over-representation. Some species indeed will achieve this over-representation, just as some species which "ought" to be represented by 1, 2, 3 or even more specimens, may fail of representation. But on the average, those relatively uncommon species whose theoretical representation is appreciably less than one specimen, will be missing from the sample. These species may themselves be of very different degrees of rarity; some "ought" to be represented perhaps by half a specimen, others by 1/20 or 1/200 of a specimen. In other words, they belong to different octaves, but these octaves are all "missing" in the sample.

There is therefore a veil line at the left end of octave A, at a representation of one specimen. The sample indicates clearly what the universe is like to the right of the veil, but we have to infer what it is like to the left of it.

Now doubling the size of the sample, other things being equal, doubles the number of individuals of each of the commoner species, in the sample. This is equivalent to moving every species one octave to the right, which in turn is equivalent to withdrawing one more octave from under the veil. The process can be repeated until the whole universe, in effect, is exposed. Thus the graph of the universe is identical with the graph of the sample, except that it is not decapitated by a veil line. The graph in fact behaves as if it were drawn complete upon a movable card, and the card slipped with its left-hand end under a veil. Continued sampling, if the universe remains constant, merely withdraws the card progressively from under that veil.

If this last statement is true, it is a very valuable property of our method of representing our results. It will be obvious to all, that it is quite accurately true for those species which, in the sample, are represented by at least a score or two of specimens. These species will be present in the sample in virtually the exact proportion in which they are present in the universe. It will not be obvious that it

will be true of those species whose representation in the sample is small, so that, by the laws of small samples, they have a good chance of being substantially over- or under-represented.

The next page or two, on the Relation between Sample and Universe, show that though any one of the less common species may very well get into the wrong octave, the probability is that every octave has very nearly the correct number of species. This is what we graph, and therefore our graph is not appreciably affected by the interchange of species among the octaves close to the veil line.

In other words, the "shape" of our universe is constant. If however we want to know the exact status of a specified species in that universe, we cannot be sure of it until the sample contains a dozen or two specimens.

Relation of sample to "universe"

Consider any universe, Gaussian or otherwise, made up of at least several hundred species, and fix attention on any one species, say a melanic, which is not one of the very common ones. Then the numbers of individuals of this species must be less than one in a thousand compared with the non-melanics. Suppose for the moment that it is one in ten thousand.

If we catch, at random, a single specimen from our universe, we have one chance in ten thousand that it is a melanic; and if we make a catch of 10,000 specimens, "perfect representation" would call for one melanic in the collection.

If we now released the specimens unharmed back into the universe, so as not to deplete it, we could make many random collections of 10,000 specimens, and when we tallied up the number of samples (collections) that showed zero, one, two, etc., melanics, we should find they conform to the several terms of the Poisson Series,

$$e^{-1}\left(1 + \frac{1}{1!} + \frac{1}{2!} + \frac{1}{3!} + \cdots\right)$$

probability for

$$0 \quad 1 \quad 2 \quad 3 \quad \cdots$$

                                      melanics.

Similarly if perfect representation calls for exactly two melanics in the collection, which would be the case if we collected 20,000 specimens all told, then the probability of our finding zero, one, two, three . . . specimens is given by the terms of the Poisson series

$$e^{-2}\left(1 + \frac{2}{1!} + \frac{2^2}{2!} + \frac{2^3}{3!} + \text{etc.}\right).$$

More generally if perfect representation calls for "p" * melanics in the collection, our series is

$$e^{-p}\left(1 + p + \frac{p^2}{2!} + \frac{p^3}{3!} + \frac{p^4}{4!} + \cdots\right).$$

The terms of this series are easily calculated if p is small or fractional, but it it is tedious if p is large, as the series then diverges at first and converges thereafter very slowly. Fortunately we do not need to use very high values of p.

What is true for melanics is of course true for any other sort of moth or species which is present to the same extent as the melanic. Thus if a hundred different species are present in identical proportions, and the sampling is carried to the point where we might expect exactly 3 specimens of each, we can tell how many of these species are in fact likely to be represented by 3 specimens, or by any other number.

Next let us suppose that, in the universe, the species in any octave of frequency are all concentrated at the midpoints of the octaves, so that if one species is present to the extent of 6 M where M is some large number—a million say—then there are a number of other species also present to the extent of 6 M, but none to the extent of 4 M, 5 M, 7 M, or 8 M. Similarly the species of the octave

2 M–4 M are all concentrated at 3 M ; * and so on.

Then we can so conduct our collecting that any species occurring in the universe ought to be present, with perfect sampling, to the extent of 3, 6, 12, 24, 48, 96, . . . individuals and not to the extent of any intermediate figure. Similarly the rarer species, if perfectly sampled, would contain fractions of specimens, and would be represented by $1\frac{1}{2}$, $\frac{3}{4}$, $\frac{3}{8}$, $\frac{3}{16}$, . . . specimens.

Now in real collections, a species has to be represented by an integral number, or be absent; and the perfect sampling has to be replaced by the most probable sampling given by the Poisson series we have just been discussing. And what is true of fractional examples is equally true of the "perfect" integral numbers. A species that ought to be represented by exactly 3 specimens will, five times out of a hundred, have zero representation, and about one time in a thousand it will have as many as ten specimens to its credit. Otherwise expressed, if there are one thousand species (in a universe containing many thousands of species) which are entitled with *perfect* sampling to a representation of three individauls, then with *random* sampling fifty of these thousand species might very likely be unrepresented, while one species might well be present to the tune of ten specimens.

Perfect sampling can be accomplished only by a conscious agent: random sampling is what happens in the absence of conscious selection.

The whole story may now be reduced to a tabulation (table V).

In this table, the left-hand column gives the midpoints of 12 octaves, six behind the veil line and six out in the open. For instance $p = 3$ represents the (arithmetical) midpoint of the octave 2–4. Suppose a given species is present in the

---

* "p," as used here, stands for "perfect representation," or "average expectation," more often denoted perhaps by c. It must not be confused with "p for probability" in the classical binomial $(p + q)^n$.

* There may be some argument that the midpoint of the octave should be $2\sqrt{2}$ M rather than 3 M, the geometric rather than the arithmetic mean: but the final numerical results will be nearly the same either way.

"universe" to such an extent that a perfect sample would contain exactly three specimens, that is, let the sampling be continued just long enough to justify an expectation of exactly three specimens of this species; then let us return the specimens unharmed to the universe and repeat our sampling many times. Then 12.5% of these many samples would show the species falling in the range 0–1, 18.7% in the octave 1–2, 42% in its own octave 2–4, and so on.

The second column from the left represents all the times in which we should expect zero representations, and half the times we should expect singletons.

The third column represents half the singletons and half the doubletons.

TABLE V. *Probabilities of representation (expressed as percentages), when "perfect" representation is p specimens*

Perfect (p) representation \ (Specimens) actual (m)	0	1	2	3	4	5	6	7	8	9	10	11	12	13	14	15	16
3/128	97.6	2.4															
3/64	95.4	4.5	0.1														
3/32	91.1	8.5	0.4														
3/16	83.0	15.5	1.4	0.1													
3/8	68.7	26.1	4.8	0.4													
3/4	47.3	35.5	13.3	3.3	0.6												
1½	22.3	33.4	25.1	12.5	4.7	1.4	0.4	0.1									
3	5.0	15.0	22.4	22.4	16.8	10.0	5.0	2.1	0.8	0.3	0.1						
6	0.2	1.5	4.5	8.9	13.4	16.2	16.2	13.9	10.4	6.9	4.1	2.2	1.1	0.5	0.2	0.1	
12					0.5	1.3	2.5	4.3	6.5	8.7	10.5	11.5	11.5	10.7	9.0	7.2	5.5
24												0.1	0.3	0.5	0.9	1.5	2.2

This table gives the percentage probability that a species will be represented, in the sample, by exactly 0, 1, 2, 3, ⋯ individuals, if its frequency or commonness in the universe (Gaussian or any other) would lead us to expect in the sample an exact representation of "p" specimens. We are interested, however, in knowing something a little less ambitious than this, *viz.*, not the probability of a particular number of specimens, but the probability of the species falling in its proper octave of the sample, or in any other octave. The table can therefore be condensed into a shorter one, covering the first four octaves of the sample. This is done below in table VI.

TABLE VI. *Probabilities of representation in each octave when "perfect" representation would be "p" specimens*

Being a condensation of the previous table (V)
(Probabilities expressed as percentages)

p	Octaves						
	Below 1	1–2	2–4	4–8	8–16	16–32	
3/128	98.8	1.2					
3/64	97.6	2.3	0.1				
3/32	95.3	4.5	0.2				
3/16	90.7	8.5	0.8				
3/8	81.7	15.5	2.8				
3/4	65.0	24.4	10.3	0.3			— Veil Line
1½	39.0	29.3	27.4	4.2			
3	12.5	18.7	42.0	25.9	0.8		
6	0.9	3.0	17.8	58.2	20.3		
12		0.2	11.6	75.1	3.1		
24				4.4			
		107.4	101.6	100.2	100.6		

321

The fourth represents half the times we should expect doubletons, all the times we should expect three specimens, and half the times we should expect four specimens.

This is in accord with our previous methods of figuring things.

The broken horizontal line is the veil line. A species theoretically represented by ¾ of a specimen should be absent. A species theoretically represented by 1½ specimens should be represented by at least one.

The sloping interrupted line indicates the percentage probability with which a species should fall in its "own" octave. A species which in a "perfect" sample would have to be represented by 1½ specimens has less than a 30% chance to be represented in its own octave, while a species theoretically represented by 12 specimens has better than a 75% chance to fall in its own octave.

We thus have (in the table) substantially exact information about the composition of the first four visible octaves of the sample (1–2, 2–4, 4–8, 8–16). Beyond this point the calculations become very tedious and the table has not been computed. But it is easy to see see that in the next octave (16–32) it is most probable that well over 90% of all species that should fall in it, *do* fall in it, and that from this point forward (32–64), and onwards, practically every species falls in its own octave, and no octave derives any adventitious support. Therefore it is the first four octaves that are important. Beyond this point the sample is a faithful replica of the "universe," whatever the distribution of commonness and rarity in that universe may be, so long as it is a reasonably smooth continuous sort of function.

We may therefore proceed to calculate what our samples should theoretically be, for two kinds of "universe." The first universe is the one in which all octaves are equally filled, the kind that results from the harmonic series of Professor Fisher and Dr. Williams. The second universe is the Gaussian one, with the coefficient "a" in the exponent set equal to 0.20.

Harmonic series: octaves in universe equally filled

This is very easily dealt with. All we have to do is to add up the columns (except the first two) in table VI, as is done at the foot of that table. It will be seen that (allowing for small errors of calculation) all columns add up to just about 100% except the first, where there seems to be about 7% excess. That is, in this column, if we had a right to expect 14 species we should most likely get 15. This error is not very serious, and we might very well say that there are no errors beyond this point. Thus for this sort of universe, the sample is a very faithful replica.

Gaussian universe

The calculations here are more tedious, and have to be made for several cases; the veil line in the sample may coincide with the crest of the curve in the universe, or it may be at any point on the ascending or descending slopes. If we consider the universe as fixed, we can bring the veil line to any point we wish by continuing our collecting. Ultimately we collect the whole universe, and then the sample coincides with the universe, and is obviously a complete Gaussian curve if we postulate that the universe also is. We are more interested in the cases where the sample's veil line is within a few octaves of the crest of the "universe."

Let the number of species falling in each octave of the universe be assumed to be

$$n = 100e^{-(0.20R)^2},$$

where R, as before, is the number of octaves to left or right of the crest of the curve.

In most of the "universes" we have encountered in practice there is a tendency for "a" to be somewhere near 0.20, so it is logical—since we must use some speci-

TABLE VII. *Distribution of species in the Gaussian universe*

R =	0	±1	±2	±3	±4	±5	±6	±7	±8	±9	±10	±11	±12
n =	100	96.1	85.2	69.8	52.7	36.8	23.7	14.1	7.7	3.9	1.8	0.8	0.3

fic value for our calculations—to use this figure. The assumption that the modal octave will have 100 species is of course entirely arbitrary, but if it has some other number, it will merely result in all our figures being multiplied by a constant, both in the universe and in the sample, and will not affect our ultimate result in any way.

Let us now take three cases as follows: (1) with the veil line placed so that the first "exposed" octave in the sample is the modal octave of the universe, (2) with the veil line four octaves to the *left* of the above position, and (3) with the veil line four octaves to the *right* of the first position.

Applying table VII to table VI, we see, in Case (1), that the modal octave ought to be made up of 29.3% of the 100 species that "belong" there, plus (24.4 + 18.7)% of the 96.1 species that theoretically belong in the immediately adjoining octaves, plus smaller contributions from more distant octaves of the universe. After completing the calculations for all the octaves, and for all three cases, we reach the following conclusions:

proximate the veil-line position of Case (1) rather than the others, it appears likely that even in the other octaves the errors will not usually exceed a few per cent. In figures 8A, 8B, 8C and 8D, we show the (species) curve for the postulated universe, and for the three samples of Cases (1), (2) and (3), respectively. In b, c, and d, the solid line is the most probable sample curve, and the broken line the corresponding curve for the universe.

This method of computing the curves is an approximation only, and the postulate that the sample curve is essentially a decapitated version of the universe curve is likewise only an approximation; but it seems that it is sufficiently close for our present purposes.

Raunkiaer's law

Some twenty years ago there was some discussion of Raunkiaer's law in the pages of ECOLOGY, (Kenoyer, '27; Gleason, '29; Romell, '30) and a much more extensive literature elsewhere. This law concerns itself with the question whether a given species occurs, or does not occur.

Sample octave	1–2	2–4	4–8	8–16	Specimens/Species
Case (1)					
Most probable representation in sample	96	96	87.2	—	Number of species in the octave
"Perfect" representation	100	96.1	85.2	—	Number of species in the octave
Case (2)					
Most probable representation	45.5	64.0	81.5	94.7	Number of species in the octave
"Perfect" representation	52.7	69.8	85.2	96.1	Number of species in the octave
Case (3)					
Most probable representation	69.3	44.4	27.3	16.0	Number of species in the octave
"Perfect" representation	52.7	36.8	23.7	14.1	Number of species in the octave

In all cases, though there are errors in the first two or three octaves, representation becomes close to perfect beyond this point: and since most samples tend to ap-

Twenty-five test areas of equal size are examined, which areas may be a square meter in size, or one-tenth of a square meter, or of some other size. The only

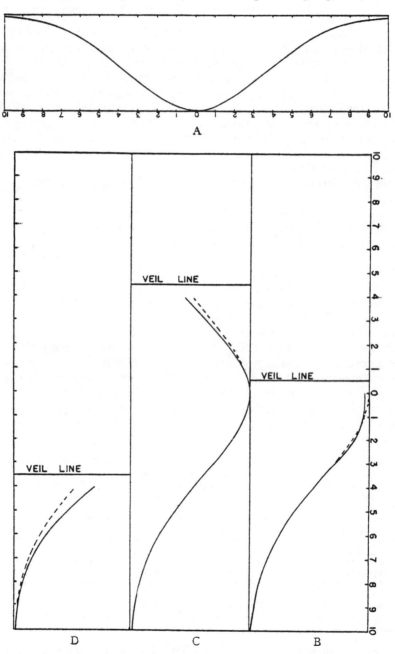

question involved is whether species S_1 (say) is present in all quadrats, or only in some, and if so in how many. It does not concern itself (in its original form)

with the question whether it is represented by thousands of specimens or only one.

The species S_1, S_2, etc., are then clas-

FIG. 8. A: a typical Gaussian curve. B, C and D, approximate relation of sample (solid line) to universe (broken line) with different degrees of decapitation (i.e., different sizes of sample). The samples follow the universe curve fairly closely.

sified into groups: Group A is represented in from 1 to 5 of the quadrats, B in 6 to 10, C in 11 to 15, D in 16 to 20, E in 21 to 25. Raunkiaer found that if the quadrat size were appropriately chosen, then group A was the largest, B less, C less still, but E tended to be larger than D. This sway-backed curve is usually described as "having two peaks."

It will be observed that what is directly measured is not commonness as understood in the present paper, where we are concerned with the numerical representation of a species in the sample or in the universe from which the sample is drawn; what Raunkiaer is concerned with is a measurement of *ubiquity*.

Gleason is quoted by Kenoyer as having called attention in 1920 to the relation existing between the actual number of individuals of the species, and the frequency index (Raunkiaer group) thereof.

But though there is a reasonable presumption that a ubiquitous species is a commoner one than a more local species, there is no *necessary* connection between the two things. We might have two species, buttercups and daisies. One buttercup occurring in each of 25 quadrats will rank the species as of a very high frequency; 25,000 daisies on a single quadrat, but missing from the other 24, will rank it as of a low frequency. Unsociable insects like the praying mantis, which on account of its cannibalism can be found only in a well-scattered arrangement, may be ranked as "frequent" but a colonial species like the bald-faced hornet, or an ant, might be infrequent, though at the nest there might be many thousands of individuals.

However, subject to biological factors of this sort, it still remains true that on the average there will be a definite relation between commonness and ubiquity, and it is easy to pass from the one to the other.

The Raunkiaer Index of Frequency is a measure of the probability that the species will occur (i.e., be represented by one or more specimens) in any one sam-

ple quadrat. Thus if the species occurs in 6 quadrats out of the 25, i.e., in 24% of them, the probability of its occurring in any one quadrat chosen at random is 0.24.

In a Raunkiaer quadrat or its equivalent, there may be one or more species to which the Poisson distribution laws do not apply. The condition for applicability is that there must be room in the quadrat for several times as many individuals as actually occur: this room will be non-existent if the species is already tightly packed in a dense stand, or if the individual bulks so large that it occupies a substantial part of a quadrat: it will also be non-existent for territorially-minded birds, and the like, where biological or ecological factors negate the possibility. The proof of the pudding, however, is in the eating, and experiment shows that normally such species, in a Raunkiaer test, are few or absent, and that therefore we can apply the Poisson distribution series with a sufficiently high degree of accuracy to all species.

In a Raunkiaer analysis, the great majority of species must be *absent* from one or more quadrats out of twenty-five, and in a random set of samples this implies that the *average* number of specimens of any one species in a *single* quadrat will rarely be as high as five.

Thus if p be the "average expectation" or "perfect representation," the series

$$e^{-p}\left(1 + p + \frac{p^2}{2!} + \frac{p^3}{3!} + \text{etc.}\right)$$

gives the probability that the species will be represented by

$$0 \quad 1 \quad 2 \quad 3$$

specimens in any one quadrat.

We are concerned now only with the first term, representation by zero specimens. The probability that the species is completely unrepresented in a sample (quadrat) chosen at random is simply e^{-p}. Therefore the probability that it is *not* unrepresented is $(1 - e^{-p})$.

This is true whatever the shape or character of the universe may be.

TABLE VIII

Veil Line
↓

Arbitrary name of octave	ζ	ε	δ	γ	β	α	A	B	C	etc.
Perfect representation p	3/128	3/64	3/32	3/16	3/8	3/4	1½	3	6	etc.
Probability that species is not unrepresented (%)	2.4	4.6	8.9	17.0	31.3	52.7	77.7	95.0	99.8	etc.
Raunkiaer index of frequency	A	A	A	A	B	C	D	E	E	etc.

Now the values of e^{-p} are given in table V above, or rather the values there given are 100 e^{-p}. Hence the values of 100 $(1 - e^{-p})$ are immediately obtainable for a series of values of perfect proportionality p. This gives us table VIII above.

The boundaries between the Raunkiaer groups A, B, etc., are incorrectly given in some treatises: the correct values, given in percentages, are A, 2–22%; B, 22–42%; C, 42–62%; D, 62–82%; E, 82–100%. It is obvious that with 25 quadrats, a species cannot occur in 5½ of

them: it may occur in either 5 or 6: if in 5 (= 20%), it falls in group A; if in 6 (= 24%), it falls in group B. If its calculated chance of being present in a single quadrat is 22%, then it is equally likely to be in A or B, so the boundary between groups is at 22%. Similarly, if its chances are 2%, it is equally likely to be represented in one solitary quadrat, or to be totally unrepresented in all.

In order to find the boundaries between groups a little more accurately, we graph the data of table VIII in figure 9. Here the circles are the calculated points,

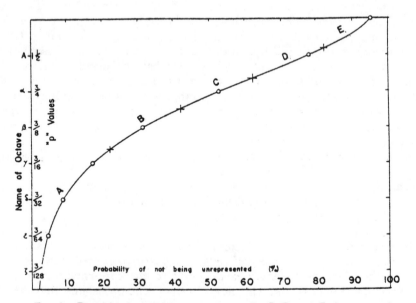

FIG. 9. Raunkiaer's frequency groups, A, B, C, D and E, in terms of octave ratings. The boundaries between groups are at 2, 22, 42, 62, 82%, as measured on the abscissa. The corresponding octave ratings are given by the ordinates. A horizontal line half way between α and A is the veil line, which cuts Raunkiaer's group D. It follows that Raunkiaer's groups are concerned for the most part with samples so small that most species in the universe are behind the veil line of the sample.

and the crosses mark the boundaries of the Raunkiaer groups. In terms of octaves, the groups are made up as follows:

The total number of species, 674.2, is of course based on the assumption that in the universe the modal octave contains

TABLE IX

<A (missing species)	All octaves to left of ζ (zeta).
A	ζ, ϵ, δ, and about 90% of γ.
B	β and about 10% of γ.
C	About 85% of α.
D	About 15% of α and 70% of A.
E	About 30% of A and all octaves to right of A.

So far the argument has been perfectly general. We have assumed nothing whatever about the "universe" from which the sample is drawn. Any universe whatever can be plotted on a logarithmic or octave base, and our discussion so far is pure mathematics. Now it becomes necessary to assume either that the octaves are all equally filled, or that they are not. The first assumption produces a universe with an infinite number of species, and of specimens: a finite universe demands unequally filled octaves. Let us therefore test the possibility that our Gaussian universe, on a logarithmic base, is the real one, and see if it agrees with the Raunkiaer findings.

Table VII above gives the ordinates of a Gaussian curve, i.e., the relative number of species in each octave if the universe is Gaussian, with an "a" value of 0.20 typical of our work with moths and birds.

Put the veil line just to the left of octave + 4, i.e., between + 3 and + 4.

Then category A contains the following number of species:

100 species. The Raunkiaer results are expressed in terms of the sample itself, i.e., 674.2 corresponds to 100%. Reducing our figures throughout in this proportion, we get table X below, where our theoretical figures and Kenoyer's experimentally-obtained frequencies are compared.

These figures agree so well that they add considerably to our confidence that we have chosen the right sort of universe, a Gaussian one, and placed the veil line at approximately the right place. If we place it somewhere else (choose larger or smaller quadrats) the Raunkiaer Frequency percentages will change, and change radically, as all the writers of twenty years ago recognized. They also recognized that it would change in a manner appropriate to that which we have found inevitable if the universe is Gaussian.

Most of the work by Raunkiaer and his followers was concerned with plant species in quadrats, but Kenoyer showed the law to hold good for sweeps of insects in a net, and for the microscopic life obtained

$$
\begin{aligned}
& 85.2 + 96.1 + 100.0 + (0.9)(96.1) &= 368.0 \\
&B\ (0.1)(96.1) + 85.2 &= 94.8 \\
&C\ (0.85)(69.8) &= 59.2 \\
&D\ (0.15)(69.8) + (0.70)(52.7) &= 47.3 \\
&E\ (0.30)(52.7) + 89.1 \text{ (all to right)} &= 104.9 \\
& \text{Total} &\overline{674.2}
\end{aligned}
$$

TABLE X

Raunkiaer group.............	A	B	C	D	E
Kenoyer's values (experimental)	53%	14%	9%	8%	16%
Our values (theoretical)	54.6	14.1	8.8	7.0	15.6

from infusions of hay. Thus the logical assumption is that these "universes" also are Gaussian.

Gleason ('29) and Romell ('30) both have excellent criticisms of the Raunkiaer law, but their perplexities seem to be removed by the present demonstration of the connection between sample and universe. Thus Gleason's concluding sentence reads: "Raunkiaer's Law is merely the expression of the fact that in any association there are more species with few individuals than with many; that the law is most apparent when quadrats are chosen of the most serviceable size to show Frequency, and that it is obscured or lost if the quadrats are either too large or too small."

This comes within an ace of correctly summarizing the situation, and yet is quite wrong. It should read, "In any *association* there are just as many very rare species as there are very common ones, but species of moderate abundance are vastly more numerous than either. In a *small sample* however, there will be more species with few individuals than with many, because those with many are the excessively common ones, and these species are few."

Romell likewise comments trenchantly ('30, p. 593): "It follows that *statistics made with different sizes of sample areas cannot be compared,* because there is no safe way of correcting the results for another size of analyzing area." (The italics are his.) Figure 10 is a chart permitting the transition, from one sample size to another, that Romell desires. On the assumption that the universe is logarithmically Gaussian with the exponent $a = 0.20$, we can calculate what the sample will be like for any position of the veil line with respect to the vertex of the Gaussian curve. This sample can then be expressed in terms of Raunkiaer groupings. Figure 10 represents the end result of the calculations.

Our chart consists of four sigmoid lines ascending from lower left to upper right. These lines mark out five domains, one each for the A, B, C, D, and E groups.

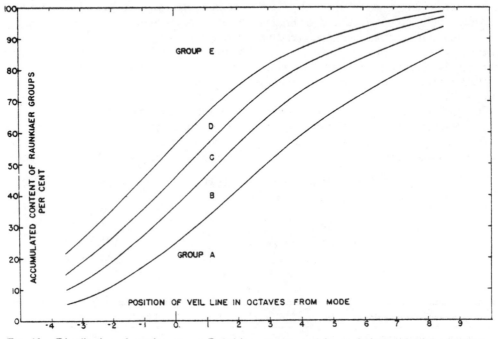

Fig. 10. Distribution of species among Raunkiaer groups and its variation with size of sample, assuming that the universe is logarithmically Gaussian.

TABLE XI

Group	A	B	C	D	E of Raunkiaer
Skead (observed)	75.4	13.1	5	5	1.6 per cent
"Theoretical," from chart	75.2	10.6	5.1	3.5	5.5 per cent

The vertical intercept between any two lines represents the percentage of species falling in one of the groups B, C, or D. On any ordinate the intercept between the lowest sigmoid and the base line represents the percentage falling in group A, and the intercept between the highest sigmoid and the ceiling line represents the percentage falling in group E.

The chart is one hundred units high. To use it, take any experimental Raunkiaer grouping, and, on the edge of a piece of paper of this same height, mark off the accumulated percentages, in sequence, A, B, C, D, and E. Move this strip of paper, with its ends in register with base and ceiling lines, from left to right till the best fit is obtained with the sigmoid lines. We have then located our sample with respect to its "universe": i.e., we have located our veil line.

Skead ('47), for instance, gives a Frequency-Grouping for birds observed over a period of a year on a small area of the Bushveld. This departs widely from Kenoyer's example, but apparently only because it was so small a sample, containing only 61 species out of the many hundred that might conceivably be observed over a much longer period. When these are separated into five groups, they average only 12 to a group, which is not enough for good statistical work. But taking the veil line at 6½ octaves to the right of the mode, we get table XI, below.

The fit is fair. It cannot be expected to be much better, for Skead's group E contains only a single species, and no group but A is large enough for good statistical work. The fit in group C is accordingly accidental, as are the slight discrepancies in the others. All that we can say is that Skead's results are not inconsistent with the hypothesis that they are drawn from a universe of the same general shape as all our other examples.

More generally, we may note the agreement of the chart with the basic conclusions of the Raunkiaer investigations. The four curves mark out a relatively narrow belt across the field, so that groups A and E together always greatly exceed B, C, and D combined. Group C is usually comparable in size with D; group B is usually greater than either; A is usually much greater than B, and E is usually greater than D, normally considerably greater (except for very small samples).

These seem to be the experimental conclusions as reported in the literature: they are immediately deducible from the chart.

The available evidence apparently does suggest that a great many biological "universes" have the logarithmically-Gaussian form, with a coefficient "a" not far from 0.20. If birds in a valley, moths in a trap, plants in a quadrat, insects in a sweep net, and micro-organisms in a suspension of hay do in fact all agree in this, it would seem that some very general law must lie behind it all.

Williams' law of collection enrichment

Williams, referring to his moths and other insects, observes: "Doubling the number of insects caught, (and hence the time of trapping) at any level, except for very small samples, always adds about 30 species to the total."

Now we have seen that doubling the sample simply means adding one more octave that was previously veiled; and so long as Williams was working over the intervals ⅛ year, 1 year, 4 years, he would be working near the crest of the curve, where one octave contains (in his

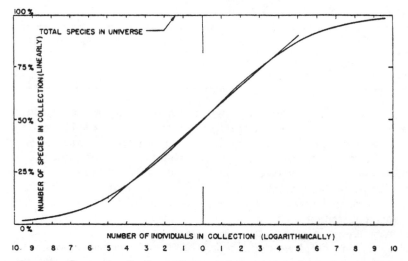

FIG. 11. Curve of collection enrichment in an unchanging universe. This "sigmoid" curve is simply the integral of the Gaussian curve. It indicates that over a wide range, doubling the size of the sample, or doubling the time of collecting, tends to add a constant increment of species to the collection (Williams' law). A compromise line as drawn sufficiently replaces the curve itself over about 8 octaves, a much longer interval than Williams tested. (But the Williams' law remains true in spite of the fact that the universe is not unchanging.)

case) about 30 species, and he would be covering about 5 octaves.

If the universe remained constant during the trapping, it is easy to show that Williams' law must necessarily be very nearly true over a range of several octaves. It is perhaps most easily seen from the integral curve. For the collection at any time consists of

$$y = n_0 \int_R^{+\infty} e^{-(aR)^2} dR \text{ species.}$$

This integral (y) plotted against R is a sigmoid curve which can be approximated very closely by a straight line over several octaves near the point of inflection.

Figure 11 shows such a curve, using a value for "a" of 0.16, close to the Saskatoon and Lethbridge figures rather than Williams' value ($a = 0.227$), for reasons that will appear later. The octave intervals are marked and it is clear that a straight line is a good approximation over 6 or 8 of the central octaves.

The surprising thing, in practice, is that Williams' law holds so well in a universe so rapidly and radically changing as the real biological universes are.

The changing "universe"

That ecological assemblages are continually changing is well understood. There is the slow business of plant "successions"; there is the remarkable build-up and subsequent crash of many mammal populations (Elton, '42); and there is the obvious fact, in a temperate region, that the moths flying tonight are not the same species that flew a month ago. Yet we might expect that the *annual* catches of moths in a given light trap would be fairly constant. This is not the case.

Dr. King tells me that the environment around his Saskatoon trap did not change in any obvious manner during 22 years, and that the trap and its efficiency did not change in any material way. Yet the catch varied considerably both in number of species and of individuals caught (fig. 12).

Even when the species remain fairly constant as to *number*, as they did over the last decade, they do not remain the same species from year to year. Species come and go from year to year to a surprising extent. Out of 277 species observed, only 38 were present every year. Table XII below shows how inconstant the fauna was.

Seamans' moths of Lethbridge confirm the picture. Figure 13 graphically summarizes the annual catches both as to number of species and as to number of individuals. As at Saskatoon the fauna seems to have become on the whole more diversified ("richer") with the passing years.

As regards the constancy of the trap itself, Mr. Seamans has this to say:

"The trap is identical, has remained in the identical spot and, except for a period of one week in 1923, has been lighted with a 60-watt clear-glass incandescent bulb. During that one week we used a 100-watt bulb and caught an entirely different series of moths, while those which normally came to the light were found roosting on the trees 15 feet away, indicating that they were susceptible to variations in light intensity. At the end of the week we switched back to our ordinary globe and have used it ever since.

"The trap is on the corner of the laboratory

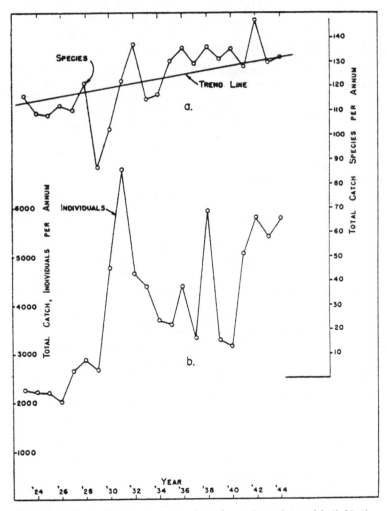

FIG. 12. Fluctuation in numbers of species and numbers of individuals in the Saskatoon trap.

331

TABLE XII. *Constancy of the Saskatoon fauna*

Present for N years out of 22	N	1	2	3	4	5	6	7	8	9	10	11	12	13	14	15	16	17	18	19	20	21	22
Number, S, of species thus present	S	49	29	9	16	8	10	12	6	11	8	11	15	3	6	6	11	8	1	7	7	6	38

TABLE XIII. *Constancy of the Lethbridge fauna*

Present for N years out of 22	N	1	2	3	4	5	6	7	8	9	10	11	12	13	14	15	16	17	18	19	20	21	22
Number, S, of species thus present	S	45	21	19	13	12	10	8	9	9	17	11	5	7	7	7	6	8	15	16	16	15	15

building, which is located on the Dominion Experimental Station grounds. The only crop changes are the normal changes which are made in rotations on irrigation plots, but none of these plots is any nearer than 70 feet from the trap.

"We have tried to maintain this trap in identical conditions in order that our yearly records may be comparable in anticipating outbreaks of the various cutworms which occur in this part of the country."

Out of the 291 species that were observed during the period, only 15 were observed in every year. This indicates how inconstant the fauna is, but the tabulation below illustrates the matter somewhat more completely.

The changing universe and Williams' law

We have seen that in a fixed universe, doubling the sample moves the curve one octave to the right, but does not change the height of the crest, or the dispersion (value of "a"). If the universe changes, it is not difficult to show that the most probable result is to increase the height of the mode, to withdraw the curve less than one octave to the right, and perhaps to leave the dispersion but little changed. These in fact are the results we also find in practice, as mentioned previously in connection with the Saskatoon and Lethbridge annual and quadrennial curves.

Lethbridge and Saskatoon data and the Williams' law

The pronounced "trends" towards enrichment in the annual count of specimens at both the above places have to be eliminated before we can do much towards testing the Williams' law. This elimination can be done by an artifice. Let us assume that at some time or other the fauna will show a trend towards im-

TABLE XIV. *Lethbridge collection enrichment with trend eliminated*
(new species appearing each year after first)

Forwards 1921–43	(102)	42	27	4	2	9	13	11	10	7	3	1	6	10	4	7	5	2	5	12	7	2
Backwards 1943–21	(155)	31	32	10	2	7	9	3	8	6	1	4	2	3	1	4	2	0	1	5	2	3
Trendless	(257)	73	59	14	4	16	22	14	18	13	4	5	8	13	5	11	7	2	6	17	9	5

TABLE XV. *Saskatoon collection enrichment with trend eliminated*

Forwards 1923–44	(116)	22	11	8	10	11	1	4	9	11	6	9	7	7	11	7	4	6	1	8	2	2
Backwards 1944–23	(132)	29	33	10	11	13	10	5	9	3	6	2	4	1	0	1	1	0	2	1	1	3
Trendless	(248)	51	44	18	21	24	11	9	18	14	12	11	11	8	11	8	5	6	3	9	3	5

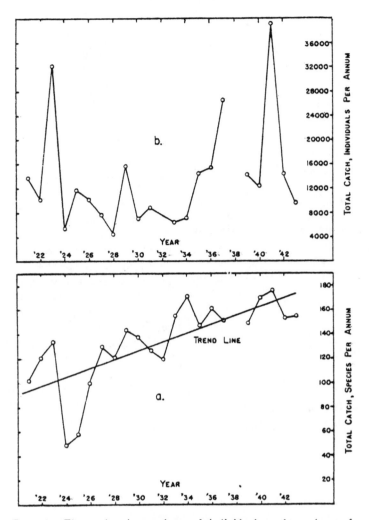

Fig. 13. Fluctuation in numbers of individuals and numbers of species in the Lethbridge trap.

poverishment exactly equal to its present trend towards enrichment. Then we might collect a set of data exactly like the present one, with the years reversed. That is, we could pretend that 1943 was really 1921, 1942 was really 1922 and so on. Adding the two sets together, we should obtain a series from which "trend" has been eliminated.

We can analyze such a series to see how many new forms appear each year in a trendless world. This is the same thing as finding how many species appear each year, for the *first* time, in the real series,

and how many appear each year, for the *last* time in the real series. This gives us two sets of data, the first of which we write "forwards" and the second "in reverse." It is most easily understood by carrying out the process as is done below. In the Lethbridge case, the year 1938 is missing, so the series is "telescoped" in order not to leave a gap.

The Williams' law says that doubling the period of observation always adds a constant increment of "new" species. We may test the law somewhat crudely as follows:

TABLE XVI. *"New" species resulting from doubling the period of observation*

End of year	1	2	4	8	16
Increment at Lethbridge	(257)	73	73	56	77
Increment at Saskatoon	(248)	51	62	65	93

It would seem that the law is by no means a bad approximation.

A more accurate method of testing the law involves putting it into a slightly different form, which might read: "The number of species in the collection increases as the logarithm of the time of collecting."

This may be expressed mathematically in either of two forms, which are equivalent:

$$N - N_1 = (N_{10} - N_1) \log_{10} t$$

or

$$N = N_1 + \frac{\Delta N}{\log_{10} 2} \log_{10} t$$
$$= N_1 + (3.3 \Delta N) \log_{10} t.$$

Here N is the number in the collection after t years; N_1 is the number at the end of the first year; N_{10} the number after 10 years of collecting, and ΔN is the constant number or increment which Williams says is added as a result of doubling the collection.

More briefly, what Williams says is that if we plot the number of specimens in the collection on a natural scale, as ordinate, against the logarithm of the time of collecting as abscissa, we shall get a straight line. Inasmuch as the Canadian data are much more extensive than those from which Williams deduced his law, we tabulate below the status of the Lethbridge and Saskatoon collections, year by year, corrected for trend. (Note: The number of species reported is finally twice the total collection of species, since we add the inverted sequence to the observed sequence, and do not halve the result to strike an average).

The results are plotted in figure 14, and it seems that they do bear out Dr. Williams' beliefs. Straight lines may be drawn fairly closely approximating the results, and what is more, the two straight

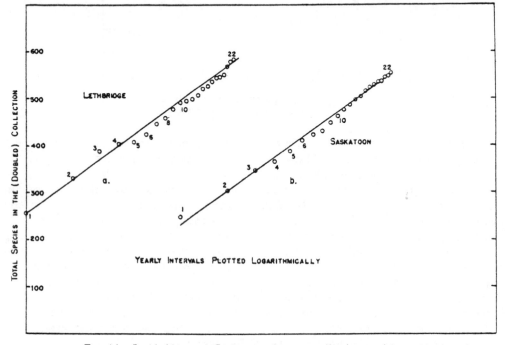

FIG. 14. Lethbridge and Saskatoon data on collection enrichment.

lines have the same slope. The number of species added by doubling the period of observation is in each case approximately 72. This is the value in each of two collections of comparable richness, having approximately 250 species in the first (composite) year. In the actual collections, which would have about 125 species in the first (or average) year, the increment ΔN to be expected from doubling the period of trapping is about 36 species, in the absence of trend. This is about 30% of an average year's catch.

TABLE XVII. *Rate at which collections (of species) grow at Lethbridge and Saskatoon (composite collections, corrected for trend)*

Years from start	Species (Leth-bridge)	Species (Saska-toon)	Years	Leth-bridge	Saska-toon
1	257	248	12	499	481
2	330	299	13	507	492
3	389	343	14	520	500
4	403	361	15	525	511
5	407	382	16	536	519
6	423	406	17	543	524
7	445	417	18	545	530
8	459	426	19	551	533
9	477	444	20	568	542
10	490	458	21	577	545
11	494	470	22	582	550

Williams' own statement is, "Doubling the number of insects caught (and hence the time of trapping) at any level, except for very small samples, always adds about 30 species to the total." His average annual catch (for four years) was 176 species. The question of "trend" is not dealt with in his report, and probably could not be usefully treated of, even if suspected. But it seems that his new species, ΔN, amount to about 17% of an average annual catch of species, instead of the 30% found at the Canadian stations.

Certain special "universes"

The universes we have been considering are mostly universes of taxonomic groups, since these are the only ones on which we appear to have adequate data. However if the correspondence between the Raunkiaer results and our present ones is accepted, then the older data on Raunkiaer groups may be considered as adding ecological assemblages to our universes, and hinting that they are Gaussian and with values of "a" not very different from those of taxonomic universes.

The universe of a light trap is not comprised in a definite area. Some species, weak fliers, like many of the geometrids, will presumably be collected only from close quarters, while strong flying noctuids or sphinges may have come from miles away—in England, for instance, they will often have come from beyond the English Channel or North Sea. The universe involved is a product of the actual commonness of the moths as we imagine an omniscient being might count them, multiplied by a factor representing relative phototropism of the various species, multiplied again by a factor representing power-of-flight, and perhaps other factors, as indicated by Seamans.

However, it will probably be conceded that the universes, whatever they are, are humped distributions approximating to a Gaussian curve on a geometric series as a base.

We should perhaps like some further evidence that the left-hand end of the curve, which is so inaccessible experimentally, actually does become asymptotic to the same axis as the right-hand end. This can only be done by a consideration of the universes themselves.

The Nearctic avifauna

As one such "universe" we might take the breeding birds of the Nearctic, which comprises, approximately, that part of the North American continent, and adjacent islands, north of the Rio Grande and the Straits of Florida. Its area is about 8.5 million square miles or 5.5×10^9 acres. Peterson ('40) estimated that there may be 5×10^9 individual birds in the continental U.S.A., so a figure of 10^{10} for the whole Nearctic, or roughly one pair per acre, may not be unreasonable. According to the American Ornithologists' Union Check-list, these individuals are distributed among 641 species, of which 35 are essentially Neotropical species whose breeding range extends slightly north of the Rio Grande. This leaves us with 606 species which may fairly be called Nearctic ones.

Now $10^{10} = 2^{33}$ approximately. Thus no species can have as many as 2^{33} individuals; and since no species can have less than one individual, the whole fauna must be comprised within less than 33 octaves, probably within 30 octaves as a maximum. In fact, since the total of all species is estimated at about 5×10^9 *pairs,* and the commonest species will not have one-fifth of all the individuals, the range of commonness will not extend over so much as from 1 to 10^9 pairs.

Let us divide this range into three equal parts (i.e., equal logarithmically, on our basis of plotting), viz. 1 to 10^3, 10^3 to 10^6, and 10^6 to 10^9 pairs.

Species falling in the range 1 to 10^3 or perhaps in the range 1 to 10^4, the first ten or twelve octaves or therabouts, appear to number about fifty-five, according to a conscientious study of the A. O. U. Check-list by a competent ornithologist friend. These are the rare species. The very common species, whose numbers exceed 10^6 pairs, or possibly 10^7, may number about sixty. Therefore the species of an intermediate level of commonness, from 10^3 to 10^6 or possibly 10^7 pairs, number about 490 and are thus in a great majority.

It is clear therefore that our Nearctic universe is bell-shaped, with a high hump in the middle, and approximates to a logarithmically-Gaussian curve. At the left-hand end we approach the axis asymptotically with some octaves blank, or occupied by only a single species, such as the Ivory-Billed Woodpecker, with only one or two pairs surviving. The

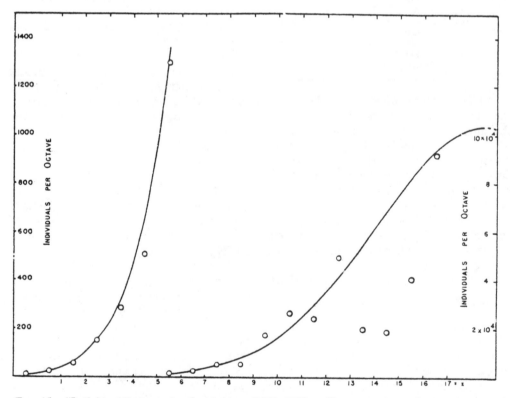

FIG. 15. "Individuals" curve for Lethbridge (1921–1943). The curve is a single curve, but the scale is changed at abscissa = 5.5. The curve is the left-hand half of a Gaussian curve.

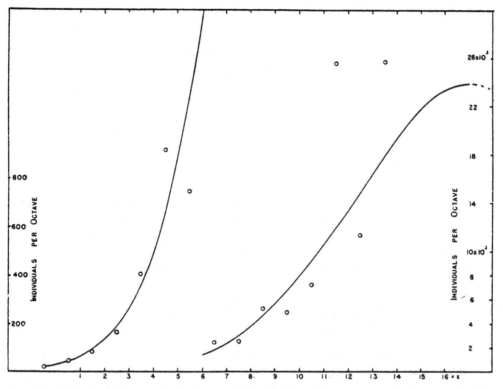

FIG. 16. "Individuals" curve for Saskatoon (1923–1944). The curve is a single curve, but the scale is changed at abscissa = 6.0.

first 10 or 12 octaves have only 55 species, or an average of 4 or 5 species per octave. The same thing is true of the last octaves to the right.

We do not know that the curve may not be "skew," but on the assumption that it is approximately symmetrical (Gaussian) we can compute its properties. The assumption is that the curve is of the form

$$n = n_0 e^{-(aR)^2};$$

the experimental information is: (1) that the total number of species

$$N = \Sigma n = \int_{-\infty}^{+\infty} ndR = 606 \text{ species,}$$

and (2) that 490 species lie within the central third of the base line.

This results in an equation

$$n = 65e^{-(0.19R)^2},$$

where as usual n is the number of species per octave and R is the number of octaves to left or right of the modal octave, which should have about 65 species.

It may be noted that the dispersion constant "a" has a value of 0.19, which is very similar to the values we have obtained in many other instances.

Thus it seems likely that a study of a whole universe would bear out the evidence of our samples, that it is in fact a logarithmically-Gaussian universe.

The "individuals" curve

If the *species* that fall in a given octave are represented by

$$n = n_0 e^{-a^2(x-b)^2} = n_0 e^{-a^2 R^2},$$

where n is the number of species in the xth octave from the veil line, n_0 the number of species in the modal octave, which is distant "b" octaves from the veil line, and "a" is the dispersion coefficient, then

337

282 F. W. PRESTON

it is not difficult to show that the *individuals* falling in any given octave (that is, the total individuals representing all the species in that octave) are given by

$$q = n_0 2^b e^{(\ln 2/2a)^2} \cdot e^{-a^2[R-(\ln 2/2a^2)]^2}.$$

This also is a Gaussian curve with the same dispersion "a" as before, with a mode displaced to the right through a distance $(\ln 2)/2a^2$ octaves, and with a height at the mode of

$$q_0 = n_0 2^b e^{(\ln 2)^2/4a^2}$$

individuals per octave.

Thus the curve can be derived directly from the "species" curves. The calculated "individuals" curves for Lethbridge and Saskatoon are given in figures 15 and 16, while figure 17 shows, for the Saskatoon data, the relative positions of of the "species" and "individuals" curves.

The individuals curve thus derived,

however, though Gaussian, is only the left-hand half of a complete Gaussian curve. Experimental points stop at or near the crest of the curve. This is because the *species* curve, though graduated by a Gaussian which theoretically extends infinitely to the right, in practice stops finitely, for an octave cannot hold less than one whole species unless it holds none. If one of the blank octaves beyond the last observed point (which is usually one species in an octave) contained species, it would hold 2, 4, or more times as many individuals as the said last observed point. Thus we cannot, with the individuals curve, observe a declining number of individuals, and are limited to the ascending portion of the curve.

It follows that the distance between the modes of the species and individuals curves ($= (\ln 2)/2a^2$ octaves) is approximately equal to the distance between the

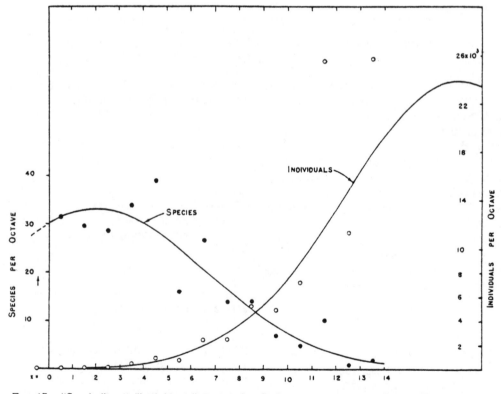

FIG. 17. "Species" and "individuals" curves for Saskatoon on the same base. The continuity of the "individuals" curve is made clear.

338

mode of the species curve and the observed last point on that curve.

The area under the individuals curve (assumed to be half a Gaussian curve) is

$$Q = \int_{-\infty}^{0} q \, dR = \frac{n_0 \sqrt{\pi} \cdot 2^{b-1}}{a} e^{(ln\,2)^2/4a^2}$$

and represents the total number of individuals observed of all species.

Table XVIII gives a comparison of some of the Q-values thus calculated and the observed totals of individuals.

TABLE XVIII. *Total individuals*

Place	Period	Calc.	Obs.
Lethbridge	22 years	579,830	303,251
Saskatoon	22 years	139,310	87,110
Orono	4 years	27,126	56,131
Rothamsted	4 years	11,246	15,609
Quaker Run		30,940	14,353

The calculated values are of the same order of magnitude as the observed ones, sometimes greater, sometimes less: the "errors" in fact are + 48, + 38, − 107, − 40, + 54%, or an average − 1%, and are due essentially to variations in the number of individuals in the two or three commonest species.

Summary

Random samples of ecological or taxonomic assemblages indicate that the universes from which they are drawn have, at least approximately, the form of an ordinary Gaussian curve drawn upon a logarithmic base (a "lognormal" curve). The sample has the same general form as the universe, but is decapitated.

The exact relationship between sample and universe is explored, and the Raunkiaer Law of Frequency explained, as is Williams' Law of Collection Enrichment. There is a remarkable tendency for the dispersion constant "a" to be not far from 0.2 in a great variety of biological universes.

Various applications of the theory are made to rather inaccessible populations, such as the Nearctic avifauna in its entirety. The findings seem reasonable in all cases.

LITERATURE CITED

Corbet, A. S. 1941. The distribution of butterflies in the Malay Peninsula. Proc. R. Ent. Soc. Lond., (A), 16: 101–16.

Dirks, C. O. 1937. Biological studies of Maine moths by light trap methods. The Maine Agriculture Experiment Station, Bulletin 389. Orono, Me.

Elton, Charles. 1942. Voles, mice and lemmings. Oxford.

Fisher, R. A., A. S. Corbet, and C. B. Williams. 1943. The relation between the number of individuals and the number of species in a random sample of an animal population. J. Anim. Ecol., 12: 42–58.

Gleason, H. A. 1929. The significance of Raunkiaer's law of frequency. Ecology, 10: 406.

Hinton, H. E. 1940. A monographic revision of the Mexican water beetles of the family Elmidae. Nov. Zool., 42: 217–396.

Kenoyer, Leslie A. 1927. A study of Raunkiaer's law of frequence. Ecology, 8: 341.

Peterson, R. T. 1940. How many birds are there? Audubon Magazine, 43: 179–87.

Romell, L. G. 1930. Comments on Raunkiaer's and similar methods of vegetation analysis and the law of frequency. Ecology, 11: 589.

Saunders, A. A. 1936. Ecology of the birds of Quaker Run Valley, Allegany State Park, New York. New York State Museum Handbook 16. Albany, N. Y.

Skead, C. J. 1947. One year's census of birds in 2½ acres of Albany Bushveld. The Ostrich, 18: 155.

ERRATUM

Page 264, column 1, line 18 from the top of the page, "it" should be deleted.

19

Patterns of Species Abundance and Diversity

Robert M. May

Contents

1. Introduction

If the relative abundances of the species in a particular plant or animal group in a given community are somehow measured, there will be found some common species, some rare species, and many species of varying intermediate degrees of rareness. These species abundance relations (the relations between abundance and the number of species possessing that abundance) are clearly of fundamental interest in the study of any ecological community. Different types of such species abundance relations have been proposed on theoretical grounds, and are observed in real situations. What these relations mean, and how they are best characterized, has been the subject of considerable discussion, much of it focused on one or another particular aspect of a specific species-abundance relation.

This chapter seeks to give an analytic review of the subject, mainly with the aim of disentangling those features that reflect the biology of the community from those features that reflect little more than the statistical law of large numbers.

This discussion of species-abundance relations also provides the basis for a consideration of species-area relations (the relation between the area of real or virtual islands, and the number of species on the island), and for some rough suggestions relating to practical problems of sampling all the species in a community.

One single number that goes a long way toward characterizing a biological community is simply the total number of species present, S_T. Another interesting single number is the total number of individuals, N_T; alternatively, if the number of individuals in the least abundant species be m, the total population can be expressed as the dimensionless ratio

$$J = N_T/m \qquad (1.1)$$

Although there may be monumental difficulties in determining S_T and J in practice (e.g., Matthew, 10:29–31), such a census is possible in principle, and the bulk of this article deals with properties of the actual species-abundance distribution. Some sketchy remarks on sampling problems are deferred to Section 6.

Going beyond this gross overview of the community in terms of S_T and J, one may ask how the individuals are distributed among the species. That is, what is N_i, the number of individuals in the ith species? Such information may be expressed as a probability distribution function, $S(N)$, where

$$S(N)\,dN = \begin{Bmatrix} \text{number of species} \\ \text{each of which} \\ \text{contains} \\ \text{between } N \text{ and} \\ N + dN \\ \text{individuals} \end{Bmatrix} \qquad (1.2)$$

That is, roughly speaking, $S(N)$ is the number of species with population N. The

quantities S_T and N_T follow immediately from this distribution function:

$$S_T = \int_0^\infty S(N)\,dN \qquad (1.3)$$

$$N_T = \int_0^\infty NS(N)\,dN \qquad (1.4)$$

[$S(N)$ has been expressed here as a continuous distribution. For a discrete distribution $S(N)$, sums replace the integrals in eqs. 1.3 and 1.4, and elsewhere. For relatively large populations this distinction between continuous and discrete distributions is generally unimportant, and it is usual for field data to be plotted as a histogram or as discrete points, and compared with the continuous curves generated by theoretical distributions. For further discussion, see Bliss (1965).]

There are a variety of currently conventional ways in which the properties of a particular species-abundance distribution, $S(N)$, may be displayed. These are surveyed in Section 2. This is largely review material, but it may be illuminating to gather together, and explicitly to relate, these superficially different ways of exhibiting the same information.

In Section 3 the most significant distributions $S(N)$ are singled out. For each distribution, we briefly review underlying theoretical ideas that lead to it, and such evidence as may be culled from appropriate field situations.

First, and most important, is the lognormal distribution (Section 3 A). Theory and observation point to its ubiquity once $S_T \gg 1$, when relative abundances must be governed by the conjunction of a variety of independent factors. In general, *two* parameters[1] are needed to characterize a specific lognormal distribution; these may be S_T and J. A further assumption as to details of the shape of the lognormal distribution reduces it to a special *one*-parameter family of "canonical" lognormal distributions; in this event S_T alone, or J alone, is sufficient to specify the distribution uniquely. This empirical assumption (Preston, 1962) fits a lot of data, but no explanation has previously been advanced as to why it may be so. Another empirical general rule relates to the width parameter in the conventional expression for the general lognormal. This rough rule, $a \sim 0.2$, has invited much speculation since first enunciated by Hutchinson in 1953. In Section 3 A it is argued in some detail that both rules derive from *mathematical* properties of the distribution, being roughly fulfilled by a wide range of general lognormal distributions, for a wide range of the values S_T and J found in nature.

MacArthur's (1957, 1960) "broken-stick" distribution (Section 3 B) may be derived in various ways; it is specified by *one* parameter, namely S_T. As discussed most fully by Webb (1973), who calls it the "proportionality space model," this distribution of relative abundance is to be

expected whenever an ecologically homogeneous group of species apportion randomly among themselves a fixed amount of some governing resource. For appropriately small and homogeneous taxa, field observations seem to fit this distribution.

Two other interesting distributions (Section 3 C) are the simple geometric series and the logseries. If the community ecology is dominated by some single factor, and if division of this niche volume proceeds in strongly hierarchical fashion with the most successful species tending to preempt a fraction k, and the next a fraction k of the remainder, and so on, we arrive at a geometric series distribution (as the ideal case), or a logseries distribution of relative abundance (as the statistically realistic expression of this underlying picture). A few natural communities, particularly simple plant communities in harsh environments, conform to these patterns. (The logseries also can often arise as a sampling distribution, as mentioned in Section 6.) Both geometric and logseries are *two*-parameter distributions: for the geometric series the parameters are usually S_T and k (but are alternatively S_T and J); and for the logseries they are conventionally S_T and "α", or "α" and "x" (but alternatively S_T and J).

In brief, if the pattern of relative abundance arises from the interplay of many independent factors, as it must once S_T is large, a lognormal distribution is both predicted by theory and usually found in nature. In relatively small and homogeneous sets of species, where a single factor can predominate, one limiting case (which

---

[1] A third parameter, namely N_T or m or N_0, would be needed if one were interested in absolute values of numbers of individuals, i.e., in the absolute position along the R axis of the peak in the distribution. By working with the ratio J, eq. 1.1, this need is circumvented, and for essentially all ecological purposes (rank-abundance curves, diversity and dominance indices, rough estimates of sampling properties, species-area curves) two parameters suffice. Similar remarks pertain to the other distributions.

may be idealized as a perfectly uniform distribution) leads to MacArthur's broken-stick distribution, whereas the opposite limit (which may be idealized as a geometric series) leads to a logseries distribution. These two extremes correspond to patterns of relative abundance which are, respectively, significantly more even, and significantly less even, than the lognormal pattern. In other words, *the lognormal distribution reflects the statistical Central Limit Theorem; conversely, in those special circumstances where broken-stick, geometric series, or logseries distributions are observed, they reflect features of the community biology.*

Many people have sought to go beyond the simple characterization of a community by the two numbers S_T and N_T (or J), yet stop short of describing the full distribution $S(N)$, by adding one further single number which will describe the "evenness" or "diversity" or "dominance" within the community. In Section 4, the theoretical relationships between such quantities and S_T are exhibited for each of the major species-abundance distributions $S(N)$ mentioned above, and their features are compared. Some of these relations have been explored for particular distributions by previous authors, but a comparative anatomy is lacking, and the lognormal distribution (which I regard as the most important) has received essentially no attention of this kind. The statistical variance to be expected in the usual Shannon-Weaver diversity index, H, is also discussed for the various distributions. We see that for relatively small S_T the various distributions $S(N)$ lead to

H-versus-S_T curves, the differences between which are of the same order as the statistical noise to be expected in any one such relation. Conversely, for $S_T \gg 1$ it is difficult to see how the distribution could be other than lognormal.

The canonical lognormal and broken-stick are one-parameter distributions. Therefore if S_T is given, J may be calculated, leading to unique relations of S_T versus J for these two species-abundance distributions. The general lognormal, geometric series, and logseries are two-parameter distributions, leading in each case to a one-dimensional family of curves of S_T versus J. Unique curves may be specified by assigning a value to the remaining parameter (i.e., by specifying γ for the lognormal, k for the geometric series, α for the logseries). Such S_T-versus-J relations may be converted to species-area relations by adding the independent biological assumption that the number of individuals is roughly proportional to the area, A:

$$J = \rho A \qquad (1.5)$$

where ρ is some constant. This assumption is of doubtful validity, but may serve as a reasonable estimate in island biogeographical contexts, with "island" interpreted in a broad sense (MacArthur and Wilson, 1967). Preston (1962) and MacArthur and Wilson have shown that the canonical lognormal distribution in conjunction with eq. 1.5 leads to a species-area curve that accords with much field data. This work begs the question of where the *canonical* lognormal came from in the first place. In Section 5 it is shown that for all reasonable lognormal distributions (the canonical lognormal being merely one special case) one gets species-area curves in rough agreement with the data. Although none of these relations are simple linear regressions of $\ln S_T$ on $\ln A$ (they have a steeper dependence of S_T on A for small A than for large A), they point to the approximate rule

$$\ln S_T \simeq x \ln A + \text{(constant)} \quad (1.6)$$

with x in the range around 0.2 to 0.3. This agrees with the data (Table 5), and suggests that the property is a rather general consequence of the lognormal species-abundance distribution (not just of the special canonical distribution). We also note the species-area relations predicted by eq. 1.5 in conjunction with the broken-stick distribution (significantly steeper than the lognormal curves), and with the geometric series or logseries distributions (significantly less steep than the lognormal curves).

Section 6 pays very brief attention to some of the problems that arise with samples that are not big enough for all species typically to be represented. Such considerations have been extensively reviewed elsewhere (e.g., Pielou, 1969; Patil, Pielou, and Waters, 1971). It is noted that sampling distributions often tend to be of logseries form (e.g., Boswell and Patil, 1971). One consequence is that if the species-area curve reflects sampling properties, with A being a simple measure of sample size (in contrast to Section 5 where S_T versus A reflected the actual species-abundance pattern along with the biological assumption eq. 1.5), the relation may

be

$$S_T \simeq \alpha \ln A + \text{(constant)} \quad (1.7)$$

Here α is a parameter of the logseries distribution (see Section 3 C). This point is briefly discussed and applied to some field data in Section 6. Estimates of the fraction of species likely to be present in small samples from the lognormal and the broken-stick distributions are also given.

A series of appendices sets out the mathematical properties of the various distributions treated in the main text. Such properties as have been discussed by earlier authors are simply listed, whereas the new work is usually developed somewhat more fully. The appendices are not exercises in mathematical pedantry, but form the backbone of the paper. They are intended to be useful to those who seek a thorough understanding of the morphology of the various distributions. On the other hand, the main text simply quotes the results as they are required, and is designed to present the main points in a self-contained and generally intelligible way, free from mathematical clutter.

Section 7 provides a summary.

2. Various Ways of Presenting $S(N)$

One way of displaying the information inherent in the species abundance distribution function $S(N)$ is to rank the populations N_i ($i = 1, 2, \ldots, S_T$) in order of decreasing abundance, with the subscript i denoting the rank in this sequence. Abundance-rank diagrams, with abundance N_i/N_T as the y axis, and rank i as the x axis may thus be drawn, and experimental findings compared with theory. Minor variants follow from the use of log (abundance) rather than abundance for the y axis (e.g., Figure 1), and/or log (rank) rather than rank for the x axis.[2]

The characteristic features of such abundance-rank presentations of lognormal, broken-stick, geometric series, and logseries distributions have been clearly and comprehensively reviewed by Whittaker (1965, 1970, 1972). He points out the distinctive character of these various curves. The choice of logarithmic or linear scale for the axes has tended to depend on the predisposition of the author. Broken-stick people tend to use abundance versus log (rank), whereupon the broken-stick $S(N)$ shows up as nearly linear [see eq. D.3, $i \simeq S_T \exp(-S_T N/N_T)$, whence the abundance-ln (rank) curve is approximately linear with slope $-S_T$]. On a log (abundance) versus rank labelling, geometric series are exactly linear [see eq. E.1, whence the ln (abundance)-rank curve is roughly linear with slope $-(1-k)$], and logseries distributions are approximately linear [see eq. F.10, $i \simeq -\alpha \ln(N/N_T)$, whence the ln (abundance)-rank curve is roughly linear with slope $-\alpha$]. The lognormal abundance versus rank curve has a shape intermediate between these two extremes. Figure 1 illustrates this.

$S(N)$ may be displayed more directly by drawing a histogram of the number of species whose populations lie in specific ranges. Williams (1964), for example,

[2]Throughout, the usual convention is employed whereby log denotes logarithms to base 10, ln denotes logarithms to base e: $\ln x = (2.303) \log x$.

gives many such "frequency plots" of $S(N)$ against N for various assemblages of data. Alternatively, a logarithmic scale may be employed for the abundance; that is, the x axis is log N. Following Preston's (1948, 1962) work, it has become customary to employ logarithms to base 2 (so that the x axis, the logarithmic abundance scale, is divided into "octaves").

Yet again, Williams (1964) has frequently analysed species-abundance data by plotting the logarithm of the abundance (x axis) against the accumulated fraction of species up to that abundance (y axis is $\int_0^N S(N') \, dN'$). By using so-called probability paper (a gaussian scale on the y axis), Williams thus arranges that a lognormal distribution will show up as a straight line on such a plot. This device for producing a straight line from a lognormal $S(N)$ may be viewed as the analogue of plotting abundance versus log (rank) to get roughly a straight line from the broken-stick $S(N)$, or log (abundance) versus rank to get roughly a straight line from geometric series or logseries $S(N)$.

The essential thing to appreciate here is that *all* such diagrams, whether one or another form of plot of abundance versus rank, or Williams- or Preston-style frequency plot, are simply interchangeable and equivalent ways of expressing the distribution $S(N)$. Intercomparison of different bodies of data would be facilitated, and a certain amount of confusion removed, if some standard format could be agreed on.

This section ends with an explicit formula relating abundance-rank plots to the distribution function $S(N)$. Define $F(N)$ to

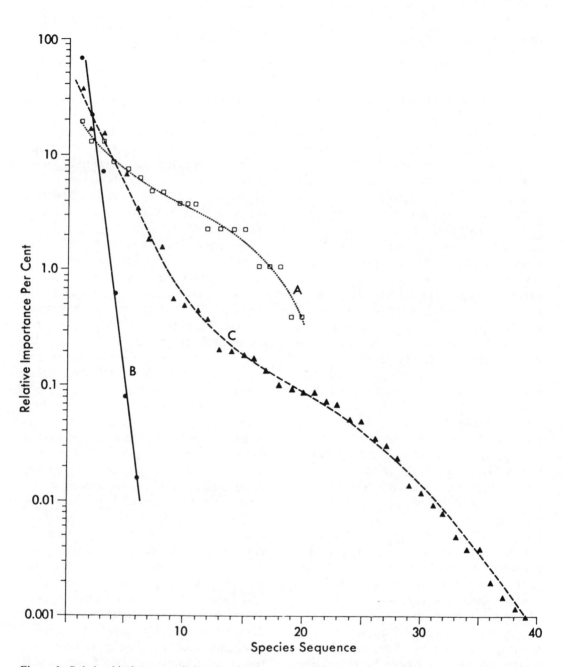

Figure 1 Relationship between relative abundance or importance (expressed as a percentage on a logarithmic scale) and rank of species for three natural communities. This figure, from Whittaker (1970), bears out the remarks in the text. Curve A is for a broken-stick distribution, and is fitted by data from a relatively small community of birds; curve B is for a geometric series distribution, and is fitted by a subalpine plant community; curve C is for a lognormal, and is fitted by the plant species in a deciduous forest.

be the total number of species with populations in excess of N:

$$F(N) = \int_N^\infty S(N')\,dN' \qquad (2.1)$$

Although $F(N)$, so defined, is a continuous function, it clearly describes the relation between rank i and abundance N_i; as N decreases from infinity, $F(N)$ attains integer values of 1, 2, etc., until $F(N) \to S_T$ as $N \to 0$. Thus a plot of N (y axis) versus $F(N)$ (x axis, scaled 1 through S_T) *is* the abundance-rank diagram. $F(N)$ may be christened the rank-order function. A geometrical prescription for converting an abundance-rank plot [N versus $F(N)$] into a frequency plot [$S(N)$ versus N] follows from the above remarks: first interchange x and y axes in the abundance-rank figure [to get $F(N)$ versus N], then calculate the slope of this curve at each point [slope $= dF/dN = -S(N)$ from eq. 2.1], and change the sign to arrive at the plot of $S(N)$ versus N. Appendix D applies this procedure explicitly in a discussion of the broken-stick distribution.

3. Specific Distributions

Some of the salient forms proposed for the species abundance distribution $S(N)$ are now reviewed, both with respect to theoretical ideas which lead to the distribution, and to corroborative evidence from field data. Note that while specific biological assumptions imply a unique $S(N)$, the converse is not true; a variety of different circumstances can imply the same $S(N)$. This lack of uniqueness in making ecological deductions from observed $S(N)$ discouraged MacArthur (1966). Even so, worthwhile distinctions can be made between properties which stem from the statistical Central Limit Theorem as opposed to broad biological features.

A: Lognormal Distribution

(i) *Theory.* At the outset, there is a need to distinguish two qualitatively different ecological regimes, commonly referred to as those of "opportunistic" and of "equilibrium" species. In the former limit, the ever-changing hazards of a randomly fluctuating environment can be all-important in determining populations, and thus relative abundances; in the latter limit, a structure of interactions within the community may, at least in principle, control all populations around steady values. One way of expressing this distinction is to define $r_i(t)$ to be the per capita instantaneous growth rate of the ith species:

$$r_i(t) = \frac{1}{N_i(t)}\frac{dN_i(t)}{dt} \qquad (3.1)$$

This growth rate may of course vary systematically or randomly from time to time, and may itself depend on the population of the ith and other species. Formally, however, eq. 3.1 integrates to

$$\ln N_i(t) = \ln N_i(0) + \int_0^t r_i(t')\,dt' \quad (3.2)$$

MacArthur (1960) discussed the opportunistic limit as being that where on the right hand side in eq. 3.2 the integral is more important than $\ln N_i(0)$, so that the population at time t is essentially unrelated to that at $t = 0$. Conversely, in the equilibrium limit the ln $N_i(0)$ term is more important than the integral; the populations are relatively unvarying, and the relative abundances form a steady pattern.

In the opportunistic regime, environmental vagaries predominate, and the $r_i(t)$ will vary randomly in time. Thus for any one population, labelled i, the accumulated integral of $r_i(t)$ in eq. 3.2 is a sum of random variables. In accord with the Central Limit Theorem (a theorem to the effect that essentially all additive statistical distributions are asymptotically gaussian, or "normal"), this integral will then in general be normally distributed (Cramer, 1948). Hence ln $N_i(t)$ is normally distributed, leading to a lognormal distribution for the population of any one species in time, and consequently a lognormal for the overall community species-abundance distribution at any one time.

The essential point here, and elsewhere throughout Section 3 A, is that populations tend to increase geometrically, rather than arithmetically, so that the natural variable is the *logarithm* of the population density. This central point has been particularly stressed by Williamson (1972, Chapter 1), Williams (1964), Montroll (1972), and others.

In the equilibrium regime, a lognormal distribution of relative abundance among the species is again most likely, once one deals with communities comprising a large number of species fulfilling diverse roles. In this event, Whittaker (1970, 1972) and others have observed that the distribution of relative abundance is liable to be gov-

erned by many more-or-less independent factors, compounded multiplicatively rather than additively, and again the Central Limit Theorem applied to such a product of factors suggests the lognormal distribution.

Alternatively, MacArthur (1960) and Williams (1964) have noted that a suggestion of Fisher's (1958) concerning community evolution can imply a lognormal distribution. Assuming roughly that beneficial genes are fixed at a rate proportional to population size, the relative abundances of species in a large community will be lognormally distributed.

In brief, the lognormal distribution is associated with products of random variables, and factors that influence large and heterogeneous assemblies of species indeed tend to do so in this fashion. Such considerations apply quite generally to multiplicative processes where, as it were, the rich grow richer (10% of 10^7 is more exciting than 10% of $10). Thus the distribution of wealth in the United States could be expected to be lognormal, and data in the *Statistical Abstract* (1971) show this to be so. Similarly, McNaughton and Wolf (1973, p. 629) have shown that the international distribution of human populations among the nations of the world, and even the distribution of the gross national products of nations, is lognormal.

(ii) *Mathematical Description of Lognormal Species-Abundance Distribution* Fuller mathematical details are given in Appendix A. Pielou (1969, Chapter 17) discusses the relation between discrete and continuous lognormal distributions.

A lognormal distribution may be written in standard form as

$$S(N) = S_0 \exp\left[-(\ln N - \ln N_0)^2/2\sigma^2\right] \quad (3.3)$$

Here N_0 is the number of individuals in the modal species (the species at the peak of the species-abundance curve in Figure 2), S_0 is the maximum value of $S(N)$ (attained at $N = N_0$), and σ is the gaussian width of the distribution. It has become conventional in much of the ecological literature, following Preston's (1948) work, to plot abundances on a scale, R, of logarithms to the base 2 (so that successive intervals or "octaves" correspond to population doublings):

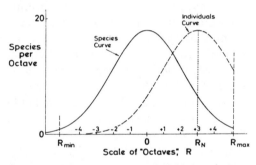

Figure 2 This figure aims to illustrate the features of a lognormal distribution, which are discussed more fully in the text. The solid curve is a lognormal species-abundance distribution, $S(R)$, after eqs. 3.4 and 3.5. The figure is specifically for $a = 0.34$, $S_0 = 18$, and consequently $\gamma = 0.6$ (see eq. 3.8). The dashed curve is the corresponding lognormal distribution in the total number of individuals, $N(R)$, eq. 3.7; the height of this dashed curve is in arbitrary units, leaving N_0 undefined (we are interested only in the shape of the curve). The species' populations are plotted in Preston's "octaves," R, which is to say as logarithms to base 2 (eq. 3.4). The boundary at R_{max} is the approximate position of the last, most abundant species; R_N is the octave in which total numbers peak. Preston's "canonical hypothesis" is that R_{max} and R_N coincide.

$$R = \ln_2 (N/N_0) \quad (3.4)$$

To avoid confusion we shall perpetuate this eccentricity, although it clutters the formulae with factors of $\ln 2$. The lognormal distribution then takes the familiar form

$$S(R)\, dR = S_0 \exp(-a^2 R^2)\, dR \quad (3.5)$$

Preston's parameter a is an inverse width of the distribution: $a = (2\sigma^2)^{-1/2}$. Figure 2 aims to illustrate this distribution. Worth mentioning are the values R_{max} and R_{min}, which mark the expected positions of the first (most abundant) and last (least abundant) species; as discussed more fully in Appendix A, $R_{max} = -R_{min}$, and both are related to a and S_0 via $aR_{max} = (\ln S_0)^{1/2}$, eq. A.2. The total number of species in the community, given in principle by eq. 1.3, is now approximately (see eqs. A.7, A.9),

$$S_T \simeq S_0 \pi^{1/2}/a$$
$$\simeq (\pi^{1/2}/a) \exp(a^2 R_{max}^2) \quad (3.6)$$

There is also a lognormal distribution in the total number of individuals in the Rth octave, $N(R)$. Combining eqs. 3.4 and 3.5 gives

$$N(R)\, dR$$
$$= S_0 N_0 2^R \exp(-a^2 R^2)\, dR$$
$$= S_0 N_0 \exp[-a^2 R^2 + R \ln 2]\, dR \quad (3.7)$$

This may be seen to be a normal distribution in the variable $(R - \ln 2/2a^2)$, that is, a gaussian with peak displaced a distance $R_N = (\ln 2)/2a^2$ to the right of that of the $S(N)$ distribution. It is now useful to define a quantity γ as the ratio between R_N and R_{max}:

$$\gamma \equiv \frac{R_N}{R_{\max}} = \frac{\ln 2}{2a\,(\ln S_0)^{1/2}} \quad (3.8)$$

The relation between the basic lognormal species-abundance distribution and the intervals R_N and $R_{\max}$ is perhaps made clear by Figure 2. Finally, noting that the above definitions imply the expected number of individuals in the least abundant species to be

$$m = N_0 2^{R_{\min}} = N_0 2^{-R_{\max}} \quad (3.9)$$

we may express the quantity $J = N_T/m$ of eqs. 1.1 and 1.4 in a form that involves any two of the interconnected parameters S_0, $R_{\max}$, a, S_T, γ. This is done in Appendix A, to arrive at eqs. A.10 and A.11.

The essential point to grasp is that the general lognormal species abundance distribution requires *two* parameters for a unique specification (see footnote 1). These two parameters may be chosen to be a and γ, in which case other interesting properties of the distribution such as S_0 (from eq. 3.8), $R_{\max}$ (from eq. A.2), and the overall S_T and J (from eqs. A.7 and A.10) follow. Alternatively, given the quantities S_T and J, which are of direct biological significance, the distribution is again uniquely described, and all other properties follow. Most commonly in Sections 4 and 5, I shall work with S_T and γ as the parameters.

Preston's (1962) "canonical hypothesis," which will be more fully discussed below, may conveniently be mentioned here, as it is the only reason for the orgy of notation leading up to the definition of the cumbersome parameter γ. This phenomenological hypothesis is that

$$\gamma = 1 \quad (3.10)$$

This is now a *one*-parameter family of canonical lognormal distributions. Given S_T, all else follows: the shape of the distribution is uniquely specified; unique values of J and of various diversity indices may be calculated; there is a unique species-area relation. The hypothesis has a purely empirical basis. In view of its predictive successes, particularly as to species-area relations, it is surprising that no theoretical justification has previously been attempted.

(iii) *Field Data.* As remarked by MacArthur (1960), the lognormal abundance distributions in communities of "opportunistic" creatures reflect nothing about the structure of the community. Dominant species are simply those that recently enjoyed a large r (eq. 3.2), and at different times different species will be most abundant. Such patterns for opportunistic species have been documented by Patrick, Hohn, and Wallace (1954) and Patrick (1968 and Chapter 15).

In steadier ("equilibrium") communities, fits to lognormal species-abundance distributions have been described for a wide variety of circumstances, including geographically diverse communities of birds, intertidal organisms, insects, and plants (Preston, 1948, 1962; Williams, 1953, 1964; Whittaker 1965, 1972; Batzli, 1969). Excellent reviews have been given by Whittaker (1970, 1972), who notes: "When a large sample is taken containing a good number of species, a lognormal distribution is usually observed, whether the sample represents a single community or more than one, whether distributions of the community fractions being combined are of geometric, lognormal or MacArthur form" (Whittaker, 1972, p. 221). Gauch and Chase (1975), who have just produced a useful computer algorithm for fitting normal distributions to ecological data, reexamined the data originally surveyed (and fitted by eye) by Preston (1948), and showed in one typical instance that 96% of the variance in the observed distribution could be accounted for by a lognormal.

In addition to these general fits between theory and data, it has been observed (originally by Preston for γ and by Hutchinson for a) that in most cases the parameters a and γ tend to have special constant values.

Preston (1962) reviewed a considerable body of material and showed that in all cases the shape of the distribution corresponded roughly to the special value $\gamma = 1$. This "canonical hypothesis" has been further discussed by MacArthur and Wilson (1967), and the ensuing unique S_T-versus-J relation applied to explain much species-area data.

Another rough rule, first noted by Hutchinson (1953) and subsequently confirmed by a growing amount of field observation, is that $a \simeq 0.2$. (The rule holds true even for the international distribution of human populations, or of the gross national products, referred to above.) Reviewing the current status of the lognormal distribution, Whittaker (1972, p. 221) observes "the constant a is usually around 0.2." This enigmatic rule has prompted many speculations, from Hutchinson's relatively cautious "it is

likely that something very important is involved here" (Hutchinson, 1953, p. 11), to one recent ecology text that indulges in the thought that "it does seem extraordinary that the constant should have the same value no matter the size or reproductive capacity of the organism, whether we are dealing with diatoms, moths or birds. Perhaps it bears some subtle relationship to the range of variation in the earth's environment, a range set by fundamental properties of the solar system or the elements in the periodic table."

A more prosaic explanation of these two rules will now be offered. They appear to be approximate mathematical properties of the lognormal distribution, once S_T is large.

(iv) *The Canonical Hypothesis and the Rule $a \simeq 0.2$.* (a) First, assume $\gamma = 1$, so that we have a canonical lognormal distribution. There is now a unique relation between a and S_T, the total number of species in the community. This relation (see eq. B.1 with the definition eq. A.6) is depicted in Figure 3. It is obvious that a depends very weakly on the actual value of S_T once there are more than ten or so species. This figure reflects the approximate fact that for $S_T \gg 1$ (see eqs. A.12, A.6)

$$a \sim \frac{\ln 2}{2\sqrt{\ln S_T}} \qquad (3.11)$$

The dependence on S_T as the square root of the logarithm is very weak indeed.

Figure 3 shows that as S_T varies from 20 to 10,000 species, a varies from 0.30 to 0.13. The rule $a \simeq 0.2$ is a mathematical property of the canonical lognormal.

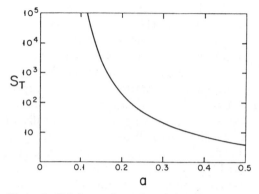

Figure 3 This figure shows the relationship between number of species, S_T, and the parameter a for the canonical lognormal distribution. Note that S_T is plotted on a logarithmic scale, emphasizing the insensitivity of a to changes in S_T.

(b) It remains to consider the canonical hypothesis itself. Without this hypothesis, the general lognormal is characterized by two parameters, conventionally a and S_0 (eq. 3.5), but equivalently a and γ, which can be determined from the total number of species, S_T, and the total number of individuals divided by the population of the rarest species, $J = N_T/m$. These relations between a, γ and S_T, J are given by eqs. A.7 and A.10 along with the definitions eqs. A.6 and A.11 in Appendix A. The form of the relation is illustrated by Figure 4, which shows the range of S_T and J values that are possible if a and γ are allowed to vary independently over the ranges a from 0.1 to 0.4 and γ from 0.5 to 1.8.

It may be observed from the figure that this enormous range of S_T and J values is roughly consistent with the rules $\gamma = 1$ and $a \simeq 0.2$, the agreement becoming more pronounced as S_T becomes larger. As in eq. 3.11, the underlying reason is

that for $S_T \gg 1$, the quantities a and γ depend on S_T and J as $\sqrt{\ln S_T}$ and $\sqrt{\ln J}$ (see eqs. A.6, A.12, A.13, A.14).

The above admittedly constitutes only an imprecise and qualitative explanation of the rules $\gamma \simeq 1$ and $a \simeq 0.2$. However, these empirical rules are themselves only rough ones, and a qualitative theory would seem to represent some advance over the prevailing total absence of any explanation as to such remarkable regularities. I see them as mathematical properties of the lognormal distribution, rather than as reflecting anything biological. This is disappointing.

The emergent moral is, presumably, to characterize lognormal distributions by indices more sensitive than a, γ and their equivalents.

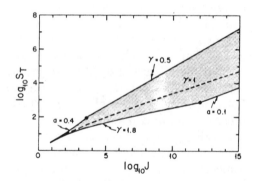

Figure 4 The shaded area illustrates the very wide range of values of S_T and J which can be described by the general lognormal distribution with the parameter a between 0.1 and 0.4 and the parameter γ between 0.5 and 1.8. This shaded area is bounded, as indicated in the figure, by the four line segments along which, respectively, $a = 0.4$ (with γ varying from 1.8 to 0.5), $\gamma = 1.8$ (with a varying from 0.4 to 0.1), $\gamma = 0.5$ (with a varying downward from 0.4), and $a = 0.1$ (with γ varying downward from 1.8). Note that both S_T and J are on logarithmic scales. The dashed line corresponds to Preston's special case, $\gamma = 1$ (and varying a).

B: MacArthur's Broken-Stick Distribution

(i) *Theory.* If attention is restricted to communities comprising a limited number of taxonomically similar species, in competitive contact with each other in a relatively homogeneous habitat, a more structured pattern of relative abundance may be expected. As pointed out by Mac-Arthur (1957, 1960), and lucidly reviewed by Whittaker (1972), if the underlying picture is one of intrinsically even division of some major environmental resource, the statistical outcome is the well-known "broken-stick" distribution. This is the distribution relevant, for example, to collecting plastic animals out of cornflakes boxes, assuming the various plastic animal species to have a uniform distribution at the factory. The most thorough treatment of the statistical properties of the relative abundances within a group of species, which apportion randomly among themselves a fixed amount of some governing resource, is due to Webb (1974). He not only shows that the familiar broken-stick distribution is the average outcome, but also considers the statistical fluctuations to be expected about this average.

The broken-stick distribution of relative abundance is considerably more even than lognormal ones. It is characterized by a single parameter, S_T. Thus once the number of species in the community is specified, diversity indices and (with eq. 1.5) species-area relations uniquely follow. The mathematical properties of the distribution are catalogued in Appendix D.

(ii) *Field Data.* For appropriately restricted samples, such broken-stick relative-abundance patterns have been found, for example, by MacArthur (1960), King (1964), and Longuet-Higgins (1971) following Tramer (1969), for birds; by Kohn (1959) for some snails; and by Goulden (1969), Deevey (1969), and Tsukada (1972) for microcrustaceans deposited in lake-bed sediments.

As pointed out by Cohen (1968), and reviewed by Pielou (1969) and others, the observation of a broken-stick distribution does not validate the very specific model initially proposed by MacArthur (1957, 1960). It does indicate, however, that some major factor is being roughly evenly apportioned among the community's constituent species (in contrast to the lognormal distribution, which suggests the interplay of many independent factors).

C: Geometric Series and Logseries Distributions

(i) *Theory.* Let us consider again a relatively small and simple community of species, whose ecology is governed by some dominant factor; the opposite extreme to an intrinsically even (or random) division of resources is one of extreme "niche preemption." In its ideal form, this limit sees the most successful species as preempting a fraction k of the niche, the next a fraction k of the remainder, and so on, to give a geometric series distribution of relative abundance (eq. E.1). An equivalent way of framing this hypothesis is to assume that all species are energetically related to the other species in the community, the magnitude of the relation being proportional to the species abundance (large populations need more energy); the addition of another species then requires the same proportional increase in the abundance of all other species (Odum, Cantlon, and Kornicher, 1960). It does not seem to be commonly appreciated that the ensuing species-abundance distribution discussed by these authors, and in particular their S_T-versus-J relation, is precisely that of the geometric series distribution. Semantically, this identity is not surprising, as their assumptions constitute a form of niche-preemption hypothesis. The formal equivalence between the mathematics of Odum, Cantlon, and Kornicher and that of the conventional geometric-series distribution is established in Appendix E.

If one considers statistically more realistic expressions of the ideas that in their ideal form lead to a geometric series distribution, one is commonly led to a logseries distribution of relative abundance. For example, suppose that the geometric series niche-preemption mechanism stems from the fact that the species arrive at successive uniform time intervals and proceed to preempt a fraction k of the remaining niche before the arrival of the next; randomization of the time intervals leads to a logseries distribution (Boswell and Patil, 1971). Kendall's (1948) remarks as to how an intrinsically geometric series distribution of species per genus is converted into a logseries distribution are also obliquely relevant here (see the discussion in Williams, 1964, Chapter 11, and Boswell and Patil, 1971). Several other ways of arriving at a logseries distribution are comprehensively reviewed by Boswell and Patil (see particularly Sections 5 and 9), and Pielou (1969, Chapter 17). The distribution often arises as a sampling distribution, in which form it was first obtained

by Fisher, Corbet, and Williams (1943). It has the elegant property that samples taken from a population distributed according to a logseries are themselves logseries.

Some mathematical properties of the logseries are listed in Appendix F. Like the geometric series, it is characterized by two parameters, which we usually choose to be S_T and α. The relation between S_T, α and J is very simple:

$$S_T = \alpha \ln (1 + J/\alpha) \qquad (3.12)$$

(ii) *Field Data.* Whittaker (1965, 1970, 1972) has reviewed data from some plant communities, generally with but a few species and either in an early successional stage or in a harsh environment, where the species-abundance distribution approximates a geometric series. The phenomenon of strong dominance may be most expected in such circumstances. McNaughton and Wolf (1973) have also reviewed a series of examples which they interpret as geometric series; however, many of their examples would seem to be fitted better, and certainly at least as well, by lognormal distributions. Some of the systems discussed by Connell (Chapter 16) are likely candidates to fit geometric or logseries distributions, as they would seem to conform to the assumptions. As we would expect, the simple models of Markovian forest succession presented by Horn (Chapter 9) give rise to an explicitly geometric series distribution of relative abundance.

D: Contrast Between these Distributions

In short, the broken-stick and geometric or logseries may be viewed as distributions characteristic of relatively simple communities whose dynamics is dominated by some single factor: the broken-stick is the statistically realistic expression of an intrinsically uniform distribution; and at the opposite extreme the logseries is often the statistical expression of the uneven niche-preemption process, of which the ideal form is the geometric series distribution. Both forms reflect dynamical aspects of the community.

However, if the environment is randomly fluctuating, or alternatively as soon as several factors become significant (as they may in general, and must if $S_T \gg 1$), we expect the statistical Law of Large Numbers to take over and produce the ubiquitous lognormal distribution. This species-abundance distribution is in most respects *intermediate* between the broken-stick and geometric series or logseries extremes, as illustrated clearly by Figure 1 and the tables and figures in Section 4. The empirical rules $a \simeq 0.2$ and $\gamma \simeq 1$ are probably no more than mathematical properties of the lognormal distribution for $S_T \gg 1$.

4. Diversity, Evenness, Dominance

I do not wish to add unnecessarily to the already voluminous literature pertaining to the meaning and relative merit of various diversity indices. Suffice it to say that many people have sought to go beyond S_T and J, yet stop short of the full $S(N)$, by using some single simple number that may characterize the distribution $S(N)$, i.e., some number that will describe whether the J individuals are roughly

evenly distributed among the S_T species, or whether they are concentrated into a few dominant species.

Among the many such indices proposed are

(i) The Shannon-Weaver diversity, which we shall refer to as H:

$$H = - \sum_{i=1}^{S_T} p_i \ln p_i \qquad (4.1)$$

Here p_i is the proportion of individuals in the ith species, $p_i = N_i/N_T$. For a given species-abundance distribution $S(N)$, the average value of H will be

$$\langle H \rangle = - \left\langle \sum_i p_i \ln p_i \right\rangle$$

$$= - \sum_i \langle p_i \ln p_i \rangle$$

That is,

$$\langle H \rangle = - \int (N/N_T) \ln (N/N_T) S(N) \, dN \qquad (4.2)$$

(ii) Also of considerable interest is the expected magnitude of the statistical fluctuations about this mean value of H. For a given distribution $S(N)$, the expected variance is

$$\sigma_H^2 = \langle (H - \langle H \rangle)^2 \rangle$$

$$= \left\langle \left(\sum_i p_i \ln p_i \right)^2 \right\rangle - \langle H \rangle^2 \qquad (4.3)$$

But we can write

$$\left\langle \left(\sum_i p_i \ln p_i \right)^2 \right\rangle = \sum_i \langle (p_i \ln p_i)^2 \rangle$$

$$+ \sum_i \sum_{j \neq i} \langle (p_i \ln p_i)(p_j \ln p_j) \rangle$$

In the second term on the right-hand side there are no correlations, and consequently the term has the value $[(S_T - 1)/S_T]\langle H\rangle^2$. Therefore

$$\sigma_H^2 = \int [(N/N_T) \ln (N/N_T)]^2 S(N)\, dN \\ - \langle H\rangle^2/S_T \quad (4.4)$$

(iii) Another dominance or diversity index is

$$C = \sum_i p_i^2 \quad (4.5)$$

The reciprocal of C,

$$D = 1/C \quad (4.6)$$

is Simpson's (1949) diversity index, which counts species, weighting them by their abundance, and can vary from $D = 1$ (one dominant species) to $D = S_T$ (completely even distribution). The index $1 - C$ is also widely used, and indeed Hurlbert (1971) has recently decried the popular index H, and has suggested that $1 - C$ may be better. For a specified distribution $S(N)$, the expected value of C is

$$\langle C\rangle = \int (N/N_T)^2 S(N)\, dN \quad (4.7)$$

In analogy to the index (ii) above, the statistical variance in C could also be calculated and displayed for any specified $S(N)$.

(iv) One simple measure of dominance in a community is (Berger and Parker, 1970)

$$d = N_{\max}/N_T \quad (4.8)$$

Here, for given $S(N)$, $N_{\max}$ is the expectation value of the most abundant population.

Various other indices of diversity, dominance, or evenness have been reviewed by Whittaker (1972), Dickman (1968), Pielou (1969), Johnson and Raven (1970), Hurlbert (1971), De Benedictis (1973), and many others.

Our attention here is confined to the four quantities H, σ_H^2, D or C, and d. For each we exhibit and compare the relationships between the index and S_T for the various distributions of Section 3. In this way we aim to clarify the relations between the various species-abundance relations as revealed by any one index, and also the relations between the various diversity indices for any one $S(N)$. Some of these diversity indices have been explored for some of the distributions listed in Section 3 (particularly for the broken-stick), but remarkably little has been done for the lognormal, which is probably the most important one.

A: The Average Value of H

For the broken-stick and the canonical lognormal distributions, unique curves of $\langle H\rangle$ versus S_T can be calculated. The general lognormal, geometric series, or logseries distributions give rise to one-parameter families of curves of $\langle H\rangle$ versus S_T, specified here by the parameters γ, k, and α respectively.

Figure 5 shows such curves for the broken-stick distribution, and for lognormal distributions with $\gamma = 0.7$, 1 (canonical), 1.3. Note that the curves are indistinguishable for relatively small S_T, but that as S_T becomes very large the broken-stick curve leads to significantly larger H values, while the lognormal curves reveal

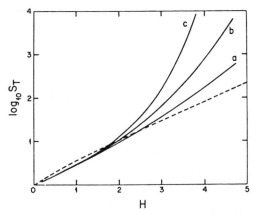

Figure 5 The expectation value of the Shannon-Weaver diversity index, H (see eq. 4.2), as a function of the total number of species, S_T (on a logarithmic scale), for various species-abundance distributions. The solid curves are lognormal distributions with: (a) $\gamma = 0.7$; (b) $\gamma = 1.0$ (canonical); (c) $\gamma = 1.3$; the dashed curve is the relation for the broken-stick distribution.

a very insensitive dependence of H on S_T.

In particular, to achieve H greater than 5 with a canonical lognormal distribution requires $S_T \sim 10^5$ species. As stressed in Section 3, for $S_T \gg 1$ a lognormal distribution is always to be expected, and this

feature of the lognormal distribution may explain the fact, remarked upon at length by Margalef (1972), that $H > 5$ is not observed in data collections.

For $S_T \gg 1$, analytic approximations may be obtained for the average value of H in the various distributions, and these are catalogued in Table 1.

These asymptotic results bear out the general remarks made in Sections 2 and 3, to the effect that uniform or broken-stick distributions are more diverse, which is to say have larger H, than lognormal distributions with $\gamma \simeq 1$, which in turn are more diverse than geometric or logseries distributions. However, these systematic differences are seen from Table 1 to scale only as $\ln S_T$ (broken-stick, uniform) compared with $\sqrt{\ln S_T}$ (lognormal with $\gamma \simeq 1$) compared with a constant (geometric or logseries), even for every large S_T. But as the logarithm of even a quite large number is of the general order of unity, one must conclude that the diversity index H is an insensitive measure of the character of $S(N)$. These comments are true a fortiori when S_T is not large (cf. Figure 5).

Table 1. Diversity, H, as a function of S_T for $S_T \gg 1$

Species-abundance distribution, $S(N)$	Average value of H, for $S_T \gg 1$	Details in appendix
lognormal (parameter γ)	$(1 - \gamma^2) \ln S_T$; for small γ $2\gamma\pi^{-1/2}\sqrt{\ln S_T}$; for $\gamma \simeq 1$ (constant); for large γ	eqs. A.18, A.19
canonical lognormal	$2\pi^{-1/2}\sqrt{\ln S_T}$	eqs. B.5, B.6
uniform	$\ln S_T$	eq. C.4
broken-stick	$\ln S_T - 0.42$	eq. D.8
geometric series (parameter k)	constant $= \dfrac{-\ln(1-k)}{k} + \ln\left(\dfrac{1-k}{k}\right)$	eq. E.6
logseries (parameter α)	constant $= \ln \alpha + 0.58$	eqs. F.5, F.6

B: The Variance of H

The variance in H may be calculated from eq. 4.4, and the details are outlined in the various mathematical appendices. Previous work along these lines consists of numerical investigations of the full statistical distribution of H for the broken-stick distribution, carried out by Bowman *et al.* (1971) and by Webb (1974).

Figure 6 shows the standard deviation to be expected in the diversity H for communities whose underlying patterns are broken-stick or canonical lognormal. (As the statistical distribution of H values is skewed, particularly at relatively small S_T values, characterizing its spread by σ_H is not an exact procedure, but should be good enough for all practical purposes; see Webb's computer simulations for the broken-stick case.)

These fluctuations in H are intrinsic properties of the distribution $S(N)$, and it is clear that their magnitude is such as to obscure any differences between the distributions unless S_T is very large.

For completeness, analytic formulae and asymptotic expressions for σ_H^2 and for the root-mean-square relative fluctuations, $\sigma_H/\langle H \rangle$, are given in the appendices. These are messy, and here we note only the asymptotic ($S_T \gg 1$) form of the relative fluctuations in the broken-stick and canonical lognormal distributions:

broken-stick:

$$\frac{\sigma_H}{\langle H \rangle} \sim \frac{1}{\sqrt{S_T}} \qquad (4.9)$$

canonical lognormal:

$$\frac{\sigma_H}{\langle H \rangle} \sim \frac{0.9}{\ln S_T} \qquad (4.10)$$

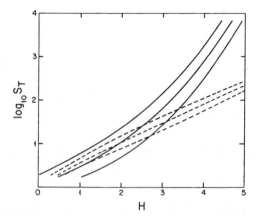

Figure 6 The three solid lines illustrate the average value of H, and the average value of one standard deviation above and below H (i.e., $\langle H \rangle$ and $\langle H \rangle \pm \sigma_H$, from eqs. 4.2 and 4.4), as a function of S_T, for the canonical lognormal distribution of species abundance. Likewise, the three dashed lines show $\langle H \rangle + \sigma_H$, $\langle H \rangle$, and $\langle H \rangle - \sigma_H$ for the broken-stick distribution.

In his computer experiments, De Benedictis (1973) noted a systematic tendency for the variance in the "evenness" (i.e., in $H/\ln S_T$) to decrease as S_T increased, albeit slowly. Figure 6 and the above formulae confirm this tendency.

Figure 7, modified from Webb's (1974) compilation of data, shows the H values for an assortment of communities of corals, copepods, plankton, benthic creatures, trees, and birds, along with the expectation values of H for the broken-stick and the canonical lognormal distributions. From a comparison of Figures 6 and 7, it may be held that the data are more consistent with a lognormal distribution, but certainly no discrimination between the two distributions is possible for small S_T.

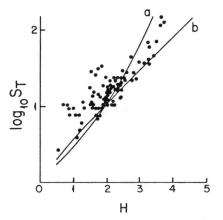

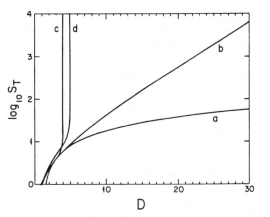

Figure 7 The theoretical curves of diversity versus species number, H versus S_T (on a logarithmic scale), for (a) the canonical lognormal distribution and (b) the broken-stick distribution, are here shown in conjunction with Webb's (1973) compilation of various authors' data (the solid dots) for birds, copepods, corals, plankton, and trees.

Figure 8 The average value of Simpson's diversity index D (from eqs. 4.6, 4.7) as a function of S_T (on a logarithmic scale) for various distributions of species abundance $S(N)$: (a) broken-stick; (b) canonical lognormal; (c) geometric series with $k = 0.4$; (d) logseries with $\alpha = 5$.

C: The Average Values of D and C

The mathematical appendices contain explicit expressions for the indices C and D of eqs. 4.6 and 4.7, for the various distributions $S(N)$ of Section 3.

Figure 8 illustrates Simpson's diversity index D as a function of S_T for the broken-stick, canonical lognormal, geometric series, and logseries distributions.

(The broken-stick and canonical lognormal curves are unique; the geometric series is for the choice $k = 0.4$, and the logseries for $\alpha = 5$.) Comparison with Figure 5 shows the index D to exhibit a stronger contrast between canonical lognormal and broken-stick distributions than does H.

Table 2 catalogues asymptotic, $S_T \gg 1$,

Table 2. Simpson's diversity, D, as a function of S_T for $S_T \gg 1$

Species-abundance distribution, $S(N)$	Average value of D, for $S_T \gg 1$	Details in appendix
lognormal (parameter γ)	see Appendix A	eqs. A.22, A.23
canonical lognormal	$\dfrac{\pi \ln S_T}{\ln 2}$	eq. B.10
uniform	S_T	eq. C.5
broken-stick	$\tfrac{1}{2} S_T$	eq. D.11
geometric series (parameter k)	$(2 - k)/k$	eq. E.8
logseries (parameter α)	α	eq. F.8

expressions for D. Both the figure and the table again illustrate the properties attributed to the distributions earlier.

D: The Average Value of d

Here Figure 9 contrasts d as a function of S_T for the canonical lognormal, broken-stick, geometric series (with $k = 0.4$), and logseries (with $\alpha = 5$) distributions. The asymptotic formulae relating d to S_T when $S_T \gg 1$ are set out in Table 3.

Yet again, all the general features are born out. As S_T becomes large, the geometric and logseries manifest their pattern of strong dominance, by having d settle to some characteristic constant value, independent of the increasing number of species. Conversely, the uniform and broken-stick distributions have d tending to zero essentially as $1/S_T$ when S_T becomes large. As ever, the canonical lognormal bestrides these extremes, with d decreasing, albeit slowly [as $(\ln S_T)^{-1/2}$], as S_T increases.

We observe that d, which is a pleasingly simple index from both conceptual and computational points of view, seems from

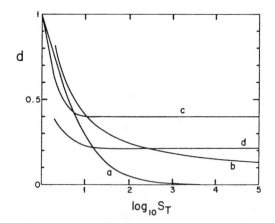

Figure 9 The average value of the simple dominance index d, eq. 4.8, as a function of S_T for various species-abundance distributions: (a) broken-stick; (b) canonical lognormal; (c) geometric series with $k = 0.4$; (d) logseries with $\alpha = 5$.

the foregoing discussion to characterize the distribution as well as any, and better than most.

E: Correlations Between H, D, d

It is obvious that, given any particular species-abundance distribution $S(N)$, one can for instance take the curves of S_T versus H and S_T versus d, and eliminate

Table 3.   Dominance, d, as a function of S_T for $S_T \gg 1$

Species-abundance distribution, $S(N)$	Average value of d, for $S_T \gg 1$	Details in appendix
lognormal (parameter γ)	see Appendix A	eq. A.24
canonical lognormal	$\dfrac{\ln 2}{\sqrt{\pi \ln S_T}}$	eq. B.11
uniform	$1/S_T$	eq. C.6
broken-stick	$(\ln S_T)/S_T$	eq. D.12
geometric series (parameter k)	k	eq. E.9
logseries (parameter α)	$\dfrac{\ln \alpha}{\alpha}$	eq. F.11, F.12

S_T to provide a unique curve of H versus d. Such relations between the diversity and dominance indices H, D, C, $1 - C$, d, etc., are already implicit in Figures 5–9 and Tables 1–3. That the various indices are correlated is a point that has been elaborated by Johnson and Raven (1970), Berger and Parker (1970), De Benedictis (1973), and others; but an exploration of the precise character of these correlations for the different species-abundance distributions $S(N)$ is lacking.

Table 4 makes this explicit, setting out the relations between H and D, and between D and d, for the various distributions. It may be noticed that (for mathematical rather than biological reasons) broken-stick and logseries distributions coincidentally tend to exhibit similar relations of H versus D and D versus d, whereas those for the lognormal distribution are qualitatively different.

5. Species—Area Relations

A subject of considerable interest, enjoying a growing literature, is the species-versus-area relation for communities of species isolated on real (see, e.g., Diamond, Chapter 14) or virtual (e.g., Cody, Chapter 10; Wilson and Willis, Chapter 18) islands. The above species-abundance distributions $S(N)$ imply predictions as to the relations between S_T and J, and these can be turned into relations between S_T and area A by addition of the biological assumption eq. 1.5, $J = \rho A$, the plausibility of which was discussed briefly in Section 1 (Preston, 1962; MacArthur and Wilson, 1967). As long as we are looking at islands or other isolated biota, this procedure may be justifiable; once A represents areas of different size from a large homogeneous mainland region, the relation of S_T versus A is likely to reflect sam-

Table 4. Asymptotic ($S_T \gg 1$) relations between H, D, d

Species-abundance distribution, $S(N)$	Asymptotic relation between H and D	Asymptotic relation between d and D
lognormal (parameter γ)	depends on γ, see Appendix A	$d^2 = \dfrac{(2\gamma - 1)\ln 2}{\gamma D}$
canonical lognormal	$H = (2/\pi)(D \ln 2)^{1/2}$	$d^2 = \dfrac{\ln 2}{D}$
uniform	$H = \ln D$	$d = 1/D$
broken-stick	$H = \ln D + 0.27$	$d = \dfrac{\ln (2D)}{2D}$
geometric series (parameter k)	$H = \ln D + 0.31 + [\text{order } D^{-2}]$	$d = \dfrac{2}{1 + D}$
logseries (parameter α)	$H = \ln D + 0.58$	$d = \dfrac{\ln D}{D}$

pling properties of the kind discussed in Section 6.

A unique relation of species versus area may be obtained from the general lognormal distribution by Preston's canonical hypothesis, $\gamma = 1$. This curve is illustrated in Figure 10. The relation has been successfully applied, initially by Preston (1962) and in more detail by MacArthur and Wilson (1967), to describe a wide range of data. These authors approximate the canonical lognormal relation S_T versus A by a simple linear regression of $\ln S_T$ on $\ln A$, a procedure that tends to overestimate the slope at large A and underestimate it at small A; the actual relation between S_T and A is not a simple function, although in the limit $S_T \gg 1$ it takes the asymptotic form given by eq. 1.6, namely

$$\ln S_T = x \ln A + \text{(constant)} \quad (5.1)$$

with $x = 0.25$. This point is discussed more fully in Appendix B. We reiterate that the more exact species-area curve illustrated in Figure 10, with its steeper slope at smaller A, tends to give a better description of real species-area data (see, e.g., Diamond, Chapter 14, Figure 3); the approximate eq. 5.1 does not flatter the theory.

The work of Preston and of MacArthur and Wilson rests on the canonical hypothesis. What happens for the general lognormal species-abundance distribution?

For each value of γ, the general lognormal gives a particular S_T-versus-A curve. Two such curves are shown in Figure 10, for $\gamma = 0.8$ and $\gamma = 1.3$. The shape is not particularly sensitive to the detailed

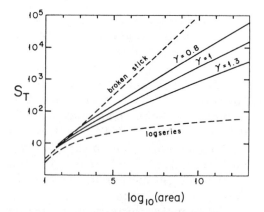

Figure 10 This figure illustrates the species-area (log S_T versus log A) relations obtained from particular species-abundance distributions $S(N)$ in conjunction with the biological assumption eq. 1.5. The solid curves are for lognormal distributions with $\gamma = 0.8$, 1.0 (Preston and MacArthur-Wilson), eq. 1.3; the dashed lines are for the broken-stick distribution, and for a logseries distribution (with $\alpha = 5$), as indicated.

value of γ, at least for the generous range of γ values around 0.6 to 1.7, which was discussed in Section 3 as pertaining to ecologically reasonable circumstances (see Figure 4). From eqs. A.15 and A.16, eq. 5.1 provides an asymptotic ($S_T \gg 1$) description of all lognormal species-area curves, with the quantity x having the values

$$x = 1/4\gamma \qquad \text{for } \gamma > 1 \quad (5.2a)$$
$$x = 1/(1 + \gamma)^2 \text{ for } \gamma < 1 \quad (5.2b)$$

Thus, roughly speaking, the above range of γ (0.6 to 1.7) leads to eq. 5.1, with x values lying between the extremes 0.39 to 0.15.

In brief, the successes of the Preston and MacArthur-Wilson species-area theories are not pathological consequences of their special canonical lognormal distribution, but rather are robust properties of any reasonable lognormal species-abundance distribution.

Table 5 summarizes this work by showing the x values obtained by various authors from fits of their field data to regressions of $\ln S_T$ versus $\ln A$, along with the asymptotic x values for lognormal distributions of various shapes (as specified by γ). An interesting next step might be to

study these data in greater detail, with the aim of understanding the biological reason why, for example, plants on the California Islands ($x = 0.37$) have a comparatively small value of γ ($\gamma \simeq 0.64$), whereas the Yorkshire nature-reserve plants ($x = 0.21$) have a comparatively large value ($\gamma \simeq 1.2$).

It may be argued that such species-area discussions are likely to involve large numbers of species, and therefore that the lognormal pattern of species abundance is

Table 5. The x values deduced from observations, and from the theoretical lognormal distribution

Observations			
Source	Flora or fauna	Location	x
Darlington, 1943	beetles	West Indies	0.34
Darlington, 1957	reptiles and amphibians	West Indies	0.30
Hamilton, Barth, and Rubinoff, 1964	birds	West Indies	0.24
Hamilton, Barth, and Rubinoff, 1964	birds	East Indies	0.28
Hamilton, Barth, and Rubinoff, 1964	birds	East-Central Pacific	0.30
MacArthur and Wilson, 1967	ants	Melanesia	0.30
Preston, 1962	land vertebrates	Lake Michigan Islands	0.24
Diamond, Chapter 14. Figure 2	birds	New Guinea Islands	0.22
Diamond, Chapter 14. Figure 3	birds	New Britain Islands	0.18
Cody, Chapter 10	birds	Mediterranean habitat gradients	0.13
Preston, 1962	land plants	Galapagos	0.32
Hamilton, Barth, and Rubinoff, 1963	land plants	Galapagos	0.33
Johnson and Raven, 1973	land plants	Galapagos	0.31
Preston, 1962	land plants	world-wide	0.22
Johnson and Raven, 1970	land plants	British Isles	0.21
Usher, 1973	land plants	Yorkshire nature reserves	0.21
Johnson, Mason, and Raven, 1968	land plants	California Islands	0.37

Theory	
Value of γ (see eq. 3.8)	x
0.6	0.39
0.8	0.31
1.0, *canonical*	0.25
1.2	0.21
1.4	0.18
1.6	0.16

the only one to work with here. Even so, the species-area consequences of the opposite extremes of broken-stick and of geometric series or logseries distributions should be mentioned.

As shown in Appendix D, eq. D.7, the broken-stick distribution in conjunction with eq. 1.5 leads *exactly* to the relation eq. 5.1 with $x = 0.5$. This relation is shown as a dashed line in Figure 10. It is not surprising to find it significantly steeper than any of the observations.

Conversely, the geometric series or logseries distributions tend to give species-area relations of the form set out in eq. 1.7: see eqs. E.5 and F.4 respectively. Such curves lead to a less steep relation than indicated from the data in Table 5. They are discussed in the next section.

6. Sampling Problems and $S(N)$

Up to this point, it has been assumed that we are dealing with situations in which the full values of S_T and N_T or J are known. All species in the community are represented in our samples. This assumption has allowed an unclouded discussion of some issues of principle. The assumption is often unrealistic, however, and in practice there will commonly be a need to work with less complete samples in which not all species are represented. To put it another way, our lognormal distributions are always (in Preston's terminology) fully unveiled; the complications introduced by distributions that are not unveiled would be distracting, and are inessential to the main points of Section 2-5.

The analysis of incomplete samples from distributions $S(N)$ constitutes a large and significant subject, to which the references given in Section 1 constitute an entry. The following discussion is confined to a few brief comments.

As first observed by Fisher, and discussed by Williams (1964), Pielou (1969), and particularly by Boswell and Patil (1971), the logseries distribution can describe sampling distributions under a diversity of circumstances. One consequence is that, if relatively small samples are taken from some large and homogeneous area, the relation between sample area (or volume), A, and the number of species represented in the sample, $S(A)$, is likely to obey the logseries eq. F.13; that is,

$$S(A) = \alpha \ln [1 + \beta A] \qquad (6.1)$$

This is approximately of the form eq. 1.7, $S \sim \alpha \ln A$. If this be the case, a regression of S on $\ln A$ will fit the data better than the $\ln S / \ln A$ regression suggested by eq. 5.1; alternatively, if an attempt is made to fit a relation actually of the form eq. 6.1 with a $\ln S / \ln A$ regression, the coefficient x thus deduced will be small, and the fit poor.

As an example, consider the elegant and much-discussed work of Sanders (1969). He studied marine benthic communities from various parts of the world, and for each particular community he plotted the number of species observed (or expected to be observed) as a function of the number of individuals in the sample. This clearly is a situation where the curves are derived from a sampling distribution. Figure 11 illustrates some of Sanders's

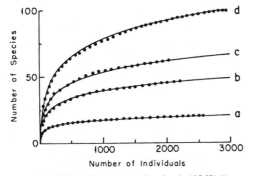

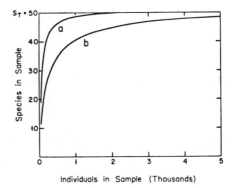

Figure 11 The solid dots are Sanders's (1969) "rare-faction curves" of number of species versus number of individuals in the sample, for data on benthic communities from: (a) boreal shallow water (IRA); (b) deep sea slope off southern New England (DR 33); (c) Walvis bay (190); (d) tropical shallow water (RH 30). The theoretical curves are for a logseries distribution, that is eq. F.13, with the single parameter α having the value: (a) 3.0; (b) 8.2; (c) 12.0 (d) 20.0.

Figure 12 An estimate of the number of species likely to be present in a relatively small sample comprising $\hat{N}$ individuals, when the underlying species-abundance distribution is (a) broken-stick, (b) canonical lognormal. In both cases, the total number of species actually present in the community is 50.

results, and the theoretical fits to them that can be obtained by assuming the sampling distribution to be a logseries, whence eq. F.13 or eq. 6.1. Each curve can be summarized by the single parameter α, which, as we saw in Section 4, is in some rough sense a measure of community diversity.

The appendices contain very crude estimates of the number of species (expressed as a fraction of the total number of species actually present in the community) likely to be observed in small samples drawn from communities whose intrinsic species abundance patterns $S(N)$ are as discussed in Section 3. Figure 12 shows characteristic results of this kind for broken-stick and canonical lognormal distributions. As is to be expected, the broken-stick distribution is more rapidly unveiled with increasing sample size than is the lognormal.

For example, if there are in fact 50 species present, the number of individuals one needs to sample to encounter half the species (25) is of the order of 73 for the broken-stick, 230 for the canonical lognormal distribution.

7. Summary

(1) For large or heterogeneous assemblies of species, a lognormal pattern of relative abundance may be expected. The intriguing rough rule $a \simeq 0.2$, and Preston's canonical hypothesis, are approximate but general mathematical properties of the lognormal distribution (e.g., Figures 3, 4). These rules thus reflect little more than the statistics of the Central Limit Theorem.

(2) For a small, ecologically homogeneous, set of species, which randomly ap-

portion among themselves a fixed amount of some governing resource, one may expect MacArthur's broken-stick distribution. Such a distribution may be thought of as the statistically realistic expression of an ideally uniform ($N_i = N_T/S_T$) distribution; the pattern of relative abundance is significantly more even than the lognormal (e.g., Figures 1, 8, 9).

(3) For a relatively small set of species where "niche-preemption" is likely to have significance, one may expect a log-series distribution. Such a distribution may be thought of as the statistically realistic expression of an ideally geometric series distribution; the pattern of relative abundance is significantly less even than the lognormal (e.g., Figures 1, 8, 9).

(4) The trends summarized in points (1), (2), and (3) are manifested in any measure of diversity or dominance in the community (Tables 1, 2, 3, 4). However, many common measures of species diversity tend not to distinguish these distributions if S_T is relatively small (e.g., Figure 6), while for large S_T we expect a lognormal distribution. Thus as one-parameter characterizations of species-abundance patterns, such indices are often of doubtful value. As a one-parameter description of the distribution, the simple dominance measure, d = (number of individuals in most abundant population)/(total number of individuals), seems as good as any.

(5) Combined with the biological assumption that $J = \rho A$, the general lognormal distribution (of which the canonical distribution of Preston and MacArthur-Wilson is a special case) leads to species-area curves that agree with the bulk of the pertinent field observations (Figure 10, Table 5). Broken-stick species-area curves are significantly steeper, and logseries or geometric series curves significantly less steep, than lognormal ones (Figure 10).

(6) Sampling problems can becloud species-abundance patterns. In particular, the statistics of sampling processes can often produce species-area curves of the form $S_T \sim \alpha \ln A$ (e.g., Figure 11).

Acknowledgments

I was drawn to this study by Robert MacArthur's seminal 1960 paper. I am grateful to many people, particularly Henry Horn, for conversations on the subject.

Mathematical Appendices

For each of the distributions $S(N)$ under consideration, we (i) make some general comments; then calculate (ii) S_T; (iii) J; (iv) species-area relations; (v) H; (vi) σ_H^2; (vii) D; (viii) d; and finally (ix) make some comments on sampling from the distribution.

A. General Lognormal Distribution

(i) *General.* To facilitate comparison with the conventional ecological literature on the lognormal distribution, we follow Preston's usage of R (see eq. 3.4) rather than N as the basic independent variable. Then the number of species in the interval R to $R + dR$ is, as discussed in the main text (eq. 3.5),

$$S(R) = S_0 \exp{(-a^2 R^2)} \qquad \text{(A.1)}$$

In practice, this distribution will extend not from R equals $-\infty$ to $+\infty$, but rather from the octave $R_{\min}$ wherein lies the least abundant species, to the octave $R_{\max}$ containing the most abundant species. As these symmetrically disposed end points are by definition those where $S(N_{min}) \simeq S(N_{max}) \simeq 1$, we have

$$1 = S_0 \exp[-a^2 R_{max}^2]$$
$$= S_0 \exp[-a^2 R_{min}^2] \quad (A.2)$$

At this point it is convenient for notational purposes to define a quantity Δ:

$$\Delta \equiv aR_{max} \equiv -aR_{min} \quad (A.3)$$

Consequently eq. A.2 can be rewritten tidily as

$$\Delta = \sqrt{\ln S_0} \quad (A.4)$$

Equation 3.9, which relates the magnitude of the smallest population, m, to that of the modal population, N_0, now takes the form

$$N_0 = m \exp[(\Delta/a)\ln 2] \quad (A.5)$$

and the definition of γ, eq. 3.8, reads

$$2a\,\Delta\gamma = \ln 2 \quad (A.6)$$

With this preliminary festival of notation disposed of, we proceed to catalogue the properties of the general lognormal distribution in terms of the three parameters a, Δ, and γ, *only two of which are independent* (see eq. A.6).

(ii) S_T. Using eq. A.4 for S_0, and taking into account that the range of octaves in practice is from R_{min} to R_{max}, we find that eq. 1.3 for the total number of species becomes

$$S_T = \exp(\Delta^2)\int_{R_{min}}^{R_{max}} \exp(-a^2 R^2)\,dR$$
$$= (\pi^{1/2}/a)\exp(\Delta^2)\,\mathrm{erf}(\Delta) \quad (A.7)$$

Here $\mathrm{erf}(\Delta)$ is the so-called error function (e.g., Abramowitz and Stegun, 1964, Chapter 7), an integral that continually appears when one deals with gaussian probability distributions, defined as

$$\mathrm{erf}(z) = 2\pi^{-1/2}\int_0^z \exp(-t^2)\,dt \quad (A.8)$$

For $\Delta > 1$, as it must be if S_T is not small,

$\mathrm{erf}(\Delta)$ is close to unity, and

$$S_T \simeq (\pi^{1/2}/a)\exp(\Delta^2) \quad (A.9)$$

If the range of integration for R is taken from $-\infty$ to $+\infty$ initially, the result eq. A.9 is exact, and it is widely quoted for the lognormal distribution. [Indeed, as the underlying assumptions leading to the lognormal species-abundance distribution usually require $S_T \gg 1$ (Preston assumes $S_T > 100$ or so), eq. A.9 should always apply; we give the result eq. A.7 only in order to treat modest values of S_T on occasion, even though the lognormal should not then be taken seriously.]

(iii) J. The total number of individuals in the community is given by eq. 1.4, which, after use of eqs. A.3, A.4, A.5, and A.6, becomes

$$N_T = m \exp[\Delta^2(1 + 2\gamma)]\int_{R_{min}}^{R_{max}}$$
$$\exp(-a^2 R^2 + R\ln 2)\,dR$$

Changing the variable of integration to $t = (aR - \gamma\Delta)$ reduces this to

$$J = (N_T/m) = (\pi^{1/2}/2a)$$
$$\exp[\Delta^2(1 + \gamma)^2]\,G(\Delta,\gamma) \quad (A.10)$$

Here we have for convenience defined G as

$$G(\Delta,\gamma) = \mathrm{erf}[\Delta(1 - \gamma)]$$
$$+ \mathrm{erf}[\Delta(1 + \gamma)] \quad (A.11)$$

Again, if S_T is not unrealistically small, $\mathrm{erf}[\Delta(1 + \gamma)]$ will be indistinguishable from unity. But as γ may in general be larger or smaller than unity, no such simplifying statement can be made about $\mathrm{erf}[\Delta(1 - \gamma)]$. Referring to Figure 2, this says biologically that the cutoff at the octave R_{max} is important for the $N(R)$ distribution, although the cutoff at R_{min} may just as well be at $-\infty$.

Given any pair of values a and γ (and hence, via eq. A.6, Δ) to characterize the shape of the lognormal distribution, the total numbers of species and of individuals, S_T and J, now follow from eqs. A.7 and A.10. Conversely, given S_T and J, the parameters a, γ, and Δ may be computed from these equations. In this way we arrive at Figure 4. Quite generally, given any two of the parameters S_T, J, a, γ, and Δ, the others may be computed.

In the limit $S_T \gg 1$, a useful analytic approximation is possible. For S_T we have eq. A.9, in which the exponential term must dominate the right-hand side for large S_T, so that roughly

$\ln S_T$
$$\simeq \Delta^2[1 + \Delta^{-2} \ln (2\pi^{1/2} \Delta\gamma/\ln 2) + \ldots]$$

That is,

$$\ln S_T \sim \Delta^2 \qquad (A.12)$$

Correspondingly, in eq. A.10 for J, we may distinguish the cases (a) $\gamma < 1$ and (b) $\gamma > 1$. In case (a), $\mathrm{erf}\,[\Delta(1 - \gamma)]$ lies between 0 and 1, generally being of order unity, and one has the asymptotic approximation

$\ln J \simeq \Delta^2(1 + \gamma)^2$
$$[1 + \Delta^{-2}(1 + \gamma)^{-2} \ln (\pi^{1/2}/a) + \ldots]$$

That is, for $\gamma < 1$,

$$\ln J \sim \Delta^2(1 + \gamma)^2 \qquad (A.13)$$

In case (b), $\mathrm{erf}\,[\Delta(1 - \gamma)]$ is negative, and the result

$$\mathrm{erf}\,(-z) = -1 + \frac{\exp(-z^2)}{\pi^{1/2}z}\left(1 - \frac{1}{2z^2} + \cdots\right)$$

leads to

$\ln J \simeq 4\gamma \Delta^2$
$$[1 - (4\gamma \Delta^2)^{-1} \ln [(\gamma - 1)(\ln 2)/\gamma] + \ldots]$$

That is, for $\gamma > 1$,

$$\ln J \sim 4\gamma \Delta^2 \qquad (A.14)$$

The intermediate case, where $\gamma \simeq 1$ so that $|\Delta(1 - \gamma)|$ is small, is covered under the canonical lognormal in Appendix B. Combining

the results eqs. A.12, A.13, and A.14, we have a rough relation between S_T and J which involves only the single parameter γ:

$$S_T \sim J^x \qquad (A.15)$$

where

$$x = (1 + \gamma)^{-2}, \text{ if } \gamma < 1 \qquad (A.16a)$$
$$x = 1/4\gamma, \qquad \text{ if } \gamma > 1 \qquad (A.16b)$$

It is strongly to be emphasized that such a linear regression of $\ln S_T$ on $\ln J$ is not an exact result, even for $S_T \gg 1$. For smallish S_T the relation is steeper than indicated by this asymptotic approximation, as is clear from Figures 4 and 10.

(iv) S_T *versus A.* The species-area curves of Figure 10 are obtained from the additional biological assumption eq. 1.5, which relates J to A. Then, given S_T and γ, A is calculated from eqs. A.7 and A.10 to give Figure 10. The analytic approximation eq. 1.6 presented in the main text follows directly from the discussion in the preceding paragraph, eqs. A.15 and A.16, along with eq. 1.5.

(v) H. To calculate the diversity index H of eq. 4.2 it is first helpful to collect the definitions eqs. 3.4, 1.1, A.4, A.5, and A.6 to write

$$p = N/N_T = J^{-1} \exp [2\gamma \Delta(\Delta + aR)] \qquad (A.17)$$

Then eq. 4.2 for the lognormal distribution takes the form

$$H = \frac{\exp [\Delta^2(1 + \gamma)^2]}{aJ} \int_{-\Delta}^{\Delta} e^{-(s-\Delta\gamma)^2}$$
$$[\ln J - 2\gamma \Delta(s + \Delta)]\, ds$$

We have written $s = aR$. After some manipulation, this reduces to

$$H = \ln \left[\frac{\pi^{1/2}\gamma \Delta}{\ln 2} G(\Delta, \gamma)\right]$$
$$+ \Delta^2(1 - \gamma^2) + \frac{2 \Delta\gamma}{\pi^{1/2}G(\Delta, \gamma)} \{\exp[-\Delta^2$$
$$(1 - \gamma)^2] - \exp[-\Delta^2(1 + \gamma)^2]\} \qquad (A.18)$$

In deriving this, use has been made of eq. A.10 for J, and the definition eq. A.11 for $G(\Delta, \gamma)$.

Once S_T and γ are specified, Δ can be computed from eqs. A.6 and A.7, and thence H from eq. A.18. In this way we obtain for any specified value of γ a unique S_T-versus-H curve, as illustrated in Figure 5.

Limiting approximations can be obtained from eq. A.18 for $S_T \gg 1$, i.e., for Δ significantly in excess of unity. The cases (a) $\gamma \simeq 1$ (where $G \simeq 1$), (b) $\gamma < 1$ (where $G \simeq 2$), and (c) $\gamma > 1$ (where G is small) are to be distinguished, as they were above in obtaining the results of eqs. A.15, A.16 (here, and elsewhere, the symbol $\mathcal{O}$ means "terms of the order of"):

(a) $\gamma \simeq 1$; $H = 2\pi^{-1/2}\Delta\gamma$
$$\left[1 - \frac{2\Delta(1-\gamma)}{\pi^{1/2}}\left\{1 - \frac{\pi(1+\gamma)}{4\gamma}\right\}\right]$$
$$+ \mathcal{O}\ln(\gamma\Delta) \quad \text{(A.19a)}$$

(b) $\gamma < 1$; $H = \Delta^2(1 - \gamma^2)$
$$+ \mathcal{O}\ln(\gamma\Delta) \quad \text{(A.19b)}$$

(c) $\gamma > 1$; $H = \left[\frac{\gamma}{\gamma-1} - \ln\left\{\frac{(\gamma-1)\ln 2}{\gamma}\right\}\right]$
$$+ \mathcal{O}\Delta^{-2} \quad \text{(A.19c)}$$

(vi) σ_H^2. In a similar way eq. 4.4 can be written down for the general lognormal distribution. The resulting expression is a mess, and we shall not set it out here. However, just as for given γ a unique H-versus-S_T curve can be computed, so here the variance in H can be calculated as a function of S_T. For the special case of the canonical distribution, $\gamma = 1$, the variance has the form given in Appendix B and displayed in Figure 6. The other asymptotic results are worth noting: (a) for $\gamma < 1$ (but still $\gamma > \frac{1}{2}$),

$$\sigma_H^2 = \frac{\Delta^2(1-\gamma)^4 \ln 2}{4\pi\gamma(2\gamma-1)}$$
$$\exp[-2\Delta^2(1-\gamma)^2][1 + \mathcal{O}(\ln S_T)^{-1}] \quad \text{(A.20a)}$$

and (b) for $\gamma > 1$,

$$\sigma_H^2 = \frac{(\gamma-1)^2 \ln 2}{\gamma(2\gamma-1)}$$
$$\{\ln[(\ln 2)(\gamma-1)/\gamma]\}\{\ln[(\ln 2)(\gamma-1)/\gamma]$$
$$- (2\gamma)/(2\gamma-1)\}[1 + \mathcal{O}(\ln S_T)^{-1}] \quad \text{(A.20b)}$$

When we combine these asymptotic results with those for H itself, the root-mean-square relative fluctuations in H are seen to have roughly the limiting forms, for $S_T \gg 1$:

(a) $\gamma \simeq 1$; $\dfrac{\sigma_H}{H} \sim \dfrac{(\text{constant})}{\ln S_T}$ \hspace{1em} (A.21a)

(b) $\gamma < 1$; $\dfrac{\sigma_H}{H} \sim (S_T)^{-(1-\gamma)^2}$ \hspace{1em} (A.21b)

(c) $\gamma > 1$; $\dfrac{\sigma_H}{H}$

$\sim$ constant (value depends on γ) \hspace{0.5em} (A.21c)

Note that, for all γ, the diversity H displays a significant amount of intrinsic statistical scatter; i.e., the r.m.s. relative variance is of order unity. This tendency is illustrated by Figure 6.

(vii) D. The diversity index D of eqs. 4.6, 4.7 may, with the aid of eqs. A.17 and A.8, be shown to be

$$1/D = \frac{\pi^{1/2}\exp[\Delta^2(1+2\gamma)^2]}{2aJ^2}$$
$$\{\text{erf}[\Delta(2\gamma+1)] - \text{erf}[\Delta(2\gamma-1)]\} \quad \text{(A.22)}$$

The case $\gamma = 1$ is discussed further in Appendix B. For any other specified value of γ, eq. A.22 may be used to calculate D as a function of S_T. Again we note the asymptotic results:

(a) $\gamma < 1$; $D \simeq [2\pi\Delta(2\gamma-1)/a]$
$$\exp[2\Delta^2(1-\gamma)^2]$$
i.e., D
$$\sim (\text{constant})(S_T)^{2(1-\gamma)^2} \quad \text{(A.23a)}$$

(b) $\gamma > 1$; $D \simeq \dfrac{(2\gamma-1)\gamma}{(\gamma-1)^2 \ln 2}$
$$+ \mathcal{O}(\ln S_T)^{-1} \quad \text{(A.23b)}$$

(viii) *d*. The dominance, defined by eq. 4.8, can be expressed immediately from eq. A.17 and the definition that N_{max} occurs in the octave labelled by R_{max} (i.e., by Δ/a):

$$d = J^{-1} \exp(4\gamma\,\Delta^2)$$

Alternatively, employing eq. A.10,

$$d = 2\pi^{-1/2}a$$
$$\exp[-\Delta^2(1-\gamma)^2]/G(\Delta,\gamma) \quad (A.24)$$

Again, it is routine to calculate the limiting forms of this expression when $S_T \gg 1$; in this limit the relation

$$D \simeq \frac{(2\gamma-1)\ln 2}{\gamma\,d^2} \quad (A.25)$$

holds for all γ.

(ix) *Sampling from the lognormal distribution.* A very simplistic approach, which glosses over statistical niceties, may be used to treat incomplete samples from the lognormal (and broken-stick) distributions. We assume that if, in the total sample of $\hat{N}$ individuals, the probable number of individuals from species *i* exceeds unity we *do* see that species, whereas if the probability corresponds to less than one animal that species is not counted.

In the sample of $\hat{N}$ individuals, let $\hat{m}$ be the average number of individuals from the least abundant species. Then $\hat{N}/\hat{m} = N_T/m$, and recalling the definition of *J*, eq. (1.1), we have $\hat{m} = \hat{N}/J$. Thus so long as

$$\hat{N} > J, \quad (A.26)$$

we have $\hat{m} > 1$, and all species are represented in our sample, at least in this naive estimate.

For $\hat{N} < J$, write

$$\hat{N} = fJ, \text{ with } f < 1 \quad (A.27)$$

In the sample, the number of individuals per species in the *R*th octave, $\hat{N}_R$, is typically a fraction ($\hat{N}/N_T$) of the actual number. With eqs. 3.4, A.27, 1.1, and 3.9, this comes to

$$\hat{N}_R = f2^{R-R_{min}} \quad (A.28)$$

If only those species that are typically represented by one or more individuals are assumed to be counted, then only those octaves with $R > R_0$ (where $\hat{N}_{R_0} = 1$) are to be counted. From eq. A.28, we have

$$R_0 = -\frac{\ln f}{\ln 2} + R_{min} \quad (A.29)$$

The total number of species, $\hat{S}_T$, represented in this sample is now

$$\hat{S}_T = S_0 \int_{R_0}^{R_{max}} \exp(-a^2R^2)\,dR$$

Alternatively, in the notation used throughout this appendix, the fraction of species represented in the sample is

$$\frac{\hat{S}_T}{S_T} = \frac{\operatorname{erf}(\Delta) + \operatorname{erf}\left[\Delta - \dfrac{\ln(1/f)}{2\gamma\,\Delta}\right]}{2\operatorname{erf}(\Delta)} \quad (A.30)$$

This sampling behavior is illustrated for one particular value of γ in Figure 12. Notice that in general 50% of the total number of species actually present in the community are represented in a sample of *fJ* individuals, where $f = \exp(-2\gamma\,\Delta^2)$. Roughly this corresponds (see eqs. A.12, A.13, A.14) to a sample of $\hat{N} \sim (S_T)^{1+\gamma^2}$ individuals if $\gamma < 1$, or $\hat{N} \sim (S_T)^{2\gamma}$ individuals if $\gamma > 1$, which agree at $\hat{N} \sim S_T^2$ if $\gamma = 1$.

B. Canonical Lognormal Distribution: $\gamma = 1$

(i) *General.* In view of Preston's (1962) empirical observations as to the ubiquity of the canonical lognormal distribution, and the present rough explanation (Section 3A) as to why this tendency is likely to be observed, we single out the case $\gamma = 1$ for special review.

(ii) S_T. As Δ and *a* are now related by eq. A.6 with $\gamma = 1$, eq. A.9 can be put in terms of the *single* parameter Δ:

$$S_T \simeq (2\pi^{1/2}/\ln 2)\,\Delta\exp(\Delta^2) \quad (B.1)$$

(iii) *J*. Equation A.10 with $\gamma = 1$ becomes

$$J = (N_T/m) = (\pi^{1/2}/\ln 2)$$
$$\Delta \exp (4\, \Delta^2)\, \mathrm{erf}\, (2\, \Delta) \quad (B.2)$$

Unless S_T has an unreasonably small value, $\mathrm{erf}\, (2\, \Delta) \simeq 1$ and one has the excellent approximation

$$J \simeq (\pi^{1/2}/\ln 2)\, \Delta \exp (4\, \Delta^2) \quad (B.3)$$

(iv) S_T *versus A*. If we make the biological assumption eq. 1.5, $J = \rho A$, the single parameter Δ may in principle be eliminated between eqs. B.1 and B.3, to give a unique species-area curve. The interesting history of such work is discussed in Section 5, and the curve illustrated in Figure 10.

Note that eqs. B.1 and B.3 do not lead to any simple functional relationship between S_T and A. For $S_T \gg 1$, Δ is significantly greater than unity, and the exponential terms dominate the right-hand side of both equations. This leads for $S_T \gg 1$ to the excellent approximation

$$\ln S_T \simeq (\tfrac{1}{4})\ln A + (\text{constant}) \quad (B.4)$$

However, for smaller values of S_T the relation is steeper than a 0.25 power law (and indeed in the unreasonable limit of $S_T \lesssim 10$ we have roughly from eqs. A.7 and A.10 a linear relation, $S_T \simeq J = \rho A$). It is for this reason that Preston's and MacArthur and Wilson's work, which fitted linear regressions to this theoretical relation $\ln S_T$ versus $\ln A$, lead to regression coefficients slightly larger than the asymptotically exact 0.25 (Preston: 0.262; MacArthur and Wilson: 0.263). It is not a good idea to fit a single such regression to a relation that systematically proceeds from a steeper to a less steep relation as S_T increases.

(v) *H*. For $\gamma = 1$, eq. A.18 reduces to

$$H = \frac{2\, \Delta[1 - \exp (-4\, \Delta^2)]}{\pi^{1/2}\, \mathrm{erf}\, (2\, \Delta)}$$
$$+ \ln [(\pi^{1/2}\, \Delta/\ln 2)\, \mathrm{erf}\, (2\, \Delta)] \quad (B.5)$$

Again the approximation $\mathrm{erf}\, (2\, \Delta) \simeq 1$ will be excellent unless S_T is small, whence

$$H = 2\pi^{-1/2}\, \Delta + \ln (\pi^{1/2}\, \Delta/\ln 2) \quad (B.6)$$

Eliminating Δ by use of eq. B.1 leads to a unique *H*-versus-S_T relationship for the canonical lognormal distribution, as illustrated in Figures 5, 6, and 7. Notice that for $S_T \gg 1$ we have the approximation

$$H \simeq 2\pi^{-1/2} (\ln S_T)^{1/2}$$
$$[1 + \mathcal{O} (\ln S_T)^{-1}] \quad (B.7)$$

As discussed in Section 4, this relation corresponds to a very slow increase in the diversity index H as S_T increases.

(vi) σ_H^2. With $\gamma = 1$, eqs. A.1 and A.17 lead to an expression for σ_H^2 for the canonical lognormal distribution: this expression generates the results displayed in Figure 6.

An excellent approximation is again obtained by assuming S_T is not small, so that terms of relative order $\exp (-\Delta^2) \sim 1/S_T$ may be neglected:

$$\sigma_H^2 = \left(\frac{2 \ln 2}{\pi\, \Delta}\right) e^{\Delta^2} \int_{\Delta}^{\infty} e^{-t^2}$$
$$\left[2\, \Delta^2 - 2\, \Delta t - \ln \left(\frac{\pi^{1/2}\, \Delta}{\ln 2}\right)\right]^2 dt \quad (B.8)$$

To a less accurate approximation, neglecting terms of relative order $1/(\ln S_T)$, this is

$$\sigma_H^2 \simeq \frac{\ln 2}{\pi\, \Delta^2} \left[\ln \left(\frac{\pi^{1/2}\, \Delta}{\ln 2}\right)\right] \left[2 + \ln \left(\frac{\pi^{1/2}\, \Delta}{\ln 2}\right)\right]$$

That is, very roughly, using "$a = 0.2$" in the logarithmic terms, we have

$$\sigma_H^2 \sim \frac{1 \cdot 1}{\ln S_T} \quad (B.9)$$

As noted in Appendix A, eq. 21a, the r.m.s. relative variance is of order $1/(\ln S_T)$, a fact that underlies the statistical dispersion in H illustrated and discussed in Figure 6.

(vii) *D*. On putting $\gamma = 1$, and neglecting

terms of relative order $\exp(-4\,\Delta^2) \sim 1/S_T^4$, we have for eq. A.22:

$$D = \pi\,\Delta^2/\ln 2 \qquad (B.10)$$

This leads to the D-versus-S_T relation illustrated in Figure 8, and has the asymptotic form given in Table 2.

(viii) d. The results displayed in Figure 9 for the canonical lognormal dominance index, d, as a function of S_T, stem from the appropriate form of eq. A.24. This is simply

$$d = \frac{\ln 2}{\pi^{1/2}\,\Delta} \qquad (B.11)$$

The asymptotic relation to S_T is clearly as given in Table 3.

(ix) *Sampling*. Here the treatment is precisely as in Appendix A, with the simplification that $\gamma = 1$. Recall that for all species to be represented we require $\hat{N} > J$, eq. A.26, which for the canonical lognormal distribution comes down to (see eqs. B.1, B.3) the requirement $\hat{N} \gtrsim S_T^4$.

C. Uniform Distribution

(i) *General*. The ideal uniform distribution,

$$N_i = N_T/S_T \qquad (C.1)$$

has the following properties.

(ii) S_T. A parameter specifying the distribution.

(iii) J. Clearly $m = N_T/S_T$, so that

$$J = (N_T/m) = S_T \qquad (C.2)$$

(iv) S_T *versus* A. Assuming eq. 1.5, we have the linear (!) relation

$$S_T = \rho A \qquad (C.3)$$

(v) H.

$$H = \ln S_T \qquad (C.4)$$

(vi) σ_H^2. For this ideally even distribution, $\sigma_H^2 = 0$.

(vii) D.

$$D = S_T \qquad (C.5)$$

(viii) d.

$$d = 1/S_T \qquad (C.6)$$

D. Broken-Stick Distribution

(i) *General*. Following the initial work of MacArthur (1957, 1960), and the numerical exploration of various aspects of the distribution by Lloyd and Ghelardi (1964), Longuet-Higgins (1971) and Webb (1974) have recently given expositions of the analytic properties of this distribution. There is also a fantastically elaborate formal and numerical exploration of "the distribution of indices of diversity" by Bowman *et al.* (1971) which, despite the generality of its title, is confined to the broken-stick distribution. These results will simply be catalogued under the appropriate headings below. There remain a few aspects of the distribution that do not seem to have been previously discussed (e.g., species-area relations, sampling aspects), and these will be developed more fully.

Unlike the lognormal or the logseries distributions, the broken-stick model is rarely presented in terms of its species-abundance distribution $S(N)$, but rather is conventionally discussed in rank-abundance form. To make concrete the abstract remarks in Section 2, we show how the broken-stick distribution $S(N)$ may be deduced from the more usual description of this distribution.

The routine broken-stick formulation gives the number of individuals in the ith most abundant of S_T species to be

$$N_i = \frac{N_T}{S_T} \sum_{n=i}^{S_T} \frac{1}{n} \qquad (D.1)$$

This is the sort of information displayed, for example, in Figure 1. For $S_T \gg 1$ this has the approximate form

$$N_i \simeq (N_T/S_T) \ln (S_T/i) \qquad \text{(D.2)}$$

Thus the rank-order function of Section 2, $F(N) \equiv i$, is simply

$$F(N) \simeq S_T \exp (-S_T N/N_T) \qquad \text{(D.3)}$$

But, as argued in eq. 2.1, $S(N)$ is now the (negative) derivative of $F(N)$:

$$S(N) = -dF/dN = (S_T^2/N_T)$$
$$\exp (-S_T N/N_T) \qquad \text{(D.4)}$$

This is exactly the asymptotic distribution function derived by Longuet-Higgins and by Webb; a thorough and elegant derivation is also presented by Cohen (1966). For a full discussion, leading to the exact result

$$S(N) = [S_T(S_T - 1)/N_T]$$
$$(1 - N/N_T)^{S_T - 2} \qquad \text{(D.5)}$$

see Webb (1973).

(ii) S_T. In the general lognormal distribution, one could choose among several alternative parameters for the two that characterized the distribution; consequently we discussed relations among these parameters. Broken-stick distributions are characterized by *one* parameter, which invariably is just S_T.

(iii) J. Although it does not seem to have been remarked previously, there is a trivial relation between J and S_T for the broken-stick distribution. From eq. D.1, the population of the least abundant species is

$$m = N_T/S_T^2$$

That is,

$$J = S_T^2 \qquad \text{(D.6)}$$

(iv) S_T *versus* A. Adding to eq. D.6 the biological assumption of eq. 1.5, we have immediately the broken-stick species-area relation

$$S_T = (\text{constant}) A^{1/2} \qquad \text{(D.7)}$$

(v) H. Numerical tables of H as a function of S_T have been given by Lloyd and Ghelardi (1964). The exact analytic result

$$H = \psi(S_T + 1) + (\gamma - 1) \qquad \text{(D.8)}$$

has been given by Webb (1974). Here $\psi(z)$ is the logarithmic derivative of the gamma function (Abramowitz and Stegun, 1964, Chapter 6), and $\gamma = 0.577 \ldots$ is Euler's constant. Neglecting terms of order $1/S_T$, one has $\psi(S_T + 1) \simeq \ln S_T$, leading to the excellent approximation given in Table 1 (Longuet-Higgins, 1971; Webb, 1974).

(vi) σ_H^2. Substitution of the exact $S(N)$ for the broken-stick distribution, eq. D.5, into eq. 4.4 for the variance of H may be shown to lead to the exact result (I omit the details: these, and any others, will be supplied on request)

$$\sigma_H^2 = \frac{2}{S_T + 1}$$
$$\{[\psi(S_T + 2) - \psi(3)]^2 + [\psi'(3) - \psi'(S_T + 2)]\}$$
$$- \frac{H^2}{S_T} \qquad \text{(D.9)}$$

Here $\psi'(z)$ is the derivative of $\psi(z)$. This expression gives the results depicted in Figure 6, which agree with Webb's numerical simulations.

For $S_T \gg 1$ one can obtain [either from eq. D.9, or directly from eq. 4.4, using eq. D.4 for $S(N)$] the asymptotic result

$$\sigma_H^2 \simeq (\ln S_T)^2/S_T \qquad \text{(D.10)}$$

The r.m.s. relative fluctuations in H then have the asymptotic form of eq. 4.9.

(vii) D. In a similar fashion, substitution of eq. D.5 for $S(N)$ into eqs. 4.6, 4.7 for D gives the exact and simple result

$$D = \tfrac{1}{2}(S_T + 1) \qquad \text{(D.11)}$$

Again the asymptotic form of this result, $D \simeq \tfrac{1}{2}S_T$, may alternatively be obtained by using the simple asymptotic formula eq. D.4 for $S(N)$.

(viii) d. From the basic eq. D.1, the dominance index of eq. 4.8 is

$$d = \frac{1}{S_T} \sum_{n=1}^{s_T} \frac{1}{n}$$

$$= \frac{1}{S_T} [\gamma + \psi(S_T + 1)] \quad (D.12)$$

Ignoring corrections of relative order $1/S_T$, we have the excellent approximation given in Table 3. Notice the neat and exact relation between H and d for the broken-stick distribution:

$$d = (H + 1)/S_T \quad (D.13)$$

(ix) *Sampling*. Referring to the discussion at the end of Appendix A as to rough estimates of the number of species represented in a sample of $\hat{N}$ individuals, we first observe that all S_T species are present once $\hat{N} > J$, eq. A.26; that is, once $\hat{N} > S_T^2$ (see eq. D.6). This confirms the features discussed in Section 6 and illustrated in Figure 12: in a sample of size $\hat{N} \sim S_T^2$, roughly 50% of species are not represented if the underlying distribution is the canonical lognormal, whereas essentially all are represented if the distribution be broken-stick.

In more detail, we see from eq. D.1 that in a sample of $\hat{N}$ total individuals, the number of individuals in the ith most abundant species, $\hat{N}_i$, is

$$\hat{N}_i = (\hat{N}/S_T)$$
$$[\psi(S_T + 1) - \psi(i + 1)] \quad (D.14)$$

That is to say a rough estimate of $\hat{S}_T$ is given implicitly by the equation

$$\psi(\hat{S}_T + 1)$$
$$= \psi(S_T + 1) - S_T/\hat{N} \quad (D.15)$$

The relation gives the curve shown in Figure 12. To an excellent approximation, $\psi(z + 1) =$ $\ln z - \gamma$, and so for $\hat{N}$ appreciably less than S_T^2 we have that the fraction of the actual species total S_T which is observed in a sample comprising $\hat{N}$ individuals is

$$\hat{S}_T/S_T = \exp(-S_T/\hat{N}) \quad (D.16)$$

Roughly 50% of all species actually present in the community are represented even in a sample as small as $S_T/\ln 2 = (1.4)S_T$ individuals.

E. Geometric Series Distribution

(i) *General*. The "niche-preemption" hypothesis, leading to a geometric series rank-abundance distribution (Motomura, 1932),

$$N_i = N_T C_k k (1 - k)^{i-1} \quad (E.1)$$

has been reviewed by Whittaker (1970, 1972) and by McNaughton and Wolf (1970). Here $k < 1$, and C_k is a normalization constant to ensure $\Sigma N_i = N_T$:

$$C_k = [1 - (1 - k)^{S_T}]^{-1} \quad (E.2)$$

Converted to a (continuous) species-abundance distribution $S(N)$, in the manner illustrated at the beginning of Appendix D, this gives

$$S(N) \, dN = \frac{dN}{N \ln [1/(1 - k)]} \quad (E.3a)$$

That is, expressed as a relation between change in the number of species ΔS and change in the number of individuals ΔN,

$$\frac{\Delta S}{\Delta N} = \frac{K}{N} \quad (E.3b)$$

This is the basic equation of Odum, Cantlon, and Kornicher (1960) and we note that their K and the conventional geometric series k are related by $K \ln (1 - k) = -1$. [As geometric series people tend to look at rank-abundance relations, whereas Odum *et al.* focus on $S(N)$, it is understandable that their relationship has not been discussed. An exception is Horn (1964), who remarked in an empirical way that the Odum *et al.* S_T-versus-J relation is similar to that for the logseries, eq. 3.12.]

In the remainder of this appendix, we emphasize some aspects of the geometric series distribution that do not appear to be widely known.

(ii) S_T. Like the general lognormal and the logseries, the geometric series distribution is characterized by *two parameters:* these are usually taken to be k, the "niche-preemption" parameter, and S_T. Note that if the product $kS_T \gg 1$, the normalization constant $C_k \simeq 1$.

(iii) J. For the least abundant species, eq. E.1 has $N_i = m$ and $i = S_T$. Consequently

$$\frac{1}{J} = \frac{m}{N_T} = \left(\frac{k}{1-k}\right)\frac{(1-k)^{S_T}}{1-(1-k)^{S_T}}$$

That is,

$$S_T = \frac{\ln\left[1 + Jk/(1-k)\right]}{\ln\left[1/(1-k)\right]} \qquad (E.4)$$

For $J \gg 1$ and any particular value of k, this takes the useful asymptotic form

$$S_T \simeq K\ln J + (\text{constant}) \qquad (E.5)$$

Here the constant K is that of eq. E.3.

(iv) S_T *versus A*. When the biological assumption of eq. 1.5 is substituted into the exact eq. E.4, or the excellent approximation eq. E.5, we obtain the species-area relations discussed in Section 5.

(v) H. It is routine to calculate the diversity index, eq. 4.1, directly from the distribution eq. E.1. We arrive at a family of curves H versus S_T, one for each value of the parameter k:

$$H = \ln\left[(1-\varepsilon)/k\right]$$
$$- \left(\frac{1-k}{k} - \frac{S_T\varepsilon}{1-\varepsilon}\right)\ln(1-k) \quad (E.6)$$

where

$$\varepsilon \equiv (1-k)^{S_T} \qquad (E.7)$$

For kS_T significantly greater than unity, $\varepsilon \simeq 0$ and eq. E.6 gives the result noted in Table 1.

(vi) σ_H^2. Interpreted strictly, the geometric series distribution of relative abundance, eq. E.1, is rigidly deterministic, and $\sigma_H^2 = 0$. Similarly the exactly even distribution, eq. C.1, has by assumption no variance. As discussed in Section 2, the precise geometric series bears the same sort of relationship to the statistical logseries distribution as the exactly evenly distributed case bears to the statistical broken-stick distribution.

(vii) D. Use of eq. E.1 in the definitions eqs. 4.5 and 4.6 leads directly to

$$D = \frac{(2-k)(1-\varepsilon)}{k(1+\varepsilon)} \qquad (E.8)$$

Here ε is as defined in eq. E.7. Again for $S_T \gg 1$ this leads to the asymptotic result quoted in Table 2.

(viii) d. From eq. E.1, the dominance measure eq. 4.8 is simply

$$d = k/(1-\varepsilon) \qquad (E.9)$$

This has the value k when S_T is large, which essentially constitutes the original definition of k.

F. The Logseries Distribution

(i) *General*. This distribution, first discussed in an ecological context by Fisher, Corbet, and Williams (1943), is reviewed in detail by Pielou (1969, Chapter 17) and, complete with numerical tables, by Williams (1964, Appendix A). The distribution is characterized by two parameters, commonly written α and x:

$$S(N) = \frac{\alpha x^N}{N} \qquad (F.1)$$

Here N runs over the positive integers. A summary of results follows; for details, see the above references.

(ii) S_T. In terms of α and x,

$$S_T = -\alpha \ln(1-x) \qquad (F.2)$$

Alternatively, S_T may be chosen as one of the two parameters characterizing the distribution.

(iii) J. The assumptions underlying the logseries distribution imply that it is never fully "unveiled" (see eq. F.1), so that the least abundant species is, or are, represented by a single individual. That is, $m = 1$ and $J = N_T$. For any specific choice of α, it may be seen that eq. 3.12 relates J to S_T.

(iv) S_T *versus A*. Given the biological assumption of eq. 1.5, eq. 3.12 corresponds to a family of species-area curves, one for each value of α:

$$S_T = \alpha \ln[1 + (\rho/\alpha)A] \qquad (F.3)$$

For $J \gg \alpha$, as it generally will be, we have

$$S_T = \alpha \ln A + (\text{constant}) \qquad (F.4)$$

This, as remarked in the main text, is of the same form as the geometric series species-area relation eq. E.5.

(v) H. From the definition eq. 4.2 and the logseries distribution eq. F.1 there follows the exact expression

$$H = (\alpha/J) \sum_{i=0}^{\infty} x^i \ln(J/i) \qquad (F.5)$$

Given S_T and α, the quantities x and J can be computed from eqs. F.2 and 3.12, and a curve H versus S_T computed for each value of α. For the usual limiting case where $J \gg \alpha$, a good approximation is

$$H \simeq (\alpha/J) \int_1^{\infty} x^t \ln(J/t)\, dt$$

That is,

$$H \simeq \ln \alpha + \gamma \qquad (F.6)$$

Here correction terms of order α/J have been neglected, and $\gamma = 0.577\ldots$ is Euler's constant.

(vi) σ_H^2. By neglect of terms of relative order $1/\ln(J/\alpha)$, it may be shown that

$$\sigma_H^2 \simeq (\ln \alpha)^2/\alpha \qquad (F.7)$$

(vii) D. The index defined by eqs. 4.6 and 4.7 is

$$1/D = (\alpha/J^2) \sum_{i=1}^{\infty} i x^i$$

Summing this series, and using eqs. F.2 and F.3, one obtains

$$D = \frac{\alpha}{1+(\alpha/J)} \simeq \alpha \qquad (F.8)$$

This makes plain the biological character of the parameter α (as did eq. F.6).

(viii) d. To calculate d, we need an estimate of the average value of the largest population. The most direct way to do this, and one with a clear biological basis, is to calculate the rank-order function $F(N)$ corresponding to eq. F.1 (see eq. 2.1), and then to estimate the maximum population as $F(N_{max}) = 1$. Replacing sums by integrals for $N > N_{max}$, one obtains

$$F(N) = \alpha E_1[N \ln(1 + \alpha/N_T)] \qquad (F.9)$$

Here E_1 is the standard exponential integral (e.g., Abramowitz and Stegun, 1964, Chapter 5). This expression is useful in itself, as it provides an analytic formula for the logseries rank-abundance computed, for example, by Whittaker (1972); in particular, for $\alpha N/N_T$ small as it will be for all but the most abundant few species, eq. F.9 has the approximate form

$$F(N) \simeq -\alpha \ln(\alpha N/N_T) - \gamma \qquad (F.10)$$

The dominance index of eq. 4.8 follows from eq. F.9 by putting $F(N_{max}) = 1$, with $N_{max} = dN_T$

In the usual case where N_T/α is large, d is given by

$$1 = \alpha E_1(\alpha d) \qquad \text{(F.11)}$$

That is, d depends only on the diversity index α, and is independent of S_T or N_T. Very roughly, eq. F.11 has the solution

$$d \sim \frac{\ln \alpha}{\alpha} [1 + \mathcal{O} (\ln \alpha)^{-1}] \qquad \text{(F.12)}$$

(ix) *Sampling.* As noted by Fisher, one of the key properties of the logseries distribution is that it is its own sampling distribution. That is, for any specified value of the diversity parameter α, the number of species $\hat{S}_T$ represented in a sample of $\hat{N}$ individuals is (see eq. 3.12)

$$\hat{S}_T = \alpha \ln (1 + \hat{N}/\alpha) \qquad \text{(F.13)}$$

References

Abramowitz, M., and I. A. Stegun. 1964. *Handbook of Mathematical Functions.* Dover, New York.

Batzli, G. O. 1969. Distribution of biomass in rocky intertidal communities on the Pacific Coast of the United States. *J. Anim. Ecol.* 38:531–546.

Berger, W. H., and F. L. Parker. 1970. Diversity of planktonic Foraminifera in deep-sea sediments. *Science* 168:1345–1347.

Bliss, C. I. 1965. An analysis of some insect trap records. *In* G. P. Patil, ed., *Classical and Contagious Discrete Distributions,* pp. 385–397. Statistical Publishing Society, Calcutta.

Boswell, M. T., and G. P. Patil. 1971. Chance mechanisms generating the logarithmic series distribution used in the analysis of number of species and individuals. *In* G. P. Patil, E. C. Pielou, and W. E. Waters, eds., *Statistical Ecology.* Vol. 3, pp. 99–130. Pennsylvania State University Press, University Park, Pa.

Bowman, K. O., K. Hutcheson, E. P. Odum, and L. R. Shenton. 1971. Comments on the distribution of indices of diversity. *In* G. P. Patil, E. C. Pielou, and W. E. Waters, eds., *Statistical Ecology.* Vol. 3., pp. 315–366.

Cohen, J. E. 1966. *A Model of Simple Competition,* pp. 77–80. Harvard University Press, Cambridge, Mass.

Cohen, J. E. 1968. Alternate derivations of a species-abundance relation. *Amer. Natur.* 102:165–172.

Cramer, H. 1948. *Mathematical Methods of Statistics.* Princeton University Press, Princeton.

Darlington, P. J. 1943. Carabidae of mountains and islands: data on the evolution of isolated faunas, and on atrophy of wings. *Ecol. Monogr.* 13:37–61.

Darlington, P. J. 1957. *Zoogeography.* John Wiley and Sons, New York.

De Benedictis, P. A. 1973. On the correlations between certain diversity indices. *Amer. Natur.* 107:295–302.

Deevey, E. S. Jr. 1969. Specific diversity in fossil assemblages. *In* G. M. Woodwell and H. H. Smith, eds., *Diversity and Stability in Ecological Systems,* Brookhaven Symposium in Biology No. 22, pp. 224–241. U.S. Department of Commerce, Springfield, Va.

Dickman, M. 1968. Some indices of diversity. *Ecol.* 49:1191–1193.

Fisher, R. A. 1958. *The Genetical Theory of Natural Selection.* Dover Publ. Inc., New York.

Fisher, R. A., A. S. Corbet, and C. B. Williams. 1943. The relation between the number of species and the number of individuals in a random sample of an animal population. *J. Anim. Ecol.* 12:42–58.

Gauch, H. G., and G. B. Chase. 1975. Fitting

the gaussian curve in ecological applications. *Ecology*. In press.

Goulden, C. E. 1969. Temporal changes in diversity. *In* G. M. Woodwell and H. H. Smith, eds., *Diversity and Stability in Ecological Systems,* Brookhaven Symposium in Biology No. 22, pp. 96–102. U. S. Department of Commerce, Springfield, Va.

Hamilton, T. H., R. H. Barth, and I. Rubinoff. 1964. The environmental control of insular variation in bird species abundance. *Proc. Nat. Acad. Sci. U.S.A.* 52: 132–140.

Hamilton, T. H., I. Rubinoff, R. H. Barth, and G. L. Bush. 1963. Species abundance: natural regulation of insular variation. *Science* 142:1575–1577.

Horn, H. S. 1964. Species diversity indices. Privately circulated manuscript, University of Washington, Seattle.

Hurlbert, S. H. 1971. The nonconcept of species diversity: a critique and alternative parameters. *Ecol.* 52:577–586.

Hutchinson, G. E. 1953. The concept of pattern in Ecology. *Proc. Acad. Nat. Sci. Philadelphia* 105:1–12.

Johnson, M. P., L. G. Mason, and P. H. Raven. 1968. Ecological parameters and plant species diversity. *Amer. Natur.* 102:297–306.

Johnson, M. P. and P. H. Raven. 1970. Natural regulation of plant species diversity. *Evol. Biol.* 4:127–162.

Johnson, M. P. and P. H. Raven. 1973. Species number and endemism: the Galapagos Archipelago revisited. *Science* 179:893–895.

Kendall, D. G. 1948. On some models of population growth leading to R. A. Fisher's logarithmic series distribution. *Biometrica* 35:6–15.

King, C. E. 1964. Relative abundance of species and MacArthur's model. *Ecology* 45:716–727.

Kohn, A. J. 1959. The ecology of Conus in Hawaii. *Ecol. Monogr.* 29:47–90.

Lloyd, M., and R. J. Ghelardi. 1964. A table for calculating the "equitability" component of species diversity. *J. Anim. Ecol.* 33:217–225.

Longuet-Higgins, M. S. 1971. On the Shannon-Weaver index of diversity, in relation to the distribution of species in bird censuses. *Theor. Pop. Biol.* 2:271–289.

MacArthur, R. H. 1957. On the relative abundance of bird species. *Proc. Nat. Acad. Sci. U.S.A.* 43:293–295.

MacArthur, R. H. 1960. On the relative abundance of species. *Amer. Natur.* 94:25–36.

MacArthur, R. H. 1966. A note on Mrs. Pielou's comments. *Ecology* 47:1074.

MacArthur, R. H. and E. O. Wilson. 1967. *The Theory of Island Biogeography.* Princeton University Press, Princeton.

McNaughton, S. J., and L. L. Wolf. 1970. Dominance and the niche in ecological systems. *Science* 167:131–139.

McNaughton, S. J., and L. L. Wolf. 1973. *General Ecology.* Holt, Rinehart, and Winston, New York.

Margalef, R. 1972. Homage to Evelyn Hutchinson, or why there is an upper limit to diversity. *Trans. Conn. Acad. Arts Sci.* 44:211–235.

Montroll, E. W. 1972. On coupled rate equations with quadratic nonlinearities. *Proc. Nat. Acad. Sci. U.S.A.* 69:2532–2536.

Motomura, I. 1932. A statistical treatment of associations [in Japanese]. *Jap. J. Zool.* 44:379–383.

Odum, H. T., J. E. Cantlon, and L. S. Kornicher. 1960. An organizational hierachy postulate for the interpretation of species-individual distributions, species entropy, ecosystem evolution, and the meaning of a species-variety index. *Ecology* 41:395–399.

Patil, G. P., E. C. Pielou, and W. E. Waters (eds.). 1971. *Statistical Ecology*. Vols. 1, 2, 3. Pennsylvania State University Press, University Park, Pa.

Patrick, R. 1968. The structure of diatom communities in similar ecological conditions. *Amer. Natur.* 102:173–183.

Patrick, R., M. Hohn, and J. Wallace. 1954. A new method of determining the pattern of the diatom flora. *Notulae Natura* 259.

Pielou, E. C. 1969. *An Introduction to Mathematical Ecology*. Wiley-Interscience, New York.

Preston, F. W. 1948. The commonness, and rarity, of species. *Ecology* 29:254–283.

Preston, F. W. 1962. The canonical distribution of commonness and rarity. *Ecology* 43:185–215, 410–432.

Sanders, H. L. 1969. Benthic marine diversity and the stability-time hypothesis. *In* G. M. Woodwell and H. H. Smith, eds., *Diversity and Stability in Ecological Systems,* Brookhaven Symposium No. 22, pp. 71–81. U.S. Department of Commerce, Springfield, Va.

Simpson, E. H. 1949. Measurement of diversity. *Nature* 163:688.

Statistical Abstract of the United States. 1971. P. 327, Tables 522, 523. U.S. Department of Commerce, Springfield, Va.

Tramer, E. J. 1969. Bird species diversity; components of Shannon's formula. *Ecology* 50:927–929.

Tsukada, M. 1972. The history of Lake Nojiri, Japan. *Trans. Conn. Acad. Arts Sci.* 44:337–365.

Usher, M. B. 1973. *Biological Management and Conservation: Ecological Theory, Application and Planning*. Chapman and Hall, London.

Webb, D. J. 1974. The statistics of relative abundance and diversity *J. Theor. Biol.* 43:277–292.

Whittaker, R. H. 1965. Dominance and diversity in land plant communities. *Science* 147:250–260.

Whittaker, R. H. 1970. *Communities and Ecosystems*. Macmillan, New York.

Whittaker, R. H. 1972. Evolution and measurement of species diversity. *Taxon* 21:213–251.

Williams, C. B. 1953. The relative abundance of different species in a wild animal population. *J. Anim. Ecol.* 22:14–31.

Williams, C. B. 1964. *Patterns in the Balance of Nature*. Academic Press, London.

Williamson, M. 1972. *The Analysis of Biological Populations*. Edward Arnold, London.

20

Reprinted from *Am. Nat.* **116**:770–787 (1980) by permission of The University of Chicago Press

MINIMAL COMMUNITY STRUCTURE: AN EXPLANATION
OF SPECIES ABUNDANCE PATTERNS

GEORGE SUGIHARA

Department of Biology, Princeton University, Princeton, New Jersey 08544

Submitted October 12, 1978; Accepted October 19, 1979

Here I attempt to provide qualitative insight into the structure of natural communities through an investigation of regularities in species abundance relations. In particular, I aim to show that the ubiquitous canonical lognormal distribution of Preston (1962) and the biogeographical species-area constant, $z \approx 1/4$ (MacArthur and Wilson 1967), are not mathematical artifacts as has been previously suggested (Connor and McCoy 1979; May 1975), but can reflect a broad and very simple underlying form of community organization.

Few propositions in ecology have as much empirical support as Preston's (1962) canonical lognormal hypothesis of species abundance (fig. 1). This statistical abundance distribution has been observed for a diverse array of organisms including diatoms (Patrick 1968; Patrick et al. 1954), soil arthropods (Hairston and Byers 1954), lepidoptera, birds, and mammal faunas (Preston 1962), and for areas ranging in size from diatometer slides to entire continents. It is the general rule for collections with large numbers of taxonomically related species, and has special importance for the theory of island biogeography (MacArthur and Wilson 1967) since it can generate the familar species-area constant, $z \approx 1/4$, for the relation, $S = cA^z$ where S and A are species count and area, respectively, and c and z are constants. (May [1975] offers the clearest derivation of this relationship.) Repeated investigations show that when z is treated as a fitted constant, it often agrees with the theoretically derived value, $z \approx 1/4$, which rests directly on the canonical hypothesis (see Preston 1962; MacArthur and Wilson 1967; May 1975; Connor and McCoy 1979 for a catalogue of examples).

By convention, a lognormal species curve is canonical if the parameter γ has a constant value of 1, where γ is the ratio of the position of the individual's curve mode, R_N, to the upper truncation point of species curve, R_{max} (fig. 1). This is essentially an empirical relationship deriving from the observation that R_N and R_{max} have a tendency to coincide. Notice that when $\gamma = 1$, the independent variables of the general lognormal distribution become coupled to yield a specific one-parameter family of lognormal curves. For example, the specific coupling between species count, S, and the variance of the lognormal, σ^2, can be written approximately as

$$S = \sigma \sqrt{\pi/2} \exp \left[\frac{(\sigma \ln 2)^2}{2} \right] \tag{1}$$

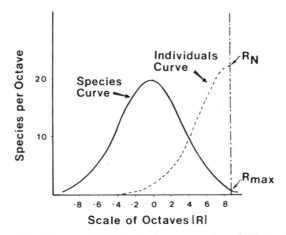

FIG. 1.—The canonical lognormal distribution for an ensemble of 178 species (after Preston 1962). By convention, the x-axis is scaled as logarithmic (base 2) abundance classes or "octaves" of individuals/species, adjusted to have a mean of zero. The species curve denotes the number of species in each octave and the individual's curve shows the number of individuals in each abundance class. This particular distribution is canonical because the mode of the individual's curve, R_N, coincides with the upper truncation point of the species curve, R_{max} (i.e., $\gamma = R_N/R_{max} = 1$). Setting $\gamma = 1$ couples species count and variance in a specific way (see fig. 2).

(after eq. [1A]). This equation contains the same information as the phenomenological rule $\gamma = 1$ and is perhaps more meaningful since it shows that under the canonical hypothesis, a large variance in the distribution accompanies a large species count. In the general lognormal distribution, on the other hand, these two parameters are completely independent. The canonical hypothesis, therefore, describes a specific positive coupling between the number of species contained under the distribution and the size of its variance.

In view of the importance and predictive success of this relationship, especially in the species-area context, it is puzzling that little serious effort has been made toward a theoretical justification. Connor and McCoy (1979) have tried to explain the island biogeographic constant, $z \simeq 1/4$, as an artifact of linear regression. They argue that z, which in practice is reckoned from the slope of a linear regression, should behave in a null model as the product of two uniform (0, 1) random variables, i.e., $z = \rho(\sigma_y/\sigma_x)$ where the correlation coefficient (ρ) and the ratio of the standard deviations of the independent and dependent variables (σ_y/σ_x) are uniform on the interval (0, 1). This treatment addresses the species-area relation out of the context of its canonical lognormal underpinnings. Although the expectation of their null model approaches $z \simeq 1/4$, the predicted distribution of z values does not agree with the distribution observed. Rather, using the empirical results that they present, one is forced to reject the null model at the 99% level (Sugihara, in press), leading to the conclusion that $z \simeq 1/4$ is not an artifact of the regression system.

An attempt at a more complete explanation, aiming directly at Preston's can-

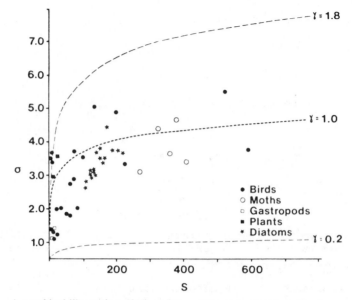

FIG. 2.—A graphical illustration of the validity of the canonical hypothesis, $\gamma \simeq 1$, for an arbitrary set of data on birds, moths (Preston 1948, 1962) gastropods (Kohn 1959), land plants (Oosting 1948), and diatoms (Patrick et al. 1954; Patrick 1968; Sugihara, unpublished data). The relationship between the standard deviation of logarithmic abundances and species count adheres to the rule $\gamma \simeq 1$, rather than to some other close value of γ. In this figure, the curves for $\gamma = 1.8$ and 0.2 were computed from the approximation given in eq. (1A), whereas the canonical relationship ($\gamma = 1$) was determined from values given in Preston (1962).

onical lognormal hypothesis, suggests that $\gamma \simeq 1$ is a robust consequence of general lognormal distributions (May 1975). Although this seems reasonable in principle, it can be shown to be insufficient. In Appendix A, for the unrestricted lognormal curve, it is shown that γ is not confined to some close neighborhood of 1 but can vary freely between 0 and $+\infty$. For ensembles containing a reasonably finite number of individuals and species, γ will range from 0.1 to 6.8. Hence, although the insensitivity of this parameter must contribute to its success, the rule $\gamma \simeq 1$ does not cover lognormal curves in general, and strictly speaking the canonical hypothesis is not an artifact of lognormal distributions. Figure 2 illustrates this result empirically. The points shown do not scatter independently as in the general lognormal distribution, but follow the canonical trend, adhering specifically to the relationship given by $\gamma \simeq 1$ (after Preston 1962) rather than to some other close value of γ (eq. [1A]). This figure shows that Preston's hypothesis reflects real regularities in the shape of the distribution and cannot be explained simply as an artifact arising from the insensitivity of γ. This is important because it means that a full explanation of the lognormal abundance curve must also account for its canonical form, ruling out many of the traditional explanations for lognormality based on random multiplicative effects (see e.g., MacArthur 1960). These traditional explanations not only fail to account for the canonical variance, but more importantly, when the multiplicative impulses are assumed to be independent one encounters the serious shortcoming that the variance of the distribution expands monotonically, and without bound, with each successive moment (Aitchison and Brown 1966). Hence, the lognormal species abundance distribution, its particular canonical form, and the species-area constant remain without a substantial theoretical basis.

In what follows, a model is developed which aims to explain these regularities as a consequence of a general form of community structure involving hierarchically related species niches. This hypothesis, which is motivated by simple evolutionary and ecological considerations for generating species diversity, proposes that a minimal form of community organization involving hierarchically related niches can explain the canonical lognormal abundance pattern. The model will be tested first against data for small assemblages containing two and three species and then extrapolated to account for the patterns observed in large ensembles. Evidence is given which suggests that niche apportionment is multidimensional and that the canonical lognormal distribution is not simply an artifact of classification.

FOUNDATION

A clue to these regularities may be culled from the fact that they seem to apply only to taxonomic collections having some measure of ecological homogeneity (Preston 1962, 1980). This demonstrates that a certain degree of evolutionary or ecological similarity is required in order for them to operate. Therefore, explanations based purely on statistical artifact, although attractive for their generality, seem less likely to apply; instead, the evidence suggests that one should seek to understand these phenomena in terms of very general biological models for generating species diversity.

A reasonable approach toward understanding patterns of species abundance is to interpret them in terms of an underlying niche structure. Indeed, because a niche translates ultimately into numbers of organisms (or biomass), observed abundance patterns can offer a useful standard measure of niches, allowing legitimate comparisons to be made between different types of species niches.

It is commonly believed that the relative abundance of a species is a reflection of the amount of limiting resources it controls (Motomura 1932; MacArthur 1957; Whittaker 1965, 1969, 1972, 1977; May 1975, 1976; Pielou 1975). Although the classical niche apportioning theories require a uniform set of limiting resources, it is also plausible to consider apportionment in a heterogeneous resource pool, involving the subdivision of several different sets of niche axes. This allows the apportionment analogy to be extended to large species ensembles which do not possess a uniform set of governing factors.

Suppose a communal niche space (the total niche requirements of a community in Whittaker's [1977] sense) is likened to a unit mass which has been sequentially split up by the component species so that each fragment denotes relative species abundance. The successive subdivisions may correspond to apportioning on different sets of niche axes, which could be driven by either ecological or evolutionary forces. This is similar in spirit to the MacArthur (1957, 1960) broken-stick model with the important exception that breakage occurs sequentially rather than simultaneously (Bulmer 1974; Pielou 1975). Kolmogoroff (1941) has shown that such sequential fractures can lead asymptotically to lognormal size-frequency distributions. The breakages themselves do not have to be random; however, the magnitude and frequency of breakage must be independent of particle size. Thus, for example, gravel fragments resulting from repeatedly crushing rocks often tend to be lognormal. Instead of smashing rocks one can imagine dividing up relative species abundance in a way that reflects a sequentially divided niche space. Therefore, large numbers of taxonomically related species should tend to have a lognormal distribution of abundance.

The biological motivation for this mechanism depends principally on two propositions. First, the underlying structure of niches should be reflected in the relative abundance pattern (Motomura 1932; MacArthur 1957; Whittaker 1965, 1969, 1972, 1977; May 1975, 1976; Pielou 1975), and second, in general, the minimal niche structure for communities should be hierarchical. This latter proposition derives from the essential differences between species niches, which allows one to sort them into natural groups according to increasing niche similarity. Such a pattern is illustrated, for example, in a niche overlap dendrogram (fig. 3) where communities are sequentially subdivided into smaller and more tightly related functional groups of species. The sequential aspect of the breakage metaphor, therefore, corresponds to this underlying niche hierarchy, with each bifurcation representing a break point (fig. 3). Although evolution is not necessary, it is a sufficient condition for generating this pattern. Whittaker (1977) has proposed that community diversity may evolve by the sequential partitioning and dispersal of species populations in a communal niche space. To some extent, speciation itself can be characterized as the successive carving up and elaboration of a taxon's niche. Such processes should lead in the end to a community structure consisting of subdivided taxonomic guilds. This inevitable tree structure or niche hierarchy is in fact the minimal kind of community structure since, apart from evolution, it could arise naturally from a gradient in niche similarities.

In terms of the breakage metaphor, this situation can be described by the simple case involving successive single fractures. Therefore, the initial unit mass is broken randomly to produce two fragments and one of these is chosen randomly and broken to yield the third, and so on. This process, where particles are sequentially chosen and broken at random, is intended to reflect a hierarchical niche pattern. According to this model, total abundance may expand or shrink proportionally for each species or may remain constant through time (cf. Van Valen's [1974] red queen hypothesis or Levinton 1979). Because the sequential aspect here corresponds to the underlying niche structure, it may be incorrect to interpret a colonization event as the subdivision of an existing hierarchy, since an entirely new set of relationships can be formed.

Although this model is similar in spirit to the MacArthur broken-stick model, it differs from it in several important respects. First, unlike MacArthur's model this mechanism involves sequential rather than simultaneous random breakages. The unit mass is split up by repeated breakage events rather than divided instantaneously with one hammer blow. It should be emphasized that simultaneous breakage

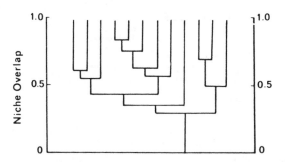

FIG. 3.—A hypothetical niche overlap dendrogram demonstrating hierarchical (minimal) community structure. In the sequential breakage model, each bifurcation in the underlying niche structure corresponds to a subdivision of the communal biomass (abundance).

does not lead to a lognormal distribution but generates a pattern which is more equitable than the one produced by sequential breakage. Second, as will be explained further in the next section, the present model incorporates complex random breakages involving the translation of several niche axes into abundance. This differs from the broken-stick model which involves the uniform breakage of a single resource axis.

Thus, this model not only leads to lognormal distributions but it is also intuitively consistent with evolutionary and ecological conditions for generating species diversity. Encouragingly, it also agrees, at least qualitatively, with Preston's canonical relationship as variance will tend to grow with the convolution of additional species to the community.

MULTIDIMENSIONAL BREAKAGE

Consider first the two-species case involving a single random breakage. At this stage, there is no difference between sequential and simultaneous breakage; therefore, expected proportional abundances can be computed from MacArthur's formula

$$E(p_i) = \frac{1}{S} \sum_{k=i}^{S} \frac{1}{S - k + 1} \qquad (2)$$

where $E(p_i)$ is the expected fraction of the ith most abundant species and S is the number of species in the ensemble. For assemblages with two species, the dominant member will assume values anywhere from 0.5 to 1.0 with uniform probability; and the expectation given by equation (2) is simply the midpoint or 0.75. As Pielou (1975) pointed out, uniform breakage means that all fractional abundances are equally likely; therefore it is essentially meaningless to seek individual examples in nature which agree with equation (2). This casts a shadow over more than a decade of such attempts (Deevey 1969; King 1964; Goulden 1969; Kohn 1959; Longuet-Higgins 1971; MacArthur 1957, 1960; Tramer 1969; Tsukada 1972). Rather, to test the hypothesis it is necessary to consider a distribution of values taken from many assemblages. Swingle's (1950) studies of bass–bluegill combi-

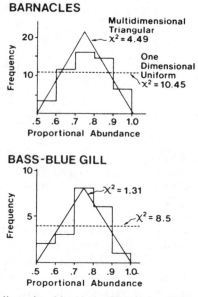

FIG. 4.—A test of one-dimensional breakage (MacArthur 1957, 1960) versus multidimensional breakage for two-species assemblages of barnacles and fish. This figure shows that the relative abundance of the most abundant species of fish (in biomass) and barnacles (in numbers of individuals) is not distributed uniformly but appears to cluster about the expected figure of 0.75. In both cases, the hypothesis of uniform breakage on one niche dimension is rejected at the 95% level.

nations are ideally suited for this as they offer many replicate data from experimental ponds which have come to equilibrium over periods of 2 to 30 yr. There was no sampling error in these studies because the ponds were either completely drained or poisoned, and total biomass could be used to calculate frequencies (fig. 4). Dayton's (1971) barnacle data, although less well suited, are also used because they contain a good number of replicates, and his total exclusion cages ensure that interaction is limited to only two species. These data, however, only represent barnacle associations in the first 3 to 5 mo after settlement, thus the temporal definition of community may be somewhat artificial. Nonetheless, there were no consistent shifts in proportional frequency (calculated from numbers of individuals) at most localities during this period. These were the only studies of two-species associations that I encountered in an arbitrary survey of the literature which had data appropriate for this analysis.

Data for the relative abundance of the dominant species of fish and barnacle are plotted in figure 4. The distributions for the less abundant members have been omitted as they are simply mirror images of the ones shown. The important result is that, rather than a uniform distribution, which is predicted if a single limiting factor is divided up randomly, there is a clear tendency in both cases for values to cluster about the expected figure of 0.75. The hypothesis of uniform breakage on a single limiting factor is rejected for both fish and barnacles at the 95% level.

Whereas individual breakages generate a flat distribution of proportions, composite breakages averaged over several dimensions can explain clustering about the mean. For example, although single throws of a fair die turn up all outcomes uniformly, pairs of throws averaged will accumulate around the expected figure of 3.5. The clustering will tighten as each point includes more throws, rising in the limit to a normal distribution with a shrinking variance. In a similar manner, the relative abundance of a species pair can be reckoned as the mean of random breakages on several dimensions, representing different resources or discrete environmental regimes which the organisms divide up. Accordingly, the fractional abundance, p_s, for k-dimensional apportioning between two species is a simple expectation,

$$p_s = \sum_{r=1}^{k} \alpha_r p_{r,s} \tag{3}$$

(where α is a weighting factor such that $\Sigma \alpha = 1$ and $p_{r,s}$ is the relative fraction that species s gets on dimension r), and the distribution of p_s will modulate around the mean value.

If, for simplicity, two equally important factors determine fractional abundance, then there are three possibilities to consider. *Case 1.* If the dominant species tends to get the larger share of both factors, then this is roughly equivalent to averaging two uniform fractures in the interval [0.5,1). The distribution will be triangular around a mean value of 0.75 (fig. 5A). *Case 2.* On the other hand, the ranked shares may be negatively correlated so that the species which gets a larger fraction on one dimension gets the smaller part of the other. In this case, a narrow triangular distribution centered about 0.5 will result (fig. 5B). *Case 3.* Intermediate between these two extremes, the rank on either dimension may be completely independent yielding a wide triangular distribution about 0.5 (fig. 5C). Data for both fish and barnacles are most consistent with case 1, and the agreement is especially good ($P < .05$) for the bass–bluegill combinations. It is not surprising that the more abundant species should get the larger share of several resources. Unfortunately, one cannot say that only two factors are involved since a similar degree of central tendency is realized with many more factors which are differentially, rather than equally, important.

Variation about the expected value of 0.75 may be further reduced if samples are large and heterogeneous so that each homogeneous local area is roughly an independent broken-stick variable. Large samples which encompass more area are simply an average of these local values, and the relative proportions should therefore cluster more tightly around the mean. This is not likely to be important, however, in the fish and barnacle data, because the experimental ponds and the small exclusion cages are essentially homogeneous.

It should be mentioned that one cannot explain clustering around the mean value in terms of sampling along a gradient. For example, in a hypothetical transect, if species A varies in abundance linearly from 0 to K and B goes from K to 0, random sampling along this interval will yield a flat distribution of proportions rather than a modal one. Similarly, a stochastic reformulation of generalized competition models having random coefficients will not generate the observed clustering of abundances (G. Sugihara and L. Nunney, in prep.). However,

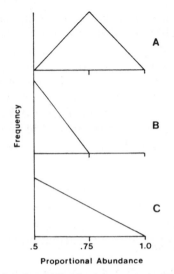

FIG. 5.—Three distinct abundance distributions are possible if two equally important factors determine fractional abundance. *A, Case 1*. If the dominant species gets the larger share of both factors this is equivalent to averaging two uniform variables in the interval [0.5, 1). *B, Case 2*. If the niche dimensions are negatively correlated so that the species which gets the larger share of one dimension, also gets the smaller share of the other, then this is equivalent to averaging two uniform variables, with one in the interval (0, 0.5] and the other in the interval [0.5, 1). *C, Case 3*. When the two dimensions are completely independent, this can be represented as the average of two uniform variables in (0, 1).

recognizing the possibility of an alternative recondite mechanism, the development that follows will depend only on the empirical fact of the pattern observed in the two-species case and only indirectly on its theoretical underpinnings.

INSTANTANEOUS VERSUS SEQUENTIAL BREAKAGE

Results observed for the two-species case will now be extended to predict ratios for assemblages with three species. Taking case 1 as the empirically validated rule for breakage, figure 6 presents the two alternative ways of generating ranked expectations depending on whether the larger or smaller fraction is broken the second time. The method used to calculate the expected sizes of the fractions generated for each breakage pathway is given in Appendix B. The final ranked expectation is simply the average of the ranked expectations for the two possible pathways.

In table 1, I compare the predictions of the sequential hypothesis and MacArthur's instantaneous breakage model for an arbitrary collection of three-species associations including molluscs (Fuller 1972), trees (Keever 1973; Jackson and Faller 1973), fish (Swingle 1950), and barnacles (Dayton 1971). Notice that these data do not represent single assemblages, but are expectations averaged over numerous collections. Although the sequential model fits more closely ($P \simeq .05$) than the MacArthur model ($P \simeq .40$), the difference is not compelling. A more

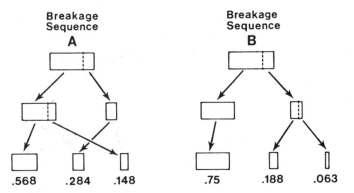

FIG. 6.—Two possible breakage sequences for generating proportional abundances for a three-species assemblage. The values at the bottom of each sequence are the expected proportional sizes (see Appendix B for the method of calculation). The solid arrows in sequence A represent the most probable sequence, and the dashed arrows indicate other possible arrangements depending on the relative sizes that pieces break into with a second breakage.

TABLE 1

COMPARISON OF THE BROKEN-STICK AND SEQUENTIAL BREAKAGE MODELS

	RANKED PROPORTIONAL ABUNDANCE			
SOURCE	p_1	p_2	p_3	χ^2
MacArthur broken-stick	.611	.277	.111	1.02
Sequential hypothesis	.659	.235	.105	.10
Observed data	.675	.227	.095	
(Molluscs, trees, barnacles, fish) $n = 57$				

NOTE.—$n = 57$ is the total number of three-species assemblages used in the calculations. p_i is the proportional abundance of the ith most abundant species. Values for the sequential hypothesis were obtained by averaging the expected values for the two possible breakage pathways (fig. 6).

powerful test was not attempted because the exact forms of the distributions are unknown.

Better evidence for sequential breakage comes from the observation of frequencies within a single community. If the breakage path is ecologically meaningful, or reflects the evolutionary partitioning of niches, then a given type of community should follow one of the two pathways shown in figure 6. Of the results surveyed, only the studies on fish (Swingle 1950), barnacles (Dayton 1971), and trees (Jackson and Faller 1973) offered a sufficient number of data to test this. These values, shown in table 2, appear to agree nicely with the sequential hypothesis: The conifers of Wizard Island follow breakage sequence A, while fish and barnacles follow B. Not only are the expected proportions in close accord with the model, but more importantly the observed variances for these p_i's are almost precisely those which would be generated by sequential breakage. For breakage

TABLE 2

DATA FROM TABLE 1 SORTED INTO SEQUENCES A AND B FROM FIGURE 6

| | | RANKED PROPORTIONAL ABUNDANCE | | |
SOURCE		p_1	p_2	p_3
Breakage Sequence A	Model prediction (approximate)	.568 $\sigma^2 = .011$	.284 $\sigma^2 = .006$	.148 $\sigma^2 = .004$
	Trees (Wizard Island) $n = 15$	.566 $\sigma^2 = .010$	.300 $\sigma^2 = .008$	.144 $\sigma^2 = .006$
Breakage Sequence B	Model prediction (exact)	.750	.188 $\sigma^2 = .004$	.063 $\sigma^2 = .001$
	Barnacles $n = 18$	.758 $\sigma^2 = .006$	.165 $\sigma^2 = .004$	.077 $\sigma^2 = .002$
	Fish $n = 12$	.757 $\sigma^2 = .006$	.174 $\sigma^2 = .004$	.069 $\sigma^2 = .002$

NOTE.—Model variances predicted for sequence A are approximate since they assume that the proportions for each successive breakage were chosen from a triangular distribution (cf. fig. 4). This assumption was not necessary in sequence B where it was possible to use the observed variance of p_1 (i.e., the observed variance for a single breakage = .006) to predict the variances of p_2 and p_3 (both resulting from a second breakage).

sequences B (fig. 6) the variances of p_2 and p_3 are conditional values predicted for the fractional abundances resulting from a second breakage. In this sequence, the larger piece, p_1, results from a first breakage and the two smaller ones are produced by a second breakage. From the observed variance in the larger fraction (first breakage) one can calculate the conditional variances for the two smaller pieces (see Appendix C for method of calculation). This conditional argument, however, does not apply to sequence A where one cannot know the exact variance of the fractions resulting from a first breakage. In this sequence, involving the breakage of the larger segment, none of the p_i's will result purely from a single fracture. For example, the largest segments, which are averaged to determine p_1, may have resulted from either a first or second breakage (fig. 6). Therefore it is necessary to approximate the variances for the ranked proportions in this sequence by assuming that the breakage fractions were chosen from a triangular distribution as observed in the two-species case (Appendix B).

GENERATION OF THE CANONICAL LOGNORMAL
AND THE SPECIES-AREA CONSTANT

Because the number of possible breakage sequences is $(S - 1)!$, it is unreasonable to discuss specific pathways for larger assemblages. Instead, the relationship

TABLE 3

COMPARISON OF THE SEQUENTIAL BREAKAGE MODEL WITH
THE CANONICAL LOGNORMAL HYPOTHESIS

S		CANONICAL HYPOTHESIS σ	SEQUENTIAL BREAKAGE HYPOTHESIS	
			$\hat{E}(\sigma)$	$\widehat{VAR}(\sigma)$
25	....	2.97	2.80	.4509
50	....	3.37	3.23	.4714
100	....	3.72	3.61	.4318
200	....	4.05	3.97	.4106
400	....	4.36	4.32	.3856
800	....	4.66	4.63	.3522
1,000	....	4.75	4.73	.3368

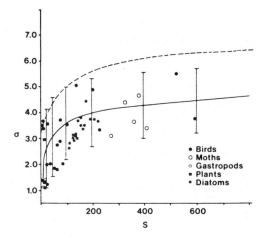

FIG. 7.—Graphical evidence for the sequential breakage hypothesis for an arbitrary assortment of taxonomic collections (see fig. 1). The solid line and error bars represent the mean and two standard deviations of the mean predicted by the sequential breakage model. The dashed line represents the trend generated by biologically trivial or arbitrary sequential breakages, i.e., where the successive breakage fractions are chosen uniformly from the interval (0,1). This latter trend is identified with the null hypothesis that species are artifacts of classification in the nominalist sense.

between the standard deviation of logarithmic abundances, σ, and species count, is examined with the aim of determining whether the coupling of these parameters agrees with the empirically based canonical relationship.

The results of a numerical simulation of the sequential breakage model are compared in table 3 to the canonical hypothesis. The correspondence is surprisingly good. An even stronger statement is made in figure 7 which demonstrates an almost perfect match between the variance in the model and the scatter of points found in nature. These numerical results are generated by assuming that the breakage fractions are precisely 0.75, 0.25, which is roughly what should be expected if the collections are large and heterogeneous, and if pairwise rank is preserved, i.e., if species A is more abundant than B in all independent localities where they occur together. Under these circumstances, because of the modulating effect of sample size, the values should peak into a spike around the expected figure. Virtually identical results can be obtained, however, by assuming that some breakages are variable as in the two-species case, and that pairwise rank is not always maintained. This can be portrayed approximately as an intergrading mixture of point fractures (0.75, 0.25, and 0.5, 0.5) and breakage distributions shown in figures 5A and 5C to yield a distribution with a peak roughly midway between 0.5 and 0.75. The relationship between S and σ generated by this regime is virtually indistinguishable from that presented in table 3. Both of these possibilities are plausible extrapolations of the two- and three-species cases previously observed. On the other hand, if one arbitrarily assumes that the successive breakages are uniform in the interval (0,1), then the canonical relationship is not obtained (fig. 7). This rules out an alternative and biologically trivial interpretation of breakage proposed by Aitchison and Brown (1967) in which arbitrary hierarchical systems of classification are conceived of as a sequential breakage operation. Clearly then, the canonical lognormal distribution cannot be explained as an artifact of classification in the nominalist sense, and species do not represent arbitrary subdivisions of biomass.

These results demonstrate that a reasonable extension of the two- and three-species cases can generate the ubiquitous canonical lognormal distribution, whereas arbitrary sequential breakages will not account for the relationship. Furthermore, because the model generates the canonical lognormal distribution, it can also explain the Preston-MacArthur-Wilson species-area constant $z \simeq 1/4$. It is surprising and encouraging that a simple induction of the patterns observed in small ensembles can account so well for such large-order regularities, and reassuring that the generality of this hypothesis corresponds to the ubiquity of the patterns which it aims to explain.

CONCLUDING REMARKS

The evidence suggests that the canonical lognormal abundance pattern and the species-area constant may be most simply explained as consequences of a hierarchically structured communal niche. This model is admittedly phenomenological; however in view of the regularities addressed, this is perhaps inevitable. It is intended strategically to capture the first-order effects of the most minimal kind of community structure, and exactly how literally it may be interpreted remains to be seen.

In constructing my argument the results of the two- and three-species cases were induced to make predictions about the multispecies case. If the hypothesis of minimal community structure is correct then it should be possible to recover roughly the pattern observed for two species from a large ensemble by constructing a dendrogram of niche overlaps and considering each bifurcation as the two-species case. If the area sampled is large enough, then the distribution of proportions thus obtained should be roughly triangular with a peak midway between 0.5 and 0.75. Therefore, an independent test of the hypothesis may be accomplished with a dendrogram containing the abundance of each species.

For the Wizard Island trees (table 2), although the breakage path was conserved, the rank for each species varied in different localities. Because these associations were from a range of altitudes and soil conditions, the lack of a fixed ordering is not surprising, and demonstrates that niche subdivision is not unique for a particular species ensemble and depends on the specific environmental context. The species rank-list for barnacles supports this idea: For tide levels under 1.3 m, the rank-list is *Balanus glandula* > *B. cariosus* > *Cthamalus dalli*, whereas at stations higher in the intertidal, *C. dalli* tends to predominate over *B. cariosus*. On the other hand, Swingle's fish associations, all from similar pond environments, tend to maintain their ordering. These examples align most consistently with ecological forces determining breakage, but this should not reduce the plausibility of evolutionary niche subdivision. It seems likely that the principle fractures which divide a large community into major functional groups will reflect evolutionary lines of niche apportionment, and that finer partitioning, within smaller more closely related species clusters, will be guided more strongly by ecological forces.

It is interesting that the breakage pathway for a given set of species was conserved (table 2). Jackson and Faller's trees cleaved the most abundant species (sequence A) while Swingle's fish and Dayton's barnacles cleaved the least abundant one (sequence B). One possible explanation, in terms of competition, is that among trees the incentive for sharing a large resource pool may have outweighed the disadvantage of competing with an abundant species, whereas the opposite may have held for fish and barnacles. The full ecological significance of this observed invariance in breakage pathway may offer interesting biological insights and deserves further consideration.

SUMMARY

Recent proposals that the canonical lognormal distribution and the resulting species-area constant, $z \simeq 1/4$, are artifacts of the general lognormal curve and regression techniques, are shown to be inadequate. An alternative hypothesis is suggested which accounts for these regularities in terms of a hierarchical community structure represented by a sequentially divided niche space. This hierarchical pattern, which can be considered to be a minimal form of community structure, derives from evolutionary and ecological considerations for generating species diversity, and is shown to account for the observed abundance structures of small ensembles as well as large natural communities. Evidence is presented which implies that niche apportionment between species may involve the random division of more than one resource, and an interesting invariance in the pattern of apportionment is observed for assemblages with three species. The possibility that the canonical lognormal distribution is a conceptual artifact resulting from arbitrary systems of classification is considered and shown to be false. Aside from its intuitive appeal, the model presented should be of interest because it offers explanations of two ubiquitous patterns in nature: the canonical lognormal and the resulting species-area constant.

ACKNOWLEDGMENTS

I thank D. A. Livingstone, R. M. May, H. S. Horn, M. Huston, G. F. Esta-brook, J. H. Vandermeer, and R. H. Whittaker for their valuable advice and discussions in preparing this manuscript. This work was supported in part by a Princeton University Prize Fellowship.

APPENDIX A

This section contains a counterargument to the proposition that $\gamma \simeq 1$ is a general property of lognormal distributions.

Following May (1975):

S = total number of species in the lognormal assemblage.

J = total number of individuals/number of individuals of the rarest species.

$a = 1/\sqrt{2\sigma^2}$ where σ^2 is the variance of the lognormal species curve.

$\gamma = R_N/R_{max}$ such that if the species curve mode is at zero, R_N = the position of the individual's curve mode, and R_{max} = the position of the upper truncation point of the species curve.

Upper and lower limits for γ are obtained conveniently from the relation

$$\gamma \simeq \frac{\ln 2}{2a\sqrt{(\ln KS)}} \tag{1A}$$

where $K = 2a/\sqrt{\pi}$. This is similar to May's eq. (3.8) with the exception that the area contained under the upper and lower limbs of the distribution is assumed to sum to 1 (after Preston 1962) rather than 2. Holding a fixed, it follows that $\gamma \to 0$ as $S \to \infty$. Because $a > 0$ and $\ln KS > 0$, γ is strictly positive. Hence, $\gamma_{min} = 0$ is a greatest lower bound. An upper limit for γ is obtained from eq. (1A) simply by setting $S = c/2a$ where $c > \sqrt{\pi}$, and allowing $a \to 0$, whence, $\gamma \to \infty$.

Collecting these remarks yields the following:

$$0 < \gamma < \infty. \tag{2A}$$

Hence, $\gamma \simeq 1$ is not a necessary property of the general lognormal.

It is now reasonable to ask whether restricting S and J to realistic limits leads to lognormal distributions with $\gamma \simeq 1$. For $50 < S < 10^4$ and $10^4 < J < 10^{12}$ the corresponding range for γ is roughly

$$0.1 < \gamma < 6.8 \tag{3A}$$

where (3A) follows from equation (1A), and the relations: $\ln J \simeq \Delta^2 (1 + \gamma)^2$ for $\gamma < 1$, and, $\ln J \simeq 4\gamma\Delta^2$ for $\gamma > 1$, where $\Delta^2 \simeq \ln KS$ (May 1975). Therefore, the property $\gamma \simeq 1$ is not inevitable for lognormal species frequency curves, even within reasonable limits for S and J.

APPENDIX B

In sequence B, which involves the division of the smaller fragment (fig. 6) the ranked expectations are easily computed as follows:

$$E(p_1) = .75$$
$$E(p_2) = (.25)(.75) = .1875$$
$$E(p_3) = (.25)(.25) = .0625$$

This simple computation is possible because there is no ambiguity in the rank ordering of fractions in relation to their sequence of generation. That is, p_1 is always generated by a single breakage and p_2 and p_3 always involve a second breakage. The situation is not as clear cut in sequence A, however, where the larger piece is divided a second time. To see this, let x_1 and x_2 be the larger fractions used in the first and second breakages, respectively, and let $(1 - x_1)$ and $(1 - x_2)$ represent the smaller fractions. Because it is the larger piece from the first breakage that is divided, three cases must be considered:

1) $(1 - x_1) > x_1 x_2 > x_1(1 - x_2)$.
2) $x_1 x_2 > (1 - x_2) > x_1(1 - x_2)$.
3) $x_1 x_2 > x_2(1 - x_2) > (1 - x_1)$.

The terms in each of these inequalities correspond to p_1, p_2, and p_3. One can see the ambiguity in p_1, for example, since it can involve pieces resulting from a first $(1 - x_1)$ and/or second $(x_1 x_2)$ breakage. Therefore, in calculating the ranked expectations and variances for sequence A the above three cases must be considered. The expectations may be computed as follows:

$$E(p_1) = \int_{\frac{1}{2}}^{1} \int_{\frac{1}{2}}^{\frac{1}{1+x_2}} (1 - x_1) dFx_1 dFx_2$$

$$+ \int_{\frac{1}{2}}^{1} \int_{\frac{1}{1+x_2}}^{1} x_1 dFx_1 dFx_2.$$

$$E(p_2) = \int_{\frac{1}{2}}^{1} \int_{\frac{1}{2}}^{\frac{1}{1+x_2}} x_1 x_2 dFx_1 dFx_2$$

$$+ \int_{\frac{1}{2}}^{1} \int_{\frac{1}{1+x_2}}^{\frac{1}{2-x_2}} (1 - x_1) dFx_1 dFx_2$$

$$+ \int_{\frac{1}{2}}^{1} \int_{\frac{1}{2-x_2}}^{1} x_1(1 - x_2) dFx_1 dFx_2.$$

$$E(p_3) = \int_{\frac{1}{2}}^{1} \int_{\frac{1}{2}}^{\frac{1}{1+x_2}} x_1(1 - x_2) dFx_1 dFx_2$$

$$+ \int_{\frac{1}{2}}^{1} \int_{\frac{1}{1+x_2}}^{\frac{1}{2-x_2}} x_1(1 - x_2) dFx_1 dFx_2$$

$$+ \int_{\frac{1}{2}}^{1} \int_{\frac{1}{2-x_2}}^{1} (1 - x_1) dFx_1 dFx_2,$$

where Fx_1 and Fx_2 are the cumulative densities of the triangular distribution on the interval $[\frac{1}{2}, 1]$. The calculation of variances follows trivially from the relations above.

APPENDIX C

This section illustrates a method for computing the conditional variances for fractions resulting from a second breakage in sequence B (see fig. 6). As a specific example, we will compute the variance of p_2 from breakage path B (see fig. 5). This is the larger fraction resulting from the second breakage.

Let $y = (1 - x_1)$ denote the size of the smallest fraction from the first breakage, and $p_2 =$ the size of the largest piece from the second fracture. The distribution of p_2, $f(p_2)$, is conditional on the distribution of y, and one can write

$$f(p_2) = \int f(p_2|y) f(y) \, dy. \tag{1B}$$

Thence, the variance of p_2, $V(p_2)$, is simply the sum of the expectation of the conditional variance plus the variance of the conditional expectation,

$$V(p_2) = E[V(p_2|y)] + V[E(p_2|y)]. \tag{2B}$$

If $V(y)$ is observed to have a value of a, then $V(p_2|y) = ay^2$ and $E(p_2|y) = (3/4)y$, whereupon

$$V(p_2) = (9/16)a + aE(y^2). \tag{3B}$$

Because, in general, $V(t) = E(t^2) - E(t)^2$, therefore $E(y^2) = a(a + 1/16)$ and equation (3B) becomes $V(p_2) = a(a + 5/8)$. For example, in the fish data $a = .006$ and $V(p_2) = .004$.

LITERATURE CITED

Aitchison, J., and J. A. C. Brown. 1966. The lognormal distribution. Cambridge University Press, Cambridge.

Bulmer, M. G. 1974. On fitting the Poisson lognormal distribution to species-abundance data. Biometrics 30:101–110.

Connor, E. F., and E. D. McCoy. 1979. The statistics and biology of the species-area relationship. Am. Nat. 113:791–833.

Dayton, P. K. 1971. Competition, distribution, and community organization: the provision and subsequent utilization of space in a rocky intertidal community. Ecol. Monogr. 41:351–389.

Deevey, E. S., Jr. 1969. Specific diversity in fossil assemblages. Pages 224–241 in Diversity and stability in ecological systems. Brookhaven Symp. Biol. No. 22.

Goulden, C. E. 1969. Temporal changes in diversity. Pages 96–102 in Diversity and stability in ecological systems. Brookhaven Symp. Biol. No. 22.

Hairston, N. G., and G. W. Byers. 1954. The soil arthropods of a field in southern Michigan: a study in community ecology. Contrib. Lab. Vertebr. Biol. Univ. Mich. 64:1–37.

Jackson, M. T., and A. Faller. 1973. Structural analysis and dynamics of the plant communities of Wizard Island, Crater Lake National Park. Ecol. Monogr. 43:441–461.

Keever, C. 1973. Distribution of major forest species in southeastern Pennsylvania. Ecol. Monogr. 54:303–327.

King, C. E. 1964. Relative abundance of species and MacArthur's model. Ecology 45:716–727.

Kohn, A. J. 1959. The ecology of Conus in Hawaii. Ecol. Monogr. 29:47–90.

Kolmogoroff, A. N. 1941. Über das logarithmisch normale Verteilungsgestz der Dimensionen der Teilchen bei Zerstückelung. C. R. (Doklady) Acad. Sci. URSS. 31:99–101.

Levington, J. S. 1979. A theory of diversity equilibrium and morphological evolution. Science 204:335–336.

Longuet-Higgins, M. S. 1971. On the Shannon-Weaver index of diversity, in relation to the distribution of species in bird censuses. Theor. Popul. Biol. 2:271–289.

MacArthur, R. H. 1957. On the relative abundance of bird species. Proc. Natl. Acad. Sci. USA 43:293–295.

———. 1960. On the relative abundance of species. Am. Nat. 94:25–36.

MacArthur, R. H., and E. O. Wilson. 1967. The theory of island biogeography. Princeton University Press, Princeton, N.J.

May, R. M. 1975. Patterns of species abundance and diversity. Pages 81–120 *in* M. L. Cody and J. M. Diamond, eds. Ecology and evolution of communities. Harvard University Press, Cambridge, Mass.

———. 1976. Patterns in multi-species communities. Pages 142–162 *in* R. M. May, ed. Theoretical ecology, principles and applications. Saunders, Philadelphia.

Motomura, I. 1932. A statistical treatment of associations (In Japanese). Jpn. J. Zool. 44:379–383.

Oosting, Henry J. 1948. The study of plant communities. Freeman, San Francisco.

Patrick, R. 1968. The structure of diatom communities in similar ecological conditions. Am. Nat. 102:173–183.

Patrick, R., M. Hohn, and J. Wallace. 1954. A new method of determining the pattern of the diatom flora. Not. Nat. 259.

Pielou, E. C. 1975. Ecological diversity. Wiley, New York.

Preston, F. W. 1948. The commonness, and rarity, of species. Ecology 29:254–283.

———. 1962. The canonical distribution of commonness and rarity. Ecology 43:185–215, 410–432.

———. 1980. Noncanonical distributions of commonness and rarity. Ecology 61:88–97.

Sugihara, G. 1981. $S = CA^z$, $z = \frac{1}{4}$: a reply to Connor and McCoy. Am. Nat. (in press).

Swingle, H. S. 1950. Relationships and dynamics of balanced and unbalanced fish populations. Agricultural Experiment Station of the Alabama Polytechnic Institute. No. 274:1–74.

Tramer, E. J. 1969. Bird species diversity; components of Shannon's formula. Ecology 50:927–929.

Tsukada, M. 1972. The history of Lake Nojiri, Japan. Trans. Conn. Acad. Arts Sci. 44:337–365.

Van Valen, L. 1973. A new evolutionary law. Evol. Theory 1:1–30.

Whittaker, R. H. 1965. Dominance and diversity in land plant communities. Science 147:250–260.

———. 1969. Evolution of diversity in plant communities. Pages 178–196 *in* Brookhaven Symp. Biol. No. 22.

———. 1972. Evolution and measurement of species diversity. Taxon 21:213–251.

———. 1977. Evolution of species diversity in land communities. Evol. Biol. 10:1–66.

AUTHOR CITATION INDEX

SUBJECT INDEX

Abundance of species, 15, 16, 17, 18, 31, 34, 35, 56, 61, 70, 74, 80, 81, 86, 107, 108, 110, 114, 115, 123, 124, 135, 152, 159, 167, 201, 204, 210, 215, 217, 218, 222, 229, 251, 289, 292, 293, 294, 296, 310–339. *See also* Diversity
  clustering, 387
  commonness, 310–339
  patterns, 340–377, 380–396
  rarity, 310–339
Africa, 33, 66, 70, 225, 229–240
Algae, 9, 46, 53, 164, 168
  abundance, 115
  benthic, 132
  competitive ability, 110
  coral reef, 54, 159–171
  effect of *L. littorea* on diversity, 117, 118, 119, 121, 124
  effect of sea urchins on diversity, 122
  ephemeral, 114, 115, 119
  intertidal, 110–124
  stream, 95, 96, 97, 98, 103, 104, 105
α-diversity, 21, 70, 98, 100, 187, 188, 189. See also Within-habitat pattern of diversity
Alpine, 188
Amazon, 34
Arctic, 71, 127, 187
Asia, 69
Australia, 33, 222–242
Autecology, 12
Autotoxins, 104

Bacteria, 288–297
β-diversity, 21, 70, 100, 186, 187, 188. *See also* Between-habitat pattern of diversity
Between-habitat pattern of diversity, 14, 20, 21, 26, 27, 33, 70, 71, 139, 151, 239, 240, 242. *See also β-*diversity
Biogeography, 296
  island, 3, 261, 380
  species-area constant, 380
Biologically accommodated community, 64, 65, 66, 78

Bird, 27, 28, 29, 32, 33, 35, 45, 71, 98, 104, 138, 139, 149, 151, 182, 189, 190, 191, 192, 197, 204, 241, 244–255, 313, 314, 315, 346, 350
  clutch size, 24
  desert, 208, 222, 238, 239
  species diversity, 19, 20, 177, 249, 251
  temperate, 139
  tropical, 139, 154, 185
Bolivia, 254
Brillouin measures of diversity, 198

Canada, 68, 335
Caribbean, 252
Carnivores, 95, 97, 105, 130, 134, 135, 136, 205
  aquatic, 110
  terminal, 128, 132
Central America, 217
Central Limit Theorem, 108
Character displacement, 22, 26, 35
Chile, 70
Climate. *See* Environmental variables
Coexisting species, 22, 24, 25, 26, 27, 40, 209, 214, 215, 216, 217, 218, 224
Colombia, 254
Colonization, 14, 27, 28, 29, 30, 31, 32, 35, 70, 95, 261, 272, 276, 279, 284, 285, 286, 384
  mechanisms, 261
Community,
  alpine, 188
  aquatic, 8, 52, 54, 98, 104, 176
  biologically accommodated, 64–66
  boreal shallow marine, 91
  coral reef, 54, 159–171
  deep-lake, 96
  deep-sea, 65, 66, 91
  desert, 21, 45, 46, 64, 208–220
  disharmonic, 27
  diversity, 53
  evolution, 348
  experimental, 289, 290, 295, 297

About the Editor

RUTH PATRICK first showed her interest in communities of aquatic organisms as a small girl under the influence of her father, who was a naturalist as well as a lawyer. This interest grew with years of collecting in the field. Her Ph.D. thesis was concerned with diatoms, and research in this field has continued to be a major interest. Her work on this group of organisms and other organisms living in rivers has led to the publication of over 100 scientific papers and two volumes on *Diatoms of the United States,* the latter written in cooperation with Dr. Charles Reimer.

Ruth Patrick founded the Limnology Department of the Academy of Natural Sciences in 1947. When it was founded the department consisted of 10 people. Over the years it has grown to a department of about 100 people scattered through three laboratories. Two of them are field laboratories — the Stroud Water Research Center and the Estuary Laboratory at Benedict, Maryland.

Ruth Patrick received the B.S. degree from Coker College, South Carolina, and the M.S. and Ph.D. degrees from the University of Virginia. This formal education was supplemented by summers spent at the Woods Hole Biological Laboratory, Cold Spring Harbor Biological Laboratory, and the Biological Laboratory at the University of Virginia, at Mountain Lake. In these summer laboratories she greatly increased her field experience and gained a lasting interest in the ecology of aquatic organisms.

She was the first person in this country to show by actual field work that the diversity of many major groups of organisms characterized natural communities. The importance of diversity in natural conditions had previously been expressed by Thienemann (1939, Grundzüge einer allgemeinen Okologie, *Arch. Hydrobiol.* **35:**267–285), but it was from her field studies in the Conestoga River Basin that the importance of pattern in the diversity of many groups of organisms in healthy aquatic communities was first emphasized.

For her scientific research Dr. Patrick has received many awards: among these are the Eminent Ecologist Award of the Ecological Society of America, the Certificate of Merit of the Botanical Society of America, and the John and Alice Taylor Ecology Award.

She is a recipient of many honorary degrees; the two most recent degrees were from Princeton University and the University of Massachusetts, awarded in 1980.